COLD SPRING HARBOR SYMPOSIA ON QUANTITATIVE BIOLOGY

VOLUME LXXIX

symposium.cshlp.org

Online access: Please visit our companion website at symposium.cshlp.org. For access questions, please contact Cold Spring Harbor Laboratory Press at cshpress@cshl.edu.

COLD SPRING HARBOR SYMPOSIA ON QUANTITATIVE BIOLOGY

VOLUME LXXIX

Cognition

symposium.cshlp.org

Symposium organizers and Proceedings editors: Cori Bargmann (*The Rockefeller University*), Daphne Bavelier (*University of Geneva, Switzerland*, and *University of Rochester*), Terrence Sejnowski (*The Salk Institute for Biological Studies*), and David Stewart and Bruce Stillman (*Cold Spring Harbor Laboratory*)

COLD SPRING HARBOR LABORATORY PRESS
2014

COLD SPRING HARBOR SYMPOSIA ON QUANTITATIVE BIOLOGY VOLUME LXXIX

International Standard Book Number 978-1-621821-26-7 (cloth)
International Standard Book Number 978-1-621821-27-4 (paper)
International Standard Serial Number 0091-7451
Library of Congress Catalog Card Number 34-8174

Printed in the United States of America

COLD SPRING HARBOR SYMPOSIA ON QUANTITATIVE BIOLOGY
Founded in 1933 by
REGINALD G. HARRIS
Director of the Biological Laboratory 1924 to 1936

Previous Symposia Volumes

I (1933) Surface Phenomena
II (1934) Aspects of Growth
III (1935) Photochemical Reactions
IV (1936) Excitation Phenomena
V (1937) Internal Secretions
VI (1938) Protein Chemistry
VII (1939) Biological Oxidations
VIII (1940) Permeability and the Nature of Cell Membranes
IX (1941) Genes and Chromosomes: Structure and Organization
X (1942) The Relation of Hormones to Development
XI (1946) Heredity and Variation in Microorganisms
XII (1947) Nucleic Acids and Nucleoproteins
XIII (1948) Biological Applications of Tracer Elements
XIV (1949) Amino Acids and Proteins
XV (1950) Origin and Evolution of Man
XVI (1951) Genes and Mutations
XVII (1952) The Neuron
XVIII (1953) Viruses
XIX (1954) The Mammalian Fetus: Physiological Aspects of Development
XX (1955) Population Genetics: The Nature and Causes of Genetic Variability in Population
XXI (1956) Genetic Mechanisms: Structure and Function
XXII (1957) Population Studies: Animal Ecology and Demography
XXIII (1958) Exchange of Genetic Material: Mechanism and Consequences
XXIV (1959) Genetics and Twentieth Century Darwinism
XXV (1960) Biological Clocks
XXVI (1961) Cellular Regulatory Mechanisms
XXVII (1962) Basic Mechanisms in Animal Virus Biology
XXVIII (1963) Synthesis and Structure of Macromolecules
XXIX (1964) Human Genetics
XXX (1965) Sensory Receptors
XXXI (1966) The Genetic Code
XXXII (1967) Antibodies
XXXIII (1968) Replication of DNA in Microorganisms
XXXIV (1969) The Mechanism of Protein Synthesis
XXXV (1970) Transcription of Genetic Material
XXXVI (1971) Structure and Function of Proteins at the Three-dimensional Level
XXXVII (1972) The Mechanism of Muscle Contraction
XXXVIII (1973) Chromosome Structure and Function
XXXIX (1974) Tumor Viruses
XL (1975) The Synapse
XLI (1976) Origins of Lymphocyte Diversity
XLII (1977) Chromatin
XLIII (1978) DNA: Replication and Recombination
XLIV (1979) Viral Oncogenes
XLV (1980) Movable Genetic Elements
XLVI (1981) Organization of the Cytoplasm
XLVII (1982) Structures of DNA
XLVIII (1983) Molecular Neurobiology
XLIX (1984) Recombination at the DNA Level
L (1985) Molecular Biology of Development
LI (1986) Molecular Biology of Homo sapiens
LII (1987) Evolution of Catalytic Function
LIII (1988) Molecular Biology of Signal Transduction
LIV (1989) Immunological Recognition
LV (1990) The Brain
LVI (1991) The Cell Cycle
LVII (1992) The Cell Surface
LVIII (1993) DNA and Chromosomes
LVIX (1994) The Molecular Genetics of Cancer
LX (1995) Protein Kinesis: The Dynamics of Protein Trafficking and Stability
LXI (1996) Function & Dysfunction in the Nervous System
LXII (1997) Pattern Formation during Development
LXIII (1998) Mechanisms of Transcription
LXIV (1999) Signaling and Gene Expression in the Immune System
LXV (2000) Biological Responses to DNA Damage
LXVI (2001) The Ribosome
LXVII (2002) The Cardiovascular System
LXVIII (2003) The Genome of Homo sapiens
LXIX (2004) Epigenetics
LXX (2005) Molecular Approaches to Controlling Cancer
LXXI (2006) Regulatory RNAs
LXXII (2007) Clocks and Rhythms
LXXIII (2008) Control and Regulation of Stem Cells
LXXIV (2009) Evolution: The Molecular Landscape
LXXV (2010) Nuclear Organization and Function
LXXVI (2011) Metabolism and Disease
LXXVII (2012) The Biology of Plants
LXXVIII (2013) Immunity and Tolerance

Front cover (*paperback*): Image by James Whitaker, Cold Spring Harbor Laboratory. Watermark inspired by Santiago Ramón y Cajal (1852–1934), a distinguished Spanish neuroscientist who, together with Italian Camillo Golgi, won the Nobel Prize in Physiology or Medicine in 1906 "in recognition of their work on the structure of the nervous system."

For a complete catalog of all Cold Spring Harbor Laboratory Press publications, visit our website www.cshlpress.org.

Online access: Please visit our companion website at symposium.cshlp.org. For access questions, please contact Cold Spring Harbor Laboratory Press at cshpress@cshl.edu.

Symposium Participants

ABDUS-SABOOR, ISHMAIL, Weill Cornell Medical College in Qatar, Doha, Qatar
ABRAHAM, ANNA, Kuwait University, Safat, Kuwait
AGETSUMA, MASAKAZU, Osaka University, Ibaraki, Japan
AHRENS, MISHA, Janelia Farm, Howard Hughes Medical Institute, Ashburn, Virginia
AHRENS, SANDRA, Cold Spring Harbor Laboratory, Cold Spring Harbor, New York
AKRAMI, ATHENA, Princeton University, Howard Hughes Medical Institute, Princeton, New Jersey
ANDERSON, DAVID, Caltech/Howard Hughes Medical Institute, Pasadena, California
ANGELAKI, DORA, Baylor College of Medicine, Houston, Texas
ANOKHIN, KONSTANTIN, Kurchatov Institute National Research Center, Moscow, Russia
ANSELMI, FRANCESCA, Cold Spring Harbor Laboratory, Cold Spring Harbor, New York
APERGIS-SCHOUTE, ANNEMIEKE, University of Cambridge, Cambridge, United Kingdom
ARAI, YOSHIO, University of Pittsburgh, Pittsburgh, Pennsylvania
AVERY, MICHAEL, Salk Institute, San Diego, California
AXEL, RICHARD, Columbia University, New York, New York
BANDYOPADHYAY, ARKARUP, Cold Spring Harbor Laboratory, Cold Spring Harbor, New York
BAO, ZHIRONG, Sloan-Kettering Institute, New York, New York
BARBASH, SHAHAR, Hebrew University of Jerusalem, Jerusalem, Israel
BARGMANN, CORI, The Rockefeller University, New York, New York
BATCHELOR, ALEXANDRA, Harvard University, Cambridge, Massachusetts
BATSCHING, SOPHIE, University Würzburg, Würzburg, Germany
BAVELIER, DAPHNE, University of Geneva, Geneva, Switzerland and University of Rochester, Rochester, New York
BENDESKY, ANDRES, Harvard University, Cambridge, Massachusetts
BHANDAWAT, VIKAS, Duke University, Durham, North Carolina
BIRNBAUM, REBECCA, Lieber Institute for Brain Development, Johns Hopkins, Baltimore, Maryland
BOBOILA, CRISTIAN, Columbia University, New York, New York
BOLDING, KEVIN, Duke University, Durham, North Carolina
BONNER, MICHAEL, University of Pennsylvania, Philadelphia, Pennsylvania
BORGES MONROY, REBECA, Cold Spring Harbor Laboratory, Cold Spring Harbor, New York
BORST, ALEXANDER, Max Planck Institute of Neurobiology, Martinsried, Germany
BRAATEN, DOUGLAS, The New York Academy of Sciences, New York, New York
BRODY, CARLOS, Princeton University, Princeton, New Jersey
BRUSH, ELEANOR, Princeton University, Princeton, New Jersey
BUZSÁKI, GYÖRGY, New York University, Langone Medical Center, New York, New York
CARCEA, IOANA, New York University Medical School, New York, New York
CARDOSO-LEITE, PEDRO, University of Geneva, Geneva, Switzerland
CARNIOL, KAREN, *Cell*, Cambridge, Massachusetts
CARRASCO, MARISA, New York University, New York, New York
CAZAKOFF, BRITTANY, Cold Spring Harbor Laboratory, Cold Spring Harbor, New York
CHAE, HONG GOO, Cold Spring Harbor Laboratory, Cold Spring Harbor, New York
CHANG, DONG-SEON, Max Planck Institute for Biological Cybernetics, Tübingen, Germany
CHARTARIFSKY, LITAL, Cold Spring Harbor Laboratory, Cold Spring Harbor, New York
CHEN, XIUYE, Harvard University, Cambridge, Massachusetts
CHOUARD, TANGUY, *Nature*, London, United Kingdom
CHUN, MARVIN, Yale University, New Haven, Connecticut
CHURCHLAND, ANNE, Cold Spring Harbor Laboratory, Cold Spring Harbor, New York
CHURCHLAND, MARK, Columbia University, New York, New York
CHURCHLAND, PATRICIA, University of California, San Diego, La Jolla, California
COEN, PHILIP, Princeton University, Princeton, New Jersey
DAN, YANG, University of California, Berkeley, Berkeley, California
DANG, CHUNG, Fred Hutchinson Cancer Research Center, Seattle, Washington
DANIELS, BRYAN, University of Wisconsin–Madison, Madison, Wisconsin
DASGUPTA, SHAMIK, University of Oxford, Oxford, United Kingdom
DAVID, KAREN, National Institutes of Health/National Institute of Neurological Disorders and Stroke, Bethesda, Maryland
DEISSEROTH, KARL, Stanford University, Stanford, California
DELEVICH, KRISTEN, Cold Spring Harbor Laboratory, Cold Spring Harbor, New York
DESIMONE, ROBERT, Massachusetts Institute of Technology/McGovern Institute for Brain Research, Cambridge, Massachusetts
DEVINENI, ANITA, Columbia University, New York, New York
DICARLO, JAMES, McGovern Institute for Brain Research/Massachusetts Institute of Technology, Cambridge, Massachusetts
DICKERSON, MICHAEL, Southwest Baptist University, Springfield, Missouri
DOBOSIEWICZ, MAY, The Rockefeller University, New York, New York
DUAN, CHUNYU, Princeton University, Princeton, New Jersey
DUBNAU, JOSH, Cold Spring Harbor Laboratory, Cold Spring Harbor, New York
DUBOWITZ, LILI, London, United Kingdom
DUBOWITZ, VICTOR, University of London, London, United Kingdom
DULAC, CATHERINE, Howard Hughes Medical Institute/Harvard University, Cambridge, Massachusetts
EBIHARA, AKINORI, The Rockefeller University, New York, New York
ECKMEIER, DENNIS, Cold Spring Harbor Laboratory, Cold Spring Harbor, New York
EMERY, VIRGINIA OLGA, Dartmouth Medical School, Bow, New Hampshire
EMMONS, SCOTT, Albert Einstein College of Medicine, Bronx, New York
ERLICH, JEFFREY, Princeton University, Princeton, New Jersey
FARNSWORTH, BRYN, Uppsala University, Uppsala, Sweden
FASSIHI, ARASH, International School for Advanced Studies (ISAS/SISSA), Trieste, Italy
FEE, MICHALE, Massachussets Institute of Technology, Cambridge, Massachusetts
FERNALD, RUSSELL, Stanford University, Stanford, California
FISCHER, JAN, University of Zurich, Zollikon, Switzerland
FISCHLER, WALTER, Columbia University, New York, New York
FISHMAN, ZVI, Columbia University, New York, New York
FLAVELL, STEVEN, Howard Hughes Medical Institute/Rockefeller University, New York, New York
FRANK, LOREN, University of California, San Francisco, San Francisco, California
FRANKS, KEVIN, Duke University, Durham, North Carolina
FREEMAN, JEREMY, Howard Hughes Medical Institute, Ashburn, Virginia
FREIWALD, WINRICH, The Rockefeller University, New York City, New York

FUKUNAGA, IZUMI, National Institute for Medical Research, London, United Kingdom
FULLER, GREGORY, Johns Hopkins University, Baltimore, Maryland
GANN, ALEX, Cold Spring Harbor Laboratory, Cold Spring Harbor, New York
GARBE, DAVID, University of Pennsylvania, Philadelphia, Pennsylvania
GARCIA DA SILVA, PEDRO, Cold Spring Harbor Laboratory, Cold Spring Harbor, New York
GARDNER, DANIEL, Weill Cornell Medical College, New York, New York
GARDNER, ESTHER, New York University School of Medicine, New York, New York
GAZZALEY, ADAM, University of California, San Francisco, San Francisco, California
GHOSH, SANCHARI, Cold Spring Harbor Laboratory, Cold Spring Harbor, New York
GLIMCHER, PAUL, New York University, New York, New York
GOARD, MICHAEL, Massachusetts Institute of Technology, Cambridge, Massachusetts
GOLDEN, JAMES, Cornell University, Ithaca, New York
GOLDSTEIN, ANN, *Neuron*, Cambridge, Massachusetts
GRAYBIEL, ANN, Massachusetts Institute of Technology, Cambridge, Massachusetts
GREENE, JOSHUA, Rockefeller University, New York, New York
GROPPE, DAVID, Feinstein Institute for Medical Research, Manhasset, New York
GU, YIRAN, New York University, New York, New York
GU, ZIRONG, Cincinnati Childrens Hospital Medical Center, Cincinnati, Ohio
GUTNISKY, DIEGO, Janelia Farm Research Campus, Ashburn, Virginia
HANGYA, BALAZS, Cold Spring Harbor Laboratory, Cold Spring Harbor, New York
HANSEN, CELIA, University of California, San Francisco, San Francisco, California
HARRIS, HILA, Weizmann Institute of Science, Rehovot, Israel
HATTORI, DAISUKE, Columbia University, New York, New York
HECK, DETLEF, University of Tennessee Health Science Center, Memphis, Tennessee
HIROKAWA, JUNYA, Cold Spring Harbor Laboratory, Cold Spring Harbor, New York
HOCEVAR BREZAVSCEK, ANA, The Rockefeller University, New York, New York
HOEKSTRA, HOPI, Harvard University, Cambridge, Massachusetts
HONG, INGIE, Johns Hopkins Medical Institute, Baltimore, Maryland
HOUSE, PATRICK, Stanford University, Stanford, California
HUGHES, VIRGINIA, *National Geographic, Popular Science, MATTER*, Brooklyn, New York
HURTADO-LÓPEZ, JULIÁN, Universidad Autónoma de Occidente, Cali, Colombia
IWASAKI, KENICHI, University of Michigan, Ann Arbor, Michigan
JAKIMO, ALAN, Hofstra University, Hempstead, New York
JARAMILLO, SANTIAGO, Cold Spring Harbor Laboratory, Cold Spring Harbor, New York
JESSELL, THOMAS, Columbia University/Hammer Health Sciences Center, New York, New York
JIANG, PING, University of Helsinki, Helsinki, Finland
JIN, XIN, Rockefeller University, New York, New York
JITSUKI, SUSUMU, Yokohama City University, Yokohama, Japan
JONES, ADAM, National Institute of Mental Health/National Institutes of Health, Bethesda, Maryland
KAHN, KEVIN, Johns Hopkins University, Baltimore, Maryland
KANWISHER, NANCY, Massachusetts Institute of Technology, Cambridge, Massachusetts
KARSH, NOAM, University of Haifa, Haifa, Israel
KAUFMAN, MATTHEW, Cold Spring Harbor Laboratory, Cold Spring Harbor, New York
KEANE, BRIAN, Rutgers University, Piscataway, New Jersey
KEBSCHULL, JUSTUS, Cold Spring Harbor Laboratory, Cold Spring Harbor, New York
KENNEDY, ANN, Columbia University, New York, New York
KEPECS, ADAM, Cold Spring Harbor Laboratory, Cold Spring Harbor, New York
KERR, MATTHEW, Johns Hopkins University, Baltimore, Maryland
KIM, JINEUN, Korea Advanced Institute of Science and Technology, Daejeon-si, South Korea
KIM, JOSHUA, Massachusetts Institute of Technology, Boston, Massachusetts
KIM, YONGSOO, Cold Spring Harbor Laboratory, Cold Spring Harbor, New York
KLEINFELD, DAVID, University of California, San Diego, La Jolla, California
KOENIG, SEBASTIAN, Universität Würzburg, Würzburg, Germany
KORFF, RICKI, Cornell University, Ithaca, New York
KOTTMANN, ANDREAS, City College of New York, New York, New York
KRISHNAN, KEERTHI, Cold Spring Harbor Laboratory, Cold Spring Harbor, New York
KRISTAN, JR., WILLIAM, University of California, San Diego, La Jolla, California
KRUG, LISA, Cold Spring Harbor Laboratory, Cold Spring Harbor, New York
KUBIE, JOHN, State University of New York Downstate Medical Center, Brooklyn, New York
KUCHIBHOTLA, KISHORE, New York University School of Medicine, New York, New York
KUHL, PATRICIA, University of Washington, Seattle, Washington
KUMAR, VIVEK, University of Texas Southwestern Medical Center, Dallas, Texas
LARA, ANTONIO, Columbia University, New York, New York
LARGE, ADAM, University of Pittsburgh, Pittsburgh, Pennsylvania
LAU, BILLY, Cold Spring Harbor Laboratory, Cold Spring Harbor, New York
LEDOUX, JOSEPH, New York University, New York, New York
LEE, EUNJEONG, National Institute of Mental Health/National Institutes of Health, Bethesda, Maryland
LEE, HYE YOUNG, University of California, San Francisco, San Francisco, California
LEE, RAY, Princeton University, Princeton, New Jersey
LI, BO, Cold Spring Harbor Laboratory, Cold Spring Harbor, New York
LI, JENNIFER, Harvard University, Cambridge, Massachusetts
LIU, HAIXIN, State University of New York at Stony Brook, Stony Brook, New York
LO, SHIH-CHING, Genentech Inc., South San Francisco, California
LOCKERY, SHAWN, University of Oregon, Eugene, Oregon
LOEHFELM, ALINE, University of Saarland, Saarbrücken, Germany
LONDON, BRIAN, Columbia University, New York, New York
LOSITSKY, OLGA, Princeton University, Princeton, New Jersey
LUO CLAYTON, ALICE, Simons Foundation, New York, New York
LVOVSKAYA, SVETLANA, University of Texas Southwestern Medical Center, Dallas, Texas
LYON, GHOLSON, Cold Spring Harbor Laboratory, Cold Spring Harbor, New York
MACHADO, TIMOTHY, Columbia University, New York, New York
MAFFEI, ARIANNA, State University of New York at Stony Brook, Stony Brook, New York
MAINEN, ZACHARY, Champalimaud Neuroscience Programme, Lisbon, Portugal
MALAIA, EVIE, University of Texas at Arlington, Arlington, Texas
MARASCO, PAUL, Lerner Research Institute, Cleveland Clinic, Cleveland, Ohio
MARDER, EVE, Brandeis University, Waltham, Massachusetts
MARLIN, BIANCA, New York University School of Medicine, New York, New York
MASSET, PAUL, Cold Spring Harbor Laboratory, Cold Spring Harbor, New York
MATHURU, AJAY, Institute of Molecular and Cell Biology, Singapore, Singapore
MAZZUCATO, LUCA, State University of New York at Stony Brook, Stony Brook, New York
MCCOMBIE, RICHARD, Cold Spring Harbor Laboratory, Cold Spring Harbor, New York
MCKINSEY, GABRIEL, University of California, San Francisco, San Francisco, California
MEHTA, ASHESH, Feinstein Institute for Medical Research, Great Neck, New York
MEISTER, MARKUS, California Institute of Technology, Pasadena, California

Michely, Julia, University of Saarland, Saarbrücken, Germany
Mizuno, Keiko, Kings College London, London, United Kingdom
Moore, Jeffrey, University of California, San Diego, La Jolla, California
Movshon, J. Anthony, New York University, New York, New York
Mu, Yu, Janelia Farm Research Campus, Howard Hughes Medical Institute, Ashburn, VIrginia
Murayama, Masanori, RIKEN, Wako, Japan
Murray, Andrew, Columbia University, New York, New York
Nagayama, Shin, University of Texas, Houston, Houston, Texas
Nakajima, Miho, The Rockefeller University, New York, New York
Nakayama, Hirofumi, The Rockefeller University, New York, New York
Nalbantian, Suzanne, Brookville, New York
Narain, Charvy, *Nature Neuroscience*, London, United Kingdom
Neftci, Emre, University of California, San Diego, La Jolla, California
Nemes, Adriana, Columbia University, New York, New York
Nestler, Eric, Mount Sinai University, New York, New York
Neunuebel, Joshua, Howard Hughes Medical Institute, Janelia Farms Research Campus, Ashburn, Virginia
Niv, Yael, Princeton University, Princeton, New Jersey
Nizami, Lance, Independent Research Scholar, Palo Alto, California
Norovich, Amy, Columbia University, New York, New York
Nunes Raposo, David, Cold Spring Harbor Laboratory, Cold Spring Harbor, New York
Oboti, Livio, Children's National Health System, Washington, D.C.
Odoemene, Kachi, Cold Spring Harbor Laboratory, Cold Spring Harbor, New York
Oka, Yuki, Columbia University, New York, New York
O'rawe, Jason, Cold Spring Harbor Laboratory, Cold Spring Harbor, New York
Oswald, Anne-Marie, University of Pittsburgh, Pittsburgh, Pennsylvania
Otazu, Gonzalo, Cold Spring Harbor Laboratory, Cold Spring Harbor, New York
Pachitariu, Marius, University College London, London, United Kingdom
Patel, Gaurav, Columbia University/New York State Psychiatric Institute, New York, New York
Peikon, Ian, Cold Spring Harbor Laboratory, Cold Spring Harbor, New York
Penzo, Mario, Cold Spring Harbor Laboratory, Cold Spring Harbor, New York
Pi, Hyun Jae, Cold Spring Harbor Laboratory, Cold Spring Harbor, New York
Pollock, Mila, Cold Spring Harbor Laboratory, Cold Spring Harbor, New York
Porciello, Giuseppina, Sapienza University, Rome, Italy
Pouget, Alexandre, University of Geneva, Geneva, Switzerland and University of Rochester, Rochester, New York
Preuschoff, Kerstin, École Polytechnique Fédérale de Lausanne, Lausanne, Switzerland
Punzi, Giovanna, Lieber Institute for Brain Development, Baltimore, Maryland
Qiao, Mu, Harvard University, Cambridge, Massachusetts
Ramírez-Moreno, David, Universidad Autónoma de Occidente, Cali, Colombia
Ranade, Sachin, Cold Spring Harbor Laboratory, Cold Spring Harbor, New York
Reardon, Thomas, Columbia University, New York, New York
Redondo, Roger, Howard Hughes Medical Institute, Massachusetts Institute of Technology, Cambridge, Massachusetts
Reid, Ashlan, Cold Spring Harbor Laboratory, Cold Spring Harbor, New York
Reimers, Mark, Virginia Commonwealth University, Richmond, Virginia
Reynolds, John, The Salk Institute, La Jolla, California
Robert, Jason, Arizona State University, Tempe, Arizona
Robson, Drew, Harvard University, Cambridge, Massachusetts
Romo, Ranulfo, Universidad Nacional Autónoma de México (UNAM), Mexico City, Mexico
Roos, Annerine, Stellenbosch University, Cape Town, South Africa
Root, Cory, Columbia University, New York, New York
Rothschild, Gideon, University of California, San Francisco, San Francisco, California
Rushworth, Matthew, University of Oxford, Oxford, United Kingdom
Sabes, Phillip, University of California, San Francisco, San Francisco, California
Sanders, Joshua, Cold Spring Harbor Laboratory, Cold Spring Harbor, New York
Sasse, Sofia, University of Münster, Münster, Germany
Schaefer, Andreas, MRC National Institute for Medical Research, London, United Kingdom
Schaefer, Anne, Icahn School of Medicine at Mount Sinai, New York, New York
Schuck, Nicolas, Princeton University, Princeton, New Jersey
Schwartz, Edmund, Columbia University, New York, New York
Seely, Jeffrey, Columbia University, New York, New York
Sejnowski, Terrence, The Salk Institute for Biological Studies, La Jolla, California
Selimi, Fekrije, INSERM, Paris, France
Shadlen, Michael, Columbia College of Physicians and Surgeons, New York, New York
Shankar, Shruti, Temasek Life Sciences, Singapore, Singapore
Shea, Stephen, Cold Spring Harbor Laboratory, Cold Spring Harbor, New York
Sheppard, John, Cold Spring Harbor Laboratory, Cold Spring Harbor, New York
Shykind, Benjamin, Weill Cornell Medical College in Qatar, Doha, Qatar
Simoncelli, Eero, New York University, New York, New York
Singer, Emily, Simons Foundation/*Quanta Magazine*, New York, New York
Slifkin, Andrew, Cleveland State University, Cleveland, Ohio
Sosulski, Dara, University College London, London, United Kingdom
Sridharan, Devarajan, Stanford University School of Medicine, Stanford, California
Stephenson-Jones, Marcus, Cold Spring Harbor Laboratory, Cold Spring Harbor, New York
Stewart, David, Cold Spring Harbor Laboratory, Cold Spring Harbor, New York
Stillman, Bruce, Cold Spring Harbor Laboratory, Cold Spring Harbor, New York
Stix, Gary, *Scientific American*, New York, New York
Stringer, Carsen, University College London, London, United Kingdom
Stryker, Michael, University of California, San Francisco, San Francisco, California
Svoboda, Karel, Howard Hughes Medical Institute, Ashburn, Virginia
Swartz, Jerome, Huntington, New York
Swartz gordon, Shanah Lloyd Harbor, New York
Tada, Hirobumi, Yokohama City University, School of Medicine, Yokohama, Japan
Talley, Edmund, National Institutes of Health, Rockville, Maryland
Toepfer, Franziska, University of Würzburg, Würzburg, Germany
Tonegawa, Susumu, RIKEN Brain Science Institute and Massachusetts Institute of Technology, Saitama, Japan
Tucciarone, Jason, Cold Spring Harbor Labs, Cold Spring Harbor, New York
Underwood, Emily, Science, Washington, D.C.
Unnikrishnan, M.K., Manipal College of Pharmaceutical Sciences, Manipal, India
Upson, Sandra, Scientific American Mind, New York, New York
Ursini, Gianluca, Lieber Institute for Brain Development, Baltimore, Maryland
Vaughan, Alexander, Cold Spring Harbor Laboratory, Cold Spring Harbor, New York
Ventimiglia, Donovan, Rockefeller University, New York, New York
Venturini, Rita, Champalimaud Foundation, Lisbon, Portugal
Vilankar, Kedarnath, Cornell University, Ithaca, New York
Wang, Bor-Shuen, Cold Spring Harbor Laboratory, Cold Spring Harbor, New York
Weber, Franz, University of California, Berkeley, Berkeley, California
Wekselblatt, Joseph, University of Oregon, Eugene, Oregon
Wigler, Michael, Cold Spring Harbor Laboratory, Cold Spring Harbor, New York

WITKOWSKI, JAN, Cold Spring Harbor Laboratory, Cold Spring Harbor, New York
WOLFE, GLENN, University of Toronto, Toronto, Canada
WOLPERT, DANIEL, University of Cambridge, Cambridge
WONG, LI CHIN, New York University School of Medicine, New York, New York
XIONG, QIAOJIE, Cold Spring Harbor Laboratory, Cold Spring Harbor, New York
YAMAZAKI, DAISUKE, University of Tokyo, Bunkyo-ku, Japan
YANG, ZHENGHONG, University of Würzburg, Würzburg, Germany
YATES, DARRAN, *Nature Reviews Neuroscience*, London, United Kingdom
YU, C. RON, Stowers Institute for Medical Research, Kansas City, Missouri
YU, KAI, Cold Spring Harbor Laboratory, New York, New York
ZADOR, TONY, Cold Spring Harbor Laboratory, Cold Spring Harbor, New York
ZANDBELT, BRAM, Vanderbilt University, Nashville, Tennessee
ZHANG, BYOUNGTAK, Seoul National University, Seoul, South Korea
ZHAO, BUYUN, University of Cambridge, Cambridge, United Kingdom
ZHONG, CHUN, Johns Hopkins University, Baltimore, Maryland
ZHOU, CHUAN, Howard Hughes Medical Institute, Janelia Farms, Ashburn, Virginia

Row 1: A. Loehfelm, S. Batsching; K. Svoboda, B. Stillman; G. Buzsáki, J. Kubie, S. Tonegawa
Row 2: D. Stewart, C. Bargmann; Y. Mu, K. Kuchibhotla, K. Chen, B. Marlin; P. Churchland
Row 3: C. Narain, D. Yates; A.F. Ebihara, G. Otazu; C. Duan, A. Akrami
Row 4: R. Axel; T. Zador; B. Marlin; G. Turner

Row 1: A. Kennedy, A. Batchelor; M. Carrasco, Y. Niv; D. Yates, N. Karsh
Row 2: A. Goldstein, M. Shadlen; Y. Dan, J.A. Movshon; J. Witkowski, P. Churchland
Row 3: E. Brush, V. Dubowitz; A. Loehfelm, S. Batsching, J. Michely; M. Stryker, D. Bavelier
Row 4: S. Lockery, I. Carcea; R. Fernald, G. Mckinsey; V. Kumar, I. Abdus-Saboor

Row 1: R. Axel, M. Shadlen, P. Glimcher; P. House; K. Vilankar, J. Golden
Row 2: S. Dasgupta, H. Harris; G. Stix; A. Graybiel; M. Rushworth
Row 3: J. DiCarlo; A. Pouget, D. Bavelier, P. Churchland, J. DiCarlo; J. Reynolds, M. Giao
Row 4: F. Toepfer, S. Koenig, L. Oboti; N. Kanwisher, M. Carrasco; E. Nestler

Row 1: G. Rothschild, M. Agetsuma; C. Brody, K. Svoboda, A. Pouget
Row 2: S. Jaramillo, I. Carcea; G. Fuller, J. Wekselblatt, D. Heck; D. Anderson
Row 3: A. Schaefer, F. Selimi; D. Gardner, D. Kleinfeld, R. Axel; M. Shadlen, D. Angelaki
Row 4: D. Gardner; R. Lee, J. Kubie; J. Watson, H. Hoekstra, A. Gann

Row 1: J. Freeman, D. Gutnisky; A. Slifkin, K. David; D. Ramirez-Moreno, J. Hurtado-Lopez
Row 2: A. Apergis-Schoute, B. Zhao, K. Deisseroth; O. Lositsky, P. Coen
Row 3: Z. Fishman; Z. Mainen; H. Furukawa; J. LeDoux
Row 4: R. Desimone, T. Sejnowski; D. Stewart, D. Kleinfeld

Row 1: G. Lyon, Y. Gu; A. Mathuru, E. Underwood; D. Wolpert
Row 2: Poster Session; Symposium Picnic; K. Deisseroth
Row 3: L. Frank, G. Buzsáki; A. Mathuru, A. Schaefer; A. Mehta, D. Groppe
Row 4: D. Sridharan; C. Narain, A. Gazzaley; P. Kuhl

Foreword

The Laboratory selected the theme of Cognition for the 79th Symposium in this historic series largely because of the tremendous advances being made by neuroscientists and psychologists working on cognitive processes at scales varying from the molecular to whole-brain and theoretical studies. Previous Symposia that addressed the brain included the fourth Symposium on Excitation Phenomena (1936), The Neuron (1952), Sensory Receptors (1965), The Synapse (1975), Molecular Neurobiology (1983), The Brain (1990), and Function & Dysfunction in the Nervous System (1996), so a return to the general theme of neuroscience was long overdue. Furthermore, a number of national projects on brain function have been initiated over the last few years, including President Barack Obama's announcement of the BRAIN (Brain Research through Advancing Innovative Neurotechnologies) Initiative in April 2013, a collaborative research project akin to the Human Genome Project, with the ultimate goal of mapping the activity of every neuron in the brain by 2025. These projects and initiatives reflect the fact that many neuroscientists are increasingly optimistic that major obstacles in understanding brain function may be overcome through the use of new technologies and approaches, and that these advances may shed light on novel approaches to treat psychiatric, neurological, and neurodegenerative diseases.

The 2014 Symposium was designed to span a broad range of approaches toward increased understanding of cognitive processes in the brain including cell/molecular biology; developmental neurobiology; genetics and genomics; electrophysiological approaches; functional neuroimaging at cellular and whole-brain resolutions; computational neuroscience; behavioral, ethological, and psychophysical studies; and evolutionary/comparative neuroscience. The program of invited speakers was arranged to ensure that fundamental discoveries were balanced with approaches relevant to societal well-being, including a variety of stratagems for harnessing our increased understanding of brain function to improve treatment of mental illness and brain disorders. Speakers responded to the challenge of putting their own research, whether based on single-neuron studies, circuits, or whole-brain studies, into a broader context, allowing for a tremendous amount of cross talk across disciplines with much insight gained from this unusually broad meeting. This is one of the strengths of the Symposium, and the 2014 meeting was particularly successful in highlighting emerging connections between molecular/cellular networks and higher brain functions such as language and decision-making.

Opening night speakers included Richard Axel (Columbia University) on innate and learned responses to odors, Allison Doupe (University of California, San Francisco) on vocal motor plasticity, Matthew Rushworth (University of Oxford) on decision-making, and Patricia Kuhl (University of Washington) on language acquisition. One of the three Churchland neuroscientists who spoke at the Symposium, Patricia Churchland (University California, San Diego) addressed "The Brains Behind Morality" in an excellent Dorcas Cummings lecture for Laboratory friends, neighbors, and Symposium participants in advance of the annual dinner parties. Terry Sejnowski delivered a compelling and comprehensive summary that captured much of the research and discussions presented at the meeting.

This Symposium was attended by almost 300 scientists from both U.S. and international universities, and the program included 60 invited presentations, six short talks selected from the openly submitted abstracts on the basis of scientific merit, and 130 poster presentations. To disseminate the latest results and discussion of the Symposium to a wider audience, attendees were able to share many of the Symposium talks with their colleagues who were unable to attend using the Leading Strand video archive. A collection of interviews by Karen Carniol (*Cell*), Ann Goldstein (*Neuron*), Charvy Narain (*Nature Neuroscience*), Gary Stix (*Scientific American*), and Jan Witkowski (CSHL Banbury Center) with leading experts in the field were arranged during the Symposium, distributed from the Cold Spring Harbor Symposium interviews website, and available in edited form in this volume.

We thank Val Pakaluk, Mary Smith, Ed Campodonico, and his staff in the Meetings & Courses Program for their assistance in organizing and running the Symposium, and John Inglis and his staff at Cold Spring Harbor Laboratory Press, particularly Inez Sialiano and Jan Argentine.

Cori Bargmann
Daphne Bavelier
Terrence Sejnowski
David Stewart
Bruce Stillman

Sponsors

Funded in part by Forest Laboratories and The Swartz Foundation.

Contributions from the following companies provide core support for the Cold Spring Harbor meetings program.

Corporate Sponsors

Agilent Technologies
Bristol-Myers Squibb Company
Genentech
Life Technologies (part of Thermo Fisher Scientific)
New England BioLabs

Plant Corporate Associates

Monsanto Company
DuPont/Pioneer Hi-Bred International

Foundations

HudsonAlpha Institute for Biotechnology

Contents

A Neural Circuit That Controls Cortical State, Plasticity, and the Gain of Sensory Responses in Mouse

MICHAEL P. STRYKER
Center for Integrative Neuroscience, Department of Physiology, University of California, San Francisco, California 94143-0444
Correspondence: stryker@phy.ucsf.edu

Neurons in the visual cortex were first found to be exquisitely selective for particular properties of visual stimuli in anesthetized animals, including mice. Studies of alert mice in an apparatus that allowed them to stand or run revealed that locomotion causes a change in cortical state that dramatically increases the magnitude of responses in neurons of the visual cortex without altering selectivity, effectively changing the gain of sensory responses. Locomotion also dramatically enhances adult plasticity in the recovery from long-term visual deprivation. We have studied the elements and operation of the neural circuit responsible for the enhancement of activity and shown that it enhances plasticity even in mice not free to run. The circuit consists of projections ascending from the midbrain locomotor region (MLR) to the basal forebrain, activating cholinergic and perhaps other projections to excite inhibitory interneurons expressing vasoactive intestinal peptide (VIP) in the visual cortex. VIP cells activated by locomotion inhibit interneurons that express somatostatin (SST), thereby disinhibiting the excitatory principal neurons and allowing them to respond more strongly to effective visual stimuli. These findings reveal in alert animals how the ascending reticular activating system described in anesthetized animals 50 years ago operates to control cortical state.

Excitatory neurons in the upper layers of the primary visual cortex (V1) of the mouse, like those in the cat and monkey that were studied 50 years ago (Hubel and Wiesel 1962), are highly selective for particular visual stimuli (Niell and Stryker 2008). Most are classic simple cells, with extended regions in the visual field responding to light increment and others responding to light decrement and generating responses at a limited range of orientations of a bar or edge stimulus and over a limited range of spatial frequencies of a grating. The preferred orientation and spatial frequency can be predicted from the arrangement of the ON and OFF regions as mapped with flashing spots (Figs. 1 and 2).

LOCOMOTION INCREASES THE GAIN OF VISUAL RESPONSES

Studying receptive fields of alert mice whose heads were fixed in an apparatus that allowed them to stand or run freely on a large Styrofoam ball floating on air (Dombeck et al. 2007) revealed a new phenomenon: Neural responses to preferred visual stimuli became dramatically larger when the mice started to walk or run but were no less selective than when the mice were stationary or anesthetized (Niell and Stryker 2010). Locomotion appeared simply to multiply the tuning functions of response as a function of orientation or spatial frequency by a constant factor for each neuron, like changing the gain of an amplifier. No similar changes were evident in simultaneous recordings from the lateral geniculate nucleus, the source of visual input to V1, revealing that the gain changes were central, taking place in the cortex and not in the eyes or the thalamus (Fig. 3).

LOCOMOTION ENHANCES ADULT PLASTICITY

In the mouse, as in the human and the carnivores and primates studied earlier, visual responses to the two eyes can be dramatically and lastingly perturbed by altered visual experience during a brief critical period in early life (Fig. 4) (Gordon and Stryker 1996). Blurring or occluding the vision of one eye during this critical period of susceptibility to the effects of monocular visual deprivation leads to a permanent reduction of visual acuity and impairment of other visual functions, a condition referred to as *amblyopia ex anopsia*. This form of amblyopia is associated with a pruning of the terminal arbors of thalamocortical afferents to V1 serving the deprived eye. In the cat, this results in a loss of 50% of the synaptic boutons and arbor length after a week of deprivation (Antonini and Stryker 1993). Not surprisingly, given the loss of much of the deprived-eye input to V1, longstanding amblyopia induced during the critical period is generally refractory to therapy in humans and in carnivores and primates. In contrast to the dramatic plasticity observed during the critical period, monocular visual deprivation in adulthood, or indeed at any time after the critical period, has little lasting effect in these species. Although visual deprivation does produce changes in visual responses in adult mice, the adult plasticity is significantly slower, smaller, and qualitatively different from critical period plasticity (Sato and Stryker 2008).

 doi: 10.1101/sqb.2014.79.024927

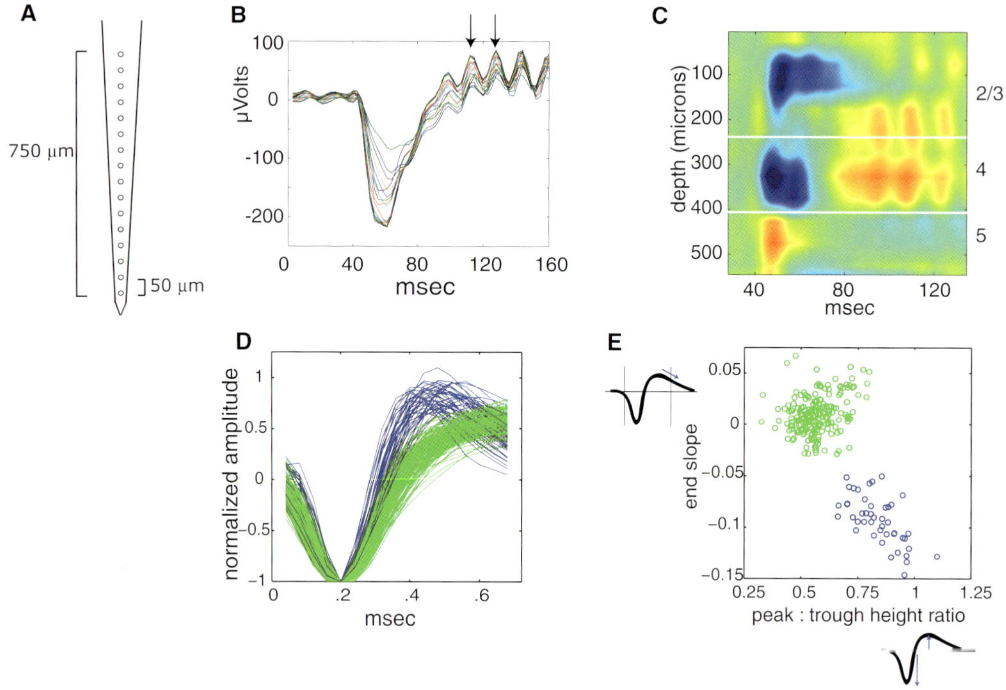

Figure 1. Multisite recording and analysis technology that made characterization of mouse visual cortex feasible by studying an average of more than 10 neurons simultaneously, determining cortical layer containing each recording site, and classifying spike waveforms as presumptive excitatory or inhibitory with no overlap. (*A*) Schematic of linear multisite probe. (*B*) Average local field potential (LFP) responses for 16 sites through the depth of cortex. Arrows show consecutive peaks of the high-frequency oscillation. (*C*) Current source density (CSD) analysis of traces in *B* reveals positions of cortical layers. (*D*) Average spike waveforms for all units analyzed ($n = 231$), normalized by trough depth, demonstrating narrow-spiking (blue), and broad-spiking (green) units. (*E*) Scatter plot of spike waveform parameters for all units. (Data from Niell and Stryker 2008.)

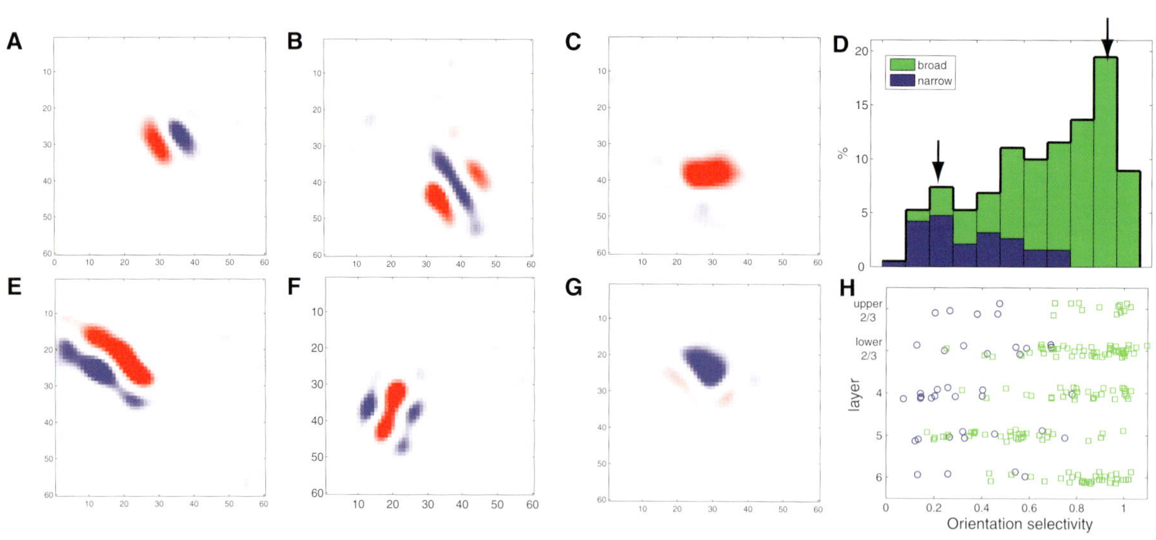

Figure 2. Receptive fields in normal mouse visual cortex measured by spike-triggered averaging (*A–C*, *E–H*). Examples of spatial receptive fields with two, three, and one subfield, respectively, showing varying orientation, ON/OFF centers, and spacing of subfields. Red indicates ON responses; blue indicates OFF responses. (*D*) Distribution of orientation selectivity in broad-spiking (presumptive excitatory) and narrow-spiking (presumptive inhibitory) cells. (*H*) Orientation selectivity of individual neurons in relation to cortical layer. Note that nearly all broad-spiking cells in the upper layers are highly selective. (*A–D*,*F*, Reprinted from Niell and Stryker 2008, with permission from the Society for Neuroscience.)

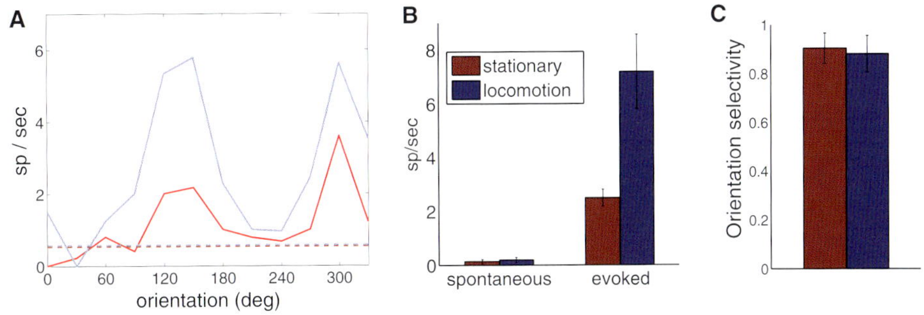

Figure 3. Enhancement of visual responses by locomotion in primary visual cortex of alert mice. (*A*) Visual response to a drifting grating as a function of direction of movement during locomotion (blue) and while stationary (red). (*B*) Average spontaneous and visually evoked responses in all broad-spiking neurons in layers 2–4. (*C*) Average selectivity for stimulus orientation in all broad-spiking neurons in layers 2–4. Note that locomotion increases magnitude of response without changing spontaneous activity or stimulus selectivity. (Data from Niell and Stryker 2010.)

In mice and rats, a number of manipulations were reported to enhance plasticity in adult animals after the end of critical period. Some manipulations produced rapid changes in balance of V1 responses to the two eyes as a result of monocular deprivation during adulthood. Others were reported to facilitate partial or complete recovery in adults from the effects of longstanding amblyopia induced during the critical period. The effective manipulations included housing or rearing in an enriched environment (containing toys, other mice, and space to climb and jump and run) and administration of antidepressants or other drugs (Sale et al. 2007; Maya Vetencourt et al. 2008; Greifzu et al. 2014). It seemed to us plausible that a common feature of these manipulations might be the increased locomotor activity of the subjects. We speculated that the larger visual responses during locomotion would, under any theory of activity-dependent plasticity, give rise to greater plasticity.

To test this hypothesis, we studied the effect of locomotion on recovery from amblyopia in the mouse (Kaneko and Stryker 2014). Imaging the intrinsic optical signal of visual responses through the intact skull is a noninvasive means for precise, quantitative measurement of neural response to the left and right eyes that can be

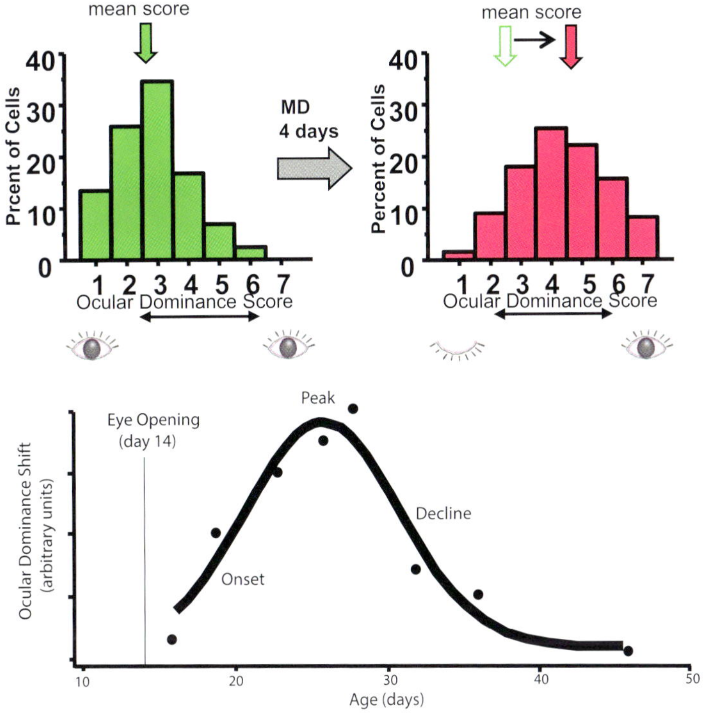

Figure 4. Ocular dominance plasticity during the critical period in mouse visual cortex. (*Top*) Ocular dominance histograms showing numbers of single neurons as a function of their relative responses to the two eyes before (*left*, green) and after (*right*, red) 4 d of occlusion of vision in the contralateral eye. (*Bottom*) Time course of the critical period of sensitivity to the effects of monocular visual deprivation (MD) in the primary visual cortex of the mouse. (Adapted from Gordon and Stryker 1996, with permission from the Society for Neuroscience.)

performed repeatedly on individual animals, making it ideal for tracking recovery of function after deprivation (Kalatsky and Stryker 2003; Kaneko et al. 2008). Mice were deprived of vision in one eye by lid suture through the critical period and continuing into the fifth month of life, leaving them amblyopic. The deprived eye was then opened and the restoration of visual responses was measured. As expected, and as is the case with human amblyopia, responses never recovered more than half way to normal levels in control mice housed in their conventional home cages for the next 3 wk. Experimental mice were allowed to run on the Styrofoam ball for 4 h/d while viewing high-contrast visual stimuli on a monitor in front of them. These animals rapidly recovered normal levels of intrinsic signal response to the stimuli to which they were exposed. Two additional control groups were studied to determine whether recovery would be produced by locomotion alone, without high-contrast visual stimulation, or by high-contrast visual stimulation alone, without the opportunity for locomotion. Neither of these additional control groups recovered significantly more than those housed in their home cages without visual stimulation. These findings were consistent with our hypothesis that high-gain visual responses might be the common explanation for many phenomena of adult plasticity (Fig. 5).

The recovery from deprivation stimulated by locomotion did not, however, produce a normal visual cortex. Instead, the recovery was highly specific to the stimuli that the animal had viewed during locomotion. Responses to other stimuli that are equally effective in intact mice failed to recover. These findings suggested that plasticity took place only in the specific neural circuits that were activated during the high-gain state.

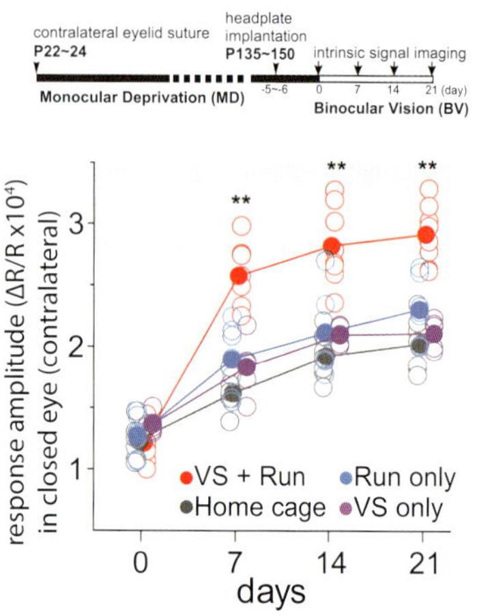

Figure 5. High-contrast visual stimulation during locomotion enhances recovery of visual responses from long-term monocular deprivation. (*Top*) Experimental timeline. (*Bottom*) Amplitude of response to contralateral eye, measured by intrinsic signal imaging, as a functions of time after reopening following ~5 mo of monocular visual deprivation beginning during the critical period. Gray bar indicates the range of response amplitudes in intact mice. Experimental condition (red) with high-contrast visual stimulation during locomotion for 4 h/d. Control conditions in the home cage (black), with locomotion 4 h/d but no high-contrast visual stimulation (blue), or with 8 h/d visual stimulation without locomotion in the home cage (magenta). (Adapted from Kaneko and Stryker 2014.)

We wondered how the responses of the vast majority of excitatory neurons in V1 could recover so well, when so much of the excitatory anatomical input to the visual cortex from the lateral geniculate nucleus had been lost (Antonini et al. 1999) and seemed unlikely to have been restored. Recordings before and after recovery from large numbers of presumptive excitatory and inhibitory single neurons, which can be distinguished reliably on the basis of the duration of the waveforms of their action potentials, suggested an answer. Although the visual responses of excitatory neurons showed recovery specific to the stimulus that accompanied locomotion, as described, the firing rates of all the inhibitory neurons were dramatically depressed (Kaneko and Stryker 2014). It appears that the reduction produced by deprivation in thalamocortical excitation is compensated for during recovery by a corresponding reduction in the magnitude of inhibition. The reduced inhibition permits the excitatory-cell responses to increase to normal levels despite reduced excitatory visual input. Such a neural-circuit-dependent homeostatic process may be quite general, supplementing cellular mechanisms for homeostatic regulation of neural activity levels.

THE SUBCORTICAL CIRCUIT ACTIVATED BY LOCOMOTION

What produces the high-gain state in visual cortex? Because it is turned on by locomotion, exploring the neural circuitry that gives rise to locomotion seemed a promising place to start. Nearly 50 yr ago, research from Orlovskii's laboratory in Moscow revealed a midbrain locomotor region (MLR), within which electrical stimulation induces locomotion at short latencies (Orlovskii et al. 1966). Anatomically, this region loosely coincides with the pedunculopontine tegmental nucleus and the cuneiform nucleus (Garcia-Rill 1991). This area has both descending projections to motor regions and ascending projections, including those to the basal forebrain cholinergic regions that have been shown to modulate behavioral state (Nauta and Kuypers 1958; Skinner et al. 2004). The same region of the brain has been considered to be part of the "ascending reticular activating system," as stimulating this area in anesthetized animals induces physiological changes resembling those of alertness (Fig. 6) (Moruzzi and Magoun 1949).

We injected a virus to transfect excitatory neurons within the MLR with a channelrhodopsin construct, enabling us to activate them optogenetically by shining a blue light onto them through an optical fiber (Lee et al. 2014). Trains of light pulses could reliably make the mice run. We decreased the light pulse intensity until 20 Hz stimulation was just sufficient to make the mice run at short latency. Not surprisingly, the cortex shifted to the high-gain state during optogenetic simulation, showing visual responses as strong as those elicited during natural running. Most informatively, when the optogenetic stim-

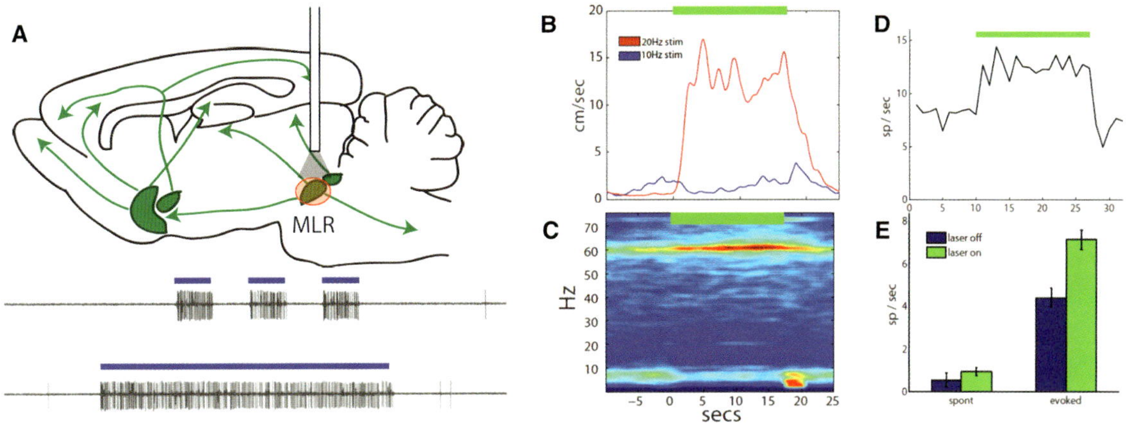

Figure 6. Optogenetic stimulation of excitatory neurons in the midbrain locomotor region (MLR) increases the gain of visual responses in primary visual cortex even without locomotion. (*A*) Drawing of sagittal section showing MLR, region of channelrhodopsin transfection and pathways projecting from MLR. Illumination through optical fiber excites MLR neuron below. (*B*) Locomotion velocity elicited by optogenetic stimulation (green bar) at 20 Hz (red) and failure of locomotion at 10 Hz (blue). (*C*) Increase in γ-band local field potential in V1 with 10-Hz MLR stimulation. (*D*) Example of increase in visual response with 10-Hz MLR stimulation, which did not elicit locomotion. (*E*) Group data showing average increase in visual response with 10-Hz MLR stimulation. (Data from Lee et al. 2014.)

ulus frequency was lowered to 10 Hz, stimulation no longer evoked locomotion, but still drove the cortex into the high-gain state. This finding that MLR stimulation below the threshold for eliciting movement could mimic the effect of locomotion on the visual cortex in stationary animals suggests that change in the gain of visual cortical responses is produced by an ascending projection of the MLR to the forebrain, not by the descending projections of the MLR to motor regions that make the legs move. Consistent with this idea, optogenetic stimulation of in the basal forebrain of these same animals showed that activating the terminals of the MLR cells there could partially mimic the effect of locomotion on cortical gain (Lee et al. 2014).

THE DISINHIBITORY CIRCUIT IN VISUAL CORTEX THAT ENHANCES RESPONSES DURING LOCOMOTION

To locate the cortical portion of the circuit by which locomotion increases the responses of visual cortical neurons, we searched for neurons that responded to locomotion in the dark using a variety of mouse lines in which particular types of neurons were altered to express Cre recombinase (Fu et al. 2014). When these Cre lines were crossed with the Ai14 mouse line that conditionally expresses the fluorescent protein tdTomato, the selected cell type was made visible by its red fluorescence. Using the calcium indicator Oregon Green BAPTA-AM (OGB), we could use two-photon microscopy in alert animals to search the upper cortical layers for a cell type that responded to locomotion in the dark. A minor type of inhibitory neuron that expresses vasoactive intestinal peptide (VIP) had a strong, consistent, and sustained response to locomotion (Fig. 7). More than 97% of the non-VIP cells did not respond to locomotion in the dark. This finding suggested that the VIP cells, which were known to have nicotinic cholinergic receptors (Porter et al. 1999), were the starting point within the cortex for the regulation of cortical gain by locomotion.

This finding was initially puzzling, for why would activating an inhibitory cell cause the responses of excitatory cells to increase? Findings in vitro from other laboratories (Pfeffer et al. 2013), however, had shown the major output of VIP cells to be directed not to excitatory cells but to another class of inhibitory neurons, the somatostatin (SST) cells. Activating VIP cells would therefore be thought to reduce the activity of SST cells and thereby relieve the majority of excitatory cells from inhibition, allowing them to respond more strongly to visual stimuli. To test this idea, we made calcium recordings from SST cells in the dark. As predicted, most of them were suppressed by locomotion (Fig. 7).

Two experiments provided additional evidence that the VIP-SST disinhibitory circuit was responsible for putting the cortex into the high-gain state during locomotion (Fu et al. 2014). First, we transfected superficial layer VIP neurons with a form of channelrhodopsin. Optogenetic stimulation of the VIP neurons in stationary mice mimicked the effect of locomotion on responses of excitatory cells to visual stimuli. This finding showed that activating the VIP-SST circuit was sufficient to account for the effect of locomotion on cortical gain. To block the response of VIP cells to locomotion, we used a cruder technique: damaging labeled cells in VIP-Cre; Ai14 mice, in which the VIP cells are red. We first measured the effect of locomotion on the visual responses of a number of excitatory cells made visible using OGB in a single two-photon microscopic field and observed the normal increase in gain. We then focused the laser, one by one, on each VIP cell in the field, figuratively "blowing it away" and causing it to swell. After such photolytic damage to the VIP cells, we measured the visual responses of the same excitatory cells a second time during locomo-

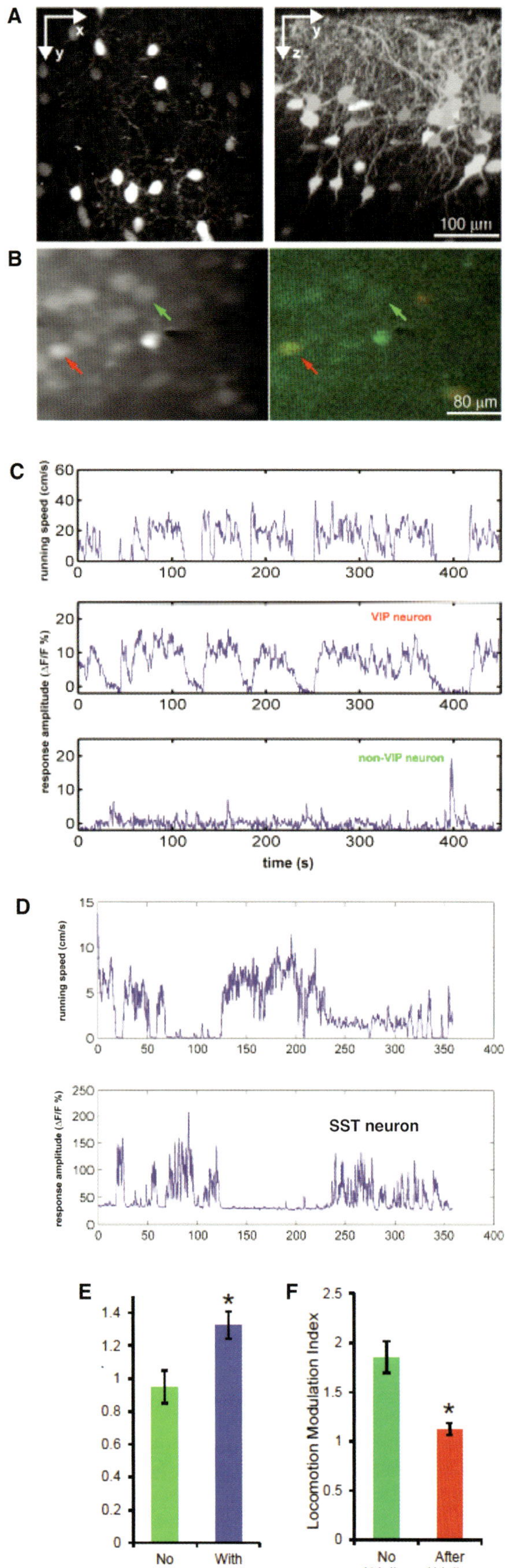

tion and while the mouse was stationary. The visual responses remained strong and selective, just as they were before the damage, but they were no longer enhanced by locomotion, so that the gain of the cortex was no longer increased. This finding shows that the VIP-SST circuit is necessary for the effect of locomotion on cortical gain.

To determine the source of the input conveying the locomotion signal to the VIP cells in visual cortex, we used a technique from the Callaway Laboratory that takes advantage of a modified rabies virus for tracing the monosynaptic inputs to a defined population of cells (Wickersham et al. 2007). We transfected the superficial layer VIP cells in a small region of visual cortex using the tracer virus and naively expected to see the retrograde label in a large population of cells in the motor cortex. Instead, very few cells in the motor cortex were labeled, and the most prominent projection was from the horizontal limb of the nucleus of the diagonal band of Broca (nDBB), a basal forebrain area containing cholinergic neurons that is within or near the region receiving projections from the MLR (not yet confirmed by double-tracer experiments). These cholinergic cells were likely candidates for conveying the locomotor input to the VIP cells. To test this notion further, we measured the effect of infusing nicotinic cholinergic blockers on the responses of VIP cells to locomotion and on the effect of locomotion on excitatory visual responses. Nicotinic blockade reduced both effects by more than two-thirds (Fu et al. 2014).

These findings are consistent with a circuit in which the locomotor signal from the MLR is conveyed via the nDBB to activate the nicotinic receptors of superficial VIP inhibitory cells of the primary visual cortex, which then inhibit the local SST cells and thereby disinhibit the excitatory cells of the visual cortex, making their visual responses larger but not altering their selectivity. This circuit can explain the influence of locomotion on cortical state and is consistent with many previous reports of cholinergic modulation of neural responses (Metherate and Weinberger 1989; Pinto et al. 2013). Because the nicotinic antagonists did not completely block the effect of locomotion in visual cortex, other neuromodulators may also play roles that have yet to be demonstrated in mouse visual cortex.

Figure 7. VIP cells carry the signal of locomotion to visual cortex to regulate cortical gain through inhibition of SST cells. (*A*) VIP cells in surface view of layer 2 (*left*) and in coronal view (*right*) visualized using two-photon microscopy in vivo of VIP-Cre-Ai14 mice. (*B*) VIP and non-VIP cells filled with Oregon Green BAPTA-AM to permit calcium recording, shown in green channel (*left*) and merge red and green channels (*right*). (*C*) Calcium responses of a VIP- and a non-VIP-cell in relation to locomotion in the dark. (*D*) Calcium responses of an SST cell expressing the indicator GCaMP6 in relation to locomotion in the dark. (*E*) Optogenetic activation of VIP cells expressing a form of channelrhodopsin increases cortical gain (With ChETA, blue), whereas similar stimulation in VIP cells without channelrhodopsin (No ChETA, green) does not. (*F*) Photodynamic damage to VIP cells (After ablation, red) blocks increase in cortical gain produced by locomotion before damage (No ablation, green). (Data from Fu et al. 2014.)

IS ENHANCED PLASTICITY THE PRODUCT OF AEROBIC EXERCISE OR THE HIGH-GAIN STATE OF VISUAL CORTEX?

Is this VIP-SST disinhibitory circuit responsible for the dramatic effects of locomotion on adult plasticity described above (Kaneko and Stryker 2014)? Locomotion, in common with other forms of aerobic exercise, might conceivably affect cortical plasticity via multiple mechanisms, including those associated with increases in metabolism, such as blood flow and transport through astrocytes. Indeed, VIP cells themselves might participate in signaling relevant to metabolic activity because they release not only the inhibitory neurotransmitter GABA, but also the peptide VIP. The most powerful effect of VIP on the brain appears to be on glycogenolysis in astrocytes, producing lactate that acts as an energy source (Magistretti et al. 1981). To determine whether locomotion enhances visual cortical plasticity via its aerobic exercise effects or via the increased visual responses produced by the VIP-SST disinhibitory circuit, we sought to block this circuit in running mice and to activate it in stationary mice and measure the resulting effects on cortical plasticity (Fu et al. 2015). We blocked this circuit in running mice using a virus that conditionally expressed tetanus toxin (a generous gift from the Sudhof Laboratory) in VIP-Cre mice. Tetanus toxin blocks the release of neurotransmitter in the cells that express it; it did not alter the survival of the VIP cells during the course of these experiments. Experimental and control mice were subjected to exactly the same regimen of more than 4 mo of monocular visual deprivation beginning in the critical period, as described above (Kaneko and Stryker 2014). After opening the deprived eye, the mice were allowed to run on the trackball while viewing high-contrast visual stimuli. As previously, the control mice rapidly recovered normal levels of visual responsiveness to the stimuli to which they were exposed. In contrast, the VIP-tetanus toxin mice did not recover responses any more than the mice kept in their home cages. This finding indicates that the release of neurotransmitter specifically from VIP cells is essential for the enhancement of adult plasticity by running in mouse visual cortex (Fig. 8) (Fu et al. 2015).

We also explored whether locomotion could enhance plasticity in a short-term assay. In normal adult mice, occluding the vision of the contralateral for 4–5 d has no effect on visual responses to stimulation of either eye; however, monocular deprivation continued past 7 d causes visual cortical responses to stimulation through the open, nondeprived eye to increase while responses through the deprived eye remain the same. This form of adult plasticity is distinct from the plasticity observed during the critical period, where the earliest and most dramatic change is the reduction in responses to deprived-eye stimulation. In normal adult animals that were monocularly deprived and then viewed high-contrast visual stimuli through the ipsilateral eye during locomotion, an increase in response to the open eye was evident after 4–5 d. Tetanus toxin expression in VIP cells prevented this increase. When we did the complementary experiment of optogenetically activating VIP cells in stationary adult mice during short-term monocular deprivation, an increase in open-eye response similar to that in running mice was observed, demonstrating that VIP cell activation without aerobic exercise could mimic the effects of locomotion. Finally, the VIP-SST disinhibitory circuit is hypothesized to alter plasticity via the suppression of SST cell activity by VIP cells during locomotion. To test this hypothesis, we expressed tetanus toxin in SST cells in adult mice and then performed short-term monocular deprivation while the mice were stationary. Blocking SST-cell output by this treatment, like optogenetically activating VIP cells, produced the adult form of plasticity, mimicking the effect of locomotion, but without aerobic activation. Taken together, these findings indicate that the effect of running on adult plasticity is principally mediated through the VIP-SST disinhibitory circuit rather than by some other effect of aerobic activity (Fu et al. 2015).

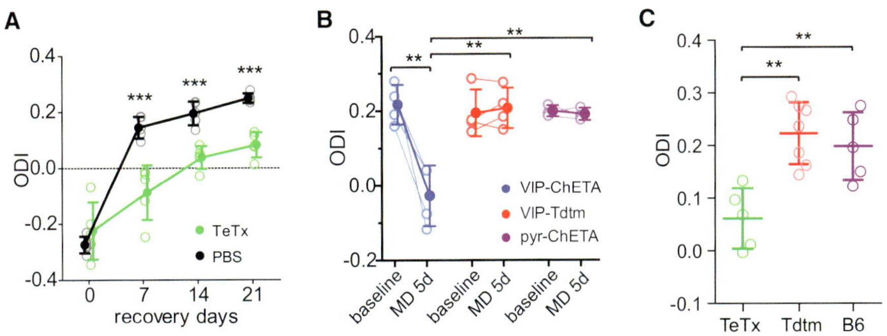

Figure 8. Blockade and activation of VIP-SST disinhibitory circuit regulate cortical plasticity. (*A*) Expression of tetanus toxin to block synaptic release from VIP cells (green) prevents the enhancement of recovery from long-term monocular deprivation by locomotion (black). Compare with Figure 5. (*B*) Optogenetic activation of VIP cells (VIP-ChETA, blue) during 5 d of monocular deprivation in stationary adult mice produces plasticity; similar stimulation of excitatory neurons (pyr-ChETA, magenta) or in VIP cells expressing tdTomato without channelrhodopsin (VIP-Tdtm, red) does not. (*C*) Expression of tetanus toxin in SST cells (TeTx, green) during 5 d of monocular deprivation in stationary adult mice produces plasticity; expression of tdTomato (Tdtm, red) does not, yielding results similar to 5-d MD in C57BL6/J control mice (B6, magenta). (ODI) Ocular dominance index of relative response to the two eyes. (Data from Fu et al. 2015.)

The fact that VIP cells rather than some other type of inhibitory neurons increase visual cortical gain and enhance adult plasticity in a manner similar to the aerobic activity of locomotion may be a coincidence, or it may suggest that the action of VIP peptide in mobilizing lactate metabolism through glycogenolysis is a normal part of the effect of aerobic exercise on the brain. It is plausible that evolution has adapted this form of signaling to increase the metabolic capacity of the brain during the high-gain state, even when sedentary, or perhaps VIP peptide operates as a rapid signal that anticipates other consequences of exercise. It will be interesting to see whether the VIP peptide alone, in the absence of VIP-SST disinhibition of excitatory pathways, might affect cortical plasticity, as experiments in the hippocampus suggest (Donato et al. 2013). Experiments to illuminate this issue in the visual cortex are in progress.

CONCLUSION

Anatomical elements of the VIP-SST disinhibitory circuit exist in many cortical areas and in many mammalian species including humans. At the present time, we do not know how this circuit operates in other cortical areas and species or whether it enhances neural activity or plasticity. Evidence for cholinergic modulation of plasticity is widespread, although some pharmacological studies implicate muscarinic rather than the nicotinic actions that we see in the VIP cells of the superficial layers of the mouse visual cortex.

Neuromodulators other than acetylcholine are also thought to influence neural plasticity. There is long-standing evidence for effects of norepinephrine and serotonin on plasticity in the visual cortex (Bear and Singer 1986; Imamura and Kasamatsu 1989; Kojic et al. 2001) and evidence for actions of dopamine on plasticity in many brain areas (Jay 2003). Indeed, early studies involving stimulation or lesions to the midbrain reticular formation, in areas near or overlapping with the MLR, were thought to implicate actions of these other neuromodulatory systems on plasticity (Singer and Rauschecker 1982). Whether or how these other neuromodulatory systems interact with the VIP-SST disinhibitory circuit is not known. There may well be multiple sites in the cortical circuit that are modulated to alter plasticity.

Deep brain electrical stimulation in the human homolog of the MLR at frequencies similar to those we used here in mice has been explored as therapy for movement disorders such as Parkinson's disease (Stefani et al. 2007). When stimulated, patients commonly report a heightened feeling of alertness (Stefani et al. 2013). It is plausible that alertness would be the subjective feeling associated with the high-gate state of cortex; however, we do not know whether locomotion induces a high-gain state in human visual cortex. In the mouse, a dramatic and sustained increase in the magnitude of γ oscillations near 60 Hz accompanies the high-gain state produced by locomotion or MLR stimulation. Experiments are under way to measure whether γ power in human visual cortex is increased by locomotion. Those, in addition to psychophysical experiments to determine whether perception is enhanced by locomotion, should resolve whether the VIP-SST disinhibitory circuit is activated in humans as it is in the mouse.

Considering the operation of the visual cortex from an information processing point of view, modulation of cortical gain by the VIP-SST disinhibitiory circuit has the tremendously desirable feature that it separates the channels that convey the stimulus specificity of cortical neurons from those that merely modulate their gain. The visually driven inputs are unchanged by locomotion, except insofar as the visual field is moved across the retinae. The filter properties of the cortical network may remain intact, but the responses of neurons become greater because of their relief from nonvisual inhibition. Other circuits in the visual cortex appear to operate similarly by disinhibition. For example, the modulation of cortical gain by activity in layer 6, thought to be responsible for some of the spatial properties of cortical neurons, has recently been shown to be achieved through the action of a special class of parvalbumin-containing inhibitory neurons with somata in layer 6 and axons that terminate in the other cortical layers (Olsen et al. 2012; Bortone et al. 2014).

The notion of the ascending reticular activating system that emerged from studies of anesthetized animals was a useful concept for many years, bringing together phenomena of neuromodulation, rhythmic activity, plasticity, movement, and alertness (Moruzzi and Magoun 1949; Skinner et al. 2004). We are now in a position, in mice at least, to dissect the cell types and circuitry responsible for this wide range of phenomena that come together in a small area of the midbrain and have such wide effects. It is not yet clear which of these phenomena are necessarily associated by common circuitry and which by the happenstance of proximity or common pharmacology that leads to apparent association by common effects of electrical stimulation, lesions, or drugs. One can be optimistic that the tools needed to decide these issues are in place or will be soon.

ACKNOWLEDGMENTS

I gratefully acknowledge the people who have performed most of the research described here, especially Megumi Kaneko, Yu Fu, Cristopher Niell, Moses Lee, and Sebastian Espinosa, and the support, principally from the National Eye Institute under R01EY02874 and from the Simons Collaboration on the Global Brain, that made it possible.

REFERENCES

Antonini A, Stryker MP. 1993. Rapid remodeling of axonal arbors in the visual cortex. *Science* **260:** 1819–1821.

Antonini A, Fagiolini M, Stryker MP. 1999. Anatomical correlates of functional plasticity in mouse visual cortex. *J. Neurosci* **19:** 4388–4406.

Bear MF, Singer W. 1986. Modulation of visual cortical plasticity by acetylcholine and noradrenaline. *Nature* **320:** 172–176.

Bortone DS, Olsen SR, Scanziani M. 2014. Translaminar inhibitory cells recruited by layer 6 corticothalamic neurons suppress visual cortex. *Neuron* **82:** 474–485.

Dombeck DA, Khabbaz AN, Collman F, Adelman TL, Tank DW. 2007. Imaging large-scale neural activity with cellular resolution in awake, mobile mice. *Neuron* **56:** 43–57.

Donato F, Rompani SB, Caroni P. 2013. Parvalbumin-expressing basket-cell network plasticity induced by experience regulates adult learning. *Nature* **504:** 272–276.

Fu Y, Tucciarone JM, Espinosa JS, Sheng N, Darcy DP, Nicoll RA, Huang ZJ, Stryker MP. 2014. A cortical circuit for gain control by behavioral state. *Cell* **156:** 1139–1152.

Fu Y, Kaneko M, Tang Y, Alvarez-Buylla A, Stryker MP. 2015. A cortical disinhibitory circuit for enhancing adult plasticity. *Elife* **4:** e05558.

Garcia-Rill E. 1991. The pedunculopontine nucleus. *Prog Neurobiol* **36:** 363–389.

Gordon JA, Stryker MP. 1996. Experience-dependent plasticity of binocular responses in the primary visual cortex of the mouse. *J Neurosci* **16:** 3274–3286.

Greifzu F, Pielecka-Fortuna J, Kalogeraki E, Krempler K, Favaro PD, Schluter OM, Lowel S. 2014. Environmental enrichment extends ocular dominance plasticity into adulthood and protects from stroke-induced impairments of plasticity. *Proc Natl Acad Sci* **111:** 1150–1155.

Hubel DH, Wiesel TN. 1962. Receptive fields, binocular interaction and functional architecture in the cat's visual cortex. *J Physiol* **160:** 106–154.

Imamura K, Kasamatsu T. 1989. Interaction of noradrenergic and cholinergic systems in regulation of ocular dominance plasticity. *Neurosci Res* **6:** 519–536.

Jay TM. 2003. Dopamine: A potential substrate for synaptic plasticity and memory mechanisms. *Prog Neurobiol* **69:** 375–390.

Kalatsky VA, Stryker MP. 2003. New paradigm for optical imaging: Temporally encoded maps of intrinsic signal. *Neuron* **38:** 529–545.

Kaneko M, Stryker MP. 2014. Sensory experience during locomotion promotes recovery of function in adult visual cortex. *Elife* **3:** e02798.

Kaneko M, Hanover JL, England PM, Stryker MP. 2008. TrkB kinase required for recovery but not loss of cortical responses following monocular deprivation. *Nat Neurosci* **11:** 497–504.

Kojic L, Gu Q, Douglas RM, Cynader MS. 2001. Laminar distribution of cholinergic- and serotonergic-dependent plasticity within kitten visual cortex. *Brain Res Dev Brain Res* **126:** 157–162.

Lee AM, Hoy JL, Bonci A, Wilbrecht L, Stryker MP, Niell CM. 2014. Identification of a brainstem circuit regulating visual cortical state in parallel with locomotion. *Neuron* **83:** 455–466.

Magistretti PJ, Morrison JH, Shoemaker WJ, Sapin V, Bloom FE. 1981. Vasoactive intestinal polypeptide induces glycogenolysis in mouse cortical slices: A possible regulatory mechanism for the local control of energy metabolism. *Proc Natl Acad Sci* **78:** 6535–6539.

Maya Vetencourt JF, Sale A, Viegi A, Baroncelli L, De Pasquale R, O'Leary OF, Castren E, Maffei L. 2008. The antidepressant fluoxetine restores plasticity in the adult visual cortex. *Science* **320:** 385–388.

Metherate R, Weinberger NM. 1989. Acetylcholine produces stimulus-specific receptive field alterations in cat auditory cortex. *Brain Res* **480:** 372–377.

Moruzzi G, Magoun HW. 1949. Brain stem reticular formation and activation of the EEG. *Electroencephalogr Clin Neurophysiol* **1:** 455–473.

Nauta WJH, Kuypers H. 1958. Some ascending pathways in the brain stem reticular formation. In *Reticular formation of the brain* (ed. Jasper HH, Proctor LD, Knighton RS, et al.), pp. 3–30. Little, Brown, Boston.

Niell CM, Stryker MP. 2008. Highly selective receptive fields in mouse visual cortex. *J Neurosci* **28:** 7520–7536.

Niell CM, Stryker MP. 2010. Modulation of visual responses by behavioral state in mouse visual cortex. *Neuron* **65:** 472–479.

Olsen SR, Bortone DS, Adesnik H, Scanziani M. 2012. Gain control by layer six in cortical circuits of vision. *Nature* **483:** 47–52.

Orlovskii GN, Severin FV, Shik ML. 1966. Locomotion induced by stimulation of the mesencephalon. *Dokl Akad Nauk SSSR* **169:** 1223–1226.

Pfeffer CK, Xue M, He M, Huang ZJ, Scanziani M. 2013. Inhibition of inhibition in visual cortex: The logic of connections between molecularly distinct interneurons. *Nat Neurosci* **16:** 1068–1076.

Pinto L, Goard MJ, Estandian D, Xu M, Kwan AC, Lee SH, Harrison TC, Feng G, Dan Y. 2013. Fast modulation of visual perception by basal forebrain cholinergic neurons. *Nat Neurosci* **16:** 1857–1863.

Porter JT, Cauli B, Tsuzuki K, Lambolez B, Rossier J, Audinat E. 1999. Selective excitation of subtypes of neocortical interneurons by nicotinic receptors. *J Neurosci* **19:** 5228–5235.

Sale A, Maya Vetencourt JF, Medini P, Cenni MC, Baroncelli L, De Pasquale R, Maffei L. 2007. Environmental enrichment in adulthood promotes amblyopia recovery through a reduction of intracortical inhibition. *Nat Neurosci* **10:** 679–681.

Sato M, Stryker MP. 2008. Distinctive features of adult ocular dominance plasticity. *J Neurosci* **28:** 10278–10286.

Singer W, Rauschecker JP. 1982. Central core control of developmental plasticity in the kitten visual cortex: II. Electrical activation of mesencephalic and diencephalic projections. *Exp Brain Res* **47:** 223–233.

Skinner RD, Homma Y, Garcia-Rill E. 2004. Arousal mechanisms related to posture and locomotion: 2. Ascending modulation. *Prog Brain Res* **143:** 291–298.

Stefani A, Lozano AM, Peppe A, Stanzione P, Galati S, Tropepi D, Pierantozzi M, Brusa L, Scarnati E, Mazzone P. 2007. Bilateral deep brain stimulation of the pedunculopontine and subthalamic nuclei in severe Parkinson's disease. *Brain* **130:** 1596–1607.

Stefani A, Peppe A, Galati S, Bassi MS, D'Angelo V, Pierantozzi M. 2013. The serendipity case of the pedunculopontine nucleus low-frequency brain stimulation: Chasing a gait response, finding sleep, and cognition improvement. *Front Neurol* **4:** 68.

Wickersham IR, Lyon DC, Barnard RJ, Mori T, Finke S, Conzelmann KK, Young JA, Callaway EM. 2007. Monosynaptic restriction of transsynaptic tracing from single, genetically targeted neurons. *Neuron* **53:** 639–647.

Diverse Effects of Conditioned Threat Stimuli on Behavior

Justin M. Moscarello[1] and Joseph LeDoux[1,2]
[1]*Center for Neural Science, New York University, New York, New York 10003*
[2]*Emotional Brain Institute, Nathan Klein Institute, Orangeburg, New York 10962*
Correspondence: *ledoux@cns.nyu.edu*

Aversive Pavlovian memory coordinates the defensive behavioral response to learned threats. The amygdala is a key locus for the acquisition and storage of aversive associations. Information about conditioned and unconditioned stimuli converge in the lateral amygdala, which is a hot spot for the plasticity induced by associative learning. Central amygdala uses Pavlovian memory to coordinate the conditioned reaction to an aversive conditioned stimulus. Aversive associations can also access the brain networks of instrumental action. The offset of an aversive conditioned stimulus can reinforce behavior, recruiting a pathway that includes the lateral and basal amygdala, as opposed to the lateral and central amygdala circuit for Pavlovian reactions. Aversive conditioned stimuli can also modulate ongoing behavior, suppressing appetitive actions and facilitating aversive actions. Facilitation depends on an amygdalar network involving the lateral and central, as well as medial, nuclei. Thus, aversive Pavlovian memory has wide-reaching effects on defensive behavior, coordinating reactive to active responses to environmental threats.

Innate and learned threats activate defensive circuits in the brain (Fanselow 1994; Gross and Canteras 2012; LeDoux 2012). In people with fear/anxiety disorders, these mechanisms can be dysregulated, resulting in hypersensitive threat detection and exaggerated and prolonged defensive responses (Charney 2004; Milad and Quirk 2012; Parsons and Ressler 2013). Understanding the brain mechanisms underlying adaptive and pathological reactions to threats is important.

Much work on circuits that detect and respond to threats has been done in the context of aversive Pavlovian conditioning, a procedure in which a biologically neutral conditioned stimulus (CS) is paired with a noxious unconditioned stimulus (US). Thereafter, the CS has the ability to elicit innate defensive reactions, such as freezing behavior (Blanchard and Blanchard 1969), and supporting physiological responses (Kapp et al. 1979; LeDoux et al. 1982), and to potentiate reflexes such as startle (Davis 1992). This form of associative learning is most commonly known as fear conditioning, but has also been called threat conditioning (LeDoux 2014). Here, we use the term *Pavlovian threat conditioning* (PTC; Fig. 1). PTC is useable across a wide range of species, from invertebrates (Carew et al. 1981; Glanzman 2010) to vertebrates (Davis 1992; Fanselow 1994; LeDoux 2000, 2012), including humans (Ohman and Mineka 2001; Phelps 2006). Further, the fact that changes in systems level functions are correlated with symptomatology in people with fear/anxiety disorders (Rauch et al. 2006; Liberzon and Sripada 2008), and that treatments that reduce symptoms in animals can also reduce symptoms in humans (Milad and Quirk 2012; Parsons and Ressler 2013), gives PTC a documented clinical relevance.

THE LATERAL AMYGDALA ACQUIRES AND STORES AVERSIVE PAVLOVIAN MEMORY

Decades of research show that the rodent amygdala is an important site for the acquisition and storage of aversive Pavlovian memory (for review, see Fanselow 1994; LeDoux 2000, 2012; Johansen et al. 2011). The amygdala is a complex network of interconnected subnuclei, which are thought to play distinct roles in PTC (Fig. 2; Pitkänen et al. 1997). The lateral amygdala (LA) receives convergent input from CS and US pathways (LeDoux et al. 1990), whereas the central nucleus of the amygdala (CeA) sends long-range projections to the brainstem effector sites that mediate conditioned reactions (LeDoux et al. 1988; Viviani et al. 2011). Information about aversive associations proceeds through the amygdala in a serial fashion. LA contacts the lateral subregion of CeA, which communicates with output neurons in medial CeA through a complex, disinhibitory microcircuit (Ciocchi et al. 2010; Li et al. 2013). In addition, LA and CeA are indirectly connected via pathways involving the basal nucleus, as well as the intercalated cell masses (Pitkänen et al. 1997). The data described below support a model in which LA encodes aversive associations, whereas CeA receives associative information and uses it to control the behavioral and physiological outputs of Pavlovian memory.

Because auditory PTC brings conditioned behavior under the control of a tone CS, the search for learning-related plasticity has focused on LA and its auditory inputs (LeDoux 2000; Maren and Quirk 2004). Pavlovian conditioning modifies CS-evoked activity in LA, as well as in the tertiary auditory cortex and medial geniculate nucleus of the thalamus (MGN), both of which project directly to LA (Quirk et al. 1997; Maren et al. 2001).

 doi: 10.1101/sqb.2014.79.024968

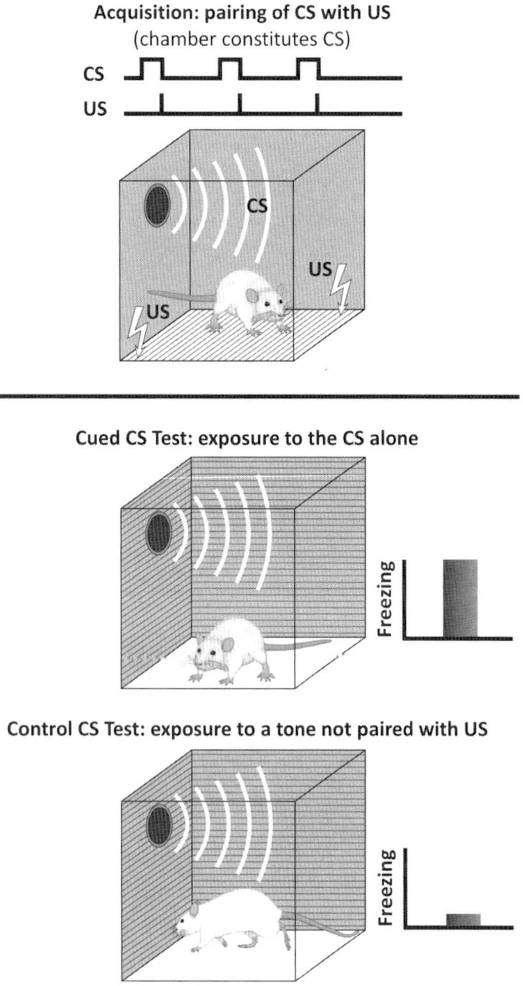

Figure 1. Pavlovian threat conditioning. This form of associative learning occurs when a previously neutral tone CS is paired with a footshock US. When the CS is presented later in the absence of the US, it evokes behavioral and physiological conditioned reactions, including freezing and changes in heart rate. These effects are not elicited by a CS that is not paired with the US.

Although these observations suggest that a widely distributed network encodes aversive associations, it is important to note that LA mediates plasticity in upstream auditory regions. CS-responsive cells in auditory cortex develop conditioned enhancements more slowly than CS-responsive cells in LA (Quirk et al. 1997), and LA lesions alter learning-related modifications in cortical activity (Armony et al. 1998). Inactivation of the basolateral complex, which includes LA, prevents conditioned neuronal responses from forming in MGN (Maren et al. 2001). It seems that feedback from the amygdala is required to maintain normal plasticity in auditory regions, suggesting a distributed representation that depends on encoding by LA neurons (Maren and Quirk 2004).

LA contains multimodal neurons that respond to both tone and shock (Romanski et al. 1993), creating a cellular substrate for the integration of CS and US information. During conditioning, enhancements in CS-evoked neuronal activity precede learned changes in behavior, highlighting the role of LA plasticity in memory formation (Quirk et al. 1995; Repa et al. 2001). Spiking also synchronizes across CS-responsive cells (Quirk et al. 1995; Paré and Collins 2000), suggesting that aversive associations are encoded by the rate and rhythm of action potentials in LA (Maren and Quirk 2004). The associative nature of these neural codes is further shown by the fact that conditioned enhancements in LA cell firing move through a "hill climbing" process that parallels the acquisition of behavioral reactions to the CS (Repa et al. 2001).

In addition to initial conditioning, extinction training has predictable effects on the activity of responsive LA neurons. This inhibitory form of learning attenuates CS-evoked activity in the dorsal LA cells that undergo rapid modifications during acquisition (Repa et al. 2001). Thus, cells in the dorsal tip of LA track changes in CS valence with corresponding changes in firing rate. In contrast, a relatively ventral subpopulation of CS-responsive LA neurons continues to display conditioned enhancements in spiking even after extinction training (Repa et al. 2001). These cells may carry an indelible element of associative memory that allows for the return of conditioned behavior following extinction. Together, these data illustrate how aversive Pavlovian memories are encoded by complex patterns of activity distributed over LA neurons participating in the memory trace.

On a synaptic level, it is theorized that conditioned modifications in LA spiking are supported by Hebbian mechanisms (Blair et al. 2001, 2003; Johansen et al. 2011). Hebb proposed a "fire together, wire together" principle by which a synapse is strengthened through concurrent activity in pre- and postsynaptic cells (Hebb 1949), suggesting a hypothetical means by which LA neurons can integrate CS and US information (Blair et al. 2001, 2003). Before Pavlovian learning, auditory synapses only weakly depolarize LA neurons. However, when inputs carrying the auditory CS occur simultaneously with robust depolarization evoked by the US, auditory synapses are strengthened through the temporal coincidence of presynaptic stimulation and postsynaptic activity. The resulting changes in synapse strength allow CS inputs to drive neuronal responses in LA, thus reactivating associative information even in the absence of the US.

NMDA receptors (NMDARs) are thought to play a key role in this Hebbian process. Postsynaptic depolarization dislodges magnesium ions from NMDARs, allowing these receptors to pass calcium when bound by glutamate (Malenka and Nicoll 1999). As such, NMDARs at auditory synapses can detect the cooccurrence of presynaptic glutamate release caused by the CS and postsynaptic depolarization caused by the US (Blair et al. 2001, 2003). The resulting influx of calcium triggers important molecular signaling cascades that underpin plasticity and memory formation (Johansen et al. 2011). This model is supported by pharmacological experiments demonstrating that intra-LA infusion of an NMDA antagonist prevents PTC (Maren et al. 1996; Rodrigues et al. 2001), as well as by in vitro experiments demonstrating that NMDARs are necessary for long-term potentiation at auditory synapses in LA (Bauer et al. 2002).

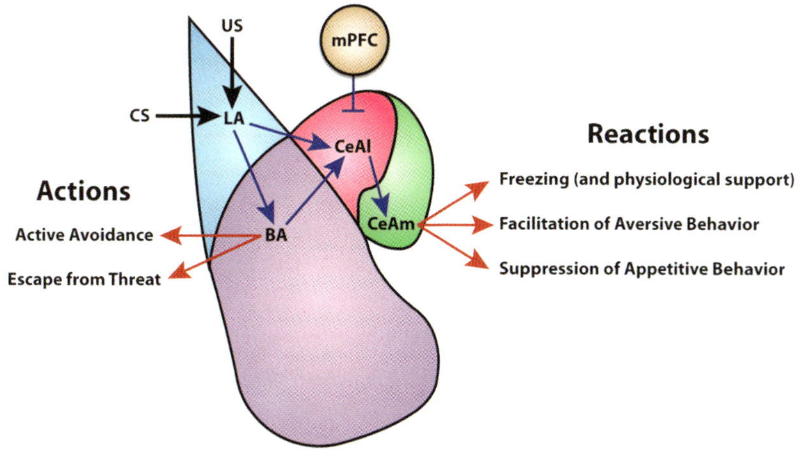

Figure 2. Role of the amygdala in aversive Pavlovian memory. CS and US converge in the lateral amygdala (LA). CS-evoked actions and reactions depend on different pathways emanating from LA. Reactions, such as freezing, occur when associative information is passed through the lateral subdivision of the central amygdala (CeA), before exiting via the medial CeA. The CS can also facilitate or suppress ongoing actions via this amygdalar pathway, though facilitation and suppression likely involve different CeA target regions. Actions, such as avoidance and escape from threat, engage the basal amygdala, but not CeA.

Along with NMDAR-dependent Hebbian mechanisms, additional factors are required for LA neurons to encode aversive Pavlovian memory, such as release of the neuromodulator norepinephrine (NE). PTC is prevented by blockade of NE β-adrenergic receptors (βARs) in LA (Bush et al. 2010). βAR signaling can mediate glutamatergic transmission via intracellular signaling cascades that lead to the phosphorylation of both NMDARs and AMPA glutamate receptors (Raman et al. 1996; Hu et al. 2007), thus regulating glutamatergic neurotransmission in a manner that could directly facilitate Hebbian learning. Other βAR-mediated mechanisms include changes in the excitability of LA principle neurons (Faber et al. 2008), as well as the suppression of inhibitory interneurons (Tully et al. 2007; Tully and Bolshakov 2010), both of which could modulate Hebbian changes in synaptic efficacy.

A recent study provides empirical support for the synergistic relationship between NE modulation and Hebbian learning mechanisms. Optogenetic techniques were used to pair an auditory CS with the depolarization of LA principle cells, in lieu of a standard footshock US. Although this treatment produced no learning on its own, combining it with pretraining intra-LA infusions of a βAR agonist produced an aversive association that was measurable the next day (Johansen et al. 2014). Thus, experimenters were able to create an artificial Pavlovian memory with a combination of βAR stimulation and an auditory CS reinforced with neuronal activations that mimic the putative effect of shock on LA neurons.

In summary, LA is a key site for the plasticity underpinning learned associations between an auditory CS and an aversive US. Aversive Pavlovian memory is encoded by the rate and rhythm of action potentials in CS-responsive LA neurons. These associative modifications in LA cell firing are supported by alterations in synaptic efficacy, which involve the complex dialogue between genome and synapse that underpins the storage of information by the nervous system (Kandel 1997; Johansen et al. 2011).

THE CENTRAL AMYGDALA CONTROLS BEHAVIORAL AND PHYSIOLOGICAL REACTIONS TO THE CS

CeA is an amygdalar output region that projects to a variety of structures mediating dissociable aspects of the response to an aversive CS (LeDoux et al. 1988). Freezing is the most commonly used readout of Pavlovian memory, though conditioned reactions also include changes in heart rate, respiration, and the hormonal milieu of the blood (LeDoux et al. 1988; Sullivan et al. 2004; Viviani et al. 2011). This constellation of behavioral and physiological responses is encoded in the activity of CeA microcircuits.

Plasticity in CeA facilitates the serial progression of information from LA, thus translating aversive associations into conditioned behavior. Inactivation of the lateral subdivision of CeA, which receives direct projections from LA (Pitkänen et al. 1997; Li et al. 2013), prevents the acquisition of aversive Pavlovian memory. In contrast, inactivation of medial CeA, which receives no direct input from LA, attenuates the expression of freezing, but has no effect on acquisition (Ciocchi et al. 2010). Electrophysiological studies show that PTC enhances the efficacy of LA synapses on lateral CeA neurons (Li et al. 2013; Penzo et al. 2014), thus creating a conduit by which associative memory can guide CS-evoked responses.

In comparison to LA, PTC evokes a combination of inhibition and excitation in different lateral CeA neurons (Duvarci et al. 2011), creating a dynamic interaction between "CS-on" and "CS-off" cell populations (Ciocchi et al. 2010). The "CS-on" population may conform to a group of neurons expressing the somatostatin (SOM^+) molecule. Optogenetic activation of SOM^+ cells causes unconditioned freezing (Li et al. 2013). The "CS-off" population conforms to a group of neurons expressing the PKCδ molecule (Ciocchi et al. 2010; Haubensak et al. 2010). Optogenetic activation of PKCδ$^+$ cells inhibits conditioned freezing (Haubensak et al. 2010). Consistent with their opposing behavioral functions, these populations mutually inhibit one another (Ciocchi et al. 2010; Haubensak et al. 2010; Li et al. 2013). PTC biases the competition between these groups by selectively enhancing the efficacy of LA synapses on SOM$^+$ neurons (Li et al. 2013). Increased activity in SOM$^+$ cells leads to the inhibition of PKCδ$^+$ cells, which form inhibitory synapses on medial CeA output neurons (Haubensak et al. 2010; Li et al. 2013). Medial CeA projects to the periaqueductal gray (PAG), an important effector site for freezing behavior (LeDoux et al. 1988; Viviani et al. 2011). PTC evokes a uniform excitation across CS-responsive cells in medial CeA (Duvarci et al. 2011). Thus, plasticity at particular LA synapses in lateral CeA leads to the disinhibition of medial CeA projections to PAG. Although aversive associations are encoded by enhancements in CS-evoked firing in LA, this conditioned activity translates into a complex pattern of excitation and inhibition in the circuitry of the CeA. It has been hypothesized that increased activity in medial CeA output neurons inhibits a network of GABA-ergic cells in PAG, facilitating CS-evoked freezing (Behbehani 1995).

Parallel channels of information through CeA control different elements of the conditioned response. These circuits are modulated by distinct neuropeptidergic inputs and have dissociable outputs from CeA. Lateral CeA contains a population of cells that express oxytocin receptors. Oxytocin agonists excite these neurons, leading to the inhibition of medial CeA neurons that drive CS-evoked freezing (Gozzi et al. 2010; Viviani et al. 2011). Although it is unknown if oxytocin receptor–expressing cells overlap with PKCδ cells in lateral CeA, they do have the same effect on medial CeA, and thus freezing, when activated. Oxytocin signaling in CeA can alter the behavioral response to the CS, but it has no effect on conditioned changes in heart rate (Viviani and Stoop 2008; Viviani et al. 2011). Conversely, vasopressin signaling in CeA can modulate the circuits for conditioned alterations in cardiac output, but has no effect on pathways that mediate freezing (Huber et al. 2005; Viviani and Stoop 2008). Conditioned changes in cardiac output are mediated by a medial CeA projection to the dorsal vagal complex (Viviani et al. 2011). These cells do not overlap with the medial CeA neurons that project to PAG. Thus, CeA contains distinct, functionally dedicated microcircuits with dissociable modulatory inputs.

In summary, plasticity at LA synapses in lateral CeA biases the competition between mutually inhibitory cell populations, creating the "CS-on" and "CS-off" dynamic that leads to the disinhibition of medial CeA projection neurons and conditioned freezing. The behavioral and physiological reactions to an aversive CS are encoded by parallel circuits in CeA that are mediated by different neuropeptides. This allows the expression of aversive Pavlovian memory to be guided by modulatory information that is specifically relevant to distinct aspects of the overall reaction to an aversive CS. Finally, the constellation of responses guided by aversive memory are driven by distinct effector regions that receive the long range projections of CeA.

THE AMYGDALA ALSO CONTRIBUTES TO AVERSIVE ACTIONS

Most work on PTC has focused on how a CS elicits innate defense reactions like freezing or alters startle reflexes. But behavior can be affected in other ways by learned threats. Animals, including humans, use cues associated with danger to prevent aversive outcomes. For example, patients with anxiety disorders often show behaviors that reduce exposure to fear-eliciting stimuli or situations. So-called avoidance responses can be effective coping strategies, but become problematic when they interfere with daily life. In comparison to *reactions* like freezing, less is known about the neural basis of aversively motivated instrumental *actions*. Escape from threat (EFT) is particularly useful for exploring how a Pavlovian CS contributes to the acquisition of new instrumental responses (Fig. 3). Although active avoidance paradigms have been more extensively studied (see below), the Pav-

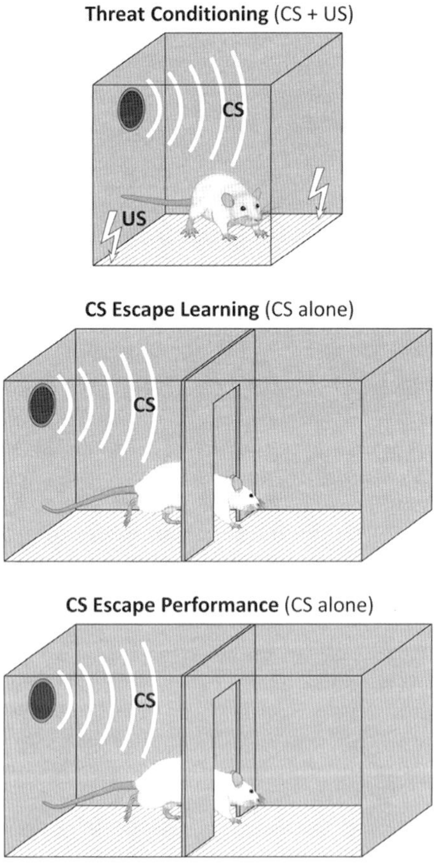

Figure 3. Escape from threat (EFT). Animals first undergo Pavlovian threat conditioning. Then, in a distinct environment, animals learn to inactivate the CS by performing a specific behavior, such as shuttling or rearing. Finally, the animal is given a reinforcement-free test consisting of a single long CS. EFT learning shifts the response to the CS from reactive to active profile of behavior.

lovian and instrumental learning components of these tasks are intermixed. EFT breaks learning into discrete phases, allowing the experimenter to isolate the role of different brain regions in the distinct learning processes that contribute to aversively motivated actions. In EFT, the CS–US association is first conditioned in a standard PTC paradigm. Then, in a different context, the animal learns to perform responses that terminate the CS. CS termination is believed to function as conditioned negative reinforcement, in which cessation of a learned cue makes the preceding behavior more likely to reoccur. This section will begin with EFT, to examine the role of amygdalar nuclei in conditioned negative reinforcement, before moving on to active avoidance, which powerfully illustrates the process by which instrumental learning fundamentally reorganizes the responses to a threatening CS.

EFT shows that conditioned negative reinforcement learning depends on the initial formation of an aversive Pavlovian memory. If the CS and US are explicitly unpaired during the PTC phase, animals will not acquire EFT, presumably because the offset of a neutral CS does not reinforce behavior (Cain and LeDoux 2007). Pretraining damage to LA prevents CS–US associations from occurring, and this treatment also prevents the acquisition of EFT (Amorapanth et al. 2000). Plasticity in LA provides an important neural substrate for the learned aversion that motivates active behavior reinforced by CS termination, suggesting that both actions and reactions overlap at this level of amygdalar processing.

The conditioned negative reinforcement phase of EFT learning allows subjects to toggle from reactive to active behavioral states by recruiting distinct pathways emanating from LA (Fig. 2). Because CeA coordinates freezing and other innate reactions to the CS, it is unsurprising that damage to CeA produces no effects on EFT learning. Instead, pretraining damage to the basal amygdala (BA) attenuates learning during the conditioned negative reinforcement phase of EFT, but has no effect on PTC (Amorapanth et al. 2000). Though no electrophysiological studies have assessed the role of BA neuronal activity in EFT, studies of BA activity during PTC suggest a complex form of processing that may be relevant to conditioned negative reinforcement. BA contains a population of cells that show conditioned enhancements in CS-evoked firing as a result of PTC, comparable to what is observed in LA. However, a distinct group of cells undergo increased spiking to the CS under extinction conditions, when the associated US is omitted (Herry et al. 2008). This suggests that BA can process information about conditioned aversion and forms of learning that mitigate the impact of a threatening CS, both of which are relevant to negative reinforcement and active coping processes.

EFT shows how active and reactive manifestations of aversive Pavlovian memory are encoded by separate pathways originating in LA. The circuits involved in conditioned negative reinforcement are also implicated in signaled active avoidance (SigAA) behavior, which intermixes Pavlovian and instrumental phases of training (Fig. 4). In SigAA, a CS is presented, followed by an aversive US. If, during the CS, the subject performs a particular

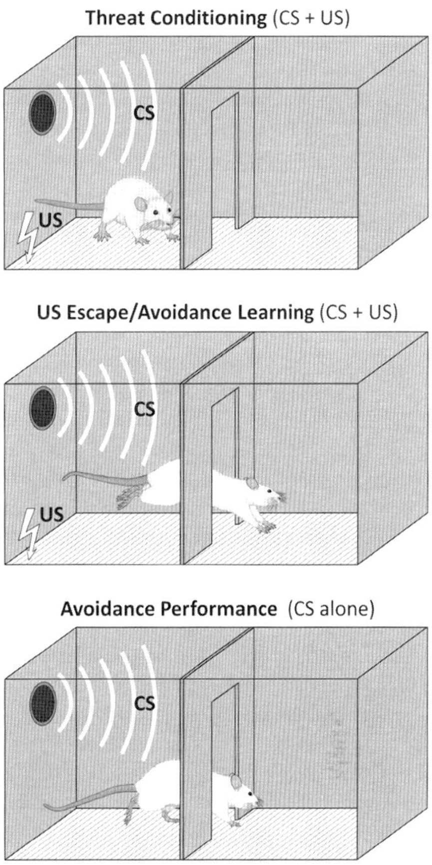

Figure 4. Signaled active avoidance. First, animals acquire an association between CS and US. Then, in subsequent trials, the animal learns that a particular behavior, such as shuttling, causes the offset of the CS and the omission of the US. In SigAA, Pavlovian and instrumental phases of learning occur in the same environment and are intermixed. Over the course of learning, animals shift from reactive freezing to active avoidance. CS-evoked freezing remains suppressed, even in alternate contexts that do not allow the avoidance response.

response (usually shuttling across a divided chamber), the CS is terminated and the US is omitted. Initially, the subject freezes to the CS. But, over trials, the avoidance response is acquired and freezing is suppressed. Similar to what has been observed using EFT, lesions of LA and BA impede the expression of SigAA behavior (Choi et al. 2010), suggesting that the two forms involve a similar mechanism in which active behavior reinforced by CS termination comes to supplant reactive freezing.

CS-evoked freezing is a more potent constraint on SigAA learning than it is on EFT. This difference is likely due to the fact that the negative reinforcement phase of EFT occurs under extinction conditions, whereas the US will always follow the CS in SigAA if the subject fails to avoid. SigAA training has a tendency to produce poor performers, a notable minority of subjects that do not acquire the avoidance response because of excessive freezing (Choi et al. 2010; Lázaro-Muñoz et al. 2010; Galatzer-Levy et al. 2014). In these animals, CeA lesion produces a substantial rebound, suggesting that some amount of latent negative reinforcement learning occurs, but is overshadowed by CS-evoked freezing (Choi et al. 2010; Lázaro-Muñoz et al. 2010). Additionally, pretraining lesions of CeA facilitate SigAA (Moscarello and LeDoux 2013), whereas they produce no significant effect on EFT (Amorapanth et al. 2000). In spite of the inherent conflict between freezing and avoidance behavior, the majority of subjects successfully transition from reactive to active responses to the CS. Freezing remains suppressed even in contexts that do not allow the subject to perform the avoidance response (Kamin et al. 1963; Moscarello and LeDoux 2013), suggesting that SigAA training is a potent means by which conditioned aversion can be managed and inhibited. This transition is mediated by the infralimbic prefrontal cortex (ilPFC), inactivation or lesion of which enhances freezing and hinders avoidance (Moscarello and LeDoux 2013). Electrophysiological studies show an inverse relationship between activity in ilPFC and brainstem-projecting neurons in medial CeA, suggesting that ilPFC exerts feed-forward inhibition on the cells that underpin freezing (Quirk et al. 2003; Likhtik et al. 2005).

The engagement of ilPFC seems to reorganize the defensive strategy evoked by an aversive CS. The decline of freezing that accompanies the acquisition of avoidance is not simply the replacement of one response with another. Not only does freezing remain inhibited in an environment that does not allow the avoidance response, but ilPFC also attenuates CS-elicited defecation, a physiological reaction that does not directly impede SigAA (Moscarello and LeDoux 2013). These data suggest the intriguing hypothesis that conditioned negative reinforcement learning leads to a fundamental reevaluation of learned threats. Because SigAA mitigates the reactive impact of an aversive stimulus by allowing the subjects to gain control of their environment, it nicely models active coping processes by which an internal locus of control can promote better mental health outcomes in humans (LeDoux and Gorman 2001).

In addition to serving as conditioned negative reinforcers, aversive CSs can influence action by modulating an ongoing instrumental behavior. A classic laboratory paradigm used to study the modulatory effect of a CS is called Pavlovian to instrumental transfer (PIT). In a PIT task, a previously conditioned CS is presented while the animal is performing a baseline instrumental behavior. Depending on the task, the CS will either facilitate or suppress action. In an early study, Estes and Skinner (1941) found that behavior maintained by food reinforcement was suppressed during the occurrence of an aversive CS. Initially called conditioned anxiety, and later conditioned emotional response (Hunt and Brady 1951), this phenomenon is most commonly referred to as conditioned suppression today (Bouton and Bolles 1979; Cardinal et al. 2002; Balleine and Killcross 2006). Conditioned suppression requires LA and CeA (Campese et al. 2014a) and has been interpreted as simple response competition between instrumental behavior and freezing (Kamin 1965). However, it has been shown that suppression is not fully accounted for by freezing. For example, lesions of the periaqueductal gray disrupt freezing but not suppression (Amorapanth et al. 1999). Suppression seems to involve broader behavioral effects than simple competition by freezing.

Conditioned suppression occurs when the Pavlovian CS is aversive and the instrumental reinforcer is appetitive. In contrast, conditioned facilitation occurs when the Pavlovian and instrumental elements have the same valence (Fig. 5). This has been studied extensively in appetitive conditioning. In a typical design, a CS that has been paired with food will enhance the performance of a behavior maintained by food reinforcement (Lovibond

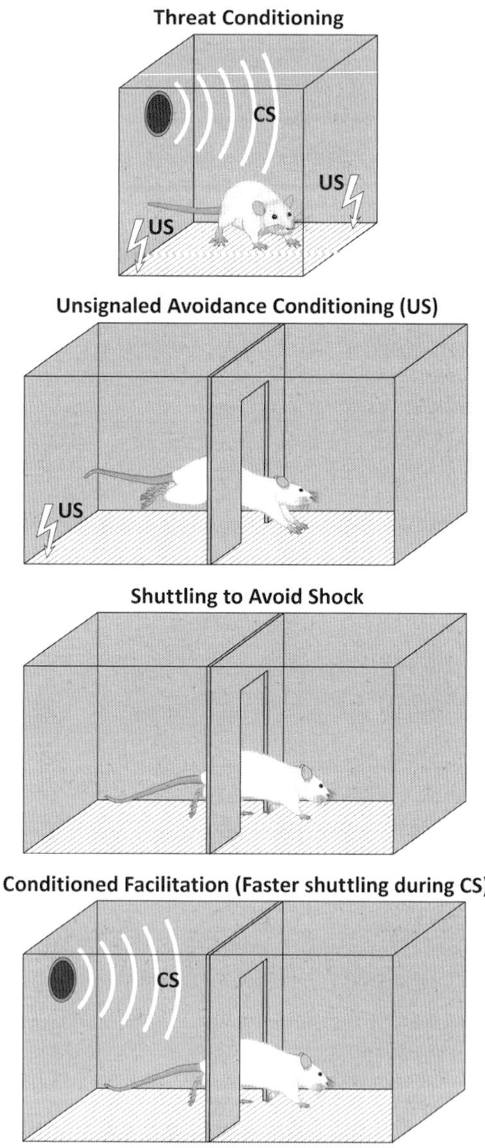

Figure 5. Pavlovian to instrumental transfer. In conditioned facilitation, animals first undergo Pavlovian threat conditioning. Then, animals learn an unsignaled active avoidance (USAA) response, in which the animal must shuttle to avoid shocks presented at a set interval. When the CS is delivered over this behavior, shuttling is strongly enhanced. This is in contrast to conditioned suppression, in which an aversive CS attenuates an appetitive action.

1983; Holland and Gallagher 2003). A comparable aversive PIT task for rodents has recently been developed (Campese et al. 2013). Rats were given Pavlovian CS–US pairings involving a tone and a shock in one context, and then trained to shuttle back and forth in a divided chamber to avoid a shock US. The CS is presented during performance of this unsignaled active avoidance behavior (USAA), facilitating the avoidance response. A CS that has been unpaired with the US does not facilitate shuttling, suggesting that the facilitation is related to the CS–US association (Campese et al. 2013).

USAA behavior initially depends on LA and BA, and is constrained by CeA-mediated freezing. However, extensive training makes USAA amygdala-independent (Lázaro-Muñoz et al. 2010), allowing the amygdalar substrates of conditioned facilitation to be examined without disrupting the baseline task. Lesions of LA and CeA disrupt PIT, but BA lesions have no effect. Baseline USAA behavior is unaffected by these lesions, demonstrating that the underlying avoidance response is intact (Campese et al. 2014b). These effects on facilitation show that LA and CeA can control opposite and incompatible responses (freezing and shuttling) via different internal circuits and/or output connections. Thus, freezing is the default response when threat is imminent (Fanselow and Lester 1988). However, if the CS occurs over a preexisting response that avoids shock, then facilitation occurs.

Although facilitation and freezing both involve LA and CeA, facilitation also requires the medial amygdala (MeA; McCue et al. 2014). MeA may play a key role as an intra-amygdala link between the LA and CeA recruited specifically to produce facilitation, because past studies found no effects of MeA lesion on freezing (Nader et al. 2001). MeA receives direct inputs from the LA (Pitkänen et al. 1997) and sends projections to both medial and lateral divisions CeA (Canteras et al. 1995). It is unknown how these pathways make the switch from freezing to conditioned facilitation, though it is reasonable to hypothesize that they exert their effect via the CeA microcircuits that encode conditioned responses to the CS. Other research suggests that opposing patterns of activity in lateral CeA microcircuits can produce mutually exclusive patterns of CS-evoked behavior. Lateral CeA contains projection neurons that disinhibit cholinergic cells in the basal forebrain. When activated, these cells cause locomotion increases in response to an aversive CS (Gozzi et al. 2010). Basal forebrain–projecting lateral CeA cells contact medial CeA cells that project to PAG, allowing this population to promote motion while inhibiting cells that underpin freezing (Gozzi et al. 2010). It is not known whether this population is involved in facilitation; however, it provides a putative mechanism by which CeA can toggle between freezing and facilitation of locomotor activation.

In summary, aversive Pavlovian memory can influence and organize a variety of behaviors beyond the reactive freezing typically used as a readout of PTC. The offset of an aversive CS can negatively reinforce behavior, transitioning the animal from a reactive to an active profile of CS-evoked responses. This transition occurs as the animal toggles between functional pathways emanating from LA, with LA–CeA connections underpinning reactive behavior and LA–BA connections underpinning active instrumental response to the CS. Aversive Pavlovian memory can also modulate ongoing actions. The CS produces freezing and suppression when set against an appetitive instrumental task and facilitation when presented over an aversively motivated behavior. This suggests that the broader context of learning and memory can dramatically influence the behavioral expression of CS–US associations. Although facilitation and freezing both require LA and CeA, MeA also plays a selective role in facilitation. The addition of MeA alters this circuit to invigorate the output of USAA behavior.

CONCLUSION

Research on PTC has revealed in great detail the system, circuit, and cellular mechanisms involved in the acquisition and storage of aversive associative memories. By following the CS through the brain, many of the pathways involved in PTC have been mapped. Importantly, a key brain region for the learning and memory in all mammals, including humans, is the amygdala. Studies in rodents have shown that synaptic plasticity occurs in LA following CS–US pairings, allowing the conditioned stimulus to flow to downstream circuitry, such as CeA. The circuitry in CeA encodes different elements of the reaction to the CS after conditioning. Aversive Pavlovian memory can also engage the substrates of instrumental action. LA stores critical aspects of the CS–US association used in the acquisition of new instrumental responses, and BA allows the CS to function as a conditioned negative reinforcer in the learning of an instrumental response. In contrast, the progression of information across an LA, MeA, and CeA pathway allows the CS to motivate an ongoing instrumental behavior. This work shows the power of using PTC to decipher processes underlying memory formation and learned changes in neural processing within a specific, behaviorally defined circuit.

REFERENCES

Amorapanth P, Nader K, LeDoux JE. 1999. Lesions of periaqueductal gray dissociate-conditioned freezing from conditioned suppression behavior in rats. *Learn Mem* **6:** 491–499.

Amorapanth P, LeDoux JE, Nader K. 2000. Different lateral amygdala outputs mediate reactions and actions elicited by a fear-arousing stimulus. *Nat Neurosci* **3:** 74–79.

Armony JL, Quirk GJ, LeDoux JE. 1998. Differential effects of amygdala lesions on early and late plastic components of auditory cortex spike trains during fear conditioning. *J Neurosci* **18:** 2592–2601.

Balleine BW, Killcross S. 2006. Parallel incentive processing: An integrated view of amygdala function. *Trends Neurosci* **29:** 272–279.

Bauer EP, Schafe GE, LeDoux JE. 2002. NMDA receptors and L-type voltage-gated calcium channels contribute to long-term potentiation and different components of fear memory formation in the lateral amygdala. *J Neurosci* **22:** 5239–5249.

Behbehani MM. 1995. Functional characteristics of the midbrain periaqueductal gray. *Prog Neurobiol* **46:** 575–605.

Blair HT, Schafe GE, Bauer EP, Rodrigues SM, LeDoux JE. 2001. Synaptic plasticity in the lateral amygdala: A cellular hypothesis of fear conditioning. *Learn Mem* **8:** 229–242.

Blair HT, Tinkelman A, Moita MA, LeDoux JE. 2003. Associative plasticity in neurons of the lateral amygdala during auditory fear conditioning. *Ann NY Acad Sci* **985:** 485–487.

Blanchard RJ, Blanchard DC. 1969. Passive and active reactions to fear-eliciting stimuli. *J Comp Physiol Psychol* **68:** 129–135.

Bouton ME, Bolles RC. 1979. Role of conditioned contextual stimuli in reinstatement of extinguished fear. *J Exp Psychol Anim Behav Process* **5:** 368–378.

Bush DE, Caparosa EM, Gekker A, Ledoux J. 2010. β-adrenergic receptors in the lateral nucleus of the amygdala contribute to the acquisition but not the consolidation of auditory fear conditioning. *Front Behav Neurosci* **4:** 154.

Cain CK, LeDoux JE. 2007. Escape from fear: A detailed behavioral analysis of two atypical responses reinforced by CS termination. *J Exp Psychol Anim Behav Process* **33:** 451–463.

Campese V, McCue M, Lázaro-Muñoz G, Ledoux JE, Cain CK. 2013. Development of an aversive Pavlovian-to-instrumental transfer task in rat. *Front Behav Neurosci* **7:** 176.

Campese VD, Gonzaga R, LeDoux JE. 2014a. Facilitation and suppression of instrumental responding require connections between the lateral and central amygdala. *Soc Neurosci Abstr*.

Campese VD, Kim J, Lázaro-Muñoz G, Pena L, LeDoux JE, Cain CK. 2014b. Lesions of lateral or central amygdala abolish aversive Pavlovian to-instrumental transfer in rats. *Front Behav Neurosci* **8:** 161.

Canteras NS, Simerly RB, Swanson LW. 1995. Organization of projections from the medial nucleus of the amygdala: A PHAL study in the rat. *J Comp Neurol* **360:** 213–245.

Cardinal RN, Parkinson JA, Hall J, Everitt BJ. 2002. Emotion and motivation: The role of the amygdala, ventral striatum, and prefrontal cortex. *Neurosci Biobehav Rev* **26:** 321–352.

Carew TJ, Walters ET, Kandel ER. 1981. Associative learning in *Aplysia*: Cellular correlates supporting a conditioned fear hypothesis. *Science* **211:** 501–504.

Charney DS. 2004. Discovering the neural basis of human social anxiety: A diagnostic and therapeutic imperative. *Am J Psychiatry* **161:** 1–2.

Choi JS, Cain CK, LeDoux JE. 2010. The role of amygdala nuclei in the expression of auditory signaled two-way active avoidance in rats. *Learn Mem* **17:** 139–147.

Ciocchi S, Herry C, Grenier F, Wolff SB, Letzkus JJ, Vlachos I, Ehrlich I, Sprengel R, Deisseroth K, Stadler MB, et al. 2010. Encoding of conditioned fear in central amygdala inhibitory circuits. *Nature* **468:** 277–282.

Davis M. 1992. The role of the amygdala in fear and anxiety. *Annu Rev Neurosci* **15:** 353–375.

Duvarci S, Popa D, Paré D. 2011. Central amygdala activity during fear conditioning. *J Neurosci* **31:** 289–294.

Estes WK, Skinner BF. 1941. Some quantitative properties of anxiety. *J Exp Psychol* **29:** 390–400.

Faber ES, Delaney AJ, Power JM, Sedlak PL, Crane JW, Sah P. 2008. Modulation of SK channel trafficking by β adrenoceptors enhances excitatory synaptic transmission and plasticity in the amygdala. *J Neurosci* **28:** 10803–10813.

Fanselow MS. 1994. Neural organization of the defensive behavior system responsible for fear. *Psychon Bull Rev* **1:** 429–438.

Fanselow MS, Lester LS. 1988. A functional behavioristic approach to aversively motivated behavior: Predatory imminence as a determinant of the topography of defensive behavior. In *Evolution and learning* (ed. Bolles RC, Beecher MD), pp. 185–212. Lawrence Erlbaum, Hillsdale, NJ.

Galatzer-Levy IR, Moscarello J, Blessing EM, Klein J, Cain CK, LeDoux JE. 2014. Heterogeneity in signaled active avoidance learning: Substantive and methodological relevance of diversity in instrumental defensive responses to threat cues. *Front Syst Neurosci* **8:** 179.

Glanzman DL. 2010. Common mechanisms of synaptic plasticity in vertebrates and invertebrates. *Curr Biol* **20:** R31–R36.

Gozzi A, Jain A, Giovannelli A, Bertollini C, Crestan V, Schwarz AJ, Tsetsenis T, Ragozzino D, Gross CT, Bifone A. 2010. A neural switch for active and passive fear. *Neuron* **67:** 656–666.

Gross CT, Canteras NS. 2012. The many paths to fear. *Nat Rev Neurosci* **13:** 651–658.

Haubensak W, Kunwar PS, Cai H, Ciocchi S, Wall NR, Ponnusamy R, Biag J, Dong HW, Deisseroth K, Callaway EM, et al. 2010. Genetic dissection of an amygdala microcircuit that gates conditioned fear. *Nature* **468:** 270–276.

Hebb DO. 1949. *The organization ofbehavior*. Wiley, New York.

Herry C, Ciocchi S, Senn V, Demmou L, Müller C, Lüthi A. 2008. Switching on and off fear by distinct neuronal circuits. *Nature* **454:** 600–606.

Holland PC, Gallagher M. 2003. Double dissociation of the effects of lesions of basolateral and central amygdala on conditioned stimulus-potentiated feeding and Pavlovian-instrumental transfer. *Eur J Neurosci* **17:** 1680–1694.

Hu H, Real E, Takamiya K, Kang MG, Ledoux J, Huganir RL, Malinow R. 2007. Emotion enhances learning via norepinephrine regulation of AMPA-receptor trafficking. *Cell* **131:** 160–173.

Huber D, Veinante P, Stoop R. 2005. Vasopressin and oxytocin excite distinct neuronal populations in the central amygdala. *Science* **308:** 245–248.

Hunt HF, Brady JV. 1951. Some effects of electro-convulsive shock on a conditioned emotional response ("anxiety"). *J Comp Phsyiol Psychol* **44:** 88–98.

Johansen JP, Cain CK, Ostroff LE, LeDoux JE. 2011. Molecular mechanisms of fear learning and memory. *Cell* **147:** 509–524.

Johansen JP, Diaz-Mataix L, Hamanaka H, Ozawa T, Ycu E, Koivumaa J, Kumar A, Hou M, Deisseroth K, Boyden ES, et al. 2014. Hebbian and neuromodulatory mechanisms interact to trigger associative memory formation. *Proc Natl Acad Sci* **111:** E5584–E5592.

Kamin LJ. 1965. Temporal and intensity characteristics of the conditioned stimulus. In *Classical conditioning: A symposium* (ed. Prokasy W), pp. 118–147. Appleton, New York.

Kamin LJ, Brimer CJ, Black AH. 1963. Conditioned suppression as a monitor of fear of the CS in the course of avoidance training. *J Comp Physiol Psychol* **56:** 497–501.

Kandel ER. 1997. Genes, synapses, and long-term memory. *J Cell Physiol* **173:** 124–125.

Kapp BS, Frysinger RC, Gallagher M, Haselton JR. 1979. Amygdala central nucleus lesions: Effect on heart rate conditioning in the rabbit. *Physiol Behav* **23:** 1109–1117.

Lázaro-Muñoz G, LeDoux JE, Cain CK. 2010. Sidman instrumental avoidance initially depends on lateral and basal amygdala and is constrained by central amygdala-mediated Pavlovian processes. *Biol Psychiatry* **67:** 1120–1127.

LeDoux JE. 2000. Emotion circuits in the brain. *Annu Rev Neurosci* **23:** 155–184.

LeDoux JE. 2012. Evolution of human emotion: A view through fear. *Prog Brain Res* **195:** 431–442.

LeDoux JE. 2014. Coming to terms with fear. *Proc Natl Acad Sci* **111:** 2871–2878.

LeDoux JE, Gorman JM. 2001. A call to action: Overcoming anxiety through active coping. *Am J Psychiatry* **158:** 1953–1955.

LeDoux JE, Del Bo A, Tucker LW, Harshfield G, Talman WT, Reis DJ. 1982. Hierarchic organization of blood pressure responses during the expression of natural behaviors in rat: Mediation by sympathetic nerves. *Exp Neurol* **78:** 121–133.

LeDoux JE, Iwata J, Cicchetti P, Reis DJ. 1988. Different projections of the central amygdaloid nucleus mediate autonomic and behavioral correlates of conditioned fear. *J Neurosci* **8:** 2517–2529.

LeDoux JE, Cicchetti P, Xagoraris A, Romanski LM. 1990. The lateral amygdaloid nucleus: Sensory interface of the amygdala in fear conditioning. *J Neurosci* **10:** 1062–1069.

Li H, Penzo MA, Taniguchi H, Kopec CD, Huang ZJ, Li B. 2013. Experience-dependent modification of a central amygdala fear circuit. *Nat Neurosci* **16:** 332–339.

Liberzon I, Sripada CS. 2008. The functional neuroanatomy of PTSD: A critical review. *Prog Brain Res* **167:** 151–169.

Likhtik E, Pelletier JG, Paz R, Paré D. 2005. Prefrontal control of the amygdala. *J Neurosci* **25:** 7429–7437.

Lovibond PF. 1983. Facilitation of instrumental behavior by a Pavlovian appetitive conditioned stimulus. *J Exp Psychol Anim Behav Process* **9:** 225–247.

Malenka RC, Nicoll RA. 1999. Long-term potentiation—A decade of progress? *Science* **285:** 1870–1874.

Maren S, Quirk GJ. 2004. Neuronal signaling of fear memory. *Nat Rev Neurosci* **5:** 844–852.

Maren S, Aharonov G, Stote DL, Fanselow MS. 1996. *N*-methyl-D-aspartate receptors in the basolateral amygdala are required for both acquisition and expression of conditional fear in rats. *Behav Neurosci* **110:** 1365–1374.

Maren S, Yap SA, Goosens KA. 2001. The amygdala is essential for the development of neuronal plasticity in the medial geniculate nucleus during auditory fear conditioning in rats. *J. Neurosci* **21:** RC135.

McCue MG, LeDoux JE, Cain CK. 2014. Medial amygdala lesions selectively block aversive Pavlovian-instrumental transfer in rats. *Front Behav Neurosci* **8:** 329.

Milad MR, Quirk GJ. 2012. Fear extinction as a model for translational neuroscience: Ten years of progress. *Annu Rev Psychol* **63:** 129–151.

Moscarello JM, LeDoux JE. 2013. Active avoidance learning requires prefrontal suppression of amygdala-mediated defensive reactions. *J Neurosci* **33:** 3815–3823.

Nader K, Majidishad P, Amorapanth P, LeDoux JE. 2001. Damage to the lateral and central, but not other, amygdaloid nuclei prevents the acquisition of auditory fear conditioning. *Learn Mem* **8:** 156–163.

Ohman A, Mineka S. 2001. Fears, phobias, and preparedness: Toward an evolved module of fear and fear learning. *Psychol Rev* **108:** 483–522.

Paré D, Collins DR. 2000. Neuronal correlates of fear in the lateral amygdala: Multiple extracellular recordings in conscious cats. *J Neurosci* **20:** 2701–2710.

Parsons RG, Ressler KJ. 2013. Implications of memory modulation for post-traumatic stress fear disorders. *Nat Neurosci* **16:** 146–153.

Penzo MA, Robert V, Li B. 2014. Fear conditioning potentiates synaptic transmission onto long-range projection neurons in the lateral subdivision of central amygdala. *J Neurosci* **34:** 2432–2437.

Phelps EA. 2006. Emotion and cognition: Insights from studies of the human amygdala. *Annu Rev Psychol* **57:** 27–53.

Pitkänen A, Savander V, LeDoux JE. 1997. Organization of intra-amygdaloid circuitries in the rat: An emerging framework for understanding functions of the amygdala. *Trends Neurosci* **20:** 517–523.

Quirk GJ, Repa C, LeDoux JE. 1995. Fear conditioning enhances short-latency auditory responses of lateral amygdala neurons: Parallel recordings in the freely behaving rat. *Neuron* **15:** 1029–1039.

Quirk GJ, Armony JL, LeDoux JE. 1997. Fear conditioning enhances different temporal components of tone-evoked spike trains in auditory cortex and lateral amygdala. *Neuron* **19:** 613–624.

Quirk GJ, Likhtik E, Pelletier JG, Paré D. 2003. Stimulation of medial prefrontal cortex decreases the responsiveness of central amygdala output neurons. *J Neurosci* **23:** 8800–8807.

Raman IM, Tong G, Jahr CE. 1996. β-adrenergic regulation of synaptic NMDA receptors by cAMP-dependent protein kinase. *Neuron* **16:** 415–421.

Rauch SL, Shin LM, Phelps EA. 2006. Neurocircuitry models of posttraumatic stress disorder and extinction: Human neuroimaging research—Past, present, and future. *Biol Psychiatry* **60:** 376–382.

Repa JC, Muller J, Apergis J, Desrochers TM, Zhou Y, LeDoux JE. 2001. Two different lateral amygdala cell populations contribute to the initiation and storage of memory. *Nat Neurosci* **4:** 724–731.

Rodrigues SM, Schafe GE, LeDoux JE. 2001. Intra-amygdala blockade of the NR2B subunit of the NMDA receptor disrupts the acquisition but not the expression of fear conditioning. *J Neurosci* **21:** 6889–6896.

Romanski LM, Clugnet MC, Bordi F, LeDoux JE. 1993. Somatosensory and auditory convergence in the lateral nucleus of the amygdala. *Behav Neurosci* **107:** 444–450.

Sullivan GM, Apergis J, Bush DE, Johnson LR, Hou M, Ledoux JE. 2004. Lesions in the bed nucleus of the stria terminalis disrupt corticosterone and freezing responses elicited by a contextual but not by a specific cue-conditioned fear stimulus. *Neuroscience* **128:** 7–14.

Tully K, Bolshakov VY. 2010. Emotional enhancement of memory: How norepinephrine enables synaptic plasticity. *Mol Brain* **3:** 15.

Tully K, Li Y, Tsvetkov E, Bolshakov VY. 2007. Norepinephrine enables the induction of associative long-term potentiation at thalamo-amygdala synapses. *Proc Natl Acad Sci* **104:** 14146–14150.

Viviani D, Stoop R. 2008. Opposite effects of oxytocin and vasopressin on the emotional expression of the fear response. *Prog Brain Res* **170:** 207–218.

Viviani D, Charlet A, ven den Burg E, Robinet C, Hurni N, Abatis M, Magara F, Stoop R. 2011. Oxytocin selectively gates fear responses through distinct outputs from the central amygdala. *Science* **333:** 104–107.

Animal-to-Animal Variability in Neuromodulation and Circuit Function

Albert W. Hamood and Eve Marder
Volen Center and Biology Department, Brandeis University, Brandeis University, Waltham, Massachusetts 02454
Correspondence: marder@brandeis.edu

Each animal alive in the world is different from all other individuals, while sharing most attributes of form and function with others of the same species. Still other attributes are shared within a phylum, and still others are common to most eukaryotic organisms. All animals have mechanisms that modulate the strength of their synapses or alter the intrinsic excitability of component neurons. What animal-to-animal variability in behavior arises from differences in neuronal structure, ion channel expression, or connectivity, and what variability arises from neuromodulation of brain states? Conversely, can robust behavior be maintained despite variability in circuit components by the action of neuromodulatory inputs? These are fundamental issues relevant to all nervous systems that have been illuminated by many years of study of the small, rhythmic motor circuits found in the crustacean stomatogastric nervous system.

In the early days of circuit study, the crustacean stomatogastric nervous system was considered a "simple system" (Maynard 1972; Fentress 1976) because the single stomatogastric ganglion (STG) found in each animal has only 27 to 30 neurons and generates stereotyped and rhythmic motor patterns. The STG quickly became a preferred preparation for the study of the mechanisms underlying the generation of rhythmic movements (Mulloney and Selverston 1974a,b; Selverston and Mulloney 1974; Hartline and Maynard 1975; Selverston et al. 1976; Hartline 1979; Hartline and Gassie 1979; Miller and Selverston 1982a,b) because when removed from the animal and placed in physiological saline, it continues to generate fictive motor patterns that resemble closely those that are seen in vivo (Fig. 1).

The pioneering work on the STG and on other "simple systems" preceded much of what we know today about the richness of ion channel diversity and neuromodulator function. Consequently, the initial goals of those seeking to understand the central pattern generating circuits that drove rhythmic movements were to (1) establish the wiring diagram, or connectivity among the neurons of the circuit, and (2) determine if a given neuron was part of the central pattern generating circuit or was merely being driven by the central pattern generating circuit. These goals were subsequently modified when it became clear that central pattern generating networks are richly modulated by tens of amines and neuropeptides (Marder 2012). Moreover, synaptic strength and intrinsic membrane properties also depend strongly on the recent history of neuronal activity (Turrigiano et al. 1996; Manor et al. 1997, 2003; Goaillard et al. 2010). Thus, the challenge is to explain how the rich dynamics of a circuit arise from its cellular mechanisms. This requires exploring the full ranges of circuit outputs that are produced by normal, healthy individuals (Bucher et al. 2005; Goaillard et al. 2009; Hamood et al. 2015). It also requires understanding how neuromodulatory inputs can alter circuit performance without disrupting appropriate physiological actions (Marder 2012).

THE RANGE OF NORMAL PYLORIC RHYTHMS

The crustacean stomatogastric nervous system (Fig. 1A) consists of four ganglia: the single STG, the midline esophageal ganglion (OG), and the paired commissural ganglia (CoGs). The STG produces two major motor patterns: the fast triphasic pyloric rhythm (Fig. 1) and the slower gastric mill rhythm. Sensory neuron activation organizes activity in sets of descending neuromodulatory inputs that reside in the CoGs and OG, which in turn influence both the pyloric and gastric mill rhythms (Harris-Warrick and Marder 1991; Nusbaum and Beenhakker 2002; Beenhakker and Nusbaum 2004; Blitz et al. 2004; Marder and Bucher 2007; Blitz and Nusbaum 2011).

In vivo recordings of the pyloric rhythm from a variety of species and temperatures show periods of ~2.5 sec in fasting animals and periods closer to 1 sec after feeding (Rezer and Moulins 1983; Clemens et al. 1998, 1999; Soofi et al. 2014). An important feature of the pyloric rhythm in vivo is that it is always present, except just at molting (Clemens et al. 1999).

The preponderance of work on the STG has been done with in vitro preparations in which fictive motor patterns are recorded from the nervous system after it has been dissected off the surface of the stomach (Fig. 1A). When the anterior modulatory inputs are left intact, the pyloric rhythm is invariably active, with characteristic pyloric

 doi: 10.1101/sqb.2014.79.024828

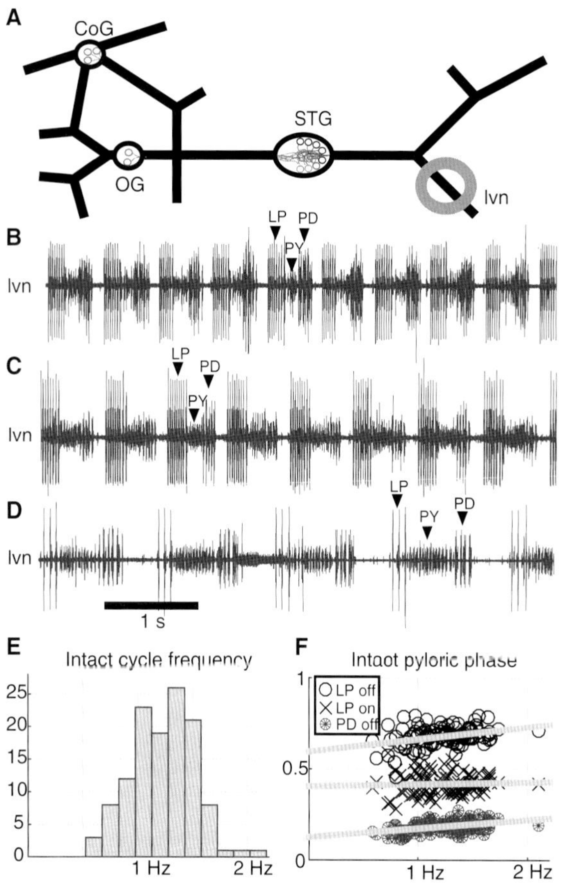

Figure 1. (*A*) Schematic diagram of the stomatogastric nervous system (STNS), as pinned out for electrophysiological recordings. The motor neurons of the pyloric rhythm reside in the stomatogastric ganglion (STG, *center*), whereas modulatory inputs arrive from anterior ganglia including the paired CoGs and the OG. Activity in the pyloric rhythm may be monitored extracellularly by recording from the lateral ventricular nerve (lvn). (*B–D*) Example lvn recordings from three preparations of the STNS, showing spontaneous pyloric rhythms. Although frequencies vary, approximate phase relationships between the pyloric (PY), pyloric dilator (PD), and lateral pyloric (LP) neurons are maintained. (*E*) Histogram shows the distribution of pyloric frequencies for intact preparations of the STNS ($N = 123$). (*F*) Pyloric phase relationships are approximately maintained across frequency for the pyloric rhythm of the same $N = 123$ intact preparations. Phase is quantified relative to PD on. Lines represent fits by linear regression. (*E,F*, Adapted from Hamood et al. 2015.)

rhythm frequencies in the 1-Hz range. Raw traces from three different preparations are shown in Figure 1B–D, with activity in the PD, LP, and PY neurons indicated. Although these preparations exhibit rhythms of different frequencies, their phase relationships (relative timing) are maintained. Figure 1E shows pooled data from 123 preparations showing the range of pyloric rhythm frequencies generated under these standard control conditions (Hamood et al. 2015). Note that there is an approximately threefold range in frequencies, but over this range of frequencies, the phases are frequency-invariant (Fig 1F). This robust phase maintenance ensures that the associated muscles at the output of the circuit are activated in a sequential manner, with their relative timing, and thus appropriate behavior, preserved.

The mechanisms that underlie the maintenance of phase over this range of frequencies have intrigued investigators, because membrane and synaptic currents have characteristic time constants, which would tend to create fixed delays, not constant phase (Hooper 1998; Manor et al. 2003; Hooper et al. 2009; Tang et al. 2010), and it is thought that phase maintenance requires the interaction of multiple synaptic and intrinsic conductances (Manor et al. 2003; Tang et al. 2010).

PYLORIC RHYTHMS IN THE ABSENCE OF NEUROMODULATION

When the descending neuromodulatory inputs are removed, the pyloric rhythm slows down or sometimes stops (Fig. 2; Russell 1979; Miller and Selverston

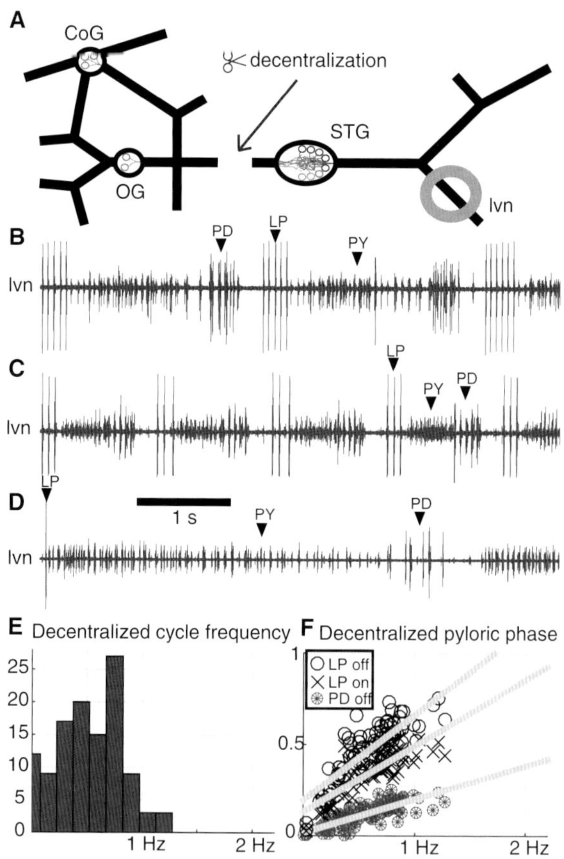

Figure 2. (*A*) Schematic diagram of the STNS including location of nerve transection known as decentralization. This perturbation severs modulatory inputs to the STG. (*B–D*) Example lvn recordings from the same three preparations as in Figure 1, 30 min following decentralization. (*E*) Histogram shows the distribution of pyloric frequencies for preparations of the STNS decentralized for 30 min ($N = 115$). (*F*) Pyloric phase relationships for the same $N = 115$ decentralized preparations show a stronger correlation with pyloric frequency than for those left intact. Phase is quantified relative to PD on. Lines represent fits by linear regression. (*E,F*, Adapted from Hamood et al. 2015.)

1982a,b; Moulins and Cournil 1982; Hamood et al. 2015). Although it has long been known that these modulatory inputs provide an excitatory drive to the network required for output in the normal frequency range, recently it has been shown that other features of the pyloric rhythm are also affected (Hamood et al. 2015). As the rhythm becomes less frequent, phase relationships between pyloric neurons advance beyond the range seen in intact controls (Fig. 2F). These phase relationships become highly correlated with pyloric frequency, compared with the weak or nonexistent correlations measured from intact preparations (Bucher et al. 2005). Additionally, the output of the pyloric rhythm becomes much more variable, reminiscent of the increased variability in the low-frequency rhythms seen in development (Richards et al. 1999; Richards and Marder 2000). Across animals, this increase in variability can be measured by the coefficient of variation (CV), which increases for all examined properties of the pyloric rhythm, including frequency, phase relationships, and the number of spikes per burst of pyloric cells (Fig. 3A). An increase in variability can also been seen within preparations, as the average cycle-to-cycle variability in the pyloric rhythm is also increased following loss of modulatory inputs (Fig. 3B,C). These changes occur rapidly within 15–30 min following decentralization, and persist as preparations are maintained in culture for up to weeks (Hamood et al. 2015).

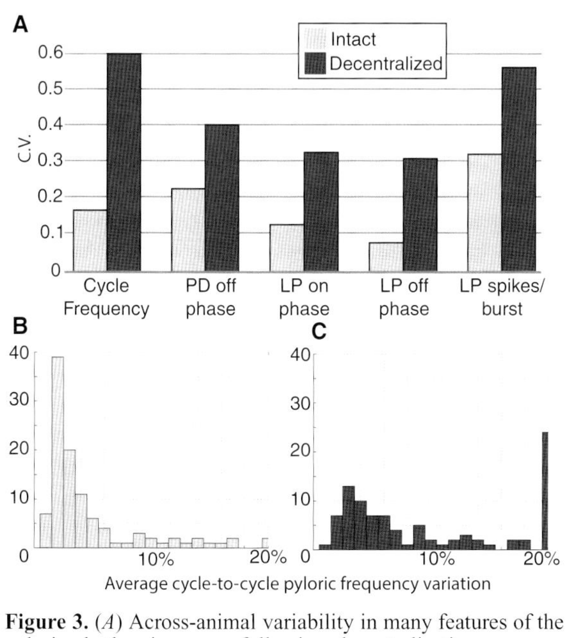

Figure 3. (*A*) Across-animal variability in many features of the pyloric rhythm increases following decentralization, as measured by the CV. (*B*) Histogram shows the average cycle-to-cycle variability in pyloric frequency, as a percentage, for $N = 123$ intact preparations of the STNS. Bar on the far *right* indicates all preparations with a >20% average pyloric cycle variability. (*C*) Histogram shows the average cycle-to-cycle variability in pyloric frequency, as a percentage, for $N = 115$ preparations of the STNS 30 min following decentralization. Bar on the far *right* indicates all preparations with a >20% average pyloric cycle variability. (*B*,*C*, Adapted from Hamood et al. 2015.)

ORIGINS OF VARIABILITY

Modeling work done on the pyloric rhythm has shown that it is possible to construct pyloric-like output from circuits with variable underlying conductance parameters (Prinz et al. 2004). Even while preserving the network connectivity, large variation in ionic and synaptic conductances can be tolerated, provided that these variables are appropriately balanced. However, neuromodulation is also likely to be highly variable across animals (Spitzer et al. 2008; Williams et al. 2013). Originally motivated by theoretical studies that showed that multiple parameter solutions can produce very similar behaviors (Goldman et al. 2001; Golowasch et al. 2002; Prinz et al. 2004; Golowasch 2014), we and others measured the variability in individual cellular and synaptic parameters in the same identified neurons in many animals (Schulz et al. 2006, 2007; Khorkova and Golowasch 2007; Goaillard et al. 2009; Grashow et al. 2010; Temporal et al. 2012, 2014; Shruti et al. 2014).

Figure 4 collects data from several studies of quantitative mRNA measurements from PD neurons for many ion channels and gap junction proteins. In all cases, a significant spread of at least twofold to sixfold is observed (Schulz et al. 2007, 2008; Temporal et al. 2012, 2014; Shruti et al. 2014). Similar results from STG neurons have been shown for features such as input resistance (Grashow et al. 2010), ionic conductances (Golowasch et al. 1999; Goldman et al. 2001), coupling coefficients of electrically coupled neurons (Shruti et al. 2014), and synaptic strengths (Goaillard et al. 2009). Similar ranges of variability are seen in measurements of synaptic strength from the leech (Norris et al. 2011; Roffman et al. 2012).

Given the measured variability in individual neuronal and synaptic properties, this raises the question of how this variability is influenced by neuromodulation. Does neuromodulation contribute to neuronal variability, or does it compensate for it? Or both?

Although this answer may depend on the system studied, in the STG the increase in variability following the loss of neuromodulatory inputs suggests that these inputs may be able to compensate for underlying neuronal variability. For example, Figure 5A,B shows pyloric rhythms from two animals that, when recorded with modulatory inputs intact, produce similar output patterns. However, following decentralization (removal of the modulatory inputs) these rhythms become very different from each other (Fig. 5C,D). A straightforward interpretation of these data is that these two pyloric rhythms are produced by networks with very different parameters, including both variable ionic conductance densities and synaptic strengths, and that the descending modulation compensates for these differences. This could arise either because the effects of the descending modulatory inputs are also variable, or because of state-dependent neuromodulation (Gutierrez and Marder 2014; Marder et al. 2014).

Certainly, the extensive array of modulatory substances released on the STG from these descending inputs (Marder and Bucher 2007) grants innumerable opportunities for such variation: in the number or relative location

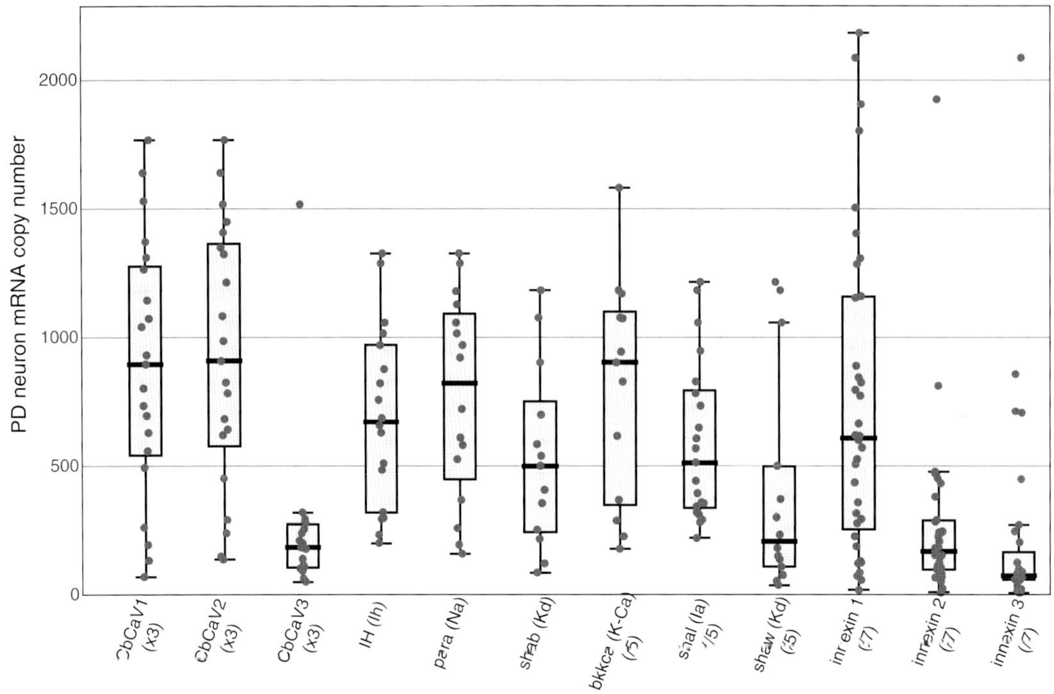

Figure 4. Quantitative polymerase chain reaction results showing measured mRNA copy number, in PD neurons, for many ion channels proteins including voltage-activated calcium channels, cation channels, and gap junctions. (Adapted from Schultz et al. 2006, Temporal et al. 2014, and Shruti et al. 2014.)

of their associated G-protein-coupled receptors (Rodgers et al. 2011a,b, 2013); in the concentrations and specific locations from which they are released (Blitz and Nusbaum 2011, 2012; Nusbaum and Blitz 2012); and in the relative abundance of various neuropeptides, which each also vary in the subset of pyloric neurons on which they converge (Swensen and Marder 2000, 2001; Thirumalai and Marder 2002). Thus, it is possible that when these modulatory inputs are removed from each animal, the direct impact of this removal is more or less extreme (Fig. 5).

NEUROMODULATION CAN INFLUENCE CORRELATIONS IN CONDUCTANCE EXPRESSION

Although variability in the measured features of pyloric neurons has now been repeatedly shown, this variability is not without structure. Many positive correlations have been demonstrated between the mRNA levels of various ion channels and conductances in the neurons of the STG, suggesting that they are regulated in a coordinated fashion (MacLean et al. 2003, 2005; Khorkova

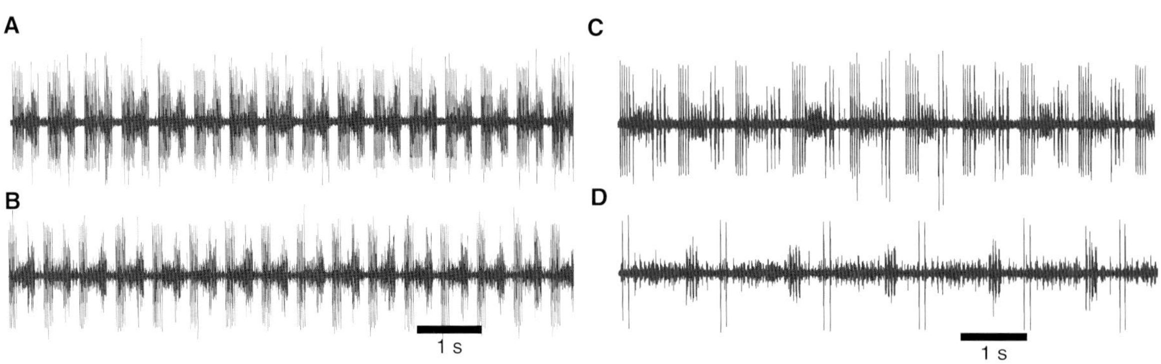

Figure 5. (*A,B*) Example lvn recordings from two intact preparations of the STNS showing highly similar pyloric rhythms. (*C,D*) Following decentralization, after 30 min these same two preparations show highly divergent responses to the perturbation, with dissimilar pyloric rhythms.

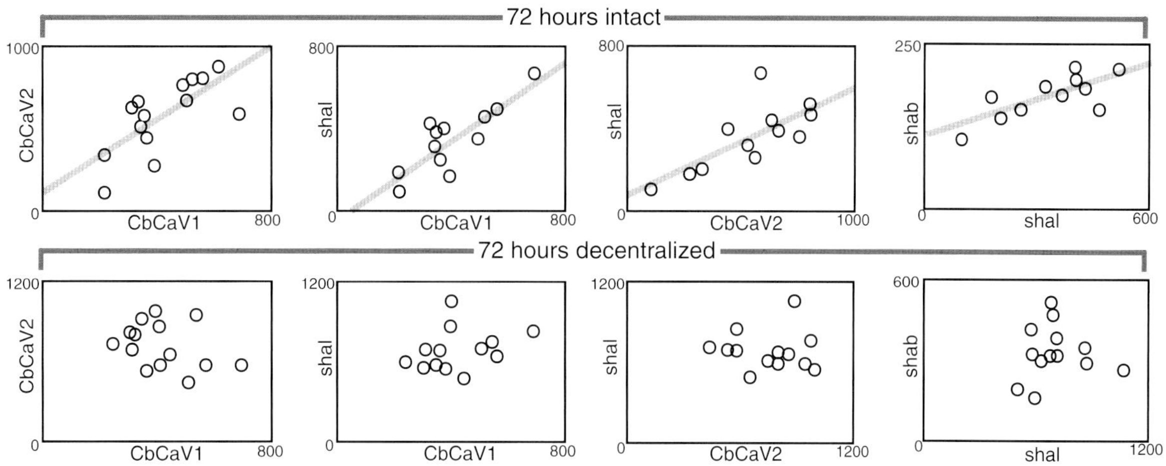

Figure 6. Correlations between mRNA levels in individual PD neurons following 72 h in culture with anterior ganglia left attached (intact, *top* row) or decentralized (*bottom* row). Lines represent linear fits when significant. Several significant positive correlations between mRNA expression levels are observed in the intact population; however, these correlations are lost in preparations maintained in the absence of neuromodulatory inputs. (Adapted from Temporal et al. 2014.)

and Golowasch 2007; Schulz et al. 2007; Temporal et al. 2012, 2014), and correlations are found in neuronal models with similar properties (Hudson and Prinz 2010). Although the measured correlations may not be necessary for the electrophysiological properties of the neurons (Taylor et al. 2009), recent theoretical work shows that neuronal-specific correlations can arise from simple homeostatic tuning rules (O'Leary et al. 2013, 2014) that function at the level of transcription and translation.

It remains largely unknown what role neuromodulation plays in the coordination of this variability, but recent experimental work suggests it might be important. In PD neurons maintained for 72 h following removal from the animal, correlations in mRNA levels seen in intact animals were lost when modulatory inputs were removed (Fig. 6; Khorkova and Golowasch 2007; Temporal et al. 2014). The mechanism by which modulatory influences might be able to produce such coordination at the molecular level is unknown.

A proof of principle that shows that variability at the level of a single neuron can be compensated by variability elsewhere comes from dynamic clamp studies in which single neurons with highly variable properties were connected by reciprocal inhibition to a simple model neuron (Grashow et al. 2010). By varying the strength of the synapses and adjusting the amount of I_H used, two-neuron circuits with similar behavior could be produced (Fig. 7). The range of the conductances needed to compensate for the variability in the individual neurons was approximately threefold, surprisingly similar to the ranges in actual measured conductances in parameters across animals (Grashow et al. 2010).

CONCLUSION

At first glance, the STG, a small circuit with well-defined connectivity (Marder and Bucher 2007), appears simple; but extensive neuromodulation, coupled with the variability of its underlying neuronal and synaptic parameters, makes it remarkable that its function is so robust (Fig. 1). More than 30 years ago, when it was first evident that neuromodulatory influences could alter STG motor patterns (Beltz et al. 1984; Eisen and Marder 1984; Hooper and Marder 1984), the notion was that neuromodulation would provide additional behavioral flexibility to the nervous system, and this has remained an important tenet of our understanding of neuromodulation across the animal kingdom (Harris-Warrick and Johnson 2010; Harris-Warrick 2011; Bargmann 2012; Marder 2012). Nonetheless, it quickly became apparent that the activity patterns of any network at a given time are specified by the modulatory environment at that time. In behaving animals, circuits are never devoid of the influences of modulators found in circulating hormones or released from specific inputs. Consequently, as important as the connectivity of a circuit is for understanding its dynamics, it is equally important to specify the neuromodulatory condition of the network that will produce the final network configuration (Brezina 2010; Harris-Warrick 2011; Bargmann 2012; Gutierrez and Marder 2014; Marder et al. 2014).

In the STG, numerous different neuromodulatory substances can strongly activate robust pyloric rhythms (Hooper and Marder 1987; Marder et al. 1987; Nusbaum and Marder 1988; Weimann et al. 1993, 1997). This raises the possibility that the presence of multiple neuromodulatory inputs with similar but not identical physiological actions can also help ensure robustness, especially as these modulatory mechanisms can counteract some of the potentially deleterious consequences of variability in channel and receptor densities over the animal's lifetime. Consistent with this interpretation, pyloric rhythms deprived of neuromodulatory inputs become more variable both within and across animals, lose their

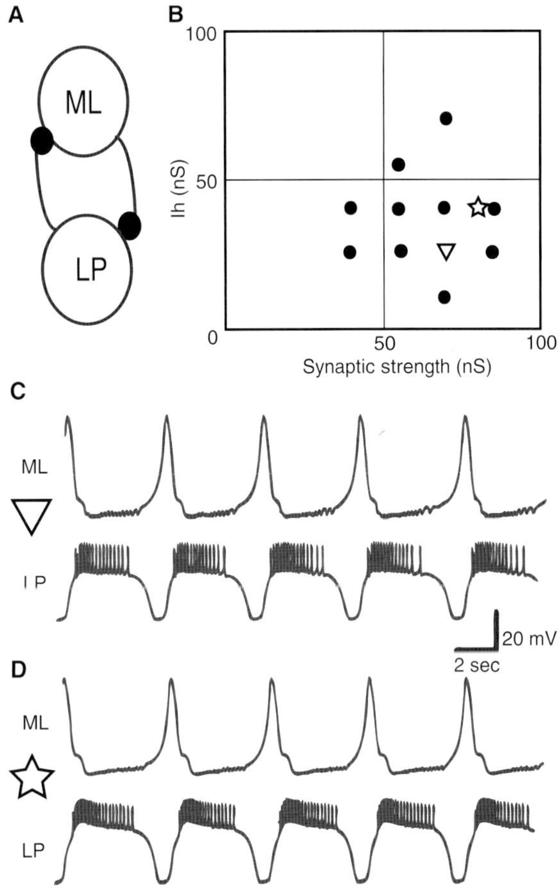

Figure 7. (*A*) Schematic diagram of the hybrid model neuron-LP neuron circuit used in Grashow et al. (2010). ML, model Morris–Lecar neuron. LP, biological LP neuron. They are coupled by mutual inhibitory synapses via dynamic clamp. Both the I_H and synaptic conductances of the model neuron were systematically varied while connected to each of 12 different LP neurons. (*B*) Points show the locations in conductance space for the model neuron where each of the 12 LP–ML circuits produced a highly similar half-center oscillator. Conductances must vary across a several-fold range in the model neuron to compensate for variability across LP neurons and produce conserved circuit function. (*C,D*) Example traces for two LP–ML circuits, marked by shapes corresponding to their locations in *B*. (Adapted from Grashow et al. 2010.)

robust phase relationships, and show fewer correlations in mRNA expression levels.

Many interesting questions remain. By what mechanism are these modulatory inputs regulated? How variable are they across animals? And what role is played by the large amount of degeneracy in these modulatory systems? Additionally, it remains to be seen to what extent the role of neuromodulation with respect to neuronal variability varies depending on the system studied. In central pattern generating networks like the STG, proper behavior demands a stereotypical output pattern that is highly similar across animals, and thus neuromodulation may be tuned to reduce variability. In other systems, such as neocortex, variability in neuronal output may increase information content and thus be a desired network feature.

ACKNOWLEDGMENTS

Research was supported by the National Institutes of Health, grant NS 17813 (E.M.) and an Individual National Research Service Award NS F31 080420 (A.W.H.).

REFERENCES

Bargmann CI. 2012. Beyond the connectome: How neuromodulators shape neural circuits. *Bioessays* **34:** 458–465.

Beenhakker MP, Nusbaum MP. 2004. Mechanosensory activation of a motor circuit by coactivation of two projection neurons. *J Neurosci* **24:** 6741–6750.

Beltz B, Eisen JS, Flamm R, Harris-Warrick RM, Hooper S, Marder E. 1984. Serotonergic innervation and modulation of the stomatogastric ganglion of three decapod crustaceans (*Panulirus interruptus, Homarus americanus* and *Cancer irroratus*). *J Exp Biol* **109:** 35–54.

Blitz DM, Nusbaum MP. 2011. Neural circuit flexibility in a small sensorimotor system. *Curr Opin Neurobiol* **21:** 544–552.

Blitz DM, Nusbaum MP. 2012. Modulation of circuit feedback specifies motor circuit output. *J Neurosci* **32:** 9182–9193.

Blitz DM, Beenhakker MP, Nusbaum MP. 2004. Different sensory systems share projection neurons but elicit distinct motor patterns. *J Neurosci* **24:** 11381–11390.

Brezina V. 2010. Beyond the wiring diagram: Signalling through complex neuromodulator networks. *Philos Trans R Soc Lond B Biol Sci* **365:** 2363–2374.

Bucher D, Prinz AA, Marder E. 2005. Animal-to-animal variability in motor pattern production in adults and during growth. *J Neurosci* **25:** 1611–1619.

Clemens S, Massabuau JC, Legeay A, Meyrand P, Simmers J. 1998. *In vivo* modulation of interacting central pattern generators in lobster stomatogastric ganglion: Influence of feeding and partial pressure of oxygen. *J Neurosci* **18:** 2788–2799.

Clemens S, Massabuau JC, Meyrand P, Simmers J. 1999. Changes in motor network expression related to moulting behaviour in lobster: Role of moult-induced deep hypoxia. *J Exp Biol* **202** (Pt 7)**:** 817–827.

Eisen JS, Marder E. 1984. A mechanism for production of phase shifts in a pattern generator. *J Neurophysiol* **51:** 1375–1393.

Fentress JC. 1976. *Simpler networks and behavior*. Sinauer Associates, Sunderland, MA.

Goaillard JM, Taylor AL, Schulz DJ, Marder E. 2009. Functional consequences of animal-to-animal variation in circuit parameters. *Nat Neurosci* **12:** 1424–1430.

Goaillard JM, Taylor AL, Pulver SR, Marder E. 2010. Slow and persistent postinhibitory rebound acts as an intrinsic short-term memory mechanism. *J Neurosci* **30:** 4687–4692.

Goldman MS, Golowasch J, Marder E, Abbott LF. 2001. Global structure, robustness, and modulation of neuronal models. *J Neurosci* **21:** 5229–5238.

Golowasch J. 2014. Ionic current variability and functional stability in the nervous system. *Bioscience* **64:** 570–580.

Golowasch J, Abbott LF, Marder E. 1999. Activity-dependent regulation of potassium currents in an identified neuron of the stomatogastric ganglion of the crab *Cancer borealis*. *J Neurosci* **19:** RC33.

Golowasch J, Goldman MS, Abbott LF, Marder E. 2002. Failure of averaging in the construction of a conductance-based neuron model. *J Neurophysiol* **87:** 1129–1131.

Grashow R, Brookings T, Marder E. 2010. Compensation for variable intrinsic neuronal excitability by circuit-synaptic interactions. *J Neurosci* **30:** 9145–9156.

Gutierrez GJ, Marder E. 2014. Modulation of a single neuron has state-dependent actions on circuit dynamics. *eNeuro* **1:** doi: 10.1523/ENEURO.0009-14.

Hamood AW, Haddad SA, Otopalik AG, Rosenbaum P, Marder E. 2015. Quantitative reevaluation of the effects of short- and long-term removal of descending modulatory inputs on the

pyloric rhythm of the crab, *Cancer borealis*: Decentralization of the STG. *eNeuro* doi: 10.1523/ENEURO.0058-14.
Harris-Warrick RM. 2011. Neuromodulation and flexibility in central pattern generator networks. *Curr Opin Neurobiol* **21:** 685-692.
Harris-Warrick RM, Johnson BR. 2010. Checks and balances in neuromodulation. *Front Behav Neurosci* **4:** 47.
Harris-Warrick RM, Marder E. 1991. Modulation of neural networks for behavior. *Annu Rev Neurosci* **14:** 39-57.
Hartline DK. 1979. Pattern generation in the lobster (*Panulirus*) stomatogastric ganglion. II. Pyloric network simulation. *Biol Cybern* **33:** 223-236.
Hartline DK, Gassie DV Jr. 1979. Pattern generation in the lobster (*Panulirus*) stomatogastric ganglion. I. Pyloric neuron kinetics and synaptic interactions. *Biol Cybern* **33:** 209-222.
Hartline DK, Maynard DM. 1975. Motor patterns in the stomatogastric ganglion of the lobster *Panulirus argus*. *J Exp Biol* **62:** 405-420.
Hooper SL. 1998. Transduction of temporal patterns by single neurons. *Nat Neurosci* **1:** 720-726.
Hooper SL, Marder E. 1984. Modulation of a central pattern generator by two neuropeptides, proctolin and FMRFamide. *Brain Res* **305:** 186-191.
Hooper SL, Marder E. 1987. Modulation of the lobster pyloric rhythm by the peptide proctolin. *J Neurosci* **7:** 2097-2112.
Hooper SL, Buchman E, Weaver AL, Thuma JB, Hobbs KH. 2009. Slow conductances could underlie intrinsic phase-maintaining properties of isolated lobster (*Panulirus interruptus*) pyloric neurons. *J Neurosci* **29:** 1834-1845.
Hudson AE, Prinz AA. 2010. Conductance ratios and cellular identity. *PLoS Comput Biol* **6:** e1000838.
Khorkova O, Golowasch J. 2007. Neuromodulators, not activity, control coordinated expression of ionic currents. *J Neurosci* **27:** 8709-8718.
MacLean JN, Zhang Y, Johnson BR, Harris-Warrick RM. 2003. Activity-independent homeostasis in rhythmically active neurons. *Neuron* **37:** 109-120.
MacLean JN, Zhang Y, Goeritz ML, Casey R, Oliva R, Guckenheimer J, Harris-Warrick RM. 2005. Activity-independent coregulation of I_A and I_h in rhythmically active neurons. *J Neurophysiol* **94:** 3601-3617.
Manor Y, Nadim F, Abbott LF, Marder E. 1997. Temporal dynamics of graded synaptic transmission in the lobster stomatogastric ganglion. *J Neurosci* **17:** 5610-5621.
Manor Y, Bose A, Booth V, Nadim F. 2003. Contribution of synaptic depression to phase maintenance in a model rhythmic network. *J Neurophysiol* **90:** 3513-3528.
Marder E. 2012. Neuromodulation of neuronal circuits: Back to the future. *Neuron* **76:** 1-11.
Marder E, Bucher D. 2007. Understanding circuit dynamics using the stomatogastric nervous system of lobsters and crabs. *Annu Rev Physiol* **69:** 291-316.
Marder E, Hooper SL, Eisen JS. 1987. Multiple neurotransmitters provide a mechanism for the production of multiple outputs from a single neuronal circuit. In *Synaptic function* (ed. Edelman GM, Gall WE, Cowan MW), pp. 305-327. John Wiley & Sons, New York.
Marder E, O'Leary T, Shruti S. 2014. Neuromodulation of circuits with variable parameters: Single neurons and small circuits reveal principles of state-dependent and robust neuromodulation. *Annu Rev Neurosci* **37:** 329-346.
Maynard DM. 1972. Simpler networks. *Ann NY Acad Sci* **193:** 59-72.
Miller JP, Selverston AI. 1982a. Mechanisms underlying pattern generation in lobster stomatogastric ganglion as determined by selective inactivation of identified neurons. II. Oscillatory properties of pyloric neurons. *J Neurophysiol* **48:** 1378-1391.
Miller JP, Selverston AI. 1982b. Mechanisms underlying pattern generation in lobster stomatogastric ganglion as determined by selective inactivation of identified neurons. IV. Network properties of pyloric system. *J Neurophysiol* **48:** 1416-1432.
Moulins M, Cournil I. 1982. All-or-none control of the bursting properties of the pacemaker neurons of the lobster pyloric pattern generator. *J Neurobiol* **13:** 447-458.
Mulloney B, Selverston AI. 1974a. Organization of the stomatogastric ganglion in the spiny lobster. I. Neurons driving the lateral teeth. *J Comp Physiol* **91:** 1-32.
Mulloney B, Selverston AI. 1974b. Organization of the stomatogastric ganglion in the spiny lobster. III. Coordination of the two subsets of the gastric system. *J Comp Physiol* **91:** 53-78.
Norris BJ, Wenning A, Wright TM, Calabrese RL. 2011. Constancy and variability in the output of a central pattern generator. *J Neurosci* **31:** 4663-4674.
Nusbaum MP, Beenhakker MP. 2002. A small-systems approach to motor pattern generation. *Nature* **417:** 343-350.
Nusbaum MP, Blitz DM. 2012. Neuropeptide modulation of microcircuits. *Curr Opin Neurobiol* **22:** 592-601.
Nusbaum MP, Marder E. 1988. A neuronal role for a crustacean red pigment concentrating hormone-like peptide-neuromodulation of the pyloric rhythm in the crab, *Cancer borealis*. *J Exp Biol* **135:** 165-181.
O'Leary T, Williams AH, Caplan JS, Marder E. 2013. Correlations in ion channel expression emerge from homeostatic tuning rules. *Proc Natl Acad Sci* **110:** E2645-E2654.
O'Leary T, Williams AH, Franci A, Marder E. 2014. Cell types, network homeostasis, and pathological compensation from a biologically plausible ion channel expression model. *Neuron* **82:** 809-821.
Prinz AA, Bucher D, Marder E. 2004. Similar network activity from disparate circuit parameters. *Nat Neurosci* **7:** 1345-1352.
Rezer E, Moulins M. 1983. Expression of the crustacean pyloric pattern generator in the intact animal. *J Comp Physiol* **153:** 17-28.
Richards KS, Marder E. 2000. The actions of crustacean cardioactive peptide on adult and developing stomatogastric ganglion motor patterns. *J Neurobiol* **44:** 31-44.
Richards KS, Miller WL, Marder E. 1999. Maturation of the rhythmic activity produced by the stomatogastric ganglion of the lobster, *Homarus americanus*. *J Neurophysiol* **82:** 2006-2009.
Rodgers EW, Fu JJ, Krenz WD, Baro DJ. 2011a. Tonic nanomolar dopamine enables an activity-dependent phase recovery mechanism that persistently alters the maximal conductance of the hyperpolarization-activated current in a rhythmically active neuron. *J Neurosci* **31:** 16387-16397.
Rodgers EW, Krenz WD, Baro DJ. 2011b. Tonic dopamine induces persistent changes in the transient potassium current through translational regulation. *J Neurosci* **31:** 13046-13056.
Rodgers EW, Krenz WD, Jiang X, Li L, Baro DJ. 2013. Dopaminergic tone regulates transient potassium current maximal conductance through a translational mechanism requiring D1Rs, cAMP/PKA, Erk and mTOR. *BMC Neurosci* **14:** 143.
Roffman RC, Norris BJ, Calabrese RL. 2012. Animal-to-animal variability of connection strength in the leech heartbeat central pattern generator. *J Neurophysiol* **107:** 1681-1693.
Russell DF. 1979. CNS control of pattern generators in the lobster stomatogastric ganglion. *Brain Res* **177:** 598-602.
Schulz DJ, Goaillard JM, Marder E. 2006. Variable channel expression in identified single and electrically coupled neurons in different animals. *Nat Neurosci* **9:** 356-362.
Schulz DJ, Goaillard JM, Marder EE. 2007. Quantitative expression profiling of identified neurons reveals cell-specific constraints on highly variable levels of gene expression. *Proc Natl Acad Sci* **104:** 13187-13191.
Schulz DJ, Temporal S, Barry DM, Garcia ML. 2008. Mechanisms of voltage-gated ion channel regulation: From gene expression to localization. *Cell Mol Life Sci* **65:** 2215-2231.
Selverston AI, Mulloney B. 1974. Organization of the stomatogastric ganglion of the spiny lobster. II. Neurons driving the medial tooth. *J Comp Physiol* **91:** 33-51.

Selverston AI, Russell DF, Miller JP. 1976. The stomatogastric nervous system: Structure and function of a small neural network. *Prog Neurobiol* **7:** 215–290.
Shruti S, Schulz DJ, Lett KM, Marder E. 2014. Electrical coupling and innexin expression in the stomatogastric ganglion of the crab *Cancer borealis*. *J Neurophysiol* **112:** 2946–2958.
Soofi W, Goeritz ML, Kispersky TJ, Prinz AA, Marder E, Stein W. 2014. Phase maintenance in a rhythmic motor pattern during temperature changes *in vivo*. *J Neurophysiol* **111:** 2603–2613.
Spitzer N, Cymbalyuk G, Zhang H, Edwards DH, Baro DJ. 2008. Serotonin transduction cascades mediate variable changes in pyloric network cycle frequency in response to the same modulatory challenge. *J Neurophysiol* **99:** 2844–2863.
Swensen AM, Marder E. 2000. Multiple peptides converge to activate the same voltage-dependent current in a central pattern-generating circuit. *J Neurosci* **20:** 6752–6759.
Swensen AM, Marder E. 2001. Modulators with convergent cellular actions elicit distinct circuit outputs. *J Neurosci* **21:** 4050–4058.
Tang LS, Goeritz ML, Caplan JS, Taylor AL, Fisek M, Marder E. 2010. Precise temperature compensation of phase in a rhythmic motor pattern. *PLoS Biol* **8:** e1000469.
Taylor AL, Goaillard JM, Marder E. 2009. How multiple conductances determine electrophysiological properties in a multicompartment model. *J Neurosci* **29:** 5573–5586.
Temporal S, Desai M, Khorkova O, Varghese G, Dai A, Schulz DJ, Golowasch J. 2012. Neuromodulation independently determines correlated channel expression and conductance levels in motor neurons of the stomatogastric ganglion. *J Neurophysiol* **107:** 718–727.
Temporal S, Lett KM, Schulz DJ. 2014. Activity-dependent feedback regulates correlated ion channel mRNA levels in single identified motor neurons. *Curr Biol* **24:** 1899–1904.
Thirumalai V, Marder E. 2002. Colocalized neuropeptides activate a central pattern generator by acting on different circuit targets. *J Neurosci* **22:** 1874–1882.
Turrigiano GG, Marder E, Abbott LF. 1996. Cellular short-term memory from a slow potassium conductance. *J Neurophysiol* **75:** 963–966.
Weimann JM, Marder E, Evans B, Calabrese RL. 1993. The effects of SDRNFLRFamide and TNRNFLRFamide on the motor patterns of the stomatogastric ganglion of the crab *Cancer borealis*. *J Exp Biol* **181:** 1–26.
Weimann JM, Skiebe P, Heinzel HG, Soto C, Kopell N, Jorge-Rivera JC, Marder E. 1997. Modulation of oscillator interactions in the crab stomatogastric ganglion by crustacean cardioactive peptide. *J Neurosci* **17:** 1748–1760.
Williams AH, Calkins A, O'Leary T, Symonds R, Marder E, Dickinson PS. 2013. The neuromuscular transform of the lobster cardiac system explains the opposing effects of a neuromodulator on muscle output. *J Neurosci* **33:** 16565–16575.

The Brainstem Oscillator for Whisking and the Case for Breathing as the Master Clock for Orofacial Motor Actions

DAVID KLEINFELD,[1,2] JEFFREY D. MOORE,[1] FAN WANG,[3] AND MARTIN DESCHÊNES[4]

[1]Department of Physics, University of California San Diego, La Jolla, California 92093

[2]Section of Neurobiology, University of California San Diego, La Jolla, California 92093

[3]Department of Neurobiology, Duke University Medical Center, Durham, North Carolina 27710

[4]Department of Psychiatry and Neuroscience, Laval University, Quebec City, Quebec GIJ 2G3, Canada

Correspondence: dk@physics.ucsd.edu

Whisking and sniffing are predominant aspects of exploratory behavior in rodents. We review evidence that these motor rhythms are coordinated by the respiratory patterning circuitry in the ventral medulla. A recently described region in the intermediate reticular zone of the medulla functions as an autonomous whisking oscillator, whose neuronal output is reset upon each breath by input from the pre-Bötzinger complex. Based on similarities between this neuronal circuit architecture and that of other orofacial behaviors, we propose that the pre-Bötzinger complex, which projects broadly to premotor regions throughout the intermediate reticular zone of the medulla, functions as a master clock to coordinate multiple orofacial actions involved in exploratory and ingestive behaviors. We then extend the analysis of whisking to the relatively slow control of the midpoint of the whisk. We conjecture, in a manner consistent with breathing as the "master clock" for all orofacial behaviors, that slow control optimizes the position of sensors whereas the breathing rhythm provides a means to perceptually bind the inputs from different orofacial modalities.

A postulate from the neuroethology movement is the presence of a hierarchical control structure for motor acts that is orchestrated by the nervous system (Weiss 1941; Tinbergen 1951). At the base of this hierarchy are individual motor "actions." For example, orofacial actions that comprise sensory exploration would correspond to a sniff, a twitch of the nose, a lick, a whisk, or a turn of the head. These actions are coordinated in nested groups to form more complex "behaviors." The nature of the interactions among the neuronal circuits that generate these actions and coordinate their output is a foundational topic in computational neuroscience. One question, which motivates the present work, concerns the possibility of using a master clock as a means to organize motor actions into behaviors. The use of such a clock could provide a computationally simple mechanism to avoid antagonistic movements and to adhere to biomechanical constraints. Presumably, the clock could function as part of a larger circuit that involves interactions between motor and premotor neuronal centers, as well as environmental and proprioceptive feedback signals. For the case of exploratory movements, the use of a common clock to actively sample the sensory environment could simplify the problem of binding inputs arising from different modalities (e.g., smell, touch, and taste) (Kleinfeld et al. 2014).

The notion of a master clock is implicit in the pioneering work of Welker (1964), published a half century ago, who observed that movements of the nose, the head, and the vibrissae of rodents appear to be coherent with rapid breathing, or sniffing. Together, these three actions are hypothesized to synchronize with the breathing cycle to form a stereotyped exploratory behavior that rodents use to rhythmically sample their surrounding olfactory and tactile environments (Fig. 1). A parsimonious conclusion is that the breathing cycle may be the master clock at the root of all rhythmic orofacial behaviors.

Here we summarize our recent work on the neuronal circuitry that controls rhythmic exploratory whisking by rats (Moore et al. 2013); this work provides evidence that whisking is subsumed by the breathing rhythm. Our summary is followed by a presentation of two threads of subsequent, and ongoing, inquiry. Motivated by anatomical similarities between this newly discovered brainstem whisking circuit and those circuits thought to control the tongue and jaw, the first thread bears directly on the issue of a master clock and concerns the potential role of breathing in the synchronization of other motor actions. The second thread addresses the mechanistic basis for controlling the slowly varying amplitude and position of whisking, which we propose may be organized similarly to the control of body posture.

It is instructive to begin with an overview of the motor plant for whisking. A Manhattan-like grid of follicles is embedded within the mystacial pad (Fig. 2), which forms the cheek. Intrinsic muscles wrap around individual vibrissa follicles and are only partially inserted into the skin (Fig. 2A). These muscles resemble slings that embrace

 doi: 10.1101/sqb.2014.79.024794

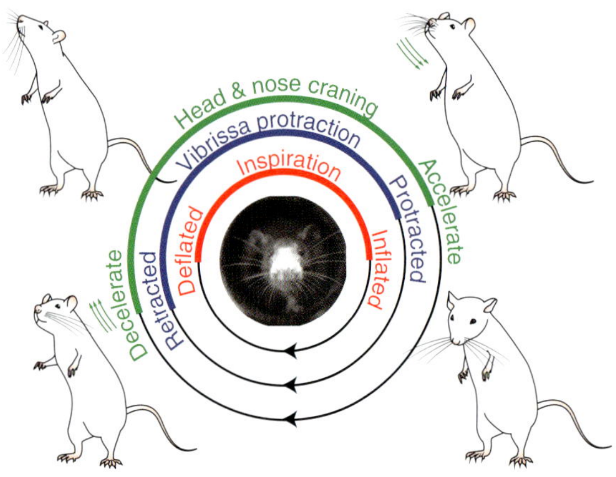

Figure 1. A cycle of exploratory behavior locked to sniffing. Three motor actions, whisking, head bobbing, and nose twitching, are linked to a fourth, sniffing, during orofacial exploration. (Abstracted from the claims of Welker 1964.)

the lower half of the follicle, and they are anchored to each of two adjacent follicles of the same row (Fig. 2D; Dörfl 1982; Wineski 1985). In addition, five superficial extrinsic muscles are inserted into either the plate or the skin of the pad, between the rows of follicles. These muscles include *m. nasalis* as a protractor and *m. nasolabialis* and *m. maxillolabialis* as retractors (Fig. 2C). Simultaneous measurements of the electromyogram (EMG) from the intrinsic and extrinsic muscles indicate the presence of three phases of muscle contraction during the whisking cycle (Fig. 2D), with different phases for the extrinsic protractor, intrinsic protractor, and extrinsic retractor muscles (Hill et al. 2008).

To verify our understanding of the muscular control of whisking, we constructed a mechanical model of the motor plant to predict the motion of the vibrissae and the mystacial pad from the measured EMGs of the intrinsic and different extrinsic muscles (Fig. 2D; Hill et al. 2008). The anatomy of the musculature, including attachment points in the skin and plate together with measurements of the viscoelastic restoring forces, provides the data that inform the model (Fig. 2E; Hill et al. 2008; Simony et al. 2010). For whisking while sniffing, we find excellent agreement between the calculated and measured motion of both the vibrissae and the pad (Fig. 2F). This implies the presence of a multiphase rhythmic pattern generator. For concreteness, we focus this presentation on the rhythm that dominates protraction through direct drive of the intrinsic muscles. Such a presentation is analogous to much of the literature that describes the motor control of breathing, which also consists of three phases of neuronal activity, yet the inspiratory phase can be driven by a single oscillator circuit (Smith et al. 2009).

Concurrent measurements of breathing and whisking in head-restrained rats reveal key aspects of their coordination (Moore et al. 2013; Ranade et al. 2013), and extend Welker's initial descriptions (Fig. 3). First, breathing over a wide range of frequencies can occur without substantial whisking (Fig. 3A). To determine whether whisking can also occur without breathing, a puff of ammonia was applied to the snout to inactivate the central inspiratory drive (Lawson et al. 1991) and temporarily inhibit respiration. Critically, rats can whisk during such a disruption in breathing (Fig. 3B). These data imply that there are separate, or separable, oscillator circuits for breathing and whisking.

Sniffing is always accompanied by a one-to-one relation with whisking (Fig. 3A,B), as first noted by Welker (1964). Yet basal breathing, which is two to five times slower than whisking, is accompanied by two forms of whisks: those that are coincident with an inspiration, which we denote "inspiratory-locked whisks," and those that occur between successive breaths and successively diminish in amplitude, which we denote "intervening whisks" (Fig. 3C,D). The "intervening whisks" are incommensurate with breathing. These data, with further support from phase-response curves (Moore et al. 2013), imply that the oscillator circuit for breathing resets the phase of the oscillator for whisking.

Where is the pattern generator for this fast, rhythmic vibrissa motion? One possibility is that rapid whisking is generated by a disynaptic feedback loop in the brainstem (Nguyen and Kleinfeld 2005; Deutsch et al. 2012; Matthews et al. 2014). That is, sensory signals generated by the motion of the vibrissa could drive the facial motoneurons, which in turn would regenerate a sensory signal, and so on. However, lesion of the vibrissa sensory nerve (i.e., the infraorbial branch of the trigeminal) has minimal effect on whisking (Welker 1964; Gao et al. 2001; Berg and Kleinfeld 2003) and thus disproves this hypothesis. Further, lesion of any of a large number of midbrain and forebrain nuclei only minimally disrupts whisking (Welker 1964; Lovick 1972; Semba and Komisaruk 1984). These observations lead to the alternative hypothesis that there is an independent oscillator that drives whisking and is located in the brainstem. But where? Given the close phase relationship between whisking and breathing, the neighborhood of the ventral respiratory column in the medulla (Smith et al. 2009) is a plausible suspect. A role for breathing in the control of whisking is also suggested by the shared facial musculature between the two behaviors (Sherrey and Megirian 1977; Haidarliu et al. 2012). We thus directed our search near the location of the pre-Bötzinger complex, which was first showed by Feldman and colleagues (Smith et al. 1991; Tan et al. 2008) to generate the inspiratory breathing rhythm.

The close, one-to-one relationship between sniffing and whisking makes it difficult to discriminate between potential breathing and whisking neuronal centers (Fig 3A). However, the difference in the frequency of basal respiration versus whisking provides a behavioral signature for such a discrimination (Fig. 3C,D; Moore et al. 2013). We recorded unit spiking activity within the pre-Bötzinger complex as well as in the rostral ventral respiratory group in head-fixed awake rats. As expected, spiking occurred in phase with inspiration and with vibrissa protraction during inspiratory-locked whisks (Fig. 3C and red dots in Fig. 3E). Critically, the spiking activity

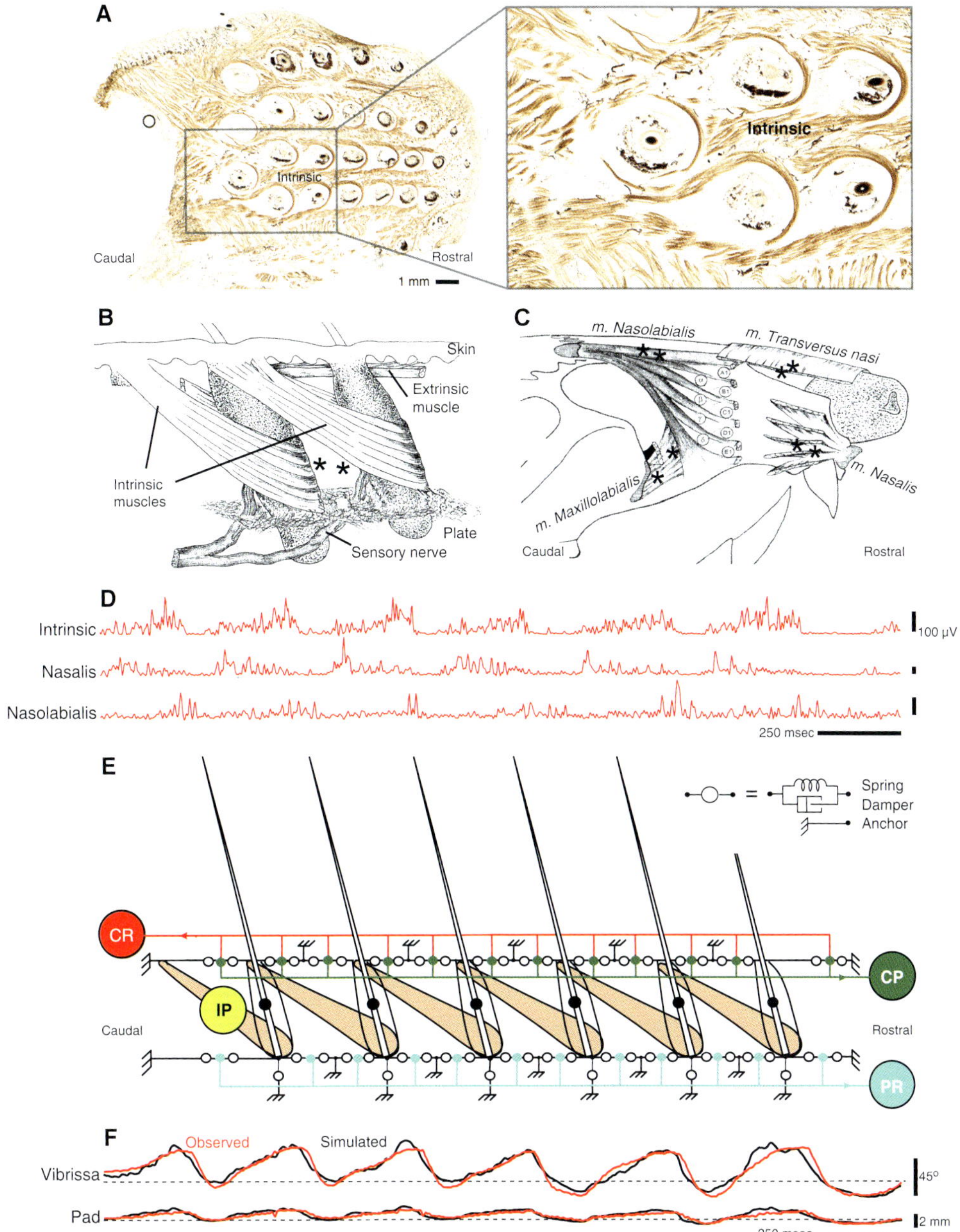

Figure 2. The nature of the motor plant for whisking. (*A*) A tangential slice containing a whole set of intrinsic muscles that form slings about the follicles that house the vibrissae. Most of the muscle slings are only partially represented in this plane. (*B*) Drawing of intrinsic musculature and follicular anatomy. The intrinsic muscles join adjacent follicles of a single row. Each muscle attaches medially and laterally to the superior part of the caudal follicle while forming a sling around the lower third of the rostral follicle. The skin on the outside and the plate on the inside provide a passive restoring force. Superficial extrinsic muscles run just below the skin. Asterisks indicate the approximate locations of the exposed tips of EMG microwires (Hill et al. 2008). (Adapted, with permission, from Dörfl 1982, © Wiley.) (*C*) Drawing of extrinsic musculature. Four extrinsic muscles invade the mystacial pad while maintaining external attachment points. The protractor *m. nasalis* attaches rostral to the pad at the nasal septum and runs under the skin as it extends caudally. The retractor *m. nasolabialis* attaches dorsal-caudal to the pad and runs superficially below the skin. The retractor *m. maxillolabialis* attaches ventral-caudal to the pad and fuses with the fibers of *m. nasolabialis* as they invade the pad. *m. transversus nasi* lies transverse to the snout and runs superficially through the pad. The recording sites for each muscle are indicated by asterisks. (Adapted from Dörfl 1982.) (*D*) Muscle activity from head-restrained rats. Electromyogram signals are abbreviated as INT (intrinsic muscles), NA (*m. nasalis*), NL (*m. nasolabialis*). The signals are high-pass filtered, full-wave rectified, and low-pass filtered. (Adapted from Hill et al. 2008.) (*E*) Schematic of a biomechanical model that represents one row of vibrissae in the maxillary compartment of the rat mystacial pad in resting position. Intrinsic whisker protractors (IP) represent intrinsic muscles. The protractor (CP) of the mystacial pad represents the Pars media inferior of the *m. nasalis* extrinsic muscle, which attached to the skin. The retractor (CR) of the mystacial pad represents the *m. maxillolabialis* and *m. nasolabialis* extrinsic muscles, which also attach to the skin. The whisker retractor (PR) represents the *Pares maxillares superficialis* and *profunda* of the *m. nasalis* extrinsic muscle, which attach to the plate. Green dots in the skin and blue dots in the plate rows represent attachment sites of the extrinsic muscles. Large black dots represent the centers of mass of the vibrissae. Empty circles represent springs coupled with dampers that symbolize the elasticity of the tissue. Anchors represent nonelastic sites in the mystacial pad. (Adapted from Haidarliu et al. 2010, after Simony et al. 2010 and Hill et al. 2008.) (*F*) Vibrissa and mystacial pad motion from head-restrained rats recorded concurrently with the EMG data of *D*. The black curve was calculated from a reduced model of the motor plant using the data of *D* as the sole input. (Adapted from Hill et al. 2008.)

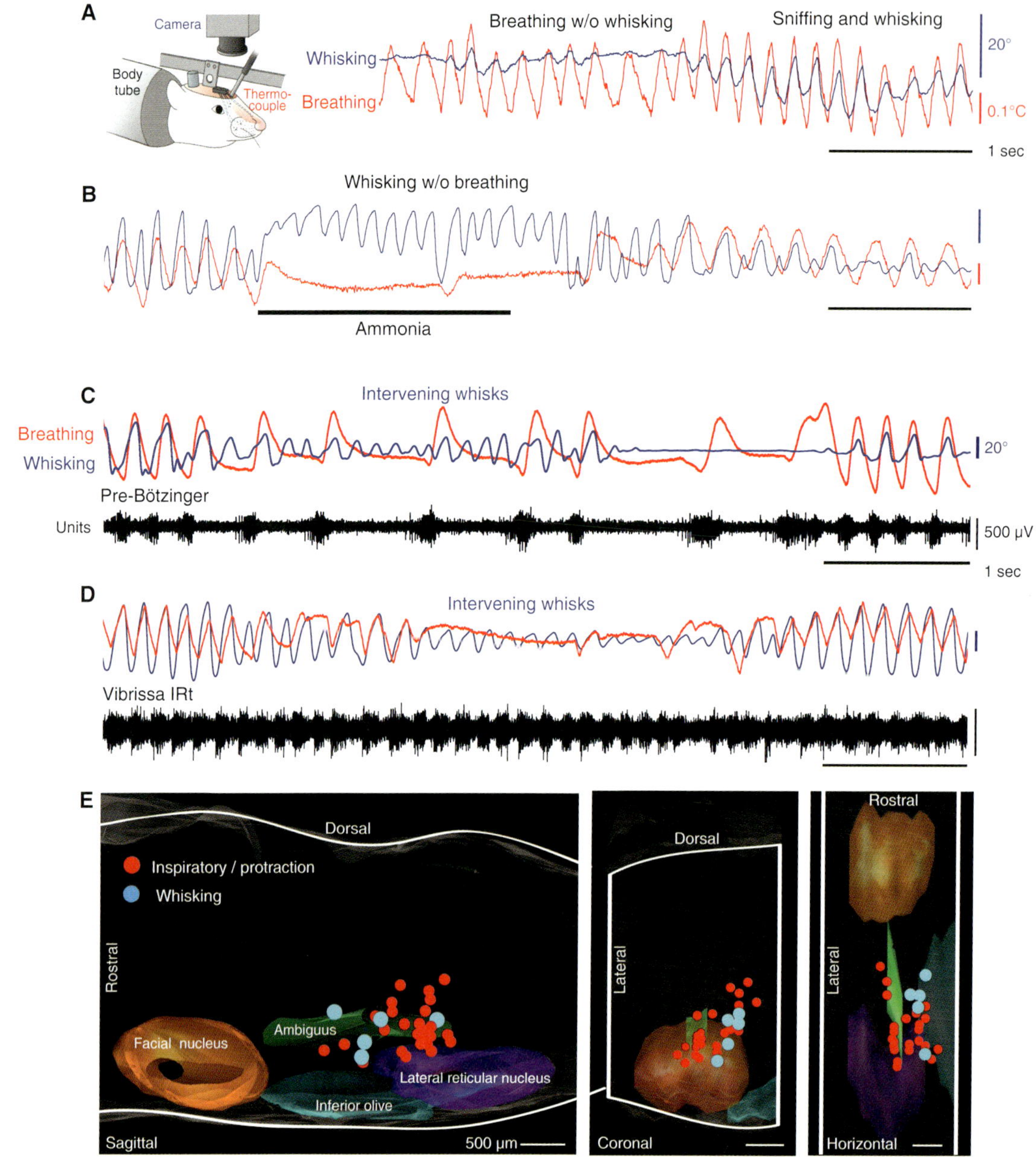

Figure 3. Simultaneous measurements of vibrissa angular position (blue) and breathing (red) in head-fixed rats. (*A*) Measurement showing epochs of breathing (red) without whisking (blue) and synchronous sniffing and whisking in a head-fixed rat. *Insert* is a schematic for simultaneous measurements of vibrissa angular position through videography and breathing through the change in temperature of a thermocouple connected to a tube inserted into the nasal passage. (*B*) Measurement showing an epoch of whisking without breathing, where inspiration was halted by introduction of ammonia into the air mixture. (*C*) Measurement showing breathing with intervening whisks between inspirations, along with concurrent recordings of multiunit activity (black) in the pre-Bötzinger complex. (*D*) Measurement showing breathing with intervening whisks between inspirations, along with concurrent recordings of multiunit activity (black) in the vibrissa zone of the intermediate reticular formation. (*E*) The recording sites for all data imposed on a three-dimensional reconstruction of the medulla. Whisking units (blue) are located in the IRt dorsomedial to inspiratory units (red) in the pre-Bötzinger complex. (Adapted from Moore et al. 2013.)

did not track the intervening whisks. In contrast, we located a subset of units in the intermediate band of the reticular formation (IRt) whose spiking was synchronous with the protraction of both inspiratory-locked and intervening whisks (Fig. 3D and blue dots in Fig. 3E). These units are potential premotor inputs to the motoneurons that drive the intrinsic muscles that protract the vibrissae (Fig. 2A,B). They are located in the ventral part of the IRt, medial to the *ambiguus* nucleus and immediately dorsomedial to the pre-Bötzinger complex. We denote this new region the vibrissa zone of the IRt (vIRt) and refer to units in this region as "whisking units."

To show the role of the vIRt in the generation of whisking, we address the necessity and sufficiency of this region for rhythmic vibrissa protraction. In support of necessity, we observe that small lesions of the vIRt abolish whisking on the ipsilateral side of the lesion, whereas whisking persists on the contralateral side (Fig. 4A). Off-site lesions, on the other hand, have minimal effects on whisking (Fig. 4B). Sufficiency of the vIRt for rhythmic motion was assessed by activation of this region through microinjection of the glutamate receptor agonist kainic acid in the vicinity of the vIRt. We observe prolonged rhythmic vibrissa movements, at normal whisking frequencies (Carvell and Simons 1990; Berg and Kleinfeld 2003), for injections in the lightly anesthetized rat (Fig. 4C) and mouse (Moore et al. 2014a). The frequency of whisking that is induced by kainic acid is observed to decrease over time, whereas the frequency of basal breathing remains at a constant rate. These data imply that the chemical activation is sufficiently strong to temporally unlock rhythmic vibrissa protraction from breathing.

The ability to induce whisking in an immobile animal through kainic acid microinjection provides a means to stably record from units whose firing times are coherent with rhythmic protraction. We identified neuronal units that spike in synchrony with protraction, as in the case of the units identified during intervening whisks in the behaving animal (Fig. 3D), as well as units that spike in antiphase (Fig. 4D). This finding is suggestive of a rhythm that results from local network dynamics as opposed to coupled single-cell oscillators (Getting 1989; Kleinfeld and Sompolinsky 1989).

The behavioral data (Fig. 3) imply that neurons in the pre-Bötzinger complex reset an oscillatory network of whisking units in the vIRt that initiate protraction of the vibrissae concurrent with each inspiration. What is the nature of the connectivity from the pre-Bötzinger complex to the vIRt? We assessed the underlying connectivity for this action by tract tracing methods (Fig. 5A; Moore et al. 2013). Injections of an anterograde tracer into the pre-Bötzinger complex, identified by the phase relation of units relative to breathing (Fig. 5B), led to dense anterograde labeling of terminals in the same vIRt region where we observed whisking units and where lesions extinguished ipsilateral whisking (Fig. 5C). These results support a direct connection from the pre-Bötzinger complex to the vIRt. In addition, injection of an anterograde tracer into the region of the vIRt does not lead to discernable labeling in the pre-Bötzinger complex (M Deschênes, unpubl.). These data imply that there is only unilateral

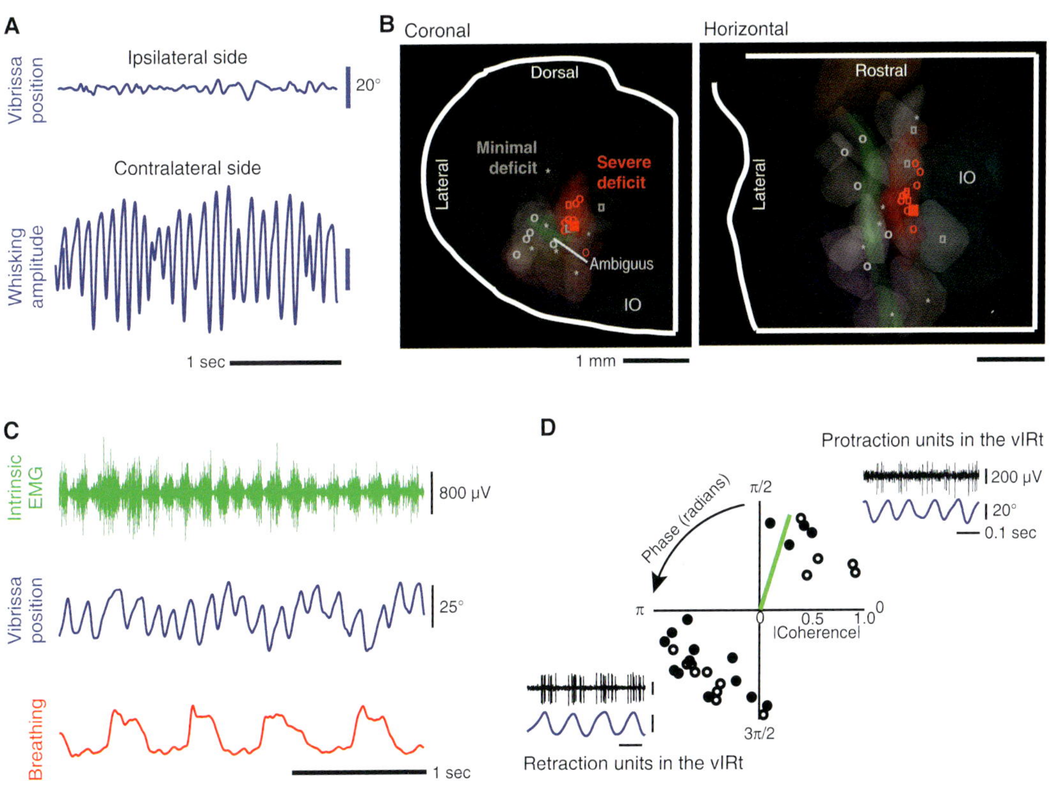

Figure 4. Necessity and sufficiency of the vIRt as the rhythm generator for whisking. Lesion of the vIRt impairs ipsilateral whisking. (*A*) An example of whisking bout following an electrolytic lesion. In *B* all lesion sites were mapped onto a three-dimensional reconstruction of the medulla and selected anatomical substructures. The lesion centroids are denoted with symbols, with circles for electrolytic lesions, triangles for lesion via transport of Sindbis virus, and squares for chemical lesion by ibotenic acid. Injection of the glutamate channel agonist kainic acid activates the vIRt, which drives facial motoneurons and induces whisking. (*C*) Vibrissa motion (blue), breathing (red), and intrinsic (green) EMG. (*D*) A polar plot of the coherence between spiking activity and vibrissa motion at the peak frequency of whisking. Open circles represent multiunit activity and closed circles represent single units. The green bar represents the coherence of the EMG for the intrinsic muscle (*B*) with vibrissa motion. (*Inserts*) Spiking activity of neuronal units in the vIRt (black) in relation to vibrissa motion (blue). (Adapted from Moore et al. 2013.)

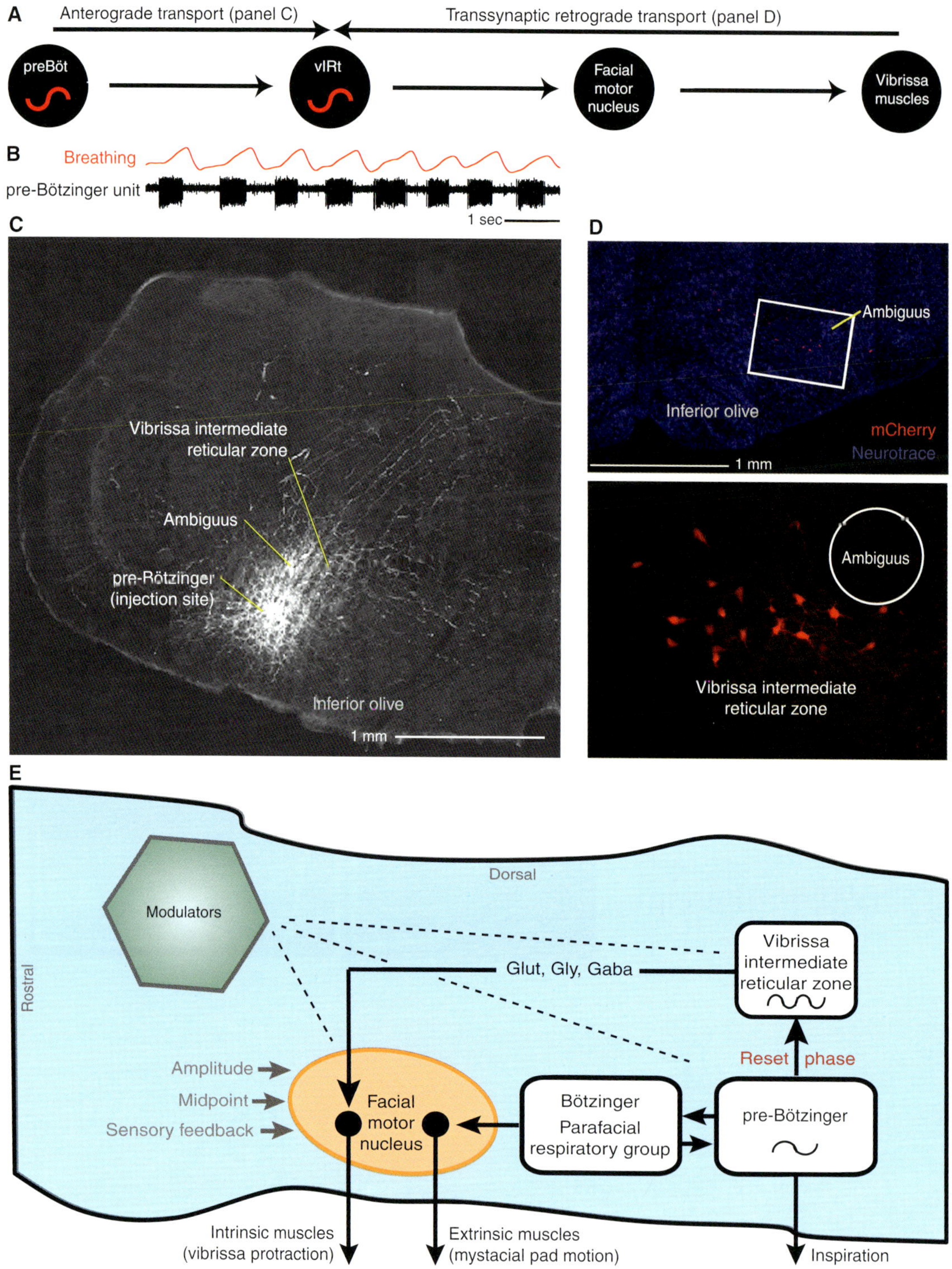

Figure 5. Anatomy and circuit for the resetting of whisking by breathing. (*A*) Proposed connectivity from the pre-Bötzinger complex, through the vIRt, and onward to the follicle muscles. Resolution of this connection pattern makes use of anterograde labeling to define the path from the pre-Bötzinger complex to the vIRt and retrograde transsynaptic labeling to define the pathway from the vIRt to the mystacial pad. (*B*) Recording of a single inspiratory unit in the pre-Bötzinger complex, together with breathing. (Adapted from Moore et al. 2013.) (*C*) Tracing via injection of the anterograde tracer biotinylated dextran amine through the same pipette used to record *A* leads to labeling of axons and terminals in the vIRt from cells in the pre-Bötzinger complex. (Adapted from Moore et al. 2013.) (*D*) Retrograde labeling of neurons in the vIRt by modified Rabies virus, ΔG-RV, that was pressure injected into the facial musculature of juvenile transgenic mice engineered to express the glycoprotein under the premotor for choline acetyltransferase. Monosynaptically connected premotor neurons in the vIRt are labeled by ΔG-RV (red) and counterstained with Neurotrace fluorescent Nissl stain (Life Technologies) (blue). Ambiguus refers to nucleus *ambiguus*. (Adapted from Takatoh et al. 2013.) (*E*) Model of the circuitry that generates whisking that is coordinated with breathing. Dashed lines indicate diffuse synaptic input from modulatory brain nuclei. GLUT, glutamate; GLY, glycine; GABA, γ-aminobutyric acid. (Adapted from Moore et al. 2013.)

connectivity from the breathing oscillator to the whisking oscillator, consistent with the reset of whisking by breathing but not vice versa (Moore et al. 2013).

We next delineated the projections from neurons in the vIRt to facial motoneurons. Here we used a retrograde scheme based on replication incompetent rabies virus to label from the muscles in the mystacial pad transynaptically back to the vIRt. A modified rabies virus, denoted ΔG-RV, in which the gene for glycoprotein, G, was deleted and the gene for red fluorescent protein inserted, was grown in cells engineered to produce G (Wickersham et al. 2007). This virus is injected into the facial musculature of juvenile transgenic mice that express G in cholinergic neurons, so that infected motoneurons can recapitulate active rabies for subsequent labeling of premotoneurons. We observe strong labeling in the location of the vIRt (Fig. 5D; Takatoh et al. 2013). These and previous (Isokawa-Akesson and Komisaruk 1987) patterns of neuronal labeling in the IRt support a direct connection from the vIRt to the facial nucleus and substantiate the role of the vIRt as the premotor pattern generator for rhythmic whisking. In toto, the vIRt is an element of a larger circuit whereby cells in or near the pre-Bötzinger complex for inspiration (Tan et al. 2008; Bouvier et al. 2010; Gray et al. 2010) reset the phase of vIRt units with each breath during exploratory whisking (Fig. 5E).

Does the pathway found for the reset of whisking by breathing generalize to the coordination of other orofacial behaviors by breathing (Fig. 1)? From a purely anatomical basis, Feldman and colleagues (Tan et al. 2010) showed that neurons that are critical to the generation of the inspiratory rhythm in the pre-Bötzinger complex extend collateral projections across the ventral to dorsal extent of the IRt (Fig. 6A). These axons are likely to intersect the dendritic fields of premotor nuclei, like that of the vIRt for whisking, that drive other rhythmic orofacial actions (Fig. 6B). The best described of these is licking, for which tongue protrusions are coordinated with the respiratory cycle (Welzl and Bures 1977), and like whisking, may be controlled by a separate oscillator (Wiesenfeld et al. 1977). A cluster of hypoglossal premotoneurons is concentrated dorsomedial to the pre-Bötzinger complex within the IRt (Fig. 6B; Travers et al. 1997; Koizumi et al. 2008; Stanek et al. 2014). The spiking output of units in this region of the IRt locks to rhythmic licking (Travers et al. 2000), whereas infusion of an inhibitory agonist into the IRt blocks licking (Chen et al. 2001). These observations provide evidence in support of sufficiency and necessity, respectively, for these units as a premotor rhythm generator. Critically, hypoglossal premotoneurons produce bursts of spikes that are locked to inspiration (Ono et al. 1998). This evidence is therefore consistent with parallel premotor patterning circuits for licking and whisking in adjacent subcompartments of the IRt that are both influenced by breathing-related neurons in the pre-Bötzinger complex.

Beyond whisking and licking, the putative premotor oscillators for chewing (i.e., the oral portion of the IRt; Travers et al. 2005, 2010), for control of the airway (i.e., the *ambiguus* nucleus; Bieger and Hopkins 1987), and for licking (i.e., the hypoglossal IRt dorsomedial to the vIRt) are located proximally to each other (Travers et al. 1997) and to the pre-Bötzinger complex (Fig. 6B). All told, the existing results are consistent with a model in which pre-Bötzinger units reset the onset of bursts in a network of hypoglossal premotor neurons in the IRt zone, and reset

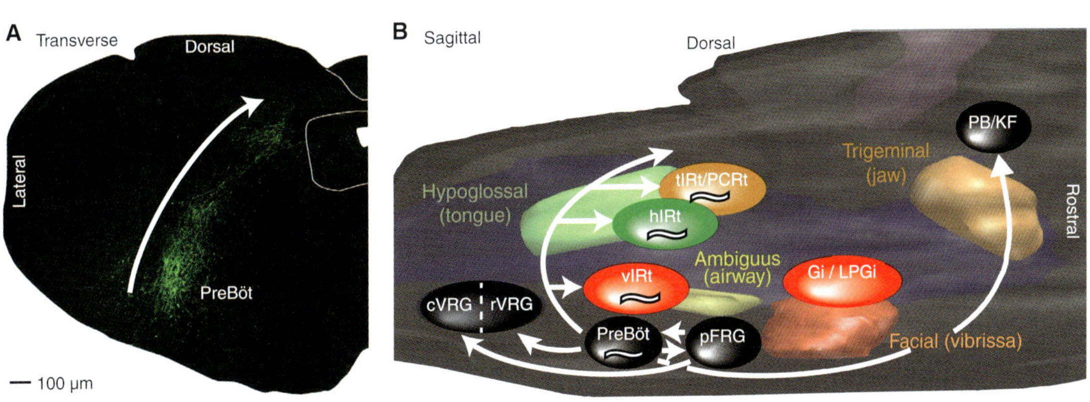

Figure 6. The pre-Bötzinger complex forms a "backplane" of collateral projections across the entire intermediate reticular zone. (*A*) Injection of adeno-associated virus 2, which labels neurons by using the somatostatin promoter to drive the expression of enhanced green fluorescent protein (EGFP) (Tan et al. 2008), forms a band of EGFP-expressing axons along the intermediate reticular zone and revealed labeled axons in parahypoglossal nucleus/nucleus of the solitary tract regions (*top*) and ventral respiratory group caudal to pre-Bötzinger complex (*bottom*). No EGFP-expressing cell bodies were observed. (Adapted, with permission, from Tan et al. 2010, © Wiley.) (*B*) Three-dimensional reconstruction of the pons and medulla, showing the pools of cranial motoneurons that control the jaws (orange), face (red), airway (yellow), and tongue (green), as well as the approximate locations of known premotor nuclei to each of the motoneuron pools. Premotor nuclei are color coded according to the primary motor nucleus that they innervate. Breathing-related regions are shown in black, and the projections of the pre-Bötzinger complex are shown as white arrows. Putative rhythmic oscillators for whisking, licking, and chewing are denoted "∼". cVRG, caudal ventral respiratory group; Gi, gigantocellular reticular formation; hIRt, hypoglossal intermediate reticular formation; LPGi, lateral paragigantocellular reticular formation; PCRt, parvocellular reticular formation; pFRG, parafacial respiratory group; PreBöt, pre-Bötzinger complex; rVRG, rostral ventral respiratory group; tIRt, trigeminal intermediate reticular formation; vIRt, vibrissa intermediate reticular formation. (Adapted from Moore et al. 2014b.)

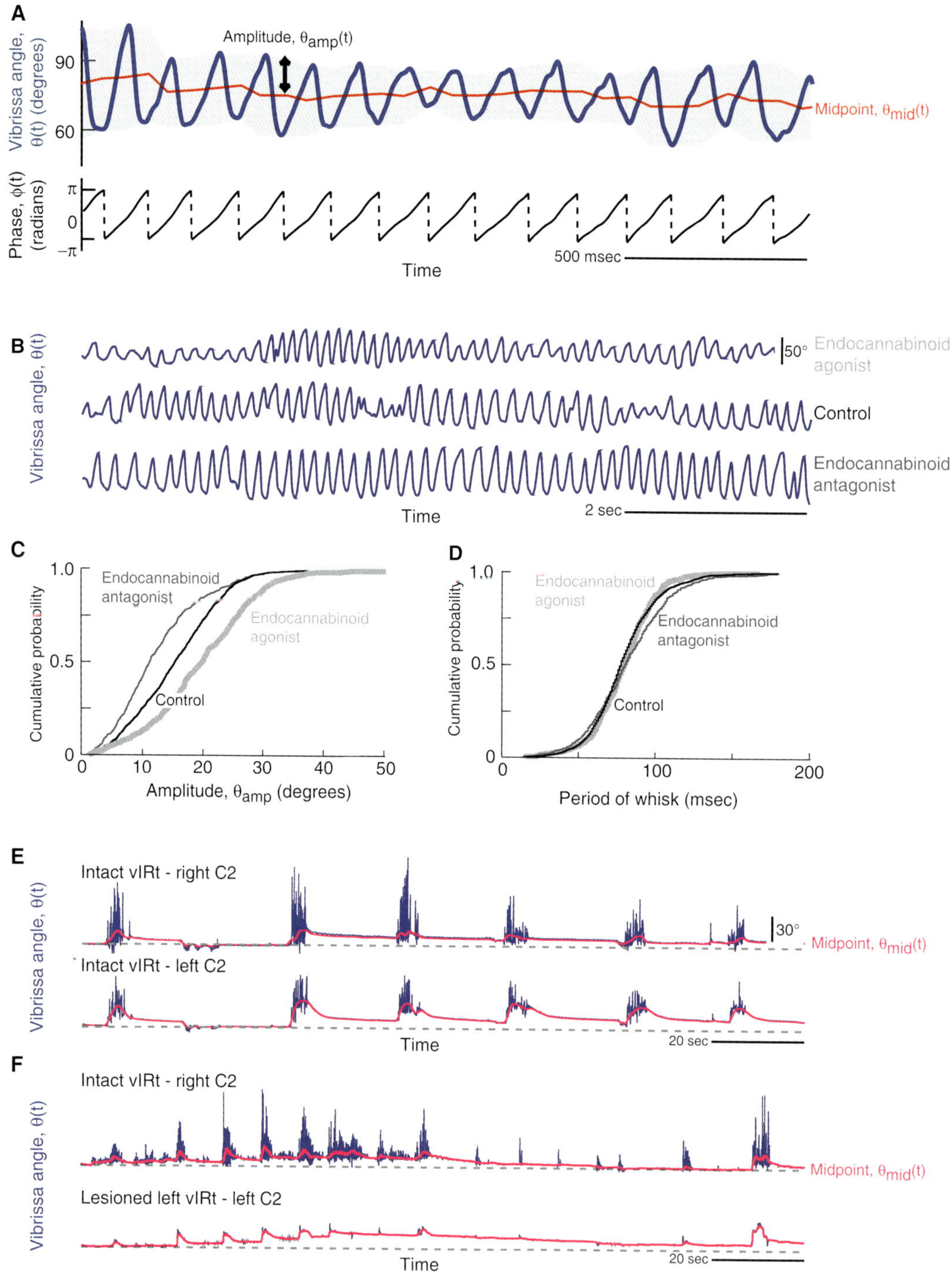

Figure 7. Evidence for the separation of slow processes (i.e., changes in amplitude and midpoint of whisking) from the fast rhythmic oscillation. (*A*) Decomposition of whisking into rapidly and slowly varying parameters via the Hilbert transform. *Top* panel shows vibrissa position along with its decomposition into the amplitude, θ_{amp}, and midpoint, θ_{mid}, of the envelope (gray band). The *lower* trace shows the phase, ϕ, along with broken vertical lines indicate wrapping of phase from π to $-\pi$. (Adapted from Hill et al. 2011.) (*B–D*) Effect of a cannabinoid receptor type 1 agonist and an antagonist on whisking amplitude but not frequency. (*B*) Typical traces of vibrissa angle executed by an animal 4 h after administration of either Δ^9-THC, an agonist (dark gray), vehicle (black), or SR141716A, an antagonist (light gray). Vertical calibration bar corresponds to 50°. (*C*) The cumulative probability distribution function of protraction amplitudes across all animals; note the significant effect of agonists and antagonist. (*D*) The cumulative probability distribution function of the period (inverse of frequency) of whisking across all animals; note the lack of a significant effect of agonists and antagonist. (*B–D*, Adapted, with permission, from Pietr et al. 2010, © The American Physiological Society.) (*E*,*F*) Effect of lesioning the vIRt in one hemisphere (Fig. 4A,B) quenches rhythmic whisking yet does not interfere with changes in midpoint. In *E*, the midpoint of the motion (red) is seen to track between the two sides of the face. In *F*, lesion of the vIRt of the left side only prevents whisking on that side. Yet the midpoint of the vibrissae of the right and left side of the face continue to track each other. (New data obtained with the methods in Moore et al. 2013.)

or otherwise influence the spiking of premotor neurons for other orofacial actions in parallel with our model of the circuit for whisking (Fig. 5E). Our ongoing work addresses the formation of behaviors through the coordinated integration of these disparate orofacial motor actions (Moore et al. 2014b).

We now step back and recall that rhythmic motor actions depend on the envelope of the rhythmic drive as well as the timing of the underlying oscillations. For the case of whisking, the envelope consists of a slowly varying amplitude, which may also gate the rhythm, and a slowly varying midpoint that sets the center of interest for exploratory whisks (Fig. 7A; Carvell et al. 1991; Berg and Kleinfeld 2003; Hill et al. 2008). A priori, the amplitude of whisking could be controlled by the vIRt or by a separate premotor nucleus that targets vibrissa motoneurons in the facial motor nucleus. The initial evidence for the latter possibility comes from a collaboration between the Ahissar and Vogel groups (Pietr et al. 2010), where it was shown that the normal variations in the amplitude of whisking are suppressed by systemic administration of an antagonist of the cannabinoid receptor but

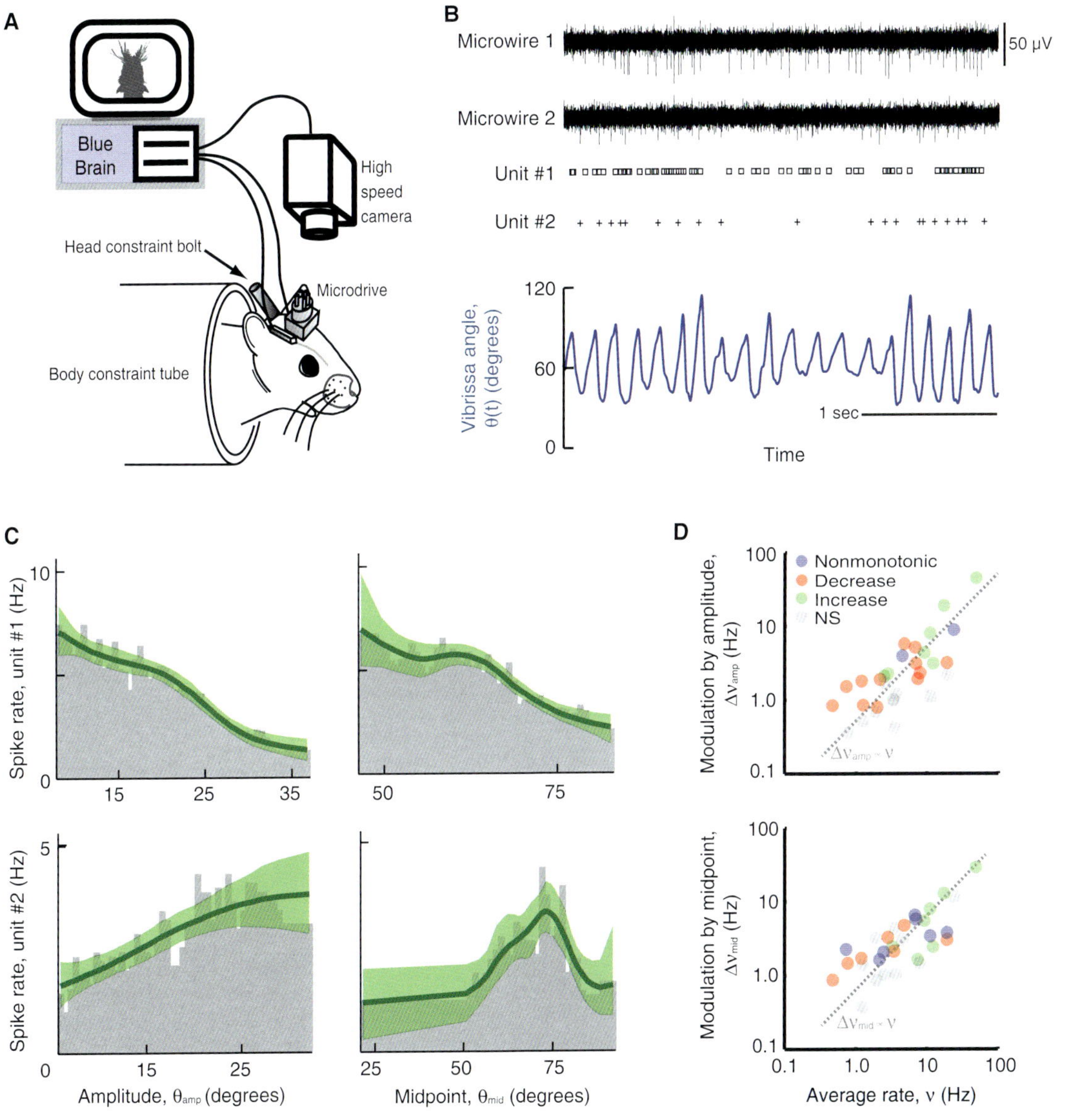

Figure 8. The vibrissa area of motor cortex codes the slowly varying midpoint and amplitude of whisking. (*A*) The scheme used to measure the spiking activity of units in agranular motor cortex as body constrained animals rhythmically whisk in air. Vibrissa position is determined from videography. Single units, indicative of spiking by an individual neuron, are extracted from extracellular microwire recordings. (*B*) Example records. (*C*) Firing rate profiles for two example units as a function of amplitude and midpoint of vibrissa motion. The two rows are profiles of units that show different relative modulation. Each plot is calculated by dividing the distribution of the respective signal at spike time by the distribution of that signal over the entire behavioral session. Green lines are fits from a smoothing algorithm along with the 95% confidence band. (*D*) Composite data across units illustrates that, on average, the rate is unaffected by whisking, consistent with the presence of units that both increase (green) and decrease (red) their rate with increasing angle. (Adapted from Hill et al. 2011.)

heightened by an agonist (Fig. 7B,C). In contrast, there is no change in the frequency of the whisk cycle (Fig. 7B,D). This result implies that the control of the slowly varying amplitude of whisking and the rapidly varying phase are independent, although the locus of the control of amplitude remains unknown.

The independence of the midpoint and the whisking rhythm was established by simultaneously monitoring the position of the vibrissae on both sides of the face before and after lesion of the vIRt in alert, head-restrained rats. In new data, we observe approximately synchronous motion of the midpoint between the right and left sides of the face during bouts of rhythmic whisking and stability of the midpoint between bouts (red lines in Fig. 7E). Lesioning the vIRt on one side abolished rhythmic whisking on that side, yet concurrent changes in the midpoint between the right and left sides of the face are maintained (Fig. 7F). This result implies the existence of a yet to be determined brainstem nucleus that controls slow changes in vibrissae position, potentially under the command of descending projections from other brain areas.

From a mathematical perspective, the independence of the slowly varying envelope of whisking from rapid changes in phase suggests that the angle of vibrissae relative to the face, $\theta(t)$, may be expressed as $\theta(t) = \theta_{amp}(t)\cdot\cos[\phi(t)] + \theta_{mid}(t)$ with $d\phi(t)/dt = 2\pi f_{whisk}$, where f_{whisk} is the instantaneous whisking frequency (Hill et al. 2011), and $\theta_{amp}(t)$ and $\theta_{mid}(t)$ are the aforementioned slowly varying amplitude and midpoint parameters. How are these slowly varying parameters coded in the brain, and how do they influence the facial motoneurons that drive whisking? The first of these questions was examined by recording unit activity in agranular motor cortex (Brecht et al. 2004), concurrent with measurements of vibrissa position (Fig. 8A), as animals whisked in air (Fig. 8B; Hill et al. 2011; Friedman et al. 2012). We observe that the firing patterns of neurons in motor cortex preferentially report the slowly varying amplitude and midpoint of a whisking bout (Fig. 8C). Across a population of neurons, the modulation is approximately linear in the basal spike rate, with both increases and decreases in rate across different units for the same change in the envelope of whisking (Fig. 8D). One possible explanation for how these signals may influence facial motoneurons, under current investigation, is that they act as commands that are relayed through trigeminal nuclei (Matthews et al. 2014) that further project to facial motor neurons (Pinganaud et al. 1999). It will be valuable to determine if similar neuronal coding applies to the cortical control of other orofacial actions (Sawczuk and Mosier 2001).

How does the motion of a "master clock" play out with cognition? Let us suppose, consistent with the anatomy and the case for whisking and most likely for licking, that the relatively tight spatial organization of orofacial oscillators in the intermediate reticular zone provides a means for their reset by breathing as well as their rapid interaction (Fig. 6A). Then every orofacial sense, be it touch through whisking, smell through sniffing, and taste and touch through licking, will have a common underlying temporal signature through breathing. Thus the breathing rhythm, in addition to its physiological role in homeostasis, may serve a perceptual role in the binding of orofacial sensory inputs that enter through the pons, medulla, and olfactory bulb. These sensory inputs converge at the level of the ventromedial prefrontal cortex (Kleinfeld et al. 2014), a site of decision-making (Rushworth et al. 2007) that presumably incorporates the convergence of evidence from multiple sensory streams. It will be of interest to determine if binding by the respiratory rhythm plays a role in the spiking patterns of units in this high-order brain region and whether these units influence motor output.

ACKNOWLEDGMENTS

Our research was funded by the Canadian Institutes of Health Research (grant MT-5877), the United States National Institutes of Health (grants NS058668, NS066664, NS047101, and NS077986), the United States National Science Foundation (grant PHY-1451026), and the US–Israeli Binational Science Foundation (grant 2003222).

REFERENCES

Berg RW, Kleinfeld D. 2003. Rhythmic whisking by rat: Retraction as well as protraction of the vibrissae is under active muscular control. *J Neurophysiol* **89:** 104–117.

Bieger D, Hopkins DA. 1987. Viscerotopic representation of the upper alimentary tract in the medulla oblongata in the rat: The nucleus ambiguus. *J Comp Neurol* **262:** 546–562.

Bouvier J, Thoby-Brisson M, Renier N, Dubreuil V, Ericson J, Champagnat J, Pierani A, Chédotal A, Fortin G. 2010. Hindbrain interneurons and axon guidance signaling critical for breathing. *Nat Neurosci* **13:** 1066–1074.

Brecht M, Krauss A, Muhammad S, Sinai-Esfahani L, Bellanca S, Margrie TW. 2004. Organization of rat vibrissa motor cortex and adjacent areas according to cytoarchitectonics, microstimulation, and intracellular stimulation of identified cells. *J Comp Neurol* **479:** 360–373.

Carvell GE, Simons DJ. 1990. Biometric analyses of vibrissal tactile discrimination in the rat. *J Neurosci* **10:** 2638–2648.

Carvell GE, Simons DJ, Lichtenstein SH, Bryant P. 1991. Electromyographic activity of mystacial pad musculature during whisking behavior in the rat. *Somatosens Mot Res* **8:** 159–164.

Chen Z, Travers SP, Travers JB. 2001. Muscimol infusions in the brain stem reticular formation reversibly block ingestion in the awake rat. *Am J Physiol Regul Intergr Comp Physiol* **280:** R1085–R1094.

Deutsch D, Pietr M, Knutsen PM, Ahissar E, Schneidman E. 2012. Fast feedback in active sensing: Touch-induced changes to whisker-object interaction. *PLoS ONE* **7:** e44272.

Dörfl J. 1982. The musculature of the mystacial vibrissae of the white mouse. *J Anat* **135:** 147–154.

Friedman WA, Zeigler HP, Keller A. 2012. Vibrissae motor cortex unit activity during whisking. *J Neurophysiol* **107:** 551–563.

Gao P, Bermejo R, Zeigler HP. 2001. Vibrissa deafferentation and rodent whisking patterns: Behavioral evidence for a central pattern generator. *J Neurosci* **21:** 5374–5380.

Getting PA. 1989. Emerging principles governing the operation of neural networks. *Annu Rev Neurosci* **12:** 185–204.

Gray PA, Hayes JA, Ling GY, Llona I, Tupal S, Picardo MC, Ross SE, Hirata T, Corbin JG, Eugenín J, et al. 2010. Developmental origin of preBötzinger complex respiratory neurons. *J Neurosci* **30:** 14883–14895.

Haidarliu S, Simony E, Golomb D, Ahissar E. 2010. Muscle architecture in the mystacial pad of the rat. *Anat Rec (Hoboken)* **293:** 1192–1206.
Haidarliu S, Golomb D, Kleinfeld D, Ahissar E. 2012. Dorsorostral snout muscles in the rat subserve coordinated movement for whisking and sniffing. *Anat Rec (Hoboken)* **295:** 1181–1191.
Hill DN, Bermejo R, Zeigler HP, Kleinfeld D. 2008. Biomechanics of the vibrissa motor plant in rat: Rhythmic whisking consists of triphasic neuromuscular activity. *J Neurosci* **28:** 3438–3455.
Hill DN, Curtis JC, Moore JD, Kleinfeld D. 2011. Primary motor cortex reports efferent control of vibrissa position on multiple timescales. *Neuron* **72:** 344–356.
Isokawa-Akesson M, Komisaruk BR. 1987. Difference in projections to the lateral and medial facial nucleus: Anatomically separate pathways for rhythmical vibrissa movement in rats. *Exp Brain Res* **65:** 385–398.
Kleinfeld D, Sompolinsky H. 1989. Associative network models for central pattern generators. In *Methods in neuronal modeling: From synapses to networks* (ed. Koch C, Segev I), pp. 195–246. MIT Press, Cambridge.
Kleinfeld D, Deschênes M, Wang F, Moore JD. 2014. More than a rhythm of life: Breathing as a binder of orofacial sensation. *Nat Neuroci* **15:** 647–651.
Koizumi H, Wilson CG, Wong S, Yamanishi T, Koshiya N, Smith JC. 2008. Functional imaging, spatial reconstruction, and biophysical analysis of a respiratory motor circuit isolated *in vitro*. *J Neurosci* **28:** 2353–2365.
Lawson EE, Richter DW, Czyzyk-Krzeska MF, Bischoff A, Rudesill RC. 1991. Respiratory neuronal activity during apnea and other breathing patterns induced by laryngeal stimulation. *J Appl Physiol* **70:** 2742–2749.
Lovick TA. 1972. The behavioural repertoire of precollicular decerebrate rats. *J Physiol* **226:** 4P–6P.
Matthews DW, Deschênes M, Furuta T, Moore JD, Wang F, Karten HJ, Kleinfeld D. 2014. Feedback in the brainstem: An excitatory disynaptic pathway for control of whisking. *J Comp Neurol*. doi: 10.1002/cne.23724.
Moore JD, Deschênes M, Furuta T, Huber D, Smear MC, Demers M, Kleinfeld D. 2013. Hierarchy of orofacial rhythms revealed through whisking and breathing. *Nature* **469:** 53–57.
Moore JD, Deschênes M, Kurnikova A, Kleinfeld D. 2014a. Activation and measurement of free whisking in the lightly anesthetized rodent. *Nat Protoc* **9:** 1792–1802.
Moore JD, Kleinfeld D, Wang F. 2014b. How the brainstem controls orofacial behaviors comprised of rhythmic actions. *Trends Neurosci* **37:** 370–380.
Nguyen QT, Kleinfeld D. 2005. Positive feedback in a brainstem tactile sensorimotor loop. *Neuron* **45:** 447–457.
Ono T, Ishiwata Y, Inaba N, Kuroda T, Nakamura Y. 1998. Modulation of the inspiratory-related activity of hypoglossal premotor neurons during ingestion and rejection in the decerebrate cat. *J Neurophysiol* **80:** 48–58.
Pietr MD, Knutsen PM, Shore DI, Ahissar E, Vogel Z. 2010. Cannabinoids reveal separate controls for whisking amplitude and timing in rats. *J Neurophysiol* **104:** 2532–2542.
Pinganaud G, Bernat I, Buisseret P, Buisseret-Delmas C. 1999. Trigeminal projections to hypoglossal and facial motor nuclei in the rat. *J Comp Neurol* **415:** 91–104.
Ranade S, Hangya B, Kepecs A. 2013. Multiple modes of phase locking between sniffing and whisking during active exploration. *J Neurosci* **33:** 8250–8256.
Rushworth MF, Buckley MJ, Behrens TE, Walton WE, Bannerman DM. 2007. Functional organization of the medial frontal cortex. *Curr Opin Neurobiol* **17:** 220–227.
Sawczuk A, Mosier KM. 2001. Neural control of tongue movement with respect to respiration and swallowing. *Crit Rev Oral Biol Med* **12:** 18–37.
Semba K, Komisaruk BR. 1984. Neural substrates of two different rhythmical vibrissal movements in the rat. *Neuroscience* **12:** 761–774.
Sherrey JH, Megirian D. 1977. State dependence of upper airway respiratory motoneurons: Functions of the cricothyroid and nasolabial muscles of the unanesthetized rat. *Electroencephalogr Clin Neurophysiol* **43:** 218–228.
Simony E, Bagdasarian K, Herfst L, Brecht M, Ahissar E, Golomb D. 2010. Temporal and spatial characteristics of vibrissa responses to motor commands. *J Neurosci* **30:** 8935–8952.
Smith JC, Ellenberger HH, Ballanyi K, Richter DW, Feldman JL. 1991. Pre-Bötzinger complex: A brainstem region that may generate respiratory rhythm in mammals. *Science* **254:** 726–729.
Smith JC, Abdala APL, Rybak IA, Paton JFR. 2009. Structural and functional architecture of respiratory networks in the mammalian brainstem. *Philos Trans R Soc Lond B Biol Sci* **364:** 2577–2587.
Stanek E IV, Cheng S, Takatoh J, Han BX, Wang F. 2014. Monosynaptic premotor circuit tracing reveals neural substrates for oro-motor coordination. *Elife* **3:** e02511.
Takatoh J, Nelson A, Zhou X, Bolton MM, Ehlers MD, Arenkiel BR, Mooney R, Wang F. 2013. New modules are added to vibrissal premotor circuitry with the emergence of exploratory whisking. *Neuron* **77:** 346–360.
Tan W, Janczewski WA, Yang P, Shao XM, Callaway EM, Feldman JL. 2008. Silencing preBötzinger complex somatostatin-expressing neurons induces persistent apnea in awake rat. *Nat Neurosci* **11:** 538–540.
Tan W, Pagliardini S, Yang P, Janczewski WA, Feldman JL. 2010. Projections of preBötzinger complex neurons in adult rats. *J Comp Neurol* **18:** 1862–1878.
Tinbergen N. 1951. *The study of instinct*. Oxford University Press.
Travers JB, Dinardo LA, Karimnamazi H. 1997. Motor and premotor mechanisms of licking. *Neurosci Biobehav Rev* **21:** 631–647.
Travers JB, DiNardo LA, Karimnamazi H. 2000. Medullary reticular formation activity during ingestion and rejection in the awake rat. *Exp Brain Res* **130:** 78–92.
Travers JB, Yoo JE, Chandran R, Herman K, Travers SP. 2005. Neurotransmitter phenotypes of intermediate zone reticular formation projections to the motor trigeminal and hypoglossal nuclei in the rat. *J Comp Neurol* **488:** 28–47.
Travers JB, Herman K, Travers SP. 2010. Supression of third ventricular NPY-elicited feeding following medullary reticular formation infusions of muscimol. *Behav Neurosci* **124:** 225–233.
Weiss PA. 1941. *Self-differentiation of the basic patterns of coordination*. Williams & Wilkins, Baltimore.
Welker WI. 1964. Analysis of sniffing of the albino rat. *Behaviour* **12:** 223–244.
Welzl H, Bures J. 1977. Lick-synchronized breathing in rats. *Physiol Behav* **18:** 751–753.
Wickersham IR, Finke S, Conzelmann KK, Callaway EM. 2007. Retrograde neuronal tracing with a deletion-mutant rabies virus. *Nat Methods* **4:** 47–49.
Wiesenfeld Z, Halpern BP, Tapper DN. 1977. Licking behavior: Evidence of hypoglossal oscillator. *Science* **196:** 1122–1124.
Wineski LE. 1985. Facial morphology and vibrissal movement in the golden hamster. *J Morphol* **183:** 199–217.

Emergence of Cognition from Action

György Buzsáki,[1,2] Adrien Peyrache,[1] and John Kubie[3]
[1]*The Neuroscience Institute, New York University, School of Medicine, New York, New York 10016*
[2]*Center for Neural Science, New York University, School of Medicine, New York, New York 10016*
[3]*Department of Cell Biology, Downstate Medical Center, State University of New York at Brooklyn, New York, New York 11203*
Correspondence: gyorgy.buzsaki@nyumc.org

Theories of brain function have evolved through multiple stages. The first proposition was that brain networks support a set of reflex responses, with current sensory inputs producing immediate motor outputs. The behaviorist paradigm suggested that actions can always be explained as a response to immediate external cues. In response to these views, the cognitive paradigm argued that behavior cannot be understood simply as input–output functions because the hidden layers of brain generate unpredictability. The central processing was termed "cognition." Here we propose a neuroscience-based model of cognition. Our core hypothesis is that cognition depends on internal models of the animal and its world, where internally generated sequences can serve to perform "what if" scenarios and anticipate the possible consequences of alternative actions without actually testing them, and aid in the decisions of overt actions. We support our hypotheses by several examples of recent experimental findings and show how externally guided cell assembly sequences become internalized to support cognitive functions.

Humans, and likely other animals, can imagine far into the future; we spend a great deal of waking time devoted to such covert activities (Killingsworth and Gilbert 2010; Immordino-Yang et al. 2012; Mar et al. 2014). The main thesis of this essay is that such ability of the brain arises from its disengagement from immediate overt actions as a result of experience. We hypothesize that brains, irrespective of size, are predictive devices that exploit regularity and recurrence as a fundamental property of the surrounding world and apply effective heuristics to make such predictions. Prediction is made possible by adaptive mechanisms that are supported by learning rules that either apply across generations (evolutionary adaptation) or within the lifetime of the organism. As a result, organisms can deal with a future occurrence of the same or similar situations more effectively. This is the fundamental organization principle of any adaptive system. Here we describe example organizational brain patterns, with increasing predictive efficiency, and present a hypothesis of how internally generated neuronal network operations, detached from outside inputs, can serve as the necessary substrate for cognition.

One aspect of the argument we will present is that most brain structures can have "dual use": at times, connected to immediate sensory input and output; at other times entrained by internal signals and brain rhythms. For many years, researchers tended to associate preexisting verbal terms, such as memory, planning, envisioning the future, volition, and decision-making, with different and distinct brain structures. Oftentimes, even studying the same structure or system generated seemingly contradictory hypotheses depending on the chosen approach, preconception, or experimental method used. A striking example is the hippocampal system. Lesion data in humans provided evidence that bilateral removal of the hippocampi produced severe and irreversible amnesia (Scoville and Milner 1957). Single unit studies in the hippocampus and entorhinal cortex of animals gave rise to the prominent theory that the fundamental function of the hippocampal system is supporting spatial navigation (O'Keefe and Nadel 1978; Hafting et al. 2005; McNaughton et al. 2006), assisted by inputs from the head-direction system (Ranck 1985; Taube 2007). In contrast, studying the collective behavior of hippocampal neurons by recording the local field potential (LFP) offered the conclusion that hippocampal theta oscillations are an unmistakable reflection of voluntary action (Vanderwolf 1969). These independent ideas of hippocampal functions persisted in parallel for decades without true interactions, apart from occasional polite gestures and references to the competing frameworks (Buzsáki 2005). As discussed below, these barriers are now being dismantled by the recognition that although neurons and networks in the memory/navigational/planning systems are primarily engaged in self-organized activity by theta oscillations, they can robustly respond to environmental or body-derived inputs (Frank et al. 2000; Wood et al. 2000; Fujisawa et al. 2008; Pastalkova et al. 2008).

In "simple" nervous systems, the connection between output and input networks is direct and immediate and the consequences of their action are mediated by the environment, as illustrated in Figure 1A. More complex brains are organized in a "multiple loop" pattern; that is, a series of interacting parallel loops are imposed between input and output, as illustrated in Figure 1B,C. For example, in mammals, new circuits are added onto the monosynaptic

 doi: 10.1101/sqb.2014.79.024679

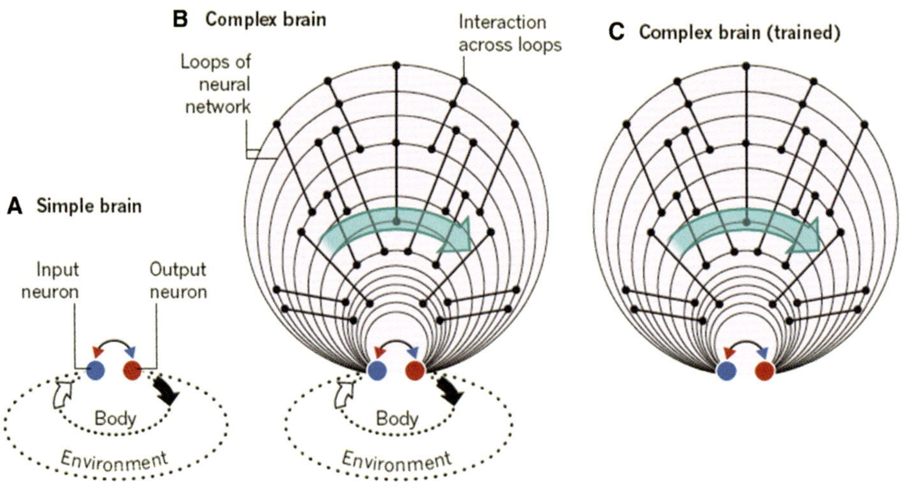

Figure 1. Externally driven and self-organized cell assemblies. (*A*) Evolutionarily simple brains contain simple neural networks. Sensory input from the body and the environment activates input neurons, which interact with output neurons to generate appropriate reflex actions in a short time window. (*B*) In more complex brains, multiple interacting loops of increasing length improve prediction of more elaborate events and can make predictions in more complex environments at longer timescales. (*C*) After extensive training, the loops can sustain self-organized, long-lasting neuronal sequences without reliance on external cues and can, therefore, support cognitive operations such as memory, planning, and imagination. Progression of neuronal operations correlates with elapsed time (green arrows) irrespective of whether the operations are driven externally or internally. (Reprinted, with permission, from Buzsáki 2013.)

spinal cord/brainstem afferent–efferent connections, including the thalamocortical system, cortico-corticortical, corticocerebellar, corticostriatal, and corticolimbic loops. As in the simple system, the parallel loops operate to predict and secure effective outcomes of actions in ever more complex environments and at much longer timescales. These operations require storage of large amounts of past experience and the ability to evaluate and weigh the importance of simultaneously experienced modalities and compare them with similar occurrences in the past and heuristics responses of the organism. Many of these operations require that brain activities continue even after transient or more durable disappearance of sensory inputs and delaying the brain's deliberations in terms of effector outputs. We speculate that such a training process, a result of both phylogenetic and ontogenetic accumulation of experience, makes the longer, more complex neuronal loops "smarter" (i.e., more effective by enabling them to interpolate and extrapolate events surrounding the organism). Our second and critical assumption is that a self-organized computation emerges as a result of training the brain, which can proceed to deliver abstract goals by disengaging the effector networks from overt (muscular or vegetative) actions.

Activity in any system, including the brain, can be maintained by only two mechanisms: externally applied energy or intrinsic resources. By definition, a system that maintains its activity without an external drive must be driven by "self-organized" mechanisms. "Self-organized" is the term most often used in physics and engineering that refers to autonomously generated patterns. In neurophysiology, the term "spontaneous" activity has been in use for decades to refer to the same process. In cognitive science, the terms "internally generated" or "internally organized" refer to the same process. These terms are synonymous. Therefore, we separate the functioning of complex brains into two modes: First, an "engaged" mode, when the immediate environment controls and feeds back on behavior even when the inputs are quite complex and actions need to be delayed, illustrated in Figure 1B. Second, a "disengaged" or detached mode, when processing continues independent of environmental inputs and behavioral responses of the animal. This input/output disengagement in the presence of continued computation of potential responses is what we refer to as "cognition." This framework may explain why damage to multiple loops of the brain has no or relatively minor impact on operations when input dimensions are limited and input–output gaps are short, but their role becomes evident when the organism needs to evaluate a large and complex parameter space and execute a response most beneficial to the organism's needs.

There are costs and advantages to the disengaged mode. The cost is that it takes greater neural resources to create the "virtual worlds"; in growing brains, these are necessary for processing internally generated patterns without reliance on immediate sensory feedback cues. In the past, Gibson and heirs developed ideas that have come to be known as "embodied cognition" (Gibson 1977; O'Regan and Noë 2001; Thompson and Varela 2001). These authors emphasize that the environment of an animal affords physical and sensory constraints (i.e., reduces the need to compute all possible circumstances) that ease the computational load (Gibson's "affordances"). Without real-world constraints, disengaged processing requires that the brain creates a well-endowed model of the world. With an internalized "virtual world" in the brain, the animal is capable of cognitive vicarious behavior; that is, it can perform "what if" scenarios in its head and anticipate the consequences of alternative actions without actually testing them. With this mechanism at hand, the brain can create new knowledge not only by interacting with the outside world, but also through "vicarious" (imagined) experience. Disengaged processing, which we call "cognition," provides dramatic advantages in predicting the consequence of actual behavior in complex environments and at long timescales.

There are numerous examples of disengagement of brain activity from afferent inputs. A striking example is the "internalization" of brain rhythms. The first organized neocortical pattern in the developing mammalian brain is the emergence of transient periods of spindle-shaped rhythmic activity at 10–25 Hz amid seconds- or tens-of-seconds-long silent periods ("tracé alternant"; Dreyfus-Brisac 1962; Anderson et al. 1985; Khazipov et al. 2004). In the newborn rat and premature human babies, all these early spindles are triggered by some kind of a movement, including isolated muscle twitches, limb jerks, and whole body startles (Khazipov et al. 2004). This organization is hypothesized to serve as a mechanism to construct an internal map of the body in the somatosensory system (Khazipov et al. 2004; Buzsáki 2006; Khazipov and Buzsáki 2010). Once the body map is formed (5–10 d postnatally in the rat and in the third trimester of pregnancy in humans), the dependence of brain activity on action-triggered reafferentation is no longer needed and spindles become part of sleep, the default organizing mode of the brain. This is a simple example of how brain patterns that are initially correlated with sensorimotor activity can disengage from external inputs, while sustaining their "spontaneous" activity patterns. Below we discuss two more examples to illustrate our disengagement hypothesis of cognition. First, we illustrate how neurons in the head direction system continue their coordinated activity during sleep. Second, we show how internally organized cell assembly sequences can support mental travel of memory and action planning, and how these functions emerge from spatial navigation in the physical environment.

SELF-ORGANIZED ATTRACTORS OF DIRECTION SENSE

The relationship between stimulus-driven and self-organized (or internally generated) activity is a recurring topic in neuroscience. It is often tacitly assumed that the information from the sensory world is "relayed" by the various thalamic nuclei to the neocortex where it is combined with top-down, internally generated cortical activity. How feed-forward, "labeled line" sensory signals in subcortical, thalamic, and cortical networks interact with self-generated ("spontaneous") activity is not well-understood. Part of the difficulty is technical, due largely to the high dimensionality of sensory signal attributes (e.g., intensity, color, orientation in case of vision; Rao and Ballard 1999; Engel et al. 2001; Varela et al. 2001; Buzsáki 2006).

Perhaps the simplest representation animals have is the sense of direction. It is simple in the sense that it has a single dimension, which is the head direction. Head-direction neurons fire robustly when the animal's head points in a specific direction (Ranck 1985; Taube et al 1990a,b; Taube 2007; McNaughton et al. 2006). Thus, the one-dimensional head-direction system offers an opportunity to understand the nature of external and internal interactions experimentally. The head-direction system consists of a largely serially connected brain network, which includes the brainstem, mammillary bodies, anterodorsal thalamic nucleus, postsubiculum, and entorhinal cortex (McNaughton et al. 2006; Sargolini et al. 2006; Taube 2007).

Numerous experiments have showed that head-direction neurons can be controlled by peripheral inputs, mainly the vestibular, visual, and ancillary afferents and the angular velocity of head rotation (Sharp et al. 2001; Taube 2007). How do these neurons behave in the absence of such inputs, when the brain is disengaged from the environment, as is the case during sleep? Computational models have long assumed that head-direction neurons with similar preferred directions fire together, while the other neurons are suppressed. According to these models, the temporally engaged group of head-direction neurons (i.e., a "hill of activity" or "an activity packet") moves on a virtual ring as the animal turns its head (Fig. 2; Skaggs et al. 1995; Redish et al. 1996; Burak and Fiete 2012; Knierim and Zhang 2012). However, experimental demonstration of the existence of internally organized neuronal populations endowed with such properties had to wait until recording technologies matured to be able to record simultaneous population of neurons and compare their dynamic interactions during both waking behavior and sleep. Recent experiments provide such evidence (Peyrache et al. 2015).

Not only do head-direction neurons continue to be active during sleep but their temporal sequential activity patterns in the waking mouse are preserved. At any time, the neuronal population was characterized by the presence of an activity packet that was consistent with its structure during waking: Neurons coding for nearby orientation continued to fire together. Therefore, a "virtual gaze" (i.e., the direction the mouse was "looking") could be reconstructed from firing of neurons in the anterodorsal thalamic nucleus and postsubiculum (Fig. 2). Although both firing rates and sequential firing patterns persisted in all brain states, the temporal dynamics of the drift of the activity packet varied. During REM sleep, when brain electrophysiology is strikingly similar to waking, the packet moved at the same speed as in the waking animal, essentially recapitulating the situation when the mouse moved its head in different direction in the waking animal. During non-REM sleep, the HD signal drifted at a 10 times faster speed, compared with waking and REM sleep, essentially following the faster dynamics of non-REM sleep described in cortical networks (Buzsáki 1989; Wilson and McNaughton 1994). Of course, no real head movement accompanied these changes in the sleeping animal; therefore, the temporal organization had to rely on internally generated mechanisms rather than external stimuli.

The observation that the neuronal code of the head-direction sense is preserved during sleep raises the possi-

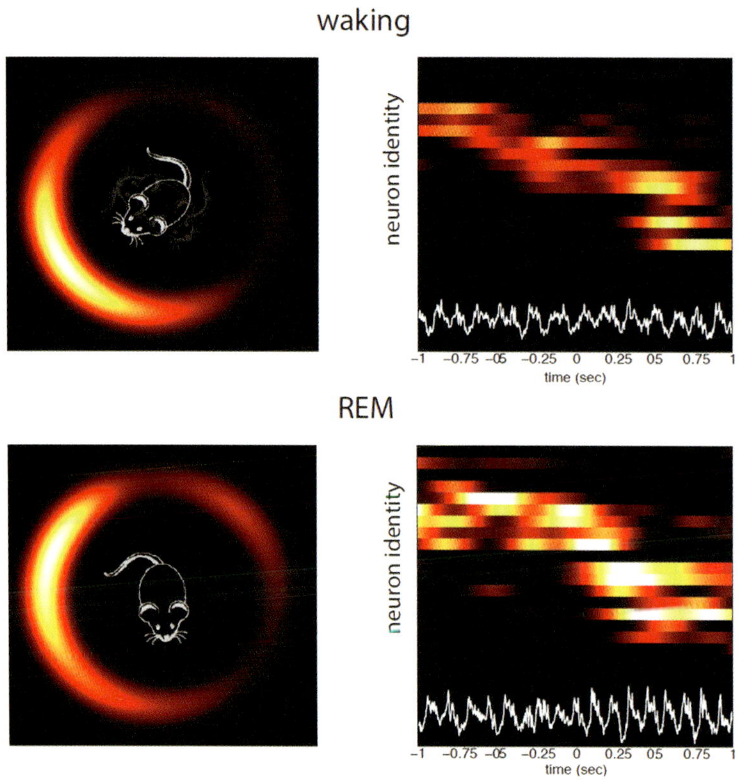

Figure 2. Persistence of head-direction cell assemblies in the anterodorsal nucleus of the thalamus during waking and sleep. Waking. Simultaneously recorded head-direction cells with different direction preferences during exploration (*right*). Each line is a neuron, increasing firing rates are represented by hot colors. Neurons are ordered according to their preferred head-direction. Note sequential activity of head-direction neurons (from *top* to *bottom*). *Bottom* trace (white) is local field potential recording from the hippocampus. (*Left*) Head-direction cell assembly activity displayed on a ring. As the head points to different directions, the hill of spiking activity moves on the ring accordingly. REM, same arrangement as in waking. Note continued sequentially changing activity of the head-direction cells despite the absence of head or body movement. (Reprinted, with permission, from Peyrache et al. 2015.)

bility that self-organized mechanisms continue to be at work also in the waking animal. Such an internal computation can amplify and enhance precision of the head-direction signal by filtering out, in the case of the head-direction, all additional information that is not compatible with a one dimensional angular signal. In the exploring animal, sensory inputs can be combined with the prediction of the internally generated head direction and rapidly adapt to reconfigured environments (Zugaro et al. 2003). In case of ambiguous or conflicting signals, self-organized mechanisms may generate the brain's "best guess" by interpolating across input signals or extrapolating the correct position of the head-direction vector from limited or ambiguous sensory information. Overall, these experimental findings in a simple but fundamentally important sense show that internally organized mechanisms are continuously used to improve the brain's interpretation of the external world (Fig. 1B).

FROM NAVIGATION IN THE PHYSICAL WORLD TO MENTAL TRAVEL

The hippocampal–entorhinal cortex has long been linked to both navigation and memory (Scoville and Milner 1957; O'Keefe and Nadel 1978). It has been proposed recently that neuronal mechanisms of memory and planning ("mental travel into the past and future"; Suddendorf and Corballis 1997, 2007) have evolved from mechanisms of navigation in the physical world (Buzsáki 2005; Buzsáki and Moser 2013). Below, we discuss that the neuronal algorithms underlying navigation in real and mental space are fundamentally the same. The key distinction is that during real world travel, external cues, such as landmarks, combined with self-motion cues aid in driving cell assembly sequences; when travel is internal, the brain must supply the supporting cues for navigating through a mental space.

To support mental travel effectively, a neural system evolved for navigation must develop the ability to self-generate temporally evolving cell assembly sequences and a capacity to store large quantities of seemingly unrelated episodes. It has been suggested that the evolution of the entorhinal–hippocampal system illustrates the phylogenetic continuity of navigation and memory and has the anatomical and physiological properties that make it especially suitable for meeting the above requirements (Buzsáki 2005; Buzsáki and Moser 2013).

SENSORY-DRIVEN ACTIVITY OF THE HIPPOCAMPAL NAVIGATION SYSTEM

The explicit spatial coordinates of an animal are encoded by a combination of a variety of neuron types, including "head-direction cells" of the thalamus and postsubiculum (Ranck 1985; Taube 2007), "place cells" in the hippocampus (O'Keefe and Nadel 1978), "grid cells" (Hafting et al. 2005), and "border cells" (Solstad et al. 2008) of the entorhinal cortex and surrounding structures. The exact functions of these cell types are yet to be determined, but it has been suggested that the periodically active fields of the grid cells provide a metric for the neural representation of space, in the same way that head-direction cells provide a directional reference frame, whereas border cells can assess allocentric distances in the discoverable environment by triangulation (Moser et al. 2014).

The "temporally evolving sequences" referred to above are, in navigational terms, paths. The paths represent a temporal sequence of neighboring locations on a map. Each map represents a specific environment. "Remapping" refers to the finding that the organization of place cells scramble from one environment to another (Muller and Kubie 1987; Kubie and Muller 1991). Because place cells that have neighboring firing fields will likely not have neighboring firing fields in a second environment, a single sequence of place cell firing not only codes a specific path, but also a specific environment. Recent findings, however, refine this strict "orthogonalized" view of remapping. Although most neurons have single place fields in a given environment, a small minority can have multiple fields and this same minority of neurons continues to fire in multiple environments. In fact, the distribution of the number of place fields per hippocampal neuron follows a lognormal distribution (Buzsáki and Mizuseki 2014). This small subset is not only more active in multiple environments, but their firing rates are higher, emit more spike bursts and their place fields are larger compared with those of the majority neurons. Importantly, the log-based rule organization applies to all cortical regions and multiple neuronal functions from synaptic strength distribution to macroscopic connectivity. The ever-active minority provides the brain's "best guess" and offers "good enough" solutions to get by in any situation. On the other hand, the majority of less active neurons comprises a large reservoir that can be mobilized to precisely distinguish one situation from another and incorporate novel ones as distinct (Buzsáki and Mizuseki 2014).

When the animal explores an environment, head-direction neurons, border cells, grid cells, and place cells continuously and coherently change their firing activity. Traditionally, two sources of information have been considered as driving factors of the temporal evolution of cell assembly sequences (Fig. 2). The first one is the constellation of the landmarks available as the animal navigates (O'Keefe and Nadel 1978). The second may derive from the ideothetic (i.e., body-derived) cues from the vestibular and proprioceptic systems and self-motion-driven optical flow from mostly local cues (McNaughton et al. 1996). Moreover, spatial firing does not require a rich set of landmarks. Landmarks can be impoverished, subsets removed, or the lights turned out, and firing will remain stable and robust (Muller and Kubie 1987; Quirk et al. 1990). These findings indicate that the brain is capable of interpolating between cues and extrapolating (generalizing or "pattern-completing") from the available external information.

However, even small insects, whose nervous systems possess many fewer neurons and are made of simpler circuits, are able to navigate using such strategies (Menzel et al. 1998). In principle, a small-sized network should be sufficient for rodents as well if navigation in a relatively fixed environment and range would be the only goal. Indeed, in laboratory experiments, the explored environment can be mapped at centimeter precision by just a dozen or so grid cells (Fyhn et al. 2007) or place cells (Wilson and McNaughton 1993), What advantage, then, do the complex networks of the entorhinal–hippocampal system of mammals offer? One can speculate that there are two advantages: first, to map specific features of multiple environments; second, that such self-organized networks (Fig. 1B) also have the ability to effectively interpolate and extrapolate from the available cues and hold information over long timescales, as well as use previous experience from similar situations in more complex environments. The ability to rapidly acquire multiple representations, to maintain separation of these representations, to hold the representations over long timescales, and to operate effectively on any, even in degraded conditions, are several of the robust navigational capacities in mammalian brains compared with simpler brains of species where navigation is accomplished by much smaller circuits.

SELF-ORGANIZED ACTIVATION OF THE HIPPOCAMPAL NAVIGATION SYSTEM

A second way of generating sequential firing of cell assemblies is by self-organization (Fig. 3), independent of cues from either the environment or the body. Such internally generated neural assembly sequences have been long assumed to be a necessary substrate of cognition for various mental operations such as memory recall, planning, and imagination (Hebb 1949). Indeed, perpetually changing hippocampal assembly sequences can be observed experimentally. For example, if a rat is trained to run in a wheel with approximately the same speed and facing in the same direction, the environmental and body-derived cues remain constant throughout the run. If such "clamped" behavior is part of a memory task to remember a previously made choice for planning for the appropriate future choice, hippocampal neurons display continuously changing assembly sequences (neuronal "trajectories"; Fig. 4). Importantly, several measures of the firing patterns of the internally organized sequences during the delay period, including the duration of the spiking activity of the neurons and the temporal relationship of their spikes relative to the reference theta oscilla-

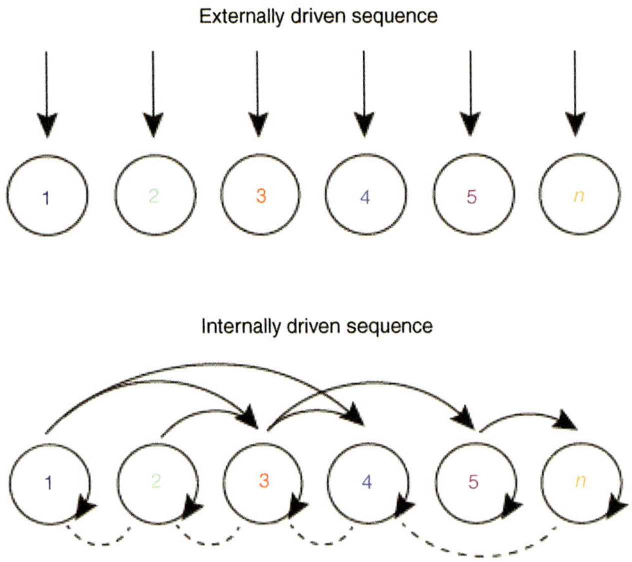

Figure 3. Cell assembly sequences. Sequential activation of neuronal assemblies neurons (1–*n*) can be controlled by the changing constellation of environmental cues and/or proprioceptive information from the body (*top*). During cognitive activity, sequential activation is supported by self-organized patterning of assembly activity (*bottom*). Not only first-order (neighbor) but also higher-order (nonneighbor) connections can be represented in strongly connected recurrent networks.

tion cycle (O'Keefe and Recce 1993), are virtually identical with those of the place cells during translational behavior (Pastalkova et al. 2008). The implication of these observations is that the physiological mechanisms that govern the progression of cell assembly sequences in the hippocampus during navigation and memory are quite similar. Although theta oscillation is an obligatory mechanism for generating internal cell assembly sequences, environmental cues can maintain sequential place cell activity even in the absence of theta rhythms (Wang et al. 2015). The cognitive relevance of self-organized sequential activity is emphasized by the observation that identical initial conditions (e.g., a left choice was rewarded) induces a similar assembly sequence each time, whereas different conditions (i.e., different memories) can give rise to uniquely different neuronal trajectories, which accurately predicts upcoming choices in the maze, including erroneous turns (Fig. 4). In accordance

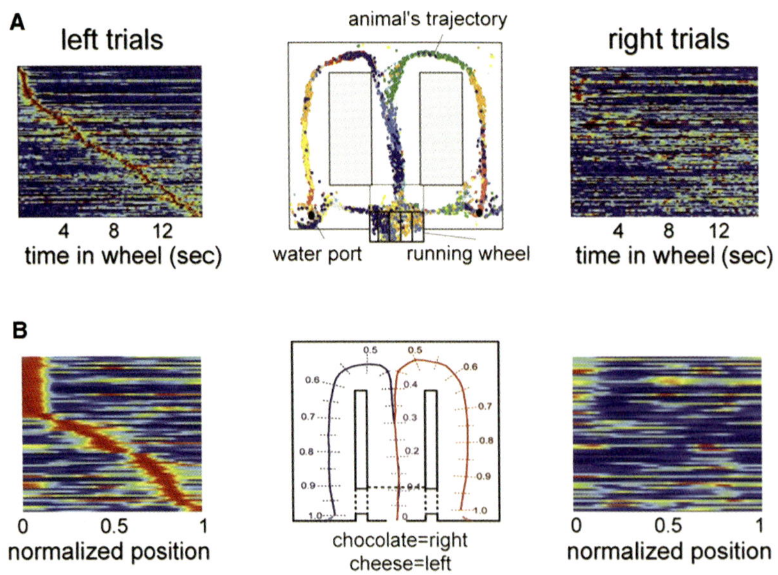

Figure 4. Internally generated assembly sequences during cognitive activity. (*A*) Sequential firing patterns of hippocampal neurons in a memory task. *Center*: Color-coded spikes (dots) of simultaneously recorded hippocampal CA1 pyramidal neurons. The rat was required to run in the wheel facing to the left during the delay between the runs in the maze. (*Left*) Normalized firing rate profiles of neurons during wheel running, ordered by the latency of their peak firing rates during left trials (each line is a single cell). (*Right*) Normalized firing rates of the same neurons during right trials. (*B*) Sequential firing patterns of prefrontal pyramidal cells in a working memory task. (*Middle*) Cheese odor or chocolate odor in the start area signals the availability of cheese or chocolate reward in the left or right goal area (position 1), respectively. Travel trajectories were linearized (0–1). (*Left*) Neurons were ordered by the location of their peak firing rates relative to the rat's position in the maze during left trials. Each row represents the position-dependent normalized firing rate of a single neuron. (*Right*) Normalized firing rates of the same neurons during right trials. (*A* and *B* are reprinted, with permission, from Pastalkova et al. 2008 and Fujisawa et al. 2008, respectively.)

with experiments in rodents, single-unit studies in human patients showed that the hippocampus and entorhinal cortex can generate numerous trajectories corresponding to different memory episodes (Gelbard-Sagiv et al. 2008). Subjects first viewed cinematic episodes that depicted short clips of famous people and characters. Several neurons responded selectively to only one or a few of the episodes. In the critical part of the experiment, subjects were asked to recall the characters and scenes without any cues. Remarkably, the neurons that fired slightly before free recall (i.e., a verbal report) corresponded to the same neurons that were activated while watching the cinematic episodes in the learning phase (Gelbard-Sagiv et al. 2008). These results support the view that the neuronal assembly sequences that encode episodes are the same that are active during the mental representation of those episodes (Tulving and Thomson 1973).

Overall, these experiments illustrate that perpetually changing cell assembly sequences can be induced in the hippocampus independent of external cues (Fig. 1C), and such disengagement can support cognitive performance. Generation of neural sentences is not confined to the hippocampal system. In the medial prefrontal cortex of the rat, long neuronal sequences can also reliably differentiate between right and left trajectories in the maze before making a choice, with individual neurons active only for a short duration (Fig. 4; Fujisawa et al. 2008). The possibility exists that choice specific neuronal trajectories in the hippocampus are selected by inputs from the medial prefrontal cortex. Similar neuronal trajectories have been also been described in the parietal cortex (Harvey et al. 2012). In summary, cortical circuits can produce multitudes of unfolding assembly sequences in two different ways: either by responding to environmental/idiothetic stimuli, when such inputs are available, or by generating them internally.

Analogous to spatial navigation, two forms of declarative memory can be distinguished (Squire 1992). Similar to the defining locations of the allocentric map, semantic memory explicitly defines living things, objects, facts, and events of the surrounding world independent from temporal context (Tulving and Thomson 1972; Squire 1992). Episodic memory, in contrast, allows learning and recalling self-referenced (first-person) experiences in the context of both space and subjective time (Tulving and Thomson 1972) for planning actions (Tulving 2002; Buckner 2010), similar to linking together location sequences during exploration. Acquiring a spatial map requires repeated exploration of the landmarks by self-referenced navigation. Similarly, semantic knowledge is generated by repeated episodic encounters until the items lose their spatiotemporal context (Fig. 5; Squire 1992; Eichenbaum et al. 1999; Buzsáki 2005).

On the one hand, declarative memories may be recalled by supplying a variety of real world cues that induce an association. This is, roughly, equivalent to an animal exposed to a few cues from an environment recalling an entire environmental representation. On the other hand, declarative memories may be recalled without real world cue triggers. For example, one may recall an episodic memory or a fact by association with another memory. Such vicarious memory recall is analogous to vicarious navigational recall. We propose they have identical mechanisms.

INTERNALLY GENERATED NEURONAL PATTERNS AND COVERT OUTPUTS IN THE CEREBRAL CORTEX

We hypothesize that internally generated patterns are a common feature of cortical organization which can function as a neuronal substrate to bridge long delays between inputs and action. Connectivity patterns found in neocortex support this notion. Superficial cortical neuronal layers may be viewed as a substrate for inducing such covert outputs. While deep layer cortical neurons project to action systems of the thalamus, colliculi, brainstem, and even the spinal cord, the superficial layer can be viewed as an extra, added loop to the deep layer with subcortical outputs (Shepherd 2013). A novel neuron loop that is added to the sensorimotor short loop is the extensive associational areas, particular the prefrontal cortex. Thus, perhaps it is not surprising that the anatomical

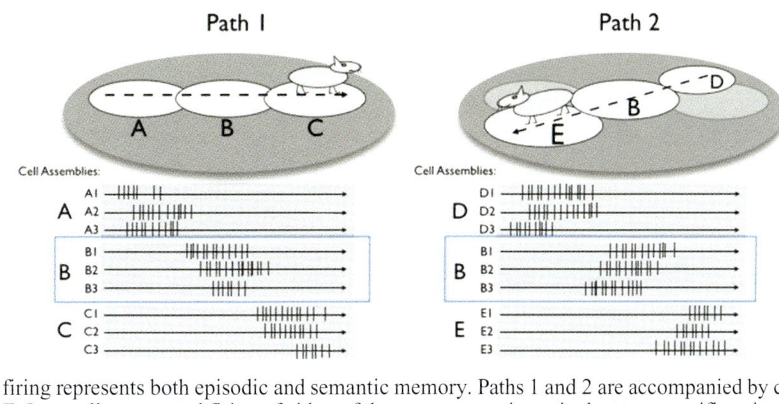

Figure 5. Place cell firing represents both episodic and semantic memory. Paths 1 and 2 are accompanied by cell assembly sequences A–B–C and D–B–E. Internally generated firing of either of these sequences is equivalent to a specific episodic memory: movement along Path 1 or Path 2. After a series of paths that activate cell assembly B, B alone is not associated with a particular path: it explicitly represents a location. In this representation, the activation of cell assembly B, alone, is a semantic fact, location B.

afferent–efferent connectivity and internal organization of the agranular motor cortex and the prefrontal cortex are so remarkably similar (Nimchinsky et al. 1995; Gabbott et al. 2005). The notion is that there are two types of motor output: the familiar "engaged" output, where motor cortical areas activate motor neurons and produce behavior; and vicarious motor output, where an internal analog of motor action is generated, with output remaining entirely within the brain. If the brain contains a sufficient model of the self and world, this type of imagined or vicarious motor action can be examined internally for potential consequences. This vicarious action is what we call a "cognitive process." Such vicarious action can be used for navigation, as in the hippocampal navigational system described earlier, or for other forms of motor behaviors. At the physiological level such cognitive operations require that the self-organized neuronal assembly sequences are propelled forward via an optimal path or by covert trial and error evaluating multiple possible outcomes, solving detours, and finding shortcuts (Gupta et al. 2012), much the same way as we navigate through locomotion in the real world in the presence of environmental feedback.

MEMORY AND THE OPTIMIZATION OF BEHAVIOR

As discussed earlier, a fundamental goal of many neuronal circuits and the brain as a whole is to predict the future. However, predicting the future is possible only after the organism has already experienced the consequences of its actions and stored the outcomes of those successful and failed actions in memory. Indeed, a barrage of recent papers document the recognition that structures that have been traditionally viewed as memory systems are inseparable parts of planning, imagining, and action systems (Buckner and Carroll 2007; Schacter and Addis 2007; Lisman and Redish 2009; Buckner 2010). What appeared to be memory of the past in the sequential activity of cell assembly sequences, equally well reflected the planned future action of the animal, including commission errors, indicating that the action outcome is the consequence of the brain's *interpretation* of the past experience (Fujisawa et al. 2008; Pastalkova et al. 2008). Numerous sleep "replay" experiments in both rodents and humans show that sleep patterns are far from random but, instead, are reminiscent of waking when the neuronal trajectories evolve under the influence of sensory inputs (Wilson and McNaughton 1994; Nádasdy et al. 1999; cf. Diekelmann and Born 2010). Transfer operations, such as the sharp wave ripples of the hippocampus, which were originally suggested to consolidate and move transient memories from the hippocampus to neocortex (Buzsáki 1989; Wilson and McNaughton 1994), have been shown to generate spike sequences that predict to-be-visited places and delayed actions or a chaining of past and expected events (Diba and Buzsáki 2007; Davidson et al. 2009; Gupta et al. 2010; cf., Carr et al. 2011; Buzsáki and Lopes da Silva 2012). Imaging experiments in humans have repeatedly found that imagining and planning invariably activated structures previously categorized as parts of the memory system (Buckner and Carroll 2007; Schacter and Addis 2007; Buckner 2010). The emerging field of "active sensing" (Katz 1989; Engel et al. 2013; Loeb and Fishel 2014) also emphasizes the primacy of action in acquiring meaning for sensation and perception (Schroeder et al. 2010). These novel (or "revisited") views break away from the tradition of "boxing" man-invented terms into particular brain structures and strategies to identify physiological boundaries for preconceived mental boundaries (Buzsáki 2010). Instead, they emphasize *interactions* of brain systems as the fundamental operations that give rise to behavioral entities that are hard to categorize strictly as memory or planning, overt or covert, conscious or "autopilot." Regarding the self-organized brain activity as the fundamental brain operation and viewing the brain's responses to external perturbations as secondary actions offers a new paradigm, in which "*meaning*" is acquired by matching preexisting neuronal patterns to action-perception (Buzsáki 2006). In support of this conjecture, experiments in developing ferrets show that the similarity between spontaneous and evoked activity increases progressively with age and is specific to responses evoked by natural scenes, suggesting that internal models result from the adaptation of brain states to the statistics of the surrounding world (Fiser et al. 2004). By extension, the construction of an internal representation of the self as a special entity can be accomplished by the reprocessing of internal data simultaneously with information about the environment.

A critical issue that has to be explored in future experiments is the significance of the duration of internally organized sequences for cognition. There are numerous brain operations, especially in the motor system, which require sequential activity of neuronal firing, yet we are not aware of the consequences of those operations and, thus, they have not been considered as cognitive. Neuronal events that are too short ($<$500 msec; Libet 2004) tend not to evoke awareness. These likely include the short ($<$100 msec) sharp wave ripple events of the hippocampus (Buzsáki 1989). Yet, these events may be regarded as "subconscious rehearsals" for future actions or priming for memory recall.

Although ample evidence exists to support the view that internally organized cell assembly sequences are the fundamental basis of mental operations, the "clutch" operation is not well understood. However, currently existing and future technologies can be deployed to test the hypothesis that mental travel and planning occur in real time, similar to navigation. By reading out brain signals and connecting them to actuators, such as a navigating robot, thought control of the robot's movements should be possible. Experiments performed within the brain–machine interface paradigm (Donoghue 2002; Fetz 2007; Nicolelis and Lebedev 2009) show the feasibility of such outcomes. Furthermore, neuronal patterns that can be read out during sleep can also be exploited to compare them with waking operations and manipulate them to establish whether internally generated patterns in the absence of conscious awareness can reflect waking-related neuronal trajectories. This paper highlights recent work in animal physiology that shows the implications of self-organized, internally generated neuronal assembly sequences. These findings support an emerging conception of the widespread importance of internal models as covert, brain-based testing grounds for efficient optimization of behavior.

ACKNOWLEDGMENTS

This work was supported by the National Institute of Health Grants MH54671 and NS074015, National Science Foundation Grant, SBE 0542013, The Mathers Foundation, Human Frontier Science Program (G.B. and A.P.), and European Molecular Biology Organization postdoctoral fellowship (A.P.).

REFERENCES

Anderson CM, Torres F, Faoro A. 1985. The EEG of the early premature. *Electroencephalogr Clin Neurophysiol* **60:** 95–105.

Buckner RL. 2010. The role of the hippocampus in prediction and imagination. *Annu Rev Psychol* **61:** 27–48.

Buckner RL, Carroll DC. 2007. Self-projection and the brain. *Trends Cogn Sci* **11:** 49–57.

Burak Y, Fiete IR. 2012. Fundamental limits on persistent activity in networks of noisy neurons. *Proc Natl Acad Sci* **109:** 17645–17650.

Buzsáki G. 1989. Two-stage model of memory trace formation: A role for "noisy" brain states. *Neuroscience* **31:** 551–570.

Buzsáki G. 2005. θ rhythm of navigation: Link between path integration and landmark navigation, episodic and semantic memory. *Hippocampus* **15:** 827–840.

Buzsáki G. 2006. *Rhythms of the brain*. Oxford University Press, New York.

Buzsáki G. 2010. Neural syntax: Cell assemblies, synapsembles, and readers. *Neuron* **68:** 362–385.

Buzsáki G. 2013. Cognitive neuroscience: Time, space and memory. *Nature* **497:** 568–569.

Buzsáki G, Lopes da Silva FL. 2012. High frequency oscillations in the intact brain. *Prog Neurobiol* **98:** 241–249.

Buzsáki G, Mizuseki K. 2014. The log-dynamic brain: How skewed distributions affect network operations. *Nat Rev Neurosci* **15:** 264–278.

Buzsáki G, Moser EI. 2013. Memory, navigation and theta rhythm in the hippocampal-entorhinal system. *Nat Neurosci* **16:** 130–138.

Carr MF, Jadhav SP, Frank LM. 2011. Hippocampal replay in the awake state: A potential substrate for memory consolidation and retrieval. *Nat Neurosci* **14:** 147–153.

Davidson TJ, Kloosterman F, Wilson MA. 2009. Hippocampal replay of extended experience. *Neuron* **63:** 497–507.

Diba K, Buzsáki G. 2007. Forward and reverse hippocampal place-cell sequences during ripples. *Nat Neurosci* **10:** 1241–1242.

Diekelmann S, Born J. 2010. The memory function of sleep. *Nat Rev Neurosci* **11:** 114–126.

Donoghue JP. 2002. Connecting cortex to machines: Recent advances in brain interfaces. *Nat Neurosci* Suppl: 1085–1088.

Dreyfus-Brisac C. 1962. The electroencephalogram of the premature infant. *World Neurol* **3:** 5–15.

Eichenbaum H, Dudchenko P, Wood E, Shapiro M, Tanila H. 1999. The hippocampus, memory, and place cells: Is it spatial memory or a memory space? *Neuron* **23:** 209–226.

Engel AK, Fries P, Singer W. 2001. Dynamic predictions: Oscillations and synchrony in top-down processing. *Nat Rev Neurosci* **2:** 704–716.

Engel AK, Maye A, Kurthen M, König P. 2013. Where's the action? The pragmatic turn in cognitive science. *Trends Cogn Sci* **17:** 202–209.

Fetz EE. 2007. Volitional control of neural activity: Implications for brain computer interfaces. *J Physiol* **579:** 571–579.

Fiser J, Chiu C, Weliky M. 2004. Small modulation of ongoing cortical dynamics by sensory input during natural vision. *Nature* **431:** 573–578.

Frank LM, Brown EN, Wilson M. 2000. Trajectory encoding in the hippocampus and entorhinal cortex. *Neuron* **27:** 169–178.

Fujisawa S, Amarasingham A, Harrison MT, Buzsáki G. 2008. Behavior-dependent short-term assembly dynamics in the medial prefrontal cortex. *Nat Neurosci* **11:** 823–833.

Fyhn M, Hafting T, Treves A, Moser MB, Moser EI. 2007. Hippocampal remapping and grid realignment in entorhinal cortex. *Nature* **446:** 190–194.

Gabbott PL, Warner TA, Jays PR, Salway P, Busby SJ. 2005. Prefrontal cortex in the rat: Projections to subcortical autonomic, motor, and limbic centers. *J Comp Neurol* **492:** 145–177.

Gelbard-Sagiv H, Mukamel R, Harel M, Malach R, Fried I. 2008. Internally generated reactivation of single neurons in human hippocampus during free recall. *Science* **322:** 96–101.

Gibson J. 1977. The theory of affordances. In *Perceiving, acting, and knowing: Toward an ecological psychology* (ed. Shaw R, Bransford J), pp. 62–82. Lawrence Erlbaum Associates, Hillsdale, NJ.

Gupta AS, van der Meer MA, Touretzky DS, Redish AD. 2010. Hippocampal replay is not a simple function of experience. *Neuron* **65:** 695–705.

Gupta AS, van der Meer MA, Touretzky DS, Redish AD. 2012. Segmentation of spatial experience by hippocampal theta sequences. *Nat Neurosci* **15:** 1032–1039.

Hafting T, Fyhn M, Molden S, Moser MB, Moser EI. 2005. Microstructure of a spatial map in the entorhinal cortex. *Nature* **436:** 801–806.

Harvey CD, Coen P, Tank DW. 2012. Choice-specific sequences in parietal cortex during a virtual-navigation decision task. *Nature* **484:** 62–68.

Hebb DO. 1949. *The organization of behavior*. Wiley, New York.

Immordino-Yang MH, Christodoulou JA, Singh V. 2012. Rest is not idleness: Implications of the brain's default mode for human development and education. *Perspect Psychol Sci* **7:** 352–364.

Katz D. 1989. *The world of touch*. Lawrence Erlbaum Associates, Hillsdale, NJ.

Khazipov R, Buzsáki G. 2010. Early patterns of electrical activity in the developing cortex. In *Oxford handbook of developmental behavioral neuroscience* (ed. Blumberg MS, Freeman JH, Robinson SR), pp. 161–177. Oxford University Press, Oxford.

Khazipov R, Sirota A, Leinekugel X, Holmes GL, Ben Ari Y, Buzsáki G. 2004. Early motor activity drives spindle bursts in the developing somatosensory cortex. *Nature* **432:** 758–761.

Killingsworth MA, Gilbert DT. 2010. A wandering mind is an unhappy mind. *Science* **330:** 932.

Knierim JJ, Zhang K. 2012. Attractor dynamics of spatially correlated neural activity in the limbic system. *Annu Rev Neurosci* **35:** 267–285.

Kubie JL, Muller RU. 1991. Multiple representations in the hippocampus. *Hippocampus* **1:** 240–242.

Libet B. 2004. *Mind time. The temporal factor in consciousness*. Harvard University Press, Cambridge, MA.

Lisman J, Redish AD. 2009. Prediction, sequences and the hippocampus. *Philos Trans R Soc Lond B Biol Sci* **364:** 1193–1201.

Loeb GE, Fishel JA. 2014. Bayesian action&perception: Representing the world in the brain. *Front Neurosci* **8:** 341.

Mar RA, Mason MF, Litvack A. 2014. How daydreaming relates to life satisfaction, loneliness, and social support: The importance of gender and daydream content. *Conscious Cogn* **21:** 401–407.

McNaughton BL, Barnes CA, Gerrard JL, Gothard K, Jung MW, Knierim JJ, Kudrimoti H, Qin Y, Skaggs WE, Suster M, et al. 1996. Deciphering the hippocampal polyglot: The hippocampus as a path integration system. *J Exp Biol* **199:** 173–185.

McNaughton BL, Battaglia FP, Jensen O, Moser EI, Moser MB. 2006. Path integration and the neural basis of the 'cognitive map'. *Nat Rev Neurosci* **7:** 663–778.

Menzel R, Geiger K, Joerges J, Muller U, Chittka L. 1998. Bees travel novel homeward routes by integrating separately acquired vector memories. *Anim Behav* **55:** 139–152.

Moser EI, Roudi Y, Witter MP, Kentros C, Bonhoeffer T, Moser MB. 2014. Grid cells and cortical representation. *Nat Rev Neurosci* **15:** 466–481.

Muller RU, Kubie JL. 1987. The effects of changes in the environment on the spatial firing of hippocampal complex-spike cells. *J Neurosci* **7:** 1951–1968.

Nádasdy Z, Hirase H, Czurkó A, Csicsvari J, Buzsáki G. 1999. Replay and time compression of recurring spike sequences in the hippocampus. *J Neurosci* **19:** 9497–9507.

Nicolelis MA, Lebedev MA. 2009. Principles of neural ensemble physiology underlying the operation of brain–machine interfaces. *Nat Rev Neurosci* **10:** 530–540

Nimchinsky EA, Vogt BA, Morrison JH, Hof PR. 1995. Spindle neurons of the human anterior cingulate cortex. *J Comp Neurol* **355:** 27–37.

O'Keefe J, Nadel L. 1978. *The hippocampus as a cognitive map.* Oxford University Press, New York.

O'Keefe J, Recce ML. 1993. Phase relationship between hippocampal place units and the EEG theta rhythm. *Hippocampus* **3:** 317–330.

O'Regan JK, Noë A. 2001. A sensorimotor account of vision and visual consciousness. *Beh Brain Sci* **25:** 883–975.

Pastalkova E, Itskov V, Amarasingham A, Buzsáki G. 2008. Internally generated cell assembly sequences in the rat hippocampus. *Science* **321:** 1322–1327.

Peyrache A, Lacroix MM, Petersen PC, Buzsáki G. 2015. Internally organized mechanisms of the head direction sense. *Nat Neurosci* **18:** 569–575.

Quirk GJ, Muller RU, Kubie JL. 1990. The firing of hippocampal place cells in the dark depends on the rat's recent experience. *J Neurosci* **10:** 2008–2017.

Ranck J. 1985. In *Electrical activity of archicortex* (ed. Buzsáki G, Vanderwolf CH). Akadémiai Kiadó, Budapest.

Rao RP, Ballard DH. 1999. Predictive coding in the visual cortex: A functional interpretation of some extra-classical receptive-field effects. *Nat Neurosci* **2:** 79–87.

Redish AD, Elga AN, Touretzky DS. 1996. A coupled attractor model of the rodent head direction system. *Netw Comput Neural Syst* **7:** 671–685.

Sargolini F, Fyhn M, Hafting T, McNaughton BL, Witter MP, Moser MB, Moser EI. 2006. Conjunctive representation of position, direction, and velocity in entorhinal cortex. *Science* **312:** 758–762.

Schacter DL, Addis DR. 2007. Constructive memory: The ghosts of past and future. *Nature* **445:** 27.

Schroeder CE, Wilson DA, Radman T, Scharfman H, Lakatos P. 2010. Dynamics of active sensing and perceptual selection. *Curr Opin Neurobiol* **20:** 172–176.

Scoville WB, Milner B. 1957. Loss of recent memory after bilateral hippocampal lesions. *J Neurol Neurosurg Psychiatry* **20:** 11–21.

Sharp PE, Blair HT, Cho J. 2001. The anatomical and computational basis of the rat head-direction cell signal. *Trends Neurosci* **24:** 289–294.

Shepherd GM. 2013. Corticostriatal connectivity and its role in disease. *Nat Rev Neurosci* **14:** 278–291.

Skaggs WE, Knierim JJ, Kudrimoti HS, McNaughton BL. 1995. A model of the neural basis of the rat's sense of direction. *Adv Neural Inf Process Syst* **7:** 173–180.

Solstad T, Boccara CN, Kropff E, Moser MB, Moser EI. 2008. Representation of geometric borders in the entorhinal cortex. *Science* **322:** 1865–1868.

Squire LR. 1992. Memory and the hippocampus: A synthesis from findings with rats, monkeys, and humans. *Psychol Rev* **99:** 195–231.

Suddendorf T, Corballis MC. 1997. Mental time travel and the evolution of the human mind. *Genet Soc Gen Psychol Monogr* **123:** 133–167.

Suddendorf T, Corballis MC. 2007. The evolution of foresight: What is mental time travel, and is it unique to humans? *Behav Brain Sci* **30:** 299–313.

Taube JS. 2007. The head direction signal: Origins and sensory-motor integration. *Annu Rev Neurosci* **30:** 181–207.

Taube JS, Muller RU, Ranck JB Jr. 1990a. Head-direction cells recorded from the postsubiculum in freely moving rats. I. Description and quantitative analysis. *J Neurosci* **10:** 420–435.

Taube JS, Muller RU, Ranck JB Jr. 1990b. Head-direction cells recorded from the postsubiculum in freely moving rats. II. Effects of environmental manipulations. *J Neurosci* **10:** 436–447.

Thompson E, Varela F. 2001. Radical embodiment: Neural dynamics and consciousness. *Trends Cogn Sci* **5:** 418–425.

Tulving E. 2002. Chronesthesia: Conscious awareness of subjective time. In *Principles of frontal lobe function* (ed. Stuss DT, Knight RC), pp. 311–325. Oxford University Press, New York.

Tulving E, Thomson D. 1973. Encoding specificity and retrieval processes in episodic memory. *Psychol Rev* **80:** 352–373.

Tulving E, Donaldson W, Bower GH. 1972. *Organization of memory.* Academic Press, New York.

Vanderwolf CH. 1969. Hippocampal electrical activity and voluntary movement in the rat. *Electroencephalogr Clin Neurophysiol* **26:** 407–418.

Varela F, Lachaux JP, Rodriguez E, Martinerie J. 2001. The brainweb: Phase synchronization and large-scale integration. *Nat Rev Neurosci* **2:** 229–239.

Wang Y, Romani S, Lustig B, Leonardo A, Pastalkova E. 2015. Theta sequences are essential for internally generated hippocampal firing fields. *Nat Neurosci* **18:** 282–288.

Wilson MA, McNaughton BL. 1993. Dynamics of the hippocampal ensemble code for space. *Science* **261:** 1055–1058.

Wilson MA, McNaughton BL. 1994. Reactivation of hippocampal ensemble memories during sleep. *Science* **265:** 676–679.

Wood ER, Dudchenko PA, Robitsek RJ, Eichenbaum H. 2000. Hippocampal neurons encode information about different types of memory episodes occurring in the same location. *Neuron* **27:** 623–633.

Zugaro MB, Arleo A, Berthoz A, Wiener SI. 2003. Rapid spatial reorientation and head direction cells. *J Neurosci* **23:** 3478–3482.

Dynamic Hippocampal Circuits Support Learning- and Memory-Guided Behaviors

EMILY B. ANDERSON, IRENE GROSSRUBATSCHER, AND LOREN FRANK

Department of Physiology and Center for Integrative Neuroscience, University of California, San Francisco, San Francisco, California 94143

Correspondence: loren@phy.ucsf.edu

The ability to learn spatial relationships and to modify stored representations when the world changes is essential for survival. In recent years, our understanding of the neural activity that underlie these learning and memory processes has improved considerably, in large part to an increase in the number of studies that use electrophysiological recordings as animals learn about their environments. In this review, we summarize some key findings from our laboratory, with a focus on recent discoveries that indicate that the rat hippocampus supports learning and decision-making behaviors via dynamic and smooth transitions in neural representation, internal processing state, and coupling with related brain structures.

Our memories are central to guiding our decisions and defining us as individuals. Historically, there have been two distinct approaches toward studying memory and the brain structures that support it: a cognitive approach and an electrophysiological approach. The first approach, perhaps best known for the seminal studies on patient H.M. (Scoville and Milner 1957; Squire and Bayley 2007; Eichenbaum 2013) uses a combination of lesion, fMRI, and behavioral methods to examine the role of different brain structures in specific learning and memory tasks. A key insight from this approach is that different brain circuits support different types of memory (Eichenbaum and Cohen 2001). From these studies, a particular region of the brain, the hippocampus (see Box 1), has emerged as particularly important in encoding episodic or declarative memories—that is, the memories of distinct, autobiographical experiences that are essential to our sense of self.

To gain insight into the specific neural mechanisms underlying memory, electrophysiologists place recording electrodes within the hippocampus and connected structures to examine how the electrical signals of individual neurons change as animals engage in tasks. Although technically challenging, the electrophysiological approach offers greater temporal precision and improved ability to identify individual circuit components. Historically, however, most electrophysiological studies of the hippocampus did not focus on learning and memory behaviors, perhaps because most classic memory tasks lack the repetitive trial structure necessary for neurophysiological data collection (Kim and Frank 2009). Instead, most electrophysiological studies have focused on characterizing the properties of individual "place cells"—hippocampal neurons that elicit action potentials at selective locations of an animal's environment—and the role of these cells in constructing an allocentric spatial map of the environment (O'Keefe and Dostrovsky 1971; Best et al. 2001). Consequently, until recently, relatively little was known about how activity at the single-neuron level could support the learning and memory functions of the hippocampus. The question of how single neurons and circuits of the hippocampus support cognitive functions like learning and memory is the main motivator for our work and the focus of this review.

ROLE OF HIPPOCAMPUS IN LEARNING

Learning is not just the formation of new memories but also a change in representation of the world. Classically, the hippocampus and neocortex had been thought to operate complementary learning systems that divide these tasks (Marr 1971; Buzsáki 1989; McClelland et al. 1995; Sutherland and McNaughton 2000; O'Reilly and Norman 2002). In this view, the hippocampus was thought to form distinct ("pattern separated") representations of the world during an initial experience. These patterns would then be essentially static within the hippocampal circuit. Once these memories were encoded by the hippocampus, reactivation of those memories during sleep was thought to drive consolidation in cortical circuits (Wilson and McNaughton 1994; Buzsáki 1996; O'Neill et al. 2010). This reactivation occurs during "sharp-wave ripple" (SWR) events, which manifest as bursts of high frequency (150–250-Hz) power in the local-field potential recorded in area CA1 (Buzsáki 1986). The neocortex would then learn slowly about the statistical regularities of the environment as the hippocampus repeatedly presented these memory sequences (i.e., replay events) during SWRs occurring in slow-wave sleep. Thus, the hippocampus was thought to represent the world while animals are awake and to send information to the neocortex for consolidation during sleep.

Although this theoretical framework has proven influential, it was developed at a time when most studies of

 doi: 10.1101/sqb.2014.79.024760

BOX 1: THE HIPPOCAMPAL CIRCUIT

The rodent hippocampus is a pair of banana-shaped structures in the medial temporal lobe (Fig. 1). As with many brain regions, the hippocampus is highly interconnected (Amaral and Witter 1995). These connections include a large number of feed-forward connections, which begin with a projection from the entorhinal cortex to all of the hippocampal subdivisions. Information also propagates along multiple internal pathways, including the Mossy fibers of the dentate gyrus that project to areas CA2 and CA3, and the Schäffer collaterals from area CA3 to area CA1. In addition, there are a number of recurrent networks within the hippocampal circuit, including one in the dentate gyrus/hilar region (not shown), a strong CA3 recurrent network, and a recently discovered CA1 recurrent network (Yang et al. 2014).

The hippocampal regions, which differ in their connectivity with subcortical structures (Amaral and Witter 1995), also vary along the dorsoventral axis. Most studies of the rodent hippocampus have focused on the more physically accessible dorsal region, which is noted for neurons that represent specific locations in space ("place cells") and is thought to be important for spatial navigation and memories involving spatial context. In contrast, the less-studied intermediate and ventral hippocampus may play an important role in anxiety and emotional memories (Fanselow and Dong 2010).

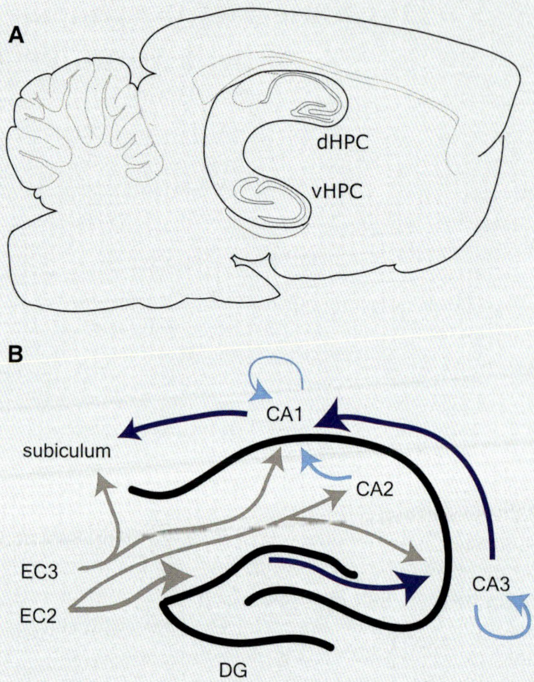

Figure 1. Hippocampal circuit. (*A*) A sagittal view of the rodent hippocampus, showing the relative locations of the dorsal (dHPC) and ventral (vHPC) hippocampus. (*B*) Simplified hippocampal circuit diagram. DG: dentate gyrus; EC2 and EC3: layers 2 and 3 of the entorhinal cortex. Outputs from CA1 and the subiculum target the deep layers of the entorhinal cortex as well as many other cortical and subcortical regions (not shown).

learning did not use electrophysiological techniques and therefore did not provide data on how learning influences the formation and maintenance of place fields, replay/reactivation events, or other aspects of hippocampal activity. Similarly, most electrophysiological studies had focused on describing patterns of activity obtained during well-learned tasks performed in familiar environments and thus also did not provide data on how patterns of neural activity support learning. In the following sections, we offer two themes that challenge the classic view and expand the hippocampus's role in learning. First, we discuss results indicating that the hippocampus does not simply form a new and stable representation following an experience. Instead, the hippocampus changes its signaling patterns as environments and objects become more familiar (Cheng and Frank 2008; Karlsson and Frank 2008; Larkin et al. 2014) and as animals learn specific tasks (Dupret et al. 2010; Singer et al. 2010). Second, we describe our findings that the memory replay events that had previously been attributed to slow-wave sleep also occur in the awake state, particularly in the context of new experiences (Cheng and Frank 2008; Karlsson and Frank 2009). Further, we discuss the role of these events in learning- and memory-guided decision-making (Jadhav et al. 2012; Yu and Frank 2014).

HIPPOCAMPAL REPRESENTATIONS CHANGE WITH EXPERIENCE

We examined the effect of novelty on place cell representations and found that the introduction of novel environments and contexts caused a novelty-dependent increase in place cell firing rate in area CA1 (Karlsson and Frank 2008; Larkin et al. 2014). This increase in firing rate was specific to CA1, as firing rates in CA3 remained stable (Karlsson and Frank 2008; Larkin et al. 2014). Over time, as the environment became more familiar and task performance improved, the distribution of firing rates across the CA1 place cell population spread: High firing rate neurons increased their firing rate, whereas low firing rate neurons decreased in rate (Karlsson and Frank 2008). The result of these changes is that the active CA1 population comes to consist of a relatively small group of cells with strong spatial tuning, creating a sparse, spatially informative population of neurons. Interestingly, we found that as this information continues from CA1 to subicular output circuits, the code is further transformed from a sparse code to a highly distributed code that may be more effective for communication with other brain regions (Kim et al. 2012).

These results complemented previous studies showing that the hippocampus is necessary for behaviors associated with contextual novelty. Measurements of evoked response potentials (ERPs) and of the BOLD response in fMRI scans have also shown an increase in activity in the hippocampus at the onset of novel stimuli (Knight 1996; Knight and Nakada 1998). These results led to the proposal that hippocampal area CA1 serves as a novelty detector (Lisman and Otmakhova 2001), as it could compare stored representations in CA3 with sensory information from the entorhinal cortex (see Box 1). Thus, we hypothesize that the increase in firing rates seen in CA1 but not upstream in CA3 could be the physiological mechanism that explains the previously observed hippocampal responses to novelty.

In a related study, we observed other long timescale patterns of change in hippocampal representations. We found that as animals learned specific behavioral tasks, coordinated hippocampal ensembles came to encode generalizations across locations (Singer et al. 2010). As the animals gained more experience with the environment, about half of the population of place cells in CA1 and CA3 developed path equivalent firing, with individual neurons firing in multiple locations with similar behavioral contexts. Such path equivalent ensembles are well-suited to encode similarities among repeating elements and potentially provide a framework for associating specific behaviors with multiple locations (Frank et al. 2000). At the same time, a population of "classic" place cells remained that could maintain specific location information. Taken together, these results argue that place cell activity not only supports the distinct representations proposed by earlier theories but also more generalized information that can support learned behaviors. We suggest that path equivalent ensembles reflect the representation of common features and behavioral associations across locations.

LEARNING MODULATES HIPPOCAMPAL REACTIVATION

In addition to these changes in hippocampal representations, we wondered whether learning modulated hippocampal communication to other brain structures. In particular, we were interested in SWR events as these are known to propagate from the hippocampus to cortical structures (Chrobak and Buzsáki 1994; Buzsáki 1996; Csicsvari et al. 2000) and are associated with a time-compressed reactivation of firing patterns associated with previous (Pavlides and Winson 1989; Wilson and McNaughton 1994; Kudrimoti et al. 1999; Nadasdy et al. 1999; Lee and Wilson 2002; Ji and Wilson 2007; O'Neill et al. 2008; Karlsson and Frank 2009) and current (Foster and Wilson 2006; Csicsvari et al. 2007; Diba and Buzsáki 2007; Carr et al. 2011) experiences. Because spiking during these ripple events is phase locked to fast oscillations similar to those used to induce LTP (Buzsáki 1986; Wilson and McNaughton 1994; Sutherland and McNaughton 2000), ripple reactivation has been proposed to drive consolidation of previously formed memories. Interestingly, we found that novelty did in fact modulate this communication: In rest sessions following exposure to a novel environment, there was an increase in the mean firing rate during SWR events (Karlsson and Frank 2008). This finding complemented parallel results from the Csicsvari laboratory (O'Neill et al. 2008). We also observed that novelty increased the correlation of neurons in the rest box with their activity in the previous behavioral session (Karlsson and Frank 2008) suggesting that novel experiences may be more faithfully consolidated during subsequent sleep sessions.

AWAKE REPLAY DURING LEARNING

Classically, reactivation of past experiences was thought to only occur during sleep, consistent with the idea of sleep as a privileged state for memory consolidation processes (Wilson and McNaughton 1994; Buzsáki 1996; O'Neill et al. 2010). Later, when sequences of place cell activity during SWRs were discovered during awake quiescence (Kudrimoti et al. 1999; Csicsvari et al. 2007 #1415; Foster and Wilson 2006; Diba and Buzsáki 2007), it was thought that although replay during sleep involved the reactivation of stored representations in the absence of specific sensory input, awake replay depended on sensory input from the current environment.

Recently, however, we discovered robust remote replay of past experiences during waking behavior (Karlsson and Frank 2009). Indeed, awake replay of a remote environment is typically a higher fidelity recapitulation of past experiences than replay during quiescence or sleep (Karlsson and Frank 2009). This led us to ask whether awake replay events might play an important role in memory retrieval and consolidation as animals learn about new environments and tasks (Carr et al. 2011).

Consistent with this hypothesis, we found an increase in the number of awake SWR events when animals were exposed to novel environments (Cheng and Frank 2008) or novel contexts (Larkin et al. 2014). We also found that novelty increased the firing rate of CA1 neurons awake during SWRs and increased the degree of coordination of neurons representing novel experiences during SWRs (Cheng and Frank 2008; Larkin et al. 2014). This suggests that in a similar fashion to SWRs occurring during sleep, awake SWRs more faithfully reflect novel experiences than familiar ones. This increase in firing rate and coordination associated with novelty was specific to activity during ripples in CA1, as no changes were seen in CA3 (Larkin et al. 2014) or outside ripple events (Cheng and Frank 2008; Larkin et al. 2014).

Parallel studies had shown that SWR events during sleep play an important role in task learning (Girardeau et al. 2009), and that the strength of awake reactivation predicts subsequent memory, suggesting that awake SWRs may indeed play an important role in learning (Dupret et al. 2010). We also observe an increase in the number of SWRs following reward, and the content of these ripples tended to reflect paths associated with reward location (Singer and Frank 2009), which could serve as a mechanism to bind rewarding outcomes to preceding experiences.

These studies did not establish a causal role for awake SWRs in learning, however. We therefore developed a real-time SWR detection system and used it to detect and then electrically interrupt awake SWRs as animals learned a hippocampus-dependent spatial alternation task (Jadhav et al. 2012). We found that the disruption of SWRs during learning caused consistent, strong, and significant deficits in performance of the task (Fig. 2). Interestingly, we found that this impairment was selective: Only outbound trials that required integration of information about immediate past and potential future options were impaired (Jadhav et al. 2012). Further, place-field activity was unaffected, indicating that the impairment was specific to the role of SWR events. As we will discuss in the next section, we hypothesize that reactivation of path trajectories during SWRs provides information about past locations and possible future options to guide decisions made in structures such as the prefrontal cortex.

AWAKE REPLAY IN RETRIEVAL, DELIBERATION AND DECISION-MAKING

We propose that one purpose of awake replay events is to contribute to internal deliberation processes whereby different routes through the animals "cognitive map" are evaluated (Yu and Frank 2014). Consistent with this idea, awake ripples frequently occur when animals pause at choice points (Karlsson and Frank 2009; Jadhav et al. 2012). Further, decoding of place cell reactivation shows that replay trajectories are biased to include the current location of the animal (Davidson et al. 2009; Karlsson and Frank 2009) and, when pertinent, can be biased to terminate at a rewarded goal location (Dupret et al. 2010; Pfeiffer and Foster 2013).

We have also shown that during learning awake SWR reactivation is more coordinated and more intense preceding correct decisions than incorrect ones (Singer et al. 2013), suggesting that replay during SWRs contributes to successful decision-making. Interestingly, however, we found that replay trajectories corresponding to both correct and incorrect choices often occur before a choice. This suggests that processing downstream (perhaps in prefrontal cortex) is critical for evaluating the content of these ripples and in making the ultimate behavioral decision (Singer et al. 2013; Yu and Frank 2014).

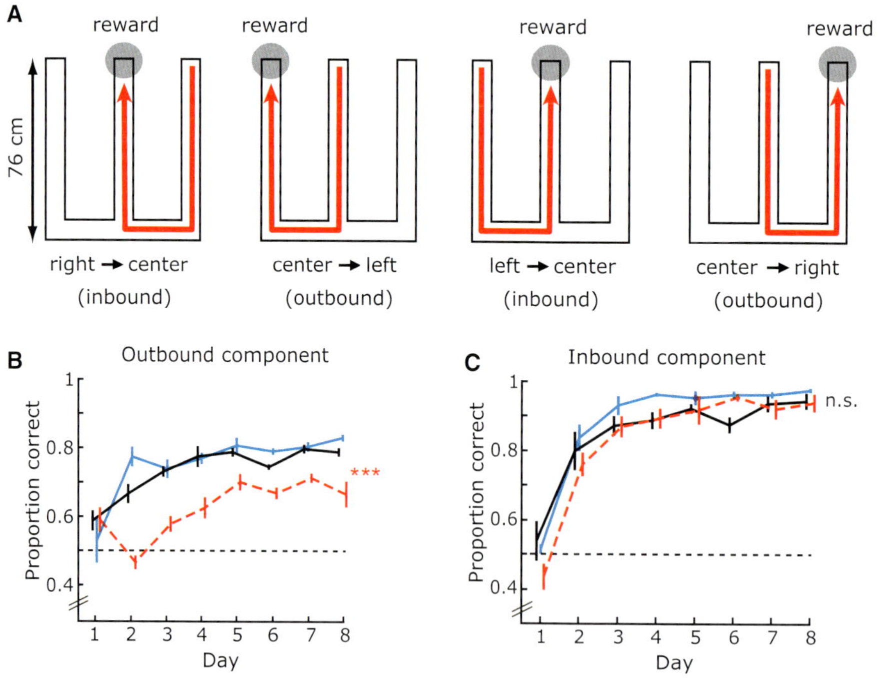

Figure 2. Disruption of SWRs during learning causes strong deficits in task performance. (*A*) Spatial alternation "W-maze" task (Kim and Frank 2009). Rats must learn two rules. When in an outer arm, animals must return to the center arm (inbound component). Thus, this "inbound" trial depends only on the rat's current location and is independent of previous history. In contrast, when leaving the center arm, the animal should not return to the most recently visited arm but instead alternate between the outer arms (outbound component). (*B*) Performance during the outbound component. Electrical disruption of SWR events significantly impaired performance ($p < 0.001$) compared with control groups. Dashed red line indicates performance with electrical stimulation, black line indicates performance in unstimulated animals, and light blue indicates control stimulation animals. (*C*) Performance during the inbound component. In contrast to the outbound behavior, disruption of SWRs did not alter performance. Colors indicate performance as in *B*. (Modified from Jadhav et al. 2012.)

FROM GEOMETRICAL TO TOPOLOGICAL "COGNITIVE MAP" IN THE HIPPOCAMPUS

In a parallel line of investigation, we examined the nature of the representation of space in the hippocampus. Place-field activity has often been characterized as representing geometric features about the environment, and the impact of some environmental manipulations can be explained in geometrical terms (O'Keefe and Burgess 1996; Eichenbaum 2000; Fenton et al. 2000a,b). In this view, place-field firing can be explained as a result of synthesizing information about distances and angles to landmarks arising from various sources, including self-motion and external, sensory cues.

Although a geometrical perspective is appealing, place cells respond to a wide variety of factors that include visual cues, head direction, goal-planning, and so on. It is not immediately clear how all of these factors can be integrated in a geometrical framework. We therefore proposed that place-field activity is better conceived as a topological representation of space. In this view, individual place cells represent regions of space and sets of place cells that fire together represent regions that are connected, but the firing of the cells does not explicitly represent distances or angles (Dabaghian et al. 2012).

We tested this hypothesis by constructing an environment where the topology was held constant while the geometry changed from trial to trial. In this dynamic environment, place-field activity was highly unstable when the specific locations, distances, and angles were considered, but stable when only the connectivity of the space was considered (Dabaghian et al. 2014). Therefore, we suggest that the hippocampal map provides information about the connectivity of spatial environments to downstream brain regions, an idea that translates naturally to the hippocampus providing a representation that would capture the connected elements of experience (Eichenbaum 2000; Dabaghian et al. 2014).

DYNAMIC ROLE FOR HIPPOCAMPUS IN INFORMATION PROCESSING

We have also examined the dynamics of hippocampal information processing states. The hippocampus is typically described as having two distinct states: one that supports encoding of new memories and a second that consolidates these memories into long-lasting distributed representations (Marr 1971; Buzsáki 1989). During periods of active exploration, processed sensory input drives coordinated neuronal activity and is associated with synchronous oscillations in the 7–10-Hz θ range (Buzsáki 2002). This "external" drive is mediated by the entorhinal cortex (Box 1) and likely supports the encoding of memories. In contrast, during slow-wave sleep or periods of quiet stillness, the hippocampus is driven by internal inputs, with stored information broadcast out of the hippocampus (via replay events) to consolidate memories in distributed hippocampal–neocortical circuits.

Our results suggest that in contrast to the two-state model, hippocampal activity can be better described as a dynamic continuum with rapid and smooth transitions in information flow and processing. Such "on the fly" reconfiguration allows the hippocampus to perform different functions at different times, depending on the behavioral state of the animal. Because relatively little was known about how the external entorhinal and the internal CA3 inputs combined to influence CA1 output, we performed a series of experiments to understand how information flow through these pathways varied as a function of behavioral state (Kemere et al. 2013). We first examined changes in γ frequencies, as recent studies had shown that power in the 30–50-Hz slow γ band of the local-field potential reflects the strength of coupling of CA3 to CA1 (Csicsvari et al. 2000; Colgin et al. 2009; Nakashiba et al. 2009), whereas power in the 50–100-Hz fast γ band reflects the strength of EC3–CA1 coupling (Colgin et al. 2009). Both ranges of γ are seen during exploration (Colgin et al. 2009), and we have shown that SWRs and replay events occur during transient increases in slow γ power and synchrony across the dorsal CA3 and CA1 networks (Carr and Frank 2012). Further, as the strength of low γ power increases, the quality of replay events also improves (Carr and Frank 2012), suggesting that the strength of CA3 input into CA1 modulates replay fidelity. Transient increases in CA3–CA1 coupling have also been seen when animals approached a position in a maze where they needed to make a memory-guided decision (Montgomery and Buzsáki 2007), consistent with the hypothesis that CA3 drive of CA1 is important for online memory retrieval.

We found that slow γ and ripple power decreased smoothly as animals moved progressively faster through their environment (Fig. 3A,B), consistent with weaker CA3 drive of CA1 at high speeds. In sharp contrast, fast γ power increased with increasing speed, suggesting greater EC drive of CA1 as animals moved more quickly (Fig. 3C). We also directly examined the propagation of activity from CA3 to CA1 by electrical stimulation and found that the slope of the field potentials recorded in CA1 varied smoothly as a function of the movement speed of the animal (Fig. 1D). These changes occurred on a timescale of <1 sec. Taken together, these results suggest a changing balance between the CA3 and EC inputs to CA1, with CA3 inputs becoming relatively less influential and EC inputs becoming relatively more influential as animals moved more quickly.

This shift from CA3 to EC drive also corresponded to a major change in the spiking output of CA1. The spiking of pairs of CA1 neurons with overlapping place fields was highly correlated at low speeds and these correlations decreased threefold with increasing speed. Thus, moment-by-moment changes in behavior that are seen as animals explore new places manifest as a transition from CA3-driven, highly correlated activity reflecting learned association (as in awake replay of past experience) at slow speeds to EC-driven, more independent representations that reflect sensory inputs available in the current location at high speeds.

How are these changes in network state regulated? Although there are likely multiple factors that sculpt the

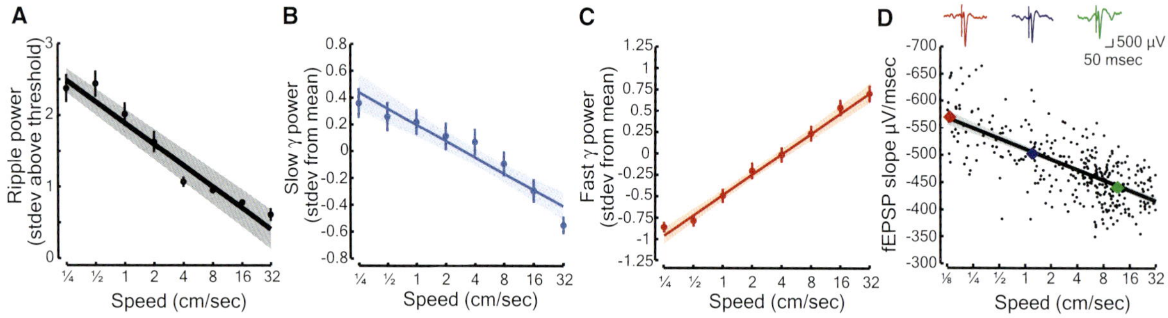

Figure 3. CA1 responses change as function of movement speed. (*A*,*B*) Measures related to CA3 drive of CA1 decline smoothly with increasing movement speed. (*A*) SWR power; (*B*) slow γ power. Note the log scale for speed. (*C*) Fast γ power, which is related to EC drive of CA1, increases smoothly with movement speed. (*D*) Linear relationship between the log of movement speed and the slope of evoked field postsynaptic potentials in CA1. Example traces for each colored point shown at right. Over 50% of the variance in the slope could be explained by the log of speed, indicating that CA1 responses to CA3 stimulation vary closely with movement speed. There was no tendency for variance of our measurements to increase at intermediate speeds as would be expected if the observation of continuous changes was a result of rapid transitions between two distinct states. Error bars represent SEM, shaded areas represent SEM for linear fits. (Modified from Kemere et al. 2013.)

relative strength of these inputs, recent work indicates that changes in cholinergic tone from the medial septum play a critical role in regulating this dynamic balance. Recent experiments have shown that optogenetic stimulation of cholinergic neurons in the medial septum results in a rapid, strong, and significant reduction in hippocampal SWR events in mice (Vandecasteele et al. 2014) and rats (Davidson et al. 2014) that likely reflects a shift away from network activity driven by the CA3–CA1 pathway. This reduction in SWRs appears to be mediated by muscarinic receptors, as application of pilocarpine (a muscarinic receptor agonist) also reduces the frequency of SWRs in vivo (Norimoto et al. 2012). Future work is needed to better understand how cholinergic modulation of hippocampal state is coordinated with other systems (such as septal GABAergic modulation of hippocampal θ activity) and how changes in hippocampal network activity supports hippocampal function.

CONCLUSION

In this review, we have highlighted how recording populations of individual hippocampal neurons as animals learn to navigate through an environment to obtain reward has changed our view of how the hippocampus supports learning. Our experiments revealed a strong effect of novelty on hippocampal representations and activity, and showed a causal role for awake SWR events in memory processes. Our work also suggests a topological framework for the hippocampal representation of space, and offers new insights into how the hippocampal circuit supports different information processing functions at different times.

These findings provide a stepping-off point for new studies that seek to understand not just the hippocampus, but the broader hippocampal–cortical and hippocampal–subcortical networks. Given the importance of awake SWRs for memory, we can now ask how these events drive information processing in downstream areas. We can also ask how these events, and network state more generally, are regulated and whether we can manipulate that regulation to alter how the hippocampus processes information. These studies, when combined with continued technology development including high-density recordings and better tools for precisely manipulating neural signals, will drive deeper insight into how interactions between the hippocampus and adjacent structures support learning and memory.

REFERENCES

Amaral DG, Witter MP. 1995. Hippocampal formation. In *The rat nervous system*, 2nd ed. (ed. Paxinos G), pp. 443–493. Academic, San Diego.

Best PJ, White AM, Minai A. 2001. Spatial processing in the brain: The activity of hippocampal place cells. *Annu Rev Neurosci* **24:** 459–486.

Buzsáki G. 1986. Hippocampal sharp waves: Their origin and significance. *Brain Res* **398:** 242–252.

Buzsáki G. 1989. Two-stage model of memory trace formation: A role for "noisy" brain states. *Neuroscience* **31:** 551–570.

Buzsáki G. 1996. The hippocampo-neocortical dialogue. *Cereb Cortex* **6:** 81–92.

Buzsáki G. 2002. Theta oscillations in the hippocampus. *Neuron* **33:** 325–340.

Carr MF, Frank LM. 2012. A single microcircuit with multiple functions: State dependent information processing in the hippocampus. *Curr Opin Neurobiol* **22:** 704–708.

Carr MF, Jadhav SP, Frank LM. 2011. Hippocampal replay in the awake state: A potential substrate for memory consolidation and retrieval. *Nat Neurosci* **14:** 147–153.

Cheng S, Frank LM. 2008. New experiences enhance coordinated neural activity in the hippocampus. *Neuron* **57:** 303–313.

Chrobak JJ, Buzsáki G. 1994. Selective activation of deep layer (V-VI) retrohippocampal cortical neurons during hippocampal sharp waves in the behaving rat. *J Neurosci* **14:** 6160–6170.

Colgin LL, Denninger T, Fyhn M, Hafting T, Bonnevie T, Jensen O, Moser MB, Moser EI. 2009. Frequency of γ oscillations routes flow of information in the hippocampus. *Nature* **462:** 353–357.

Csicsvari J, Hirase H, Mamiya A, Buzsáki G. 2000. Ensemble patterns of hippocampal CA3-CA1 neurons during sharp wave-associated population events. *Neuron* **28:** 585–594.

Csicsvari J, O'Neill J, Allen K, Senior T. 2007. Place-selective firing contributes to the reverse-order reactivation of CA1

pyramidal cells during sharp waves in open-field exploration. *Eur J Neurosci* **26:** 704–716.
Dabaghian Y, Mémoli F, Frank L, Carlsson G. 2012. A topological paradigm for hippocampal spatial map formation using persistent homology. *PLoS Comput Biol* **8:** e1002581.
Dabaghian Y, Brandt VL, Frank LM. 2014. Reconceiving the hippocampal map as a topological template. *Elife* **3:** e03476.
Davidson TJ, Kloosterman F, Wilson MA. 2009. Hippocampal replay of extended experience. *Neuron* **63:** 497–507.
Davidson TJ, Anderson EB, Lerner TN, Ramakrishnan C, Mattias J, Grosenick LM, Kauvar IV, Frank LM, Deisseroth K. 2014. Subsecond cholinergic dynamics underlying hippocampal network state in freely-behaving rats. In *Society for Neuroscience Abstracts*, Program No. 566.18. 2014 Neuroscience Meeting Planner. Society for Neuroscience, Washington, DC. Online.
Diba K, Buzsáki G. 2007. Forward and reverse hippocampal place-cell sequences during ripples. *Nat Neurosci* **10:** 1241–1242.
Dupret D, O'Neill J, Pleydell-Bouverie B, Csicsvari J. 2010. The reorganization and reactivation of hippocampal maps predict spatial memory performance. *Nat Neurosci* **13:** 995–1002.
Eichenbaum H. 2000. Hippocampus: Mapping or memory? *Curr Biol* **10:** R785–R787.
Eichenbaum H. 2013. What H.M. taught us. *J Cogn Neurosci* **25:** 14–21.
Eichenbaum H, Cohen NJ. 2001. *From conditioning to conscious recollection: Memory systems of the brain*. Oxford University Press, New York.
Fanselow MS, Dong HW. 2010. Are the dorsal and ventral hippocampus functionally distinct structures? *Neuron* **65:** 7–19.
Fenton AA, Csizmadia G, Muller RU. 2000a. Conjoint control of hippocampal place cell firing by two visual stimuli. I. The effects of moving the stimuli on firing field positions. *J Gen Physiol* **116:** 191–209.
Fenton AA, Csizmadia G, Muller RU. 2000b. Conjoint control of hippocampal place cell firing by two visual stimuli. Ii. A vector-field theory that predicts modifications of the representation of the environment. *J Gen Physiol* **116:** 211–221.
Foster DJ, Wilson MA. 2006. Reverse replay of behavioural sequences in hippocampal place cells during the awake state. *Nature* **440:** 680–683.
Frank LM, Brown EN, Wilson MA. 2000. Trajectory encoding in the hippocampus and entorhinal cortex. *Neuron* **27:** 169–178.
Girardeau G, Benchenane K, Wiener SI, Buzsáki G, Zugaro MB. 2009. Selective suppression of hippocampal ripples impairs spatial memory. *Nat Neurosci* **12:** 1222–1223.
Jadhav SP, Kemere C, German PW, Frank LM. 2012. Awake hippocampal sharp-wave ripples support spatial memory. *Science* **336:** 1454–1458.
Ji D, Wilson MA. 2007. Coordinated memory replay in the visual cortex and hippocampus during sleep. *Nat Neurosci* **10:** 100–107.
Karlsson MP, Frank LM. 2008. Network dynamics underlying the formation of sparse, informative representations in the hippocampus. *J Neurosci* **28:** 14271–14281.
Karlsson MP, Frank LM. 2009. Awake replay of remote experiences in the hippocampus. *Nat Neurosci* **12:** 913–918.
Kemere C, Carr MF, Karlsson MP, Frank LM. 2013. Rapid and continuous modulation of hippocampal network state during exploration of new places. *PLoS ONE* **8:** e73114.
Kim SM, Frank LM. 2009. Hippocampal lesions impair rapid learning of a continuous spatial alternation task. *PLoS ONE* **4:** e5494.
Kim SM, Ganguli S, Frank LM. 2012. Spatial information outflow from the hippocampal circuit: Distributed spatial coding and phase precession in the subiculum. *J Neurosci* **32:** 11539–11558.
Knight R. 1996. Contribution of human hippocampal region to novelty detection. *Nature* **383:** 256–259.
Knight RT, Nakada T. 1998. Cortico-limbic circuits and novelty: A review of EEG and blood flow data. *Rev Neurosci* **9:** 57–70.
Kudrimoti HS, Barnes CA, McNaughton BL. 1999. Reactivation of hippocampal cell assemblies: Effects of behavioral state, experience, and EEG dynamics. *J Neurosci* **19:** 4090–4101.
Larkin MC, Lykken C, Tye LD, Wickelgren JG, Frank LM. 2014. Hippocampal output area CA1 broadcasts a generalized novelty signal during an object-place recognition task. *Hippocampus* **24:** 773–783.
Lee AK, Wilson MA. 2002. Memory of sequential experience in the hippocampus during slow wave sleep. *Neuron* **36:** 1183–1194.
Lisman JE, Otmakhova NA. 2001. Storage, recall, and novelty detection of sequences by the hippocampus: Elaborating on the SOCRATIC model to account for normal and aberrant effects of dopamine. *Hippocampus* **11:** 551–568.
Marr D. 1971. Simple memory: A theory for archicortex. *Philos Trans R Soc Lond B Biol Sci* **262:** 23–81.
McClelland JL, McNaughton BL, O'Reilly RC. 1995. Why there are complementary learning systems in the hippocampus and neocortex: Insights from the successes and failures of connectionist models of learning and memory. *Psychol Rev* **102:** 419–457.
Montgomery SM, Buzsáki G. 2007. γ Oscillations dynamically couple hippocampal CA3 and CA1 regions during memory task performance. *Proc Natl Acad Sci* **104:** 14495–14500.
Nadasdy Z, Hirase H, Czurkó A, Csicsvari J, Buzsáki G. 1999. Replay and time compression of recurring spike sequences in the hippocampus. *J Neurosci* **19:** 9497–9507.
Nakashiba T, Buhl DL, McHugh TJ, Tonegawa S. 2009. Hippocampal CA3 output is crucial for ripple-associated reactivation and consolidation of memory. *Neuron* **62:** 781–787.
Norimoto H, Mizunuma M, Ishikawa D, Matsuki N, Ikegaya Y. 2012. Muscarinic receptor activation disrupts hippocampal sharp wave-ripples. *Brain Res* **1461:** 1–9.
O'Keefe J, Burgess N. 1996. Geometric determinants of the place fields of hippocampal neurons. *Nature* **381:** 425–428.
O'Keefe J, Dostrovsky J. 1971. The hippocampus as a spatial map. Preliminary evidence from unit activity in the freely-moving rat. *Brain Res* **34:** 171–175.
O'Neill J, Senior TJ, Allen K, Huxter JR, Csicsvari J. 2008. Reactivation of experience-dependent cell assembly patterns in the hippocampus. *Nat Neurosci* **11:** 209–215.
O'Neill J, Pleydell-Bouverie B, Dupret D, Csicsvari J. 2010. Play it again: Reactivation of waking experience and memory. *Trends Neurosci* **33:** 220–229.
O'Reilly RC, Norman KA. 2002. Hippocampal and neocortical contributions to memory: Advances in the complementary learning systems framework. *Trends Cogn Sci* **6:** 505–510.
Pavlides C, Winson J. 1989. Influences of hippocampal place cell firing in the awake state on the activity of these cells during subsequent sleep episodes. *J Neurosci* **9:** 2907–2918.
Pfeiffer BE, Foster DJ. 2013. Hippocampal place-cell sequences depict future paths to remembered goals. *Nature* **497:** 74–79.
Scoville WB, Milner B. 1957. Loss of recent memory after bilateral hippocampal lesions. *J Neurol Neurosurg Psychiatry* **20:** 11–21.
Singer AC, Frank LM. 2009. Rewarded outcomes enhance reactivation of experience in the hippocampus. *Neuron* **64:** 910–921.
Singer AC, Karlsson MP, Nathe AR, Carr MF, Frank LM. 2010. Experience-dependent development of coordinated hippocampal spatial activity representing the similarity of related locations. *J Neurosci* **30:** 11586–11604.
Singer AC, Carr MF, Karlsson MP, Frank LM. 2013. Hippocampal SWR activity predicts correct decisions during the initial learning of an alternation task. *Neuron* **77:** 1163–1173.
Squire LR, Bayley PJ. 2007. The neuroscience of remote memory. *Curr Opin Neurobiol* **17:** 185–196.
Sutherland GR, McNaughton B. 2000. Memory trace reactivation in hippocampal and neocortical neuronal ensembles. *Curr Opin Neurobiol* **10:** 180–186.
Vandecasteele M, Varga V, Berényi A, Papp E, Barthó P, Venance L, Freund TF, Buzsáki G. 2014. Optogenetic activation

of septal cholinergic neurons suppresses sharp wave ripples and enhances θ oscillations in the hippocampus. *Proc Natl Acad Sci* **111:** 13535–13540.
Wilson MA, McNaughton BL. 1994. Reactivation of hippocampal ensemble memories during sleep. *Science* **265:** 676–679.
Yang S, Yang S, Moreira T, Hoffman G, Carlson GC, Bender KJ, Alger BE, Tang CM. 2014. Interlamellar CA1 network in the hippocampus. *Proc Natl Acad Sci* **111:** 12919–12924.
Yu JY, Frank LM. 2014. Hippocampal-cortical interaction in decision making. *Neurobiol Learn Mem* **117:** 34–41.

Identification and Manipulation of Memory Engram Cells

Xu Liu,[1,2,3] Steve Ramirez,[1] Roger L. Redondo,[1,2] and Susumu Tonegawa[1,2]
[1]*RIKEN-MIT Center for Neural Circuit Genetics at the Picower Institute for Learning and Memory, Department of Biology and Department of Brain and Cognitive Sciences, Massachusetts Institute of Technology, Cambridge, Massachusetts 02139*
[2]*Howard Hughes Medical Institute, Massachusetts Institute of Technology, Cambridge, Massachusetts 02139*
Correspondence: tonegawa@mit.edu

How memories are formed and stored in the brain remains a fascinating question in neuroscience. Here we discuss the memory engram theory, our recent attempt to identify and manipulate memory engram cells in the brain with optogenetics, and how these methods are used to address questions such as how false memory is formed and how the valence of a memory can be changed in the brain.

How and where memory is stored in the brain network is one of the fundamental questions in brain and cognitive sciences. At the onset of the 20th century, a German biologist Richard Semon proposed the engram theory of memory (Semon 1923), but the theory was nearly completely ignored by his contemporary and subsequent brain researchers, until Daniel Schactor, James Eich, and Endel Tulving revived the theory in the late 1970s (Schacter et al. 1978). Semon's memory engram theory was built on two fundamental postulates termed the "Law of Engraphy" and the "Law of Ecphory" for memory storage and memory retrieval, respectively. The Law of Engraphy posits: "All simultaneous excitations (derived from experience) . . . with in our organisms form a connected simultaneous complex of excitations which, as such, acts engraphically, that is to say leaves behind it a connected, and to that extent, unified engram-complex" (Semon 1923). The Law of Ecphory on the other hand posits: "The partial return of an energetic situation which has fixed itself engraphically acts in an ecphoric sense upon a simultaneous engram-complex" (Semon 1923).

Semon's conceptualizations of the memory process were novel for his time and were remarkably predictive of the contemporary concepts of memory storage and retrieval. For instance, Semon's memory retrieval process contained the concept of "pattern completion," which was advanced years later (Marr 1970; Nakazawa et al. 2003; Leutgeb et al. 2004). However, Semon did not elaborate the biological basis of the "simultaneous excitations" nor "a connected, unified engram-complex." This is not surprising considering that his theory was put forward nearly a century before the development of molecular, cellular, and genetic biology and sophisticated imaging and electrophysiological technologies for the analysis of the nervous system.

Incorporating the current knowledge about neurons, synaptic connections, and neuronal circuits, Semon's Engram Theory of Memory can be rephrased as follows: When a subject undergoes or encounters an episode, a set of selected stimuli from the experience or episode activate populations of neurons to induce enduring physical and/or chemical changes (engrams) in them and their connections, each contributing to the storage of memory. Subsequently, when a part of the original stimuli returns, these cells (engram cells) are reactivated to evoke the recall of the specific memory.

A half-century after Semon's book was published, Karl Lashley pioneered a systemic hunt for engram cells in the rodent brain by introducing lesions of varying sizes into different areas of the cerebral cortex, attempting to find an engram for a maze task. However, Lashley found that memory was impaired in many of these lesioned animals, and the severity of the impairments was proportional to the sizes of the lesions. On the basis of these findings, Lashley concluded that the engrams for maze-resolving memory are spread throughout the cerebral cortex with no obvious localization (Mass Action Principle) (Lashley 1950). However, soon after Lashley's study, Wilder Penfield and Theodor Rasmussen obtained the first evidence suggesting that the engrams of episodic memories are stored in the medial temporal lobes (MTLs) (Penfield and Rasmussen 1950). This chance finding was supported several years later by William Scoville and Brenda Milner, who discovered that a patient H.M., who lost a large portion of his MTLs as a result of surgery, had severe anterograde amnesia for episodes as well as a graded retrograde amnesia. These studies were consistent with the notion that episodic memories are stored in the MTLs.

As to the nature of memory engrams—enduring physical and chemical changes induced by learning—the guiding hypothesis has been Donald Hebb's theory, which posits that neurons encoding memory stimuli undergo enduring strengthening of some of their synapses through their coactivation with presynaptic cells: neurons that "fire together wire together" (Hebb 1949). Starting with Tim Bliss and Terje Lomo's discovery of long-term

[3]Present address: Department of Neurobiology, Northwestern University, Evanston, Illinois 60208.

 doi: 10.1101/sqb.2014.79.024901

potentiation (Bliss and Lomo 1973), which supports Hebb's hypothesis, a large number of studies have been directed to the characterization of LTP and other facets of synaptic plasticity and their potential role in learning and memory. However, none of these studies could link these activity-dependent alterations of synapses and neurons directly to engram cells, which are activated by specific learning and whose reactivation by the specific recall cues elicited behavioral responses.

In this review, we shall outline our recent attempt to identify memory engram cells and to manipulate them by optogenetics to investigate several thus far unresolved issues associated with episodic memory.

OPTOGENETIC ACTIVATION OF MEMORY ENGRAM CELLS

Previous studies have linked selected neuronal populations with particular memory events by correlational evidence (Reijmers et al. 2007) and loss-of-function evidence (Han et al. 2009; Zhou et al. 2009), but a critical piece of evidence was largely missing. The most direct evidence of engram cells should come from gain-of-function experiments, where a population of neurons that are considered responsible for a particular memory are selectively labeled and activated artificially to mimic their natural activity. If such manipulation causes the recall of that memory, then this provides evidence that the selected population of neurons is sufficient for the memory, and thus argues the selected neuronal population is the neuronal basis for the engram of this particular memory (Martin and Morris 2002). However, this type of gain-of-function experiment is technically challenging, as one has to be able to correctly isolate the neurons involved in one particular memory from their seemingly indistinguishable neighbors and activate them with proper spatial and temporal precision. Recent advances in technology such as optogenetics (Fenno et al. 2011; Goshen 2014) enabled such experiments.

To achieve this goal, we combined activity-dependent, drug-regulatable expression system with optogenetics (Liu et al. 2012). We used a transgenic mouse model where the artificial tetracycline transactivator (tTA), which can be blocked by doxycycline (Dox), is driven by the promoter of immediate early gene (IEG) *c-fos* (Reijmers et al. 2007). The activity dependency of *c-fos* promoter poses a natural spatial constrain on the identities of the neurons that can be labeled, reflecting the normal biological selection process of the brain during memory formation, whereas the Dox-dependency of the system poses an artificial temporal constrain as to when these neurons can be labeled, which can be controlled by the experimenters. With these two constraints, the downstream effector of tTA can express selectively in neurons that are active during a particular behavior episode, only if the animals are off Dox diet. Using this system, we expressed channelrhodopsin-2 (ChR2) delivered by a viral vector AAV-TRE-ChR2-EYFP targeting the dentate gyrus (DG) of the hippocampus and implanted optical fibers right above the infected areas (Fig. 1A). These animals were habituated in one context A with light stimulation while on Dox, and then taken off Dox and fear conditioned in context B, where DG neurons active during the formation of this context-fear association memory were labeled by ChR2 (Fig. 1B). After which they were put back on Dox diet to stop further labeling and tested again in context A by light stimulation of the labeled neurons (Fig. 1C).

Although light had no effect on the test subjects in context A before training, these animals showed reversible, light-dependent freezing in context A after training (Fig. 1D), indicating light-induced recall of fear memory associated with fear conditioning, which happened in context B. Because these animals did not freeze in con-

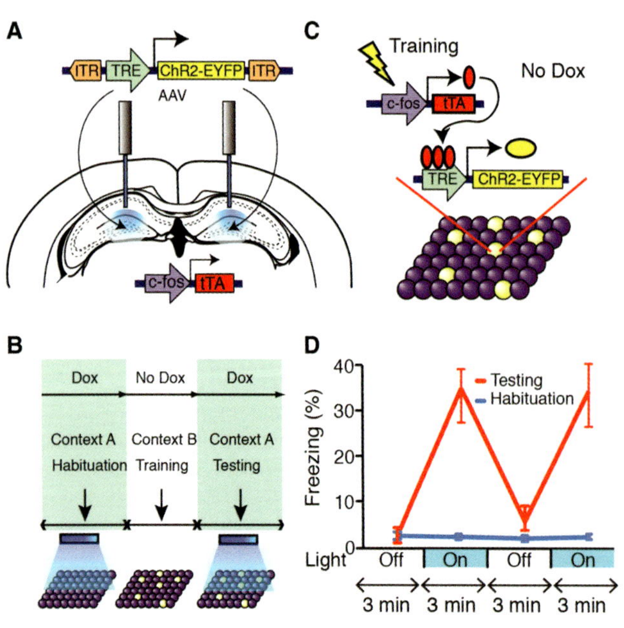

Figure 1. Optogenetic activation of memory engram cells induced memory recall. (*A*) Basic composition of the system. Virus expressing TRE-ChR2 and optic fibers are targeted bilaterally into the dentate gyrus (DG) of transgenic mouse line expressing c-fos-tTA. (*B*) In the absence of Dox, DG neurons that are active during the formation of a memory are labeled with ChR2. (*C*) Behavior schedules. Animals were habituated to context A with light stimulation while on Dox, trained in context B while off Dox, and texted again in context A with light stimulation while on Dox. (*D*) Although light had no effect during pretraining habituation sessions, the animals showed light-dependent freezing behavior post-training, indicating the light-induced recall of a fear memory.

text A in the absence of light posttraining, this ruled out the possibility that the freezing was due to generalization between context A and B. Control animals underwent similar treatments except did not receive footshock in context B did not show light-induced freezing after training, although a similar number of DG neurons were labeled after exposure to context B with or without shock (Liu et al. 2012). This indicated that simply activating a population of DG neurons not associated with a contextual fear memory by light stimulation was not the cause of freezing. Another control group underwent the same treatment as the experimental group including the footshock in B, but expressed only EYFP instead of ChR2 and also failed to show light-induced freezing after training, indicating that potential sensitization of light after fear conditioning also could not account for the behavior. These two controls reflected the requirement of both the presence of a fear memory and light activation of neurons associated with that memory for the observed light-induced freezing, and thus supported the idea that in the experimental group, activation of neurons associated with a previous fear memory by light indeed caused the ectopic recall of that memory in an otherwise neutral context (Liu et al. 2012). Taken together, these experiments provided the gain-of-function evidence for the memory engram cells, suggesting these cells are sufficient for the memory.

The artificial recall of the memory was also faithful to some extent, as it showed contextual specificity. Experiments have shown that two statistically independent (Liu et al. 2012) or even distinct (Deng et al. 2013) populations of DG neurons are active in two different contexts. We examined the cross talk between neuronal ensembles representing different contexts by testing if artificial activation of neurons representing one context could induce the recall of memory associated with another context. If the neurons active in a neutral context A were labeled while the animals were off Dox, followed by fear conditioned in a different context B while on Dox, the animals did not display freezing behavior upon light stimulation in a third context (Liu et al. 2012). This results showed that light activation of a neuronal population associated with a neutral context will not elicit the recall of another fear-conditioned context, as long as the neuronal representation of these two contexts are orthogonal, thus supporting the context specificity of the light-induced fear memory recall. This observation is also consistent with the proposed patter separation function of the DG (Leutgeb et al. 2004). This study showed that the behavior expression of a memory could be controlled from the neuronal ensemble level and opened possibilities for further memory manipulations using the engram cell method, as we will discuss below.

GENERATING FALSE MEMORIES BASED ON MEMORY ENGRAM CELLS

The experimental realization of reactivating discrete memories in the rodent brain (Liu et al. 2012) enabled subsequent studies to test long-standing hypotheses about the malleability of memory with unprecedented spatial–temporal resolution (Ramirez et al. 2013). These notions began in the early 1930s when the British psychologist Frederic Bartlett recited slightly inconsistent fables to people from several cultural backgrounds, most famously The War of the Ghosts (Bartlett 1932). While recalling the fable, many subjects unintentionally modified the contents of the story into a logical narrative that contained new elements that fit within their cultural milieu. Bartlett discovered that streams of recalled contextual information could act as a modifiable scaffold onto which information is added or distorted. Indeed, memories are not immutable video records of the experienced past that are projected onto a mental theater; they are mnemonic rivers that ebb and flow and thereby reconstruct the neuronal riverbeds that structurally support various streams of information. Bartlett had discovered the labile nature of memory.

Since Bartlett, the process of memory "updating" has been experimentally shown in both humans and rodents. Rats given electroconvulsive shocks shortly after recalling an aversive memory subsequently display profound amnesia for the original event (Misanin et al. 1968). This process of memory updating, later termed "reconsolidation," was rediscovered in 2000 and shown to be dependent on protein synthesis in the lateral amygdala (Nader et al. 2000) or hippocampus depending on the type of memory recalled (Debiec et al. 2002). Examples abound in humans that highlight the dramatic instances in which distorted memories of crime scenes, childhood events, and traumatic experiences—often recalled under interrogation in the court of law or during psychotherapy sessions—disrupt both individual well-being and modern jurisprudence (Loftus 2003; Schacter and Loftus 2013).

Of course, although the rich repertoire of human false memories is difficult to fully model in animals, a starting point is to take a Pavlovian approach and deconstruct the learning process into conditioned stimuli (CSs) and unconditioned stimuli (USs) associations. A series of recent studies have successfully showed the proof of principle of artificially linking CSs and USs to form novel associative memories. For instance, Johansen et al. (2010) showed that optically activated lateral amygdala (LA) cells were sufficient to substitute as a US during tone (CS) presentations and, upon subsequent tone presentations, animals displayed fear behavior despite the CS and US having never been naturally, or exogenously, presented. Another study showed that an activated population of pyriform cortex neurons, when paired with rewards or shocks, could drive the associated appetitive or aversive behavioral output upon stimulation of the same neurons (Choi et al. 2011). Moreover, pairing footshocks with optogenetically reactivated secondary auditory cortex and medial geniculate nucleus (MGN) inputs to the LA was also sufficient to form an associative fear memory to the optically activated terminals (Kwon et al. 2014). A more recent study elegantly showed that optically inducing long-term potentiation (LTP) or long-term depression (LTD) from MGN terminals into the LA was sufficient to promote or inhibit a previously formed memory, thus engineering the inactivation and reactivation of a specific memory and causally linking its expression to these types of synaptic plasticity (Nabavi et al. 2014).

Finally, a context-specific artificial memory was recently achieved using the "engram-labeling" technology (Ramirez et al. 2013). Animals were first taken off Dox to open a window for activity-dependent labeling and exposed to a neutral environment (context A). Thus, DG cells active during the exploration of context A became ChR2-positive (Fig. 2A). While back on Dox—and thus the window for activity-dependent labeling was closed—animals were then exposed to context C. The following day, fear conditioning occurred in context B with or without light stimulation of DG cells. In this experiment, even though the animals had a memory for both context A and C, only the neuronal populations activated by the former was ChR2-positive and presumably reactivated during fear conditioning in context B. We hypothesized that, if DG cells previously active during context A exploration defined an active neural population sufficient for memory recall, then optogenetic activation of these cells during fear conditioning in context B should form an artificial association between the light-induced context A memory and aversive information (e.g., footshocks). To test this possibility, animals were first placed back in context C (C′) and showed low basal levels of freezing (Fig. 2A). However, when placed back in context A (A′), only the animals in which DG cells were reactivated displayed robust freezing behavior, indicating that a false memory for context A had been artificially formed. Moreover, when placed in another neutral environment (context D), control animals showed basal levels of freezing during light-off and light-on epochs, whereas the experimental group in which an artificial memory had been created displayed robust freezing behavior only during light-on epochs (Fig. 2B). In other words, the expression of the false memory was behaviorally similar to the light-induced expression of a natural fear memory as previously reported (Liu et al. 2012).

Together, these data suggest that optical reactivation of hippocampal dentate gyrus cells that were previously active during context exploration is sufficient to act as an artificial, context-specific CS during fear conditioning. This thereby forms an artificial CS–US association, or a putative false memory, because the artificially constructed memory never had its contiguous experiences naturally linked (Ramirez et al. 2013). These results are consistent with the temporal context model (TCM) in humans, which posits that contextual memory reactivation can be linked to novel information that is presented at the time of reactivation (Gershman et al. 2013; St Jacques and Schacter 2013). Notably, the formation of false memories in humans often occurs as a result of recombining mnemonic elements of discrete experiences into a new, reconstructed memory that is not a veridical representation of the past. These memories are not de novo and require preexisting memories as a scaffold onto which distinct experiences can be incorporated to update the memory itself (Tse et al. 2007; Gershman et al. 2013). Similarly, in all rodent studies to date, any artificial memories generated were not de novo constructions; rather, they are results of artificially linking either a preexisting memory or concurrent learning processes with events of high valence.

CHANGING MEMORY VALENCE THROUGH MEMORY ENGRAM CELLS

After demonstrating the possibility to assign negative valence to an original neutral DG engram (Ramirez et al.

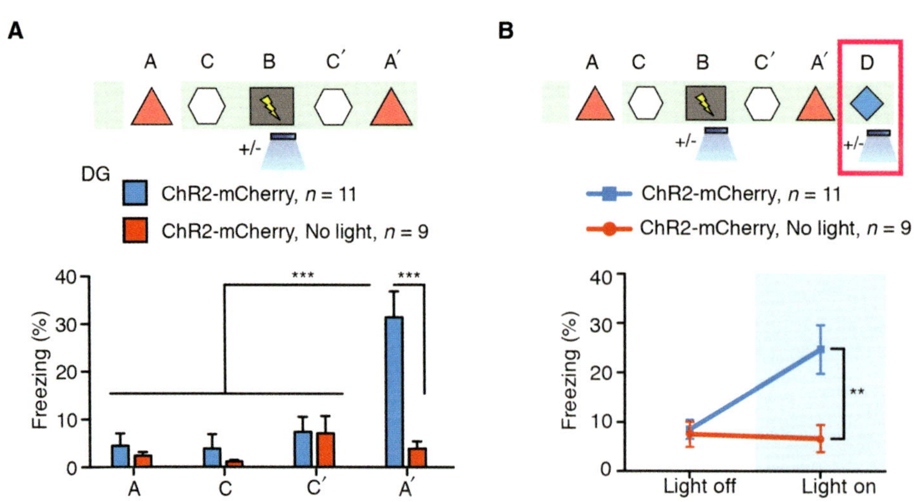

Figure 2. Creation of a false memory. (*A*) (*Top*) c-fos-tTA mice injected with AAV_9-TRE-ChR2-mCherry in the DG were taken off Dox and exposed to context A to label correspondingly activate DG cells with ChR2-mCherry and then put back on Dox and exposed to context C the following day. A control group injected with AAV_9-TRE-ChR2-mCherry did not receive light stimulation during fear conditioning (ChR2-mCherry, no light). (*Bottom*) animals' freezing levels in context A and C before and after fear conditioning with or without light revealed that a false memory had been formed only in the group in which DG stimulation occurred during fear conditioning in context B; (***$P < 0.001$). (*B*) Animals that underwent the behavioral protocol shown in the *top* panel were exposed to a novel context D, and the freezing levels were examined both in the absence and presence of light stimulation (**$P = 0.007$).

2013), we set out to identify the nodes of the circuit capable of associating valence to contextual representations. If the DG processes contextual information, the main hub for emotional learning in the brain is the amygdala, a group of nuclei deep in the temporal lobe that when lesioned impairs emotion but spares memory in tasks without emotional components (Zola-Morgan et al. 1991; Anderson and Phelps 2001). What are the contributions of the DG and the amygdala to the memory engram? In Redondo et al.'s (2014) study, neuronal ensembles that up-regulated c-fos at the time of memory encoding were manipulated in the DG or in the basolateral amygdala (BLA). Active avoidance responses were detected when mice reduced the time spent in a side of a maze where laser stimulation was delivered (OptoPA test). Active appetitive responses were identified by an increase in the time spent inside the zone targeted with laser (OptoPP test). First, we showed that both DG and BLA engrams are capable of driving an aversive response (day 5; Fig. 3A,B). Then, after pairing the optogenetic reactivation of the DG engram while mice experience a positive emotion (i.e., female contact, day 7 in Fig. 3A), the circuit changed so that subsequent activation of the DG ensemble drove an appetitive behavioral response on day 9 (Fig. 3A,B) and no longer produced an aversive response (Fig. 3C). After the induction protocol on day 7, the BLA engram failed to produce an appetitive response on day 9 (Fig. 3B) and retained the ability to drive an aversive response upon its activation (Fig. 3C). A symmetrical result was obtained when DG ensembles originally linked to an appetitive memory were optogenetically reactivated during fear learning: The output of the DG engram then drove an aversive response (Redondo et al. 2014).

To understand how the functional connectivity between the DG and the BLA had changed, the DG and the BLA engram were simultaneously labeled in a group of mice during memory encoding (Fig. 3D). On day 3, mice were divided into three groups: one experienced a US of opposite valence to that encoded on day 1 without laser stimulation (US^+, $Light^-$); another received an induction protocol pairing the presentation of a US of op-

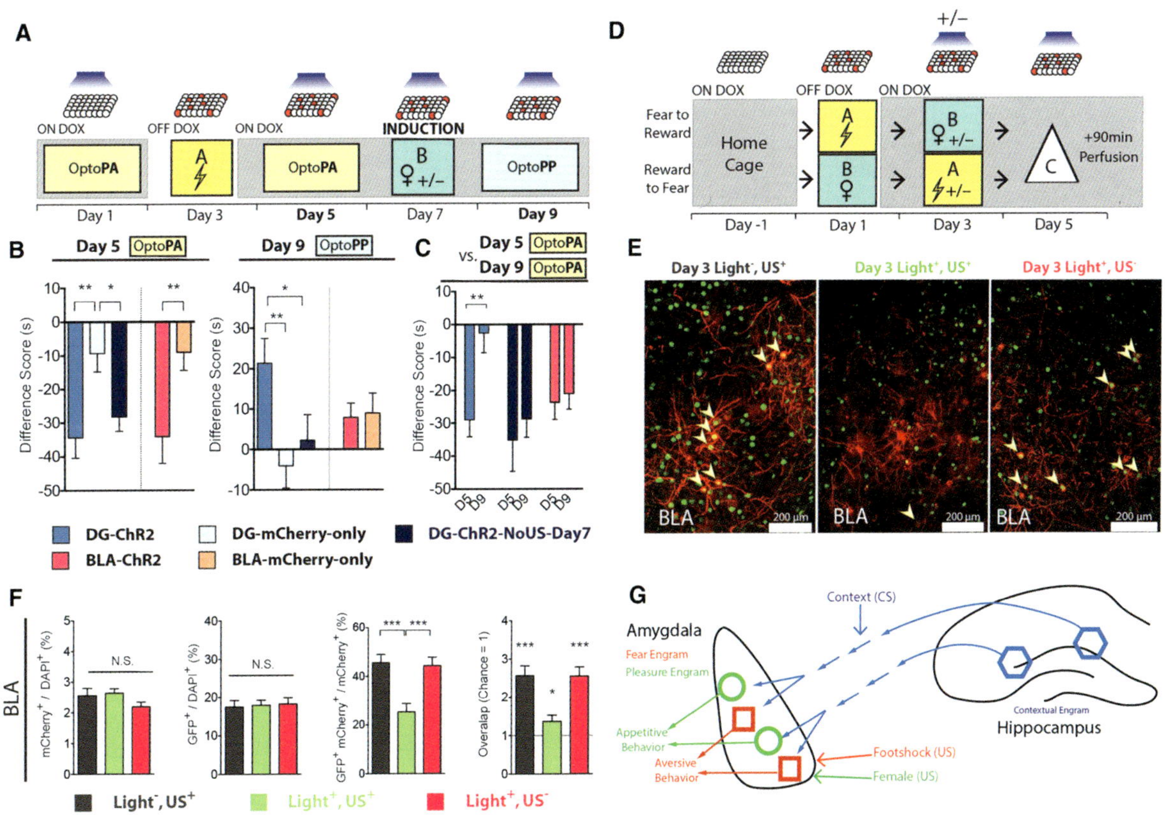

Figure 3. Reversal of the valence associated with the hippocampal memory engram. (*A*) Daily protocol with labeling of fear memory engram (day 3) and later its reactivation during a pleasurable experience (day 7). (*B*) Only in the DG engram but not in the BLA engram, the aversive response driven on Day 5 (reduction in time spent in the area where the memory engram is reactivated) is reversed into an appetitive response on day 9. (*C*) Only in the DG engram but not in the BLA engram, the aversive response on day 5 is eliminated on day 9. (*D*) Daily protocol depicting the labeling of an engram (day 1) followed by induction (day 3) and ending in the reactivation of the DG engram and brain extraction (day 5). (*E*) Representative images of the BLA showing engram neurons labeled on day 1 (red) and neurons reactivated on day 5 (green). All groups show high overlap except the group that underwent the valence reversal protocol. (*F*) Bar graphs quantifying similar levels of labeling and activation in the BLA across groups. However, only the group that underwent induction shows a reduced reactivation score that approximates to chance levels. (*G*) Summary graph depicting the potential of contextual information from the hippocampus to associate with amygdala neurons specialized in driving appetitive or aversive responses.

posite value to that encoded with optogenetic stimulation (Light$^+$, US$^+$), and the last group received optogenetic stimulation without US delivery (Light$^+$, US$^-$). On day 5 all animals had the DG ensembles reactivated optogenetically and 90 min later the brains were collected and processed for immunohistochemistry. Green fluorescence revealed those neurons that had recently up-regulated c-fos, whereas red fluorescence identified the neurons expressing ChR2-mCherry labeled on day 1 (Fig. 3E). The laser stimulation delivered to the DG was equally effective across groups at reactivating the DG engram (data not shown). Also in the BLA, the three experimental groups had the same amount of cells labeled (mCherry$^+$/DAPI$^+$) and the same levels of c-fos activation (GFP$^+$/DAPI$^+$) (Fig. 3F). However, although the control groups (US$^+$, Light$^-$ and Light$^+$, US$^-$) showed high levels of reactivation, the animals that experienced the induction protocol on day 3 had a decreased level of reactivation very close to the levels expected by chance (Fig. 3E,F). Two main conclusions can be drawn from these data: First, activating the DG ensemble reactivates the BLA neuronal population colabeled during memory encoding; and second, the induction protocol changes the connectivity between the DG and the BLA in such a way that even though the DG drives a similar number of neurons in the BLA, the identity of the activated BLA population has changed and no longer corresponds to those neurons that were active during memory encoding. That is, after reversing the valence associated with a DG ensemble, the output from the DG drove a different population of amygdala neurons.

The emerging picture of the circuit for memory valence depicts a series of neutral components of the engram (DG) free to associate with either positive or negative valences coded by nodes downstream in the circuit (BLA) (Fig. 3G). The development of new technologies capable of altering the connectivity between the nodes in the circuits of memory valence opens up the possibility of circumventing classical approaches to the treatment of emotional psychopathologies (i.e., post-traumatic stress disorder [PTSD], depression).

CONCLUSION

By combining activity-dependent gene expression system and optogenetics, we have established a system where we can identify and manipulate neurons that are active during the formation of a memory. Using this system, several important discoveries were made related to the mechanisms of memory. First, activation of these cells induced the recall of the associated memory, indicating that these cells are sufficient for the memory (Liu et al. 2012). Together with other studies with observational and loss-of-function evidence (Reijmers et al. 2007; Han et al. 2009; Zhou et al. 2009), this gain-of-function experiment pinpointed these cells as the cellular basis of memory engram. Second, artificial activation of these cells made the associated memory labile and capable of incorporating new information to form a new memory (Ramirez et al. 2013). This could be a potential mechanism whereby false memories are formed. Third, pairing the activation of a memory of certain valence while experiencing an event of opposite valence can reverse the valence originally associated with the memory. This alteration of memory valence is possibility due to changes of functional connectivity between the hippocampus and the amygdala (Redondo et al. 2014).

What we discussed here are just some examples of memory-related questions that can be answered by this new engram cell–based technology. There are ample equally exciting yet unexplored topics waiting to be addressed. For example, what plasticity changes are happening inside these cells harboring memory engrams? What are other memory engram pathways inside the brain? Can we use this technology to tackle disease models for mental disorders, such as depression and anxiety? With the fast evolving new technology, we have reason to believe that the memory engram cell–related studies will continue to bring deeper insights and exciting new discoveries in the years to come.

ACKNOWLEDGMENTS

We thank P. Pang, C. Puryear, A. Govindarajan, P. Lin, J. Suh, M. Pignatelli, T.J. Ryan, J. Kim, and A. Arons for help with the experiments and all the members of the Tonegawa laboratory for their support. This work was supported by RIKEN Brain Science Institute and Howard Hughes Medical Institute to S.T.

REFERENCES

Anderson AK, Phelps EA. 2001. Lesions of the human amygdala impair enhanced perception of emotionally salient events. *Nature* **411:** 305–309.

Bartlett FC. 1932. *Remembering: A study in experimental and social psychology*. Cambridge University Press, Cambridge.

Bliss TV, Lomo T. 1973. Long-lasting potentiation of synaptic transmission in the dentate area of the anaesthetized rabbit following stimulation of the perforant path. *J Physiol* **232:** 331–356.

Choi GB, Stettler DD, Kallman BR, Bhaskar ST, Fleischmann A, Axel R. 2011. Driving opposing behaviors with ensembles of piriform neurons. *Cell* **146:** 1004–1015.

Debiec J, LeDoux JE, Nader K. 2002. Cellular and systems reconsolidation in the hippocampus. *Neuron* **36:** 527–538.

Deng W, Mayford M, Gage FH. 2013. Selection of distinct populations of dentate granule cells in response to inputs as a mechanism for pattern separation in mice. *Elife* **2:** e00312.

Fenno L, Yizhar O, Deisseroth K. 2011. The development and application of optogenetics. *Annu Rev Neurosci* **34:** 389–412.

Gershman SJ, Schapiro AC, Hupbach A, Norman KA. 2013. Neural context reinstatement predicts memory misattribution. *J Neurosci* **33:** 8590–8595.

Goshen I. 2014. The optogenetic revolution in memory research. *Trends Neurosci* **37:** 511–522.

Han JH, Kushner SA, Yiu AP, Hsiang HL, Buch T, Waisman A, Bontempi B, Neve RL, Frankland PW, Josselyn SA. 2009. Selective erasure of a fear memory. *Science* **323:** 1492–1496.

Hebb D. 1949. *The organization of behavior*. Wiley & Sons, New York.

Johansen JP, Hamanaka H, Monfils MH, Behnia R, Deisseroth K, Blair HT, LeDoux JE. 2010. Optical activation of lateral amygdala pyramidal cells instructs associative fear learning. *Proc Natl Acad Sci* **107:** 12692–12697.

Kwon JT, Nakajima R, Kim HS, Jeong Y, Augustine GJ, Han JH. 2014. Optogenetic activation of presynaptic inputs in lateral amygdala forms associative fear memory. *Learn Mem* **21:** 627–633.

Lashley K. 1950. In search of the engram. *Symp Soc Exp Biol* **4:** 454–482.

Leutgeb S, Leutgeb JK, Treves A, Moser MB, Moser EI. 2004. Distinct ensemble codes in hippocampal areas CA3 and CA1. *Science* **305:** 1295–1298.

Liu X, Ramirez S, Pang PT, Puryear CB, Govindarajan A, Deisseroth K, Tonegawa S. 2012. Optogenetic stimulation of a hippocampal engram activates fear memory recall. *Nature* **484:** 381–385.

Loftus E. 2003. Our changeable memories: Legal and practical implications. *Nat Rev Neurosci* **4:** 2–5.

Marr D. 1970. A theory for cerebral neocortex. *Proc R Soc Lond B Biol Sci* **176:** 161–234.

Martin SJ, Morris RG. 2002. New life in an old idea: The synaptic plasticity and memory hypothesis revisited. *Hippocampus* **12:** 609–636.

Misanin JR, Miller RR, Lewis DJ. 1968. Retrograde amnesia produced by electroconvulsive shock after reactivation of a consolidated memory trace. *Science* **160:** 554–555.

Nabavi S, Fox R, Proulx C, Lin J. 2014. Engineering a memory with LTD and LTP. *Nature* **511:** 348–352.

Nader K, Schafe GE, LeDoux JE. 2000. The labile nature of consolidation theory. *Nat Rev Neurosci* **1:** 216–219.

Nakazawa K, Sun LD, Quirk MC, Rondi-Reig L, Wilson MA, Tonegawa S. 2003. Hippocampal CA3 NMDA receptors are crucial for memory acquisition of one-time experience. *Neuron* **38:** 305–315.

Penfield W, Rasmussen T. 1950. *The cerebral cortex of man: A clinical study of localization of function.* Macmillan, New York.

Ramirez S, Liu X, Lin PA, Suh J, Pignatelli M, Redondo RL, Ryan TJ, Tonegawa S. 2013. Creating a false memory in the hippocampus. *Science* **341:** 387–391.

Redondo RL, Kim J, Arons AL, Ramirez S, Liu X, Tonegawa S. 2014. Bidirectional switch of the valence associated with a hippocampal contextual memory engram. *Nature* **513:** 426–430.

Reijmers LG, Perkins BL, Matsuo N, Mayford M. 2007. Localization of a stable neural correlate of associative memory. *Science* **317:** 1230–1233.

Schacter DL, Loftus EF. 2013. Memory and law: What can cognitive neuroscience contribute? *Nat Neurosci* **16:** 119–123.

Schacter DL, Eich JE, Tulvlng E. 1978. Richard Semon's theory of memory. *J Verbal Learn Verbal Behav* **17:** 721–743.

Semon R. 1923. *Mnemic philosophy.* Allen & Unwin, Australia.

St Jacques PL, Schacter DL. 2013. Modifying memory: Selectively enhancing and updating personal memories for a museum tour by reactivating them. *Psychol Sci* **24:** 537–543.

Tse D, Langston RF, Kakeyama M, Bethus I, Spooner PA, Wood ER, Witter MP, Morris RGM. 2007. Schemas and memory consolidation. *Science* **316:** 76–82.

Zhou Y, Won J, Karlsson MG, Zhou M, Rogerson T, Balaji J, Neve R, Poirazi P, Silva AJ. 2009. CREB regulates excitability and the allocation of memory to subsets of neurons in the amygdala. *Nat Neurosci* **12:** 1438–1443.

Zola-Morgan S, Squire LR, Alvarez-Royo P, Clower RP. 1991. Independence of memory functions and emotional behavior: Separate contributions of the hippocampal formation and the amygdala. *Hippocampus* **1:** 207–220.

A Dynamical Basis Set for Generating Reaches

Mark M. Churchland[1] and John P. Cunningham[2]

[1]*Department of Neuroscience, Grossman Center for the Statistics of Mind, David Mahoney Center for Brain and Behavior Research, Kavli Institute for Brain Science, Columbia University Medical Center, New York, New York 10032*

[2]*Department of Statistics, Grossman Center for the Statistics of Mind, Center for Theoretical Neuroscience, Institute for Data Science and Engineering, Columbia University, New York, New York 10027*

Correspondence: mc3502@columbia.edu

The motor cortex was the one of the first cortical areas to be explored electrophysiologically, yet little agreement has emerged regarding its basic response properties. Often it is assumed that single-neuron responses reflect a preference for a particular movement or movement variable. It may be further assumed that movement is generated by (or at least accompanied by) a growing population-level preference for the relevant movement. This view has been attractive because it provides a canonical form for the single neuron, a link between preparatory and movement activity, a way of interpreting the population response, and a platform for designing analyses and couching hypotheses. However, this traditional view yields predictions that are at odds with basic features of the data. We discuss an alternative simplified model, in which outgoing commands are produced by dynamics that generate different output patterns as a function of the initial preparatory state. For reaching tasks, we hypothesized simple quasioscillatory dynamics because they provide a natural basis set for the empirical patterns of muscle activity. The predictions of the dynamical model match the data well at both the single-neuron and population levels, and the quasioscillatory patterns explain many of the otherwise odd features of the neural responses.

Since the pioneering work of Evarts (1968), a central goal of motor physiology has been to understand the relationship between neural activity in motor cortex and movement itself. Much of the subsequent literature has leveraged tasks in which monkeys execute reaches (or directed wrist movements) to targets at different spatial locations (Georgopoulos et al. 1986; Moran and Schwartz 1999; Kakei et al. 2001). In many versions of this task (Fig. 1) a target is presented but it is required that movement execution be delayed until after a go cue (Weinrich et al. 1984; Crammond and Kalaska 2000). The resulting "preparatory period" (between target onset and the go cue) is employed because it temporally separates neural activity related to movement preparation from neural activity related to movement itself. The key question is then: How does neural activity unfold during the preparatory and movement periods? Does activity before the go cue prepare the system for the production of movement activity? If so, how? What patterns of activity are present during movement, and what purpose do they serve?

It is widely accepted that neural activity in motor cortex is complex and heterogeneous—across time, across neurons, and across different reach types. A central goal of motor physiology is then to create simple models that capture basic response features and offer explanations for why those features exist. Consider the neural response before and during the two reaches illustrated in Figure 2A. Many models and/or analyses have assumed the framework illustrated in Figure 2B,D. The schematic single-neuron response (Fig. 2B) gives a larger response for the reach down and to the right (*red*); that neuron is thus said to "prefer" that direction. This preference becomes apparent during preparation and grows stronger during movement. Neural responses consistent with this view are often observed in population averages (Requin et al. 1988; Georgopoulos et al. 1989a; Riehle and Requin 1989; Bastian et al. 1998, 2003; Erlhagen et al. 1999; Cisek 2006a). Accordingly, it is often posited that movement onset is triggered when response strength passes a threshold or reaches a peak.

The single-neuron view in Figure 2B is readily extended to the population level (Georgopoulos et al. 1986, 1989b) by considering the response in neural "state space" (Fig. 2D). Each axis captures a "factor": a weighted sum of the neural responses. In this case x_1 might be the average activity of rightward-preferring neurons minus the average activity of leftward-preferring neurons. Accordingly, x_2 might be the average activity of upward-preferring neurons minus the average activity of downward-preferring neurons. A reach in a given direction would be generated when the "neural state" expands away from baseline and crosses a threshold. This hypothesis possesses some appealing features. First, it resembles the mechanism by which saccades are prepared and generated (e.g., Hanes and Schall 1996). Second, directions in neural state space correspond to directions in physical space. This provides a concrete interpretation of the neural state as a representation of external factors such as reach direction or velocity. For example, the neural state might be modeled as

$$\mathbf{x}(t, c) = g(\mathbf{v}(t, c)), \tag{1}$$

doi: 10.1101/sqb.2014.79.024703

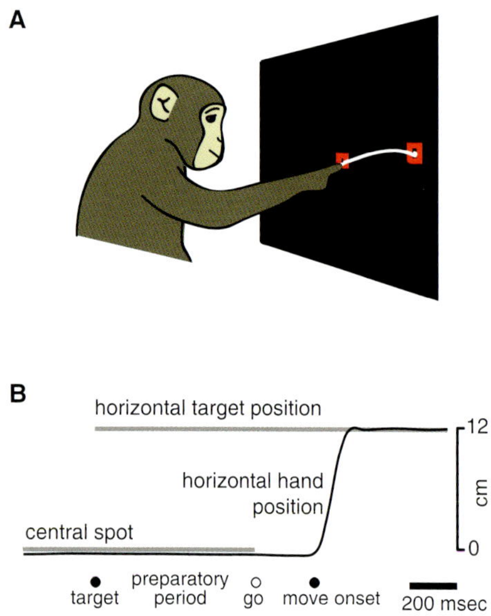

Figure 1. Illustration of behavior for a reaching task, of the type often used to study motor cortex. (*A*) Reaches were from a central spot to a target. An example trajectory is shown. (*B*) Task timeline. Target onset was separated from a go cue by a preparatory period. Movement onset occurred shortly (~300 msec) after the go cue. Reaches were typically very brisk; in this case lasting <200 msec.

where $\mathbf{v}(t,c)$ is a vector of hand velocity for time t and condition c. The function $g(\,)$ captures "tuning" of neural activity for reach velocity and may be linear or nonlinear. This traditional view remains prevalent in motor physiology (e.g., Pearce and Moran 2012) and virtually all motor prosthetic applications use models that contain relationships similar to that in Equation 1 (e.g., Gilja et al. 2012).

We recently proposed (Churchland et al. 2012; Shenoy et al. 2013) an alternative "dynamical" model. The dynamical model extends the view that motor cortex produces a muscle-like output (Evarts 1968; Cheney and Fetz 1980; Lemon et al. 1986; Todorov 2000; Sergio et al. 2005; Pohlmeyer et al. 2007; Schieber and Rivlis 2007; Lillicrap and Scott 2013), yet carries important additional implications. In this dynamical model the dominant motif during reaching is a rotation of the neural state (Fig. 2E) rather than an expansion (Fig. 2D). It is still proposed, as in the traditional model, that the preparatory response involves an expansion away from baseline. However, this preparatory state then "seeds" subsequent oscillatory dynamics, producing a rotation of the neural state that lasts ~1–1.5 cycles (Churchland et al. 2010). Rotations of the neural state imply quasioscillatory single-neuron response patterns (Fig. 2C), but the converse is not true—the central prediction is more specific than multiphasic single-neuron responses. It is hypothesized that the population neural state evolves according to dynamics of the form

$$\dot{\mathbf{x}}(t,c) = f(\mathbf{x}(t,c)), \qquad (2)$$

where $f(\,)$ shows a large rotational component. A critical feature of this model is that the dynamics $f(\,)$ are the same across all conditions. Thus, each condition is a different neural trajectory in the same dynamical flow field. Although many models imply multiphasic responses at the level of single neurons, many fewer predict population-level rotational dynamics that are consistent across conditions. Below we illustrate why such a model might make sense, given the empirical patterns of muscle activity.

MODEL AND NEURAL RESPONSE PATTERNS

A Basis Set for Muscle Activity? Motivating a Dynamical Hypothesis

The basic features of the traditional model are fairly intuitive: The factors that describe the neural state space correspond to externally measurable parameters such as direction or velocity. In contrast, the appeal of the dynamical model is not immediately obvious; what could be the mechanistic purpose of a neural state governed by internal dynamics (Eq. 2)? A possible answer presents itself if we consider the patterns of activity that can be generated by the traditional and dynamical models. The traditional model generates pulses of activity, whereas the dynamical model generates brief quasioscillatory patterns. Both pulses and oscillatory patterns are potentially useful: Each could constitute a set of building blocks (a "basis set") for outgoing commands.

To investigate the possible utility of an oscillatory basis set, we simulated a dynamical model in which the state space was greater than two dimensional, allowing for more than one plane of rotation and thus more than one frequency. Such a system produces a small basis set of oscillatory patterns (Fig. 3A), and can do so with any phase and amplitude depending on the preparatory state. If the output is a (fixed) sum of the resulting patterns, one then has a mechanism for generating a range of output patterns.

We examined the patterns of muscle activity across different reaches (Fig. 3B) and found that they could indeed be reconstructed (Fig. 3C) from a basis set of just two underlying oscillations (four total dimensions plus an additional dimension to capture offset). Oscillation frequency and duration were fixed (and thus the dynamics $f(\,)$ in Eq. 2 were fixed); only phase and amplitude varied across conditions (implied by the initial state $\mathbf{x}(0,c)$). This simple construction was sufficient to accurately reproduce muscle-like commands (Fig. 3C vs. 3B).

Quasioscillatory patterns are thus an effective basis set, in agreement with a recent modeling study (Rokni and Sompolinsky 2012). There are two further reasons to suspect such a basis set might be used. First, a common feature of large recurrent networks—even randomly connected networks—is the presence of multiple oscillatory "modes." Network dynamics can, via training, be simplified to produce a small set of well-behaved output activities that are muscle-like (Schaffer et al. 2006; Sussillo et al. 2012). Second, oscillatory activity is common during movement in many species, not only during locomotion but sometimes during brief movements as well.

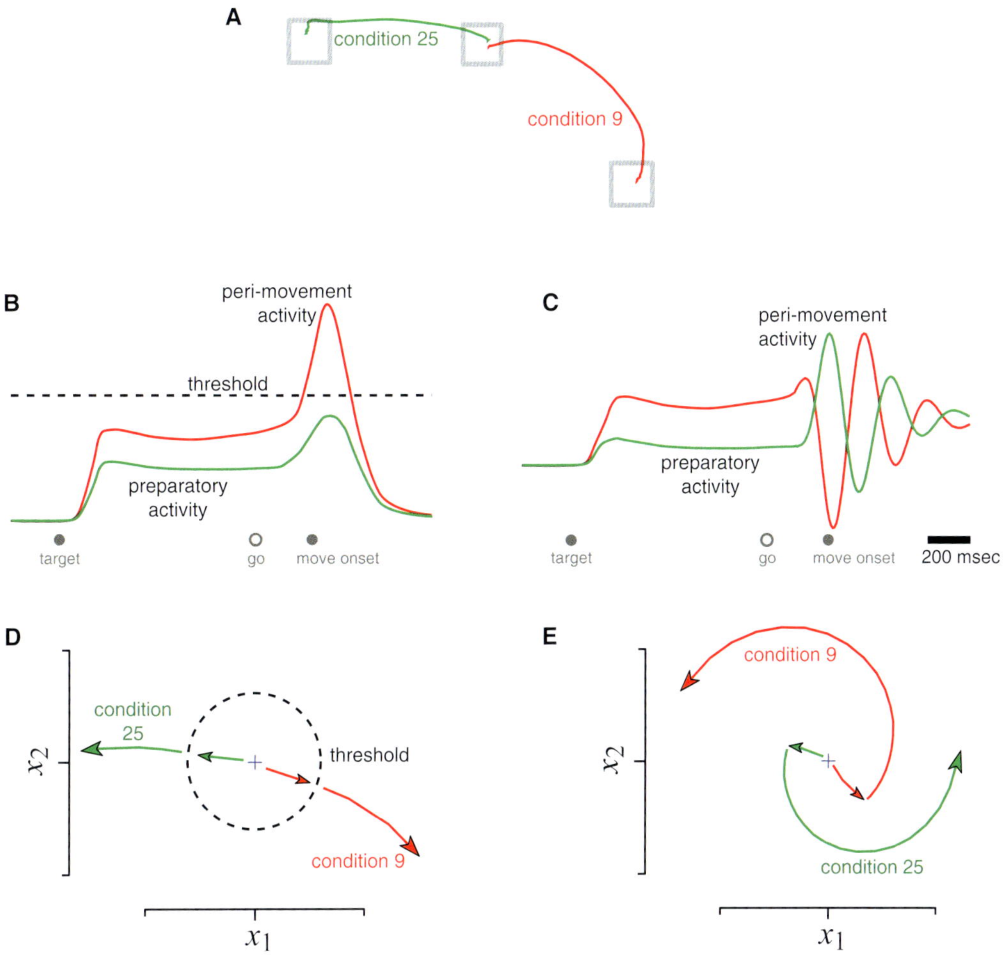

Figure 2. Two simplified models of neural activity during reach generation. (*A*) Example reach trajectories (2 of 27 total conditions). (*B*) Cartoon illustration of the traditional model showing the response of one model unit as a function of time for two conditions. The red trace shows the response for the "preferred condition" (condition 9, a roughly rightward reach). (*C*) Cartoon illustration of the dynamical model, showing the response of one model unit as a function of time for two conditions. (*D*) Illustration of the traditional model in state space. Each dimension (x_1 and x_2) captures one of the factors represented by the population. In this case x_1 and x_2 correspond to horizontal and vertical reach direction. Motion of the neural state during preparation is away from baseline, and this expansion continues during movement. This expansion corresponds, at the single-unit level, to the development of a preference during preparation that is heightened during movement. For example, a single unit whose response primarily reflects x_1 would show a rightward preference, as in *B*. (*E*) Illustration of the dynamical model in state space. The neural state moves away from baseline during preparation, as with the traditional model. During reaching, the neural state rotates with time, tracing a circular trajectory. To connect back to the single-unit level, note that a single unit whose response primarily reflects x_1 would show an oscillatory pattern with an ever-changing "preference," as in *C*.

For example, zebrafish larvae produce briefly oscillatory swimming bouts (Orger et al. 2008). Thus, rhythmic activity is a common motif across multiple motor systems that use very different neural architectures. Given the ease with which different neural systems (real and artificial) produce oscillatory patterns, and given the empirical effectiveness of an oscillatory basis set in building muscle activity, a seemingly straightforward hypothesis is that such a mechanism might be at play during reaching.

It should be stressed that a quasioscillatory basis set is not the only (or even necessarily the optimal) solution. The generation of pulsatile patterns—as in the traditional model—is also common across many motor systems and could provide a basis set. And although the above arguments may suggest a quasioscillatory set of basis patterns, by no means do they necessitate that solution. It was therefore an empirical question whether the data during reaching resembled the traditional model, the dynamical model, or neither.

Dynamical Model Responses Resemble Neural Response

We recorded neural activity from motor and premotor cortex during a variety of reaching tasks (Churchland et al. 2006; Kaufman et al. 2010). Neurons showed a broad range of response patterns (Fig. 4) and rarely resembled the predictions of the traditional model (Churchland and Shenoy 2007; Churchland et al. 2010, 2012). There was little evidence of a "preferred movement." Indeed, the

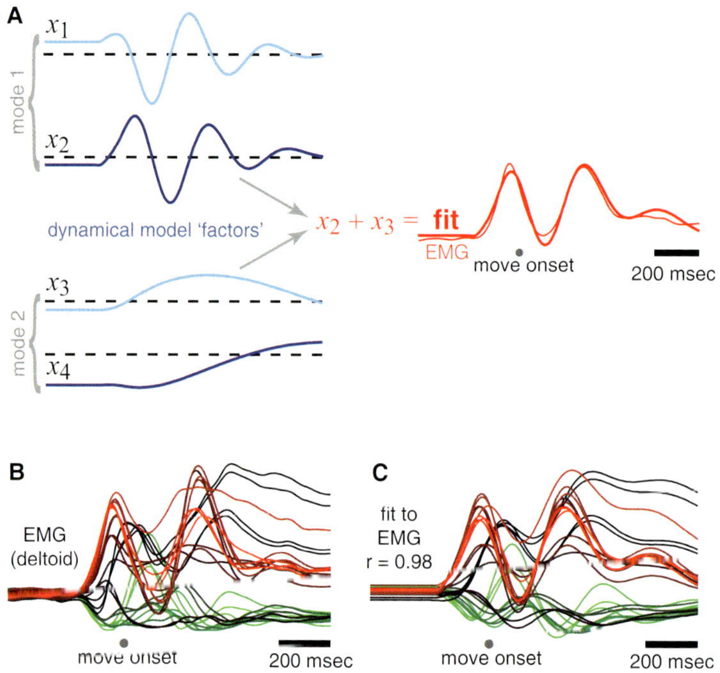

Figure 3. Constructing muscle activity from an oscillatory basis set. (*A*) The time-evolving factors proposed by the dynamical model. The first pair of factors (x_1 and x_2) capture the two phases of a 2.8-Hz oscillation. The model employs both phases because it is the most natural way for a dynamical system to generate an oscillatory response. The second pair of factors (x_3 and x_4) capture the two phases of a 0.3-Hz oscillation. For both frequencies the oscillations are short-lived; the 2.8-Hz oscillation lasts only a little over one cycle and the 0.3-Hz oscillation lasts less than one cycle. The sum of factors x_2 and x_3 produces a pattern (thick red trace) that approximates the empirical pattern of electromyography (EMG) recorded from the deltoid (thin red trace). Data is shown for one condition (9) of 27 total. Oscillation phase and amplitude were set by the "preparatory state." The preparatory state is reflected in the fact that the factors are offset from baseline (dashed line) before the onset of pattern generation. For each condition we chose the preparatory state to produce the phases and amplitudes that best fit the EMG. (*B*) EMG recorded from the deltoid for all 27 conditions. Each trace plots the intensity of the muscle activity for one condition. Red-to-green coloring of the traces is used simply to allow the eye to follow individual traces. (*C*) Fits to the EMG via the dynamical model, as in *A*, for all conditions. Correlation between model fit and recorded EMG was 0.98. (*B*,*C*, Adapted from Churchland et al. 2012.)

condition that garnered the larger response was typically different at different times during the trial (e.g., sometimes *red traces* are higher and sometimes *green traces* are higher). Thus, individual-neuron responses do not particularly resemble the predictions of the traditional model.

We then compared neural responses with responses predicted by the dynamical model. Simulated neural responses were based on the model in Figure 3A, which was optimized to generate the empirical electromyography (EMG). The response of each simulated unit was a randomly weighted sum of the patterns present in the model, plus a small amount of noise (Churchland et al. 2012). The level of preparatory activity was determined by the initial plateau (left side of Fig. 3A). This plateau of preparatory activity is critical to the model: It sets the phase and amplitude of the subsequent oscillations ($\mathbf{x}(0,c)$ in Eq. 2). The duration of simulated preparatory activity was set to roughly match that seen for the neural responses.

Dynamical model responses showed a variety of multiphasic patterns that resembled those of cortical neurons (Fig. 4). Preparatory activity was present, yet had an unclear relationship with the subsequent pattern of movement activity. For the model this occurs because the relationship between preparatory state and rotation phase becomes clear only at the level of the population neural state. In summary, the dynamical model, but not the traditional model, produces single-unit responses that resemble the data. Importantly the dynamical model units in Figure 4 are not fit to recorded neurons; the resemblance naturally arises when the model is fit to the empirical patterns of EMG (Fig. 3A). The next key question was thus: Does the similarity between the data and the dynamical model extend to the population level?

DYNAMICAL STRUCTURE IN THE POPULATION RESPONSE

Model Predictions

The dynamical model makes three key predictions at the population level. First, the population state should evolve according to dynamics described by Equation 2, which can be linearly approximated as

$$\dot{\mathbf{x}}(t,c) = M\mathbf{x}(t,c). \quad (3)$$

It is proposed that dynamics are very "smooth": They are similar across conditions and can thus be approximated with a single matrix M. The second prediction is that the dynamics in question should have a large oscillatory/

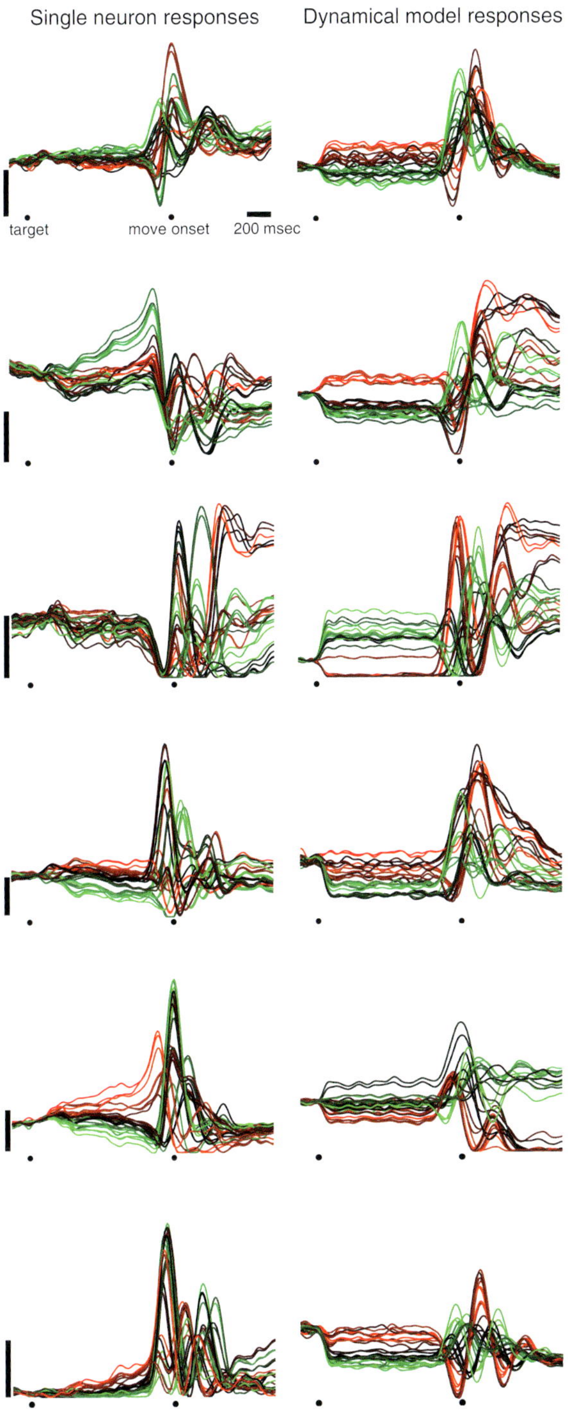

Figure 4. Example responses of real (*left* column) and simulated units from the dynamical model (*right* column). Each subpanel plots the response for one neuron/unit. Each trace plots the response for one reach condition. Vertical scale bars indicate 20 spikes/sec. Traces are colored red to green for visualization. So that the same color coding can be used for all panels, color-coding is based on the initial strength of deltoid EMG, just before movement onset. Note that although the real and simulated data share a number of interesting general features, the individual units/neurons are not meant to (and do not) map directly onto one another. The model responses were never fit to the neural responses—they are a consequence of how the model reproduced the empirical EMG.

rotational component: M should be fairly close to skew-symmetric ($M = -M^{\mathrm{T}}$). Thus, the fit to the data should be almost as good if M is constrained to be skew-symmetric. Third, it should be possible to directly view projections of the data that reveal the rotational structure illustrated in Figure 2E. If there exist factors x_1 and x_2 with the hypothesized structure, then it should be possible to recover those factors by taking an appropriately weighted sum of the activity of all the neurons. Doing so should reveal state-space patterns with clear rotations.

Before examining these predictions, we stress that the use of a weighted sum to yield interpretable "factors" is common in systems neuroscience (Cunningham and Yu 2014)—particularly when decoding represented parameters. For example, in the traditional model, one can recover a representation of reach direction by assigning each neuron a pair of weights based on its preferred direction. Thus, x_1 (x_2) would be the sum of the activities of all neurons, weighted according to the degree to which that neuron preferred rightward (upward) reaches. When examining the predictions of the dynamical model, we wished to accomplish something similar, but to recover projections that might relate to rotational dynamics.

We used principle component analysis as a preprocessing step and reduced the dimensionality of the data to a few factors (typically six). We then fit the temporal evolution of the data with Equation 3. Fit quality (R^2) was assessed using both an unconstrained matrix M, and a matrix M_{skew} that was constrained to be skew-symmetric. Finally, we recovered factors x_1 and x_2 via weights based on the eigenvectors of M_{skew} (the jPCA method; Churchland et al. 2012). We used the eigenvectors that corresponded to the highest-frequency rotations. As a technical aside, we used the eigenvectors of M_{skew} rather than M for three practical reasons. First, we wished to test for the presence of rotations in a subset of dimensions, even if other dimensions contained other patterns. Second, M_{skew} is "normal" and conveniently yields orthogonal eigenvectors. Third, in practice M and M_{skew} were typically very similar for the neural data.

Rotations Are a Dominant Motif in the Population Response

When applied to the population of dynamical model responses (e.g., Fig. 4) the jPCA method accurately recovered the true factors (compare Fig. 5A,B). As expected, the quality of the dynamical fit (R^2) was high and was nearly as good when constrained to be purely rotational (R^2_{skew}). For the neural data (Fig. 5C,D) jPCA revealed a similar rotational structure (Churchland et al. 2012). The quality of the dynamical fit was high (although certainly not as high as the idealized model) and was nearly as good when constrained to be purely rotational. Rotation phase and amplitude followed from the preparatory state for both the model and the data. Both model and data contained a second plane with lower-frequency rotations (not shown, although see Fig. 7B,C below). Thus, the three population-level predictions of the dynamical mod-

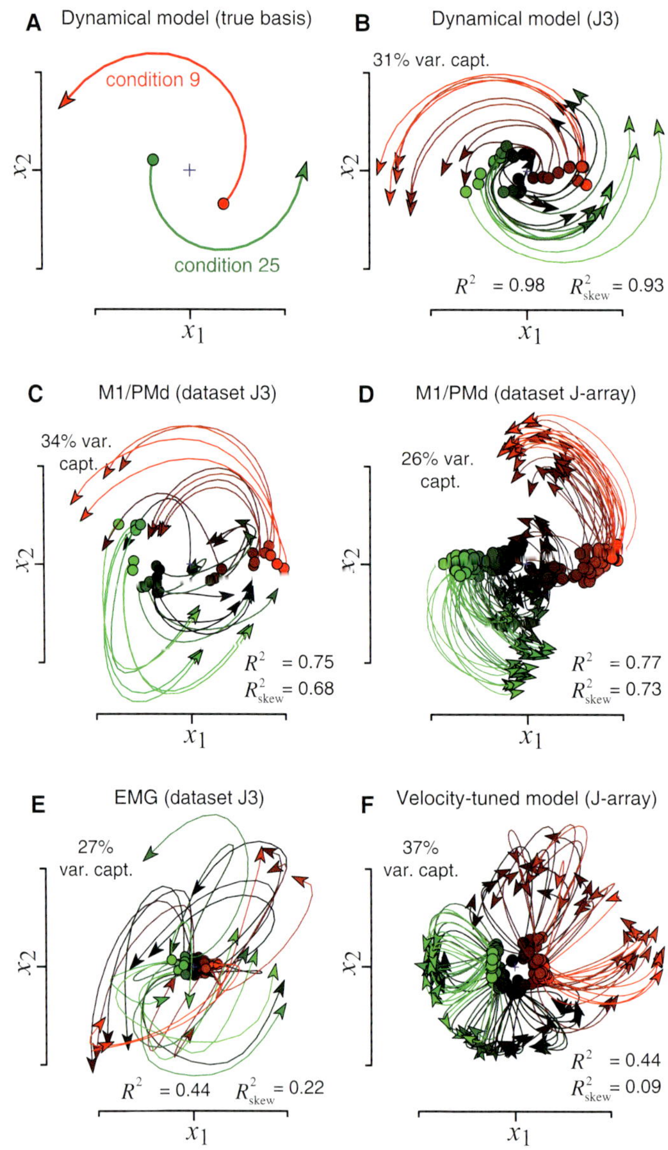

Figure 5. Population structure for real and simulated data. All plots show 200 msec of the evolution of the neural state, beginning just as the preparatory activity (dots) transitions to movement activity (subsequent trajectories). (*A*) The first two dimensions of the true basis set used by the dynamical model. For simplicity two conditions are shown. (*B*) Projection, via jPCA, of the simulated population response for the dynamical model. The model was fit to produce the EMG of the deltoid from data set J3, which was recorded across 27 conditions. For visualization, traces are colored based on the projection of the preparatory state onto x_1. The jPCA method accurately captures the underlying basis from *A*. Values of R^2 and R^2_{skew} indicate the fit quality provided by a linear model (unconstrained or skew-symmetric respectively) for the two dimensions shown. The variance captured (var. capt.) indicates the total data variance that lies within these two dimensions, as with PCA. (*C*) jPCA projection of the population response for data set J3 (27 conditions). (*D*) jPCA projection of the population response for data set J-array (108 conditions). (*E*) jPCA projection of the muscle population for data set J3. (*F*) jPCA projection of the simulated population response for a traditional model in which neural responses were tuned for end point position and upcoming peak speed during preparation and tuned for velocity and absolute speed during movement. Traditional model responses were based upon the kinematics recorded for data set J-array. (All panels adapted from Churchland et al. 2012.)

el were observed: (i) the temporal evolution of the data could be fit by a dynamical model, (ii) the fit was almost as good when dynamics were constrained to be rotational, and (iii) the population response could be projected to reveal factors with clear rotations that followed from the preparatory state.

The comparison between neural and model responses in Figure 5 is an instance of "hypothesis-guided dimensionality reduction." We hypothesized the presence of rotational structure and designed a method (jPCA) to seek factors that might display such structure. But to what degree might this constitute a self-fulfilling prophesy: Would rotational structure be seen in any data with certain basic surface-level features (e.g., smoothly changing multiphasic responses)? Might the traditional model also show such rotational structure when analyzed via jPCA?

Rotational Structure Is Absent for Muscles and Traditional Models

For most neural data sets we also recorded the corresponding population of muscle responses. Single-muscle responses were complex and multiphasic (e.g., Fig. 3B) much like single-neuron responses. Single-muscle responses varied smoothly across time and conditions, much like single-neuron responses. Yet the muscle populations did not contain factors with strong rotational structure. Projections of the EMG population (Fig. 5E) often contained looping trajectories, but these did not travel in a consistent direction: The circular flow of some conditions/times was counteracted by a different flow for other conditions/times and was thus inconsistent with the dynamical model. In agreement, the overall dynamical fit was weaker for the muscle population than for the neural populations, ($R^2 = 0.44$ vs. 0.75 and 0.77) and suffered more when constrained to be purely rotational ($R^2_{\text{skew}} = 0.22$ vs. 0.68 and 0.73). Thus, it is not the case that any population of complex responses will show rotational structure; muscle responses share most surface-level features with the neural data yet did not show consistent rotations. This relates to a key point to which we return below: Although the goal of the dynamical model is to produce muscle activity, the model contains multiple patterns other than the final output, and model neurons are therefore not expected to appear "tuned" for muscle activity.

We also simulated population responses for traditional models where activity is a function of externally measurable parameters (Eq. 1). These population responses did not show strong rotational structure (Fig. 5F), but instead tended to be dominated by the expansion of the neural state. This was true for both simple versions of the traditional model (tuned for reach end point during preparation and reach velocity and speed during movement) and for more complex versions that included tuning for position, acceleration, and jerk (Churchland et al. 2012). This analysis acts as a further control: Again, it is not the case that jPCA reveals rotations for most any data set. More importantly the analyses in Figure 5 reveal that a basic feature of the population response—rotations that follow from the preparatory state—is consistent with the dynamical model but not with a pure muscle-tuning hypothesis or with the traditional model.

In principle, it might be possible to adapt traditional models to produce rotations. For example, a model tuned purely for horizontal position and velocity would produce a partial rotation in state space, if those representations were separable into two roughly orthogonal factors. A more extended rotation could result from a representation of acceleration and jerk. Yet the neural data contain features that are not naturally accounted for by such models. First, rotations during movement follow naturally from the preparatory state, something that would not be true if plotting a "jerk representation" versus an "acceleration representation." Second, rotation frequency was ~1.5–2 Hz, considerably slower than the profile of acceleration or jerk. Some movements had durations of <150 msec, which would yield a triphasic jerk pattern at ~10 Hz—much faster than observed. Third, the observed rotation frequency did not depend on reach speed/movement duration (Churchland et al. 2012) in opposition to the expectation if jerk or acceleration were being represented. Fourth, rotation direction was the same regardless of whether the actual reach was straight, curved clockwise, or curved counterclockwise. These features all emerge naturally from the dynamical model. It might be possible to adapt the traditional model to also account for these features, but it is not clear how to do so.

TRADITIONAL AND DYNAMICAL MODELS SUGGEST DIFFERENT EXPLANATORY APPROACHES

Interpretation of "Represented" Parameters

The traditional model assumes that factors constitute "representations" of parameters such as reach velocity or speed (e.g., Pearce and Moran 2012). A neuron's response should therefore be related to the parameters with which it covaries most strongly. For example, if a given neuron's response correlates weakly with muscle activity and strongly with reach velocity, then that would be taken as evidence that the response relates principally to velocity. Under the dynamical model such inferences are fraught. Consider four hypothetical neurons whose responses respectively reflect x_1 through x_4 from the model in Figure 3A. Note that x_3 and x_4 resemble the pattern of EMG rather weakly: x_3 resembles a low-pass filtered version of velocity, and x_4 resembles position. A general feature of the dynamical model is that individual basis elements may often weakly resemble the final output and may appear to represent other incidental factors. This can be seen as the extension, into the temporal domain, of the observation that in a highly redundant system a neuron's response need not match its causal impact on the final output (Baraduc et al. 2001; Todorov 2002; Schieber and Rivlis 2007).

The difference in the interpretations suggested by the traditional and dynamical models is heightened if there are many observed response patterns. Traditionally, a plethora of patterns suggests the representation/control of many parameters (e.g., Ashe and Georgopoulos 1994; Fu et al. 1995; Wang et al. 2007; Pearce and Moran 2012). Under the dynamical hypothesis, a plethora of patterns suggests a rich basis set (Fetz 1992; Churchland and Shenoy 2007; Churchland et al. 2010, 2012; Sussillo et al. 2012; Shenoy et al. 2013; Hennequin et al. 2014).

Interpretation of Why "Tuning" Varies with Other Factors

Neural responses are often analyzed in terms of "tuning": the manner in which firing rates differ across conditions. Under the traditional framework, tuning is fundamental and reflects the hypothesized relationship between factors and represented parameters. A large literature has sought to identify those parameters to which tuning relates most reliably, on the assumption that those are the parameters that are truly represented. For example, an extensive series of elegant experiments from multiple laboratories has asked whether tuning is constant with respect to Cartesian reach direction or with respect to other factors such as force and/or muscle activity (for review, see Scott 2008; Kalaska 2009; Reimer and Hatsopoulos 2009). Almost universally, such experiments reveal that tuning is not constant with respect to any of the proposed parameters. Tuning either varies idiosyncratically or undergoes "intermediate" shifts. More generally, tuning for a given factor such as direction is rarely constant with respect to other factors such as reach speed (Churchland et al. 2006; Churchland and Shenoy 2007). For example, Figure 6A shows the mean preparatory response of a neuron whose tuning for each direction depended on reach distance (top and bottom panels). This might initially seem troubling. Is the underlying representation very complicated? Are responses simply "messy" and unreliable?

Under the dynamical model, inconstant tuning is expected. "Tuning" is merely a consequence of different movement conditions requiring the elements of the basis set to be modulated differently. For example, every trace in Figure 3C was built from the same oscillatory basis set, but with different phases and amplitudes. Because phase and amplitude depend on the preparatory state (Fig. 2E), the preparatory state necessarily varies across conditions. However, such tuning is not constant with respect to other factors. One example is shown in Figure 6B: The model unit changes its preparatory tuning with distance, much as has been regularly observed in the data. The dynamical

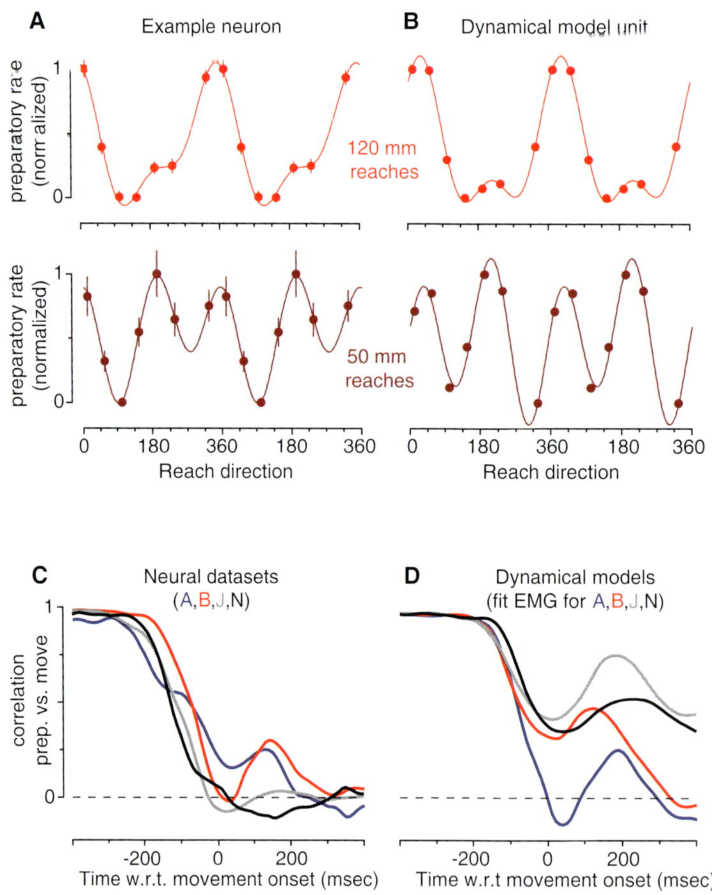

Figure 6. Both real neurons and dynamical model units undergo idiosyncratic changes in the preferred direction as a function of reach distance and as a function of time. (*A*) The average preparatory firing rate of a single neuron as a function of reach direction, recorded during a center-out reaching task. Error bars give the SEM. Fits use a cosine-tuning model employing both the fundamental (unimodal tuning) and the first harmonic (bimodal tuning). (*B*) Similar plot but for the preparatory activity of a unit simulated according to the dynamical model, which was trained to produce the EMG of the deltoid. (*C*) Quantification of the degree to which neurons "preferred" the same conditions during the movement as they did during the preparatory period. The correlation between preparatory and peri-movement tuning is plotted as a function of time. All correlations are between the time shown and a reference time during the middle of the preparatory period. The response of a given neuron at the reference time is a vector containing the firing rate for every condition. For each analyzed time, we correlated the vector of firing rates at that time with the vector at the reference time. This correlation was averaged across all neurons within the data set. Analysis was restricted to neurons robustly tuned during both epochs. Correlations are initially near unity, in agreement with the observation that tuning is stable during the preparatory period. Around 175 msec before movement onset the average correlation began to decrease sharply, indicating that the preference during the movement typically bore little resemblance to that during the preparatory period. (*D*) Same analysis as above but for data sets simulated using the dynamical model. (*C*, Adapted from Churchland et al. 2010.)

model thus provides a rather different account from that suggested by the traditional model, where preparatory tuning constitutes a representation of factors such as reach direction or target location (e.g., Shen and Alexander 1997; Pearce and Moran 2012). For the dynamical model "tuning" will rarely have a consistent surface-level relationship with any movement parameter, even those directly controlled by the network. These observations extend an important fact previously noted in the literature: Any system that controls movement will necessarily have activity that varies across conditions, resulting in directional "tuning" that may or may not be consistent (Mussa-Ivaldi 1988; Fetz 1992; Sanger 1994; Todorov 2000, 2002; Cisek 2006b; Ajemian et al. 2008; Kalaska 2009; Reimer and Hatsopoulos 2009; Lillicrap and Scott 2013).

Interpretation of Why "Tuning" Changes with Time

A striking aspect of most single-neuron responses is that tuning changes with time (in addition to the tuning inconstancy across conditions, as noted above). The "preferred condition" during the preparatory period tends not to agree with the preferred condition during movement (Crammond and Kalaska 2000; Churchland et al. 2010). Furthermore, the preferred condition early in the movement tends not to agree with the preferred condition later in the movement (Sergio and Kalaska 1998; Churchland and Shenoy 2007; Hatsopoulos et al. 2007). For example, in Figure 4 the red-to-green ordering of traces is not preserved as time passes. Figure 6C plots the average correlation of "tuning" between each time and a reference time during the preparatory interval. The correlation plummets just before movement begins and approaches zero by the time of movement onset. Thus, the cartoon neuron illustrated in Figure 2B is rarely observed. Under the traditional model these findings suggest that different movement features are represented in different ways at different times (e.g., Fu et al. 1995; Pearce and Moran 2012). An alternative explanation has been that consistent tuning is neither expected nor desired of a system whose function is to generate and control movement (Fetz 1992; Todorov 2000; Cisek 2006b; Kalaska 2009; Reimer and Hatsopoulos 2009; Lillicrap and Scott 2013; Shenoy et al. 2013).

The dynamical model accords with this latter interpretation and supplies a straightforward explanation for tuning that varies with time. The oscillatory basis intrinsically results in a "preference" that is always in flux. For example, in Figure 2C the red trace is sometimes higher and the green trace is sometimes higher. Like the data, the model also shows a declining correlation between preparatory and movement-period "tuning" (Fig. 6D).

Projections of the Population

Different perspectives on tuning lead to different perspectives regarding how to summarize the population response. In the traditional view, underlying factors relate to represented movement parameters and should be recovered via reference to those parameters. As one example, a two-dimensional population vector can be constructed by weighting each neuron's preferred direction by its response and taking the vector sum. More formally, assume the encoding model:

$$\mathbf{r} = D\mathbf{x},$$

where $\boldsymbol{r}$ is an $n \times 1$ vector of neural responses, $\boldsymbol{x}$ is a 2×1 vector of factors describing horizontal and vertical velocity, and D is an $n \times 2$ matrix of preferred directions. The underlying factors in $\boldsymbol{x}$ can be extracted via

$$\mathbf{x} = D^{\dagger}\mathbf{r},$$

where $D^{\dagger}$ is the pseudoinverse of D. When D has columns that are close to orthonormal, this can be approximated using the transpose:

$$\mathbf{x} = D^{T}\mathbf{r}.$$

A population vector produced via this method is shown in Figure 7A for 28 reach conditions (7 directions $\times$ 2 distances $\times$ 2 reach speeds). This view captures a considerable proportion of the overall variance (20%). Still, although an expansive component is present, it is somewhat disorganized and there is little clear relationship between preparatory and movement activity across conditions. Thus, although Figure 7A is a valid view of the population response, one wonders whether other reasonable views also exist and might be at least as informative.

Under the dynamical perspective, the most revealing projections of the population will be those that capture the dynamical relationship between factors. One particularly desires projections in which a flow field describes the evolution of the neural state. jPCA pursues this end by fitting with a dynamical model. The data are then projected via

$$\mathbf{x} = V\mathbf{r},$$

where each row of the matrix V is an eigenvector of the matrix that fit the dynamics. The number of rows of V can be chosen based on the number of eigenvectors with large corresponding eigenvalues. We found that the data typically contained two to three planes of rotations (4–6 total dimensions) as illustrated in Figure 7B,C (same data set as in Fig. 7A). The first two planes capture 56% of the variance and reveal rotations at different frequencies. Such structure accords with the dynamical model in which a basis set is built from a small handful of oscillatory modes

Pitfalls of the "Population PSTH"

It should be stressed that the three views in Figure 7A–C are all linear transformations of the underlying population response. One might prefer one view to

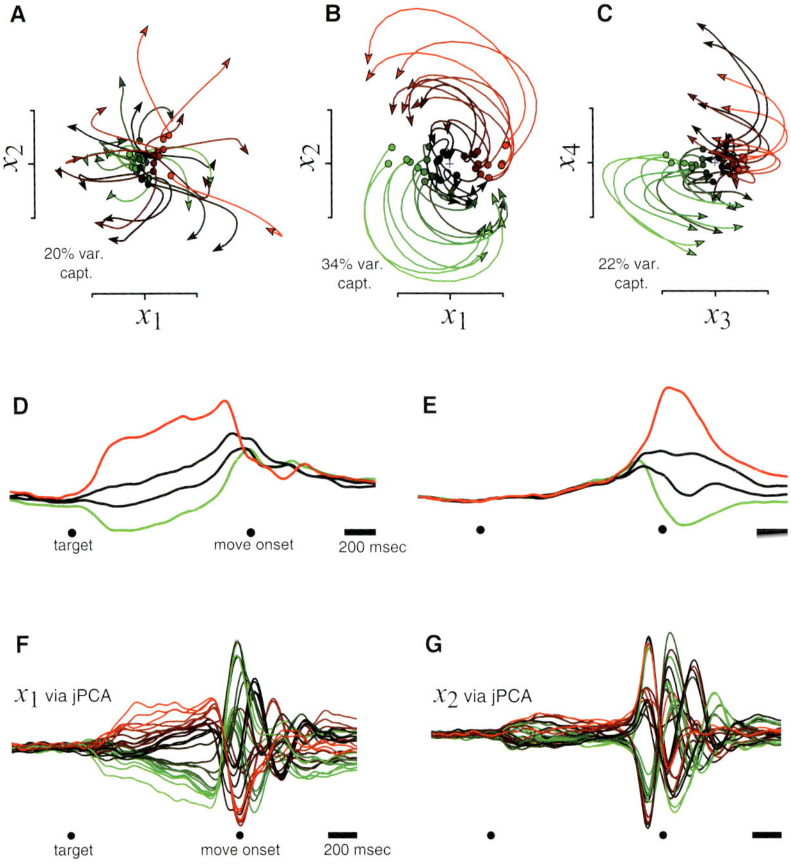

Figure 7. Various views of the population response. For state-space plots (*A–C*) all conventions are as in Figure 5. (*A*) The population vector for a data set which was recorded during straight reaches to seven radially arranged targets at two different distances and two reach speeds (data from monkey B). (*B*) The first pair of dimensions of the jPCA projection for that same data set. (*C*) The second pair of dimensions of that jPCA projection. (*D*) Population peristimulus time histograms (PSTHs) based on the preferred condition during the preparatory period (monkey J). Traces plot the average firing rate (across all neurons) for the condition with the strongest preparatory response (red), the condition with the weakest preparatory response (green), and two intermediate conditions (intermediate shades). Such conditions were found separately for each neuron. Responses were then averaged after normalizing each neuron's activity by its maximum rate (results were similar without normalization). The resulting population PSTH displays very little movement period activity, even though such activity was strongly present. Indeed, analysis was restricted to include only neurons with robust preparatory tuning and robust movement-period tuning. (*E*) A similar population PSTH but with the preferred condition found during the movement period. Despite being based on exactly the same data as in *D*, the population PSTH now looks very different: There appears to be little preparatory activity. (*F*) First dimension (x_1) of the projection found via jPCA for data set J3, plotted as a function of time. Each trace corresponds to one condition. (*G*) As in *F*, but for the second jPCA dimension (x_2). (*B*, Adapted from Churchland et al. 2012.)

another depending on one's hypothesis and on the structure and variance captured by that view. Yet all are valid views of the data, and each reveals response aspects not captured by the others. However, there exist common analysis methods, often inspired by the traditional model, that will not in general produce valid projections of the population response. For example, it is common to summarize the temporal evolution of the population response by computing the "population PSTH (peristimulus time histogram)": the average activity of every neuron, each for its own "preferred" condition. Thus, the activity of a rightward-preferring neuron for a rightward reach is averaged with the activity of an upward-preferring neuron for an upward reach. This procedure is justified only if the tuning model is accurate and behavior is symmetric (e.g., if every reach trajectory is similar excepting the direction itself). Otherwise the resulting view is not a projection of the population response and may be difficult to interpret.

To illustrate, Figure 7D plots the population PSTH based on the preferred condition during the preparatory period. The red trace is the average response of all neurons for their most-preferred condition. The green trace corresponds to the least-preferred condition, and dark traces show intermediate preferences. The population PSTH shows a small peak before movement onset, after which the four traces become similar and relatively flat. The plot thus fails to reflect the fact that most neurons are strongly "tuned" and responsive at this time. That structure is lost in the population PSTH because the preferred condition during movement and preparation are weakly correlated (Fig. 6C). The population PSTH looks very different if the preferred condition is measured later, during the movement (Fig. 7E). Indeed, peaks in tuning can be produced at different times depending on exactly when the preferred condition was assessed. Thus, the seemingly sensible and oft-used approach of averaging responses based on their preferences can produce tautological conclusions: A population will appear strongly tuned at a particular time simply because it was sorted accordingly.

The dynamical perspective suggests alternative approaches for examining the population response versus time. For example, one can produce "factor PSTHs" in which each factor is plotted versus time across all conditions (Fig. 7F,G). These projections of the neural data reveal response structure that is lost in the traditional population PSTH. In particular, these projections reveal that a rough plateau of preparatory activity leads to a strongly oscillatory response during the movement. This pattern—also often observed at the single-neuron level—is lost in the population PSTH.

The Dynamical Model: Toward a Mechanistic Hypothesis

Under the dynamical perspective, the "meaning" of the neural state can seem abstract because most factors do not relate to externally measurable parameters. However, the dynamical perspective provides a natural language for expressing high-level mechanistic hypotheses. The model in Figure 3A is not a circuit-level model, but does propose a mechanism by which the circuit as a whole might perform its function. Dynamical hypotheses can often be succinctly expressed in state space, where they become hypotheses regarding the flow field (Fig. 8B)—that is, the influence of factors on each other. Because many of those factors have no external meaning, they can be interpreted in terms of their mechanistic roles in producing the final output. This style of explanation is often useful even when the underlying connectivity is known, as is the case for many model networks (Sussillo and Barak 2013). Furthermore, this approach promotes the generation of hypotheses that might otherwise lack a clear language. Two examples follow.

A long-standing question has been how preparatory activity avoids causing movement. For the traditional model, it is usually suggested that preparatory activity lies below threshold or is held at bay by an inhibitory gate. The dynamical model suggests a different hypothesis: that preparatory activity lives in the null space of the

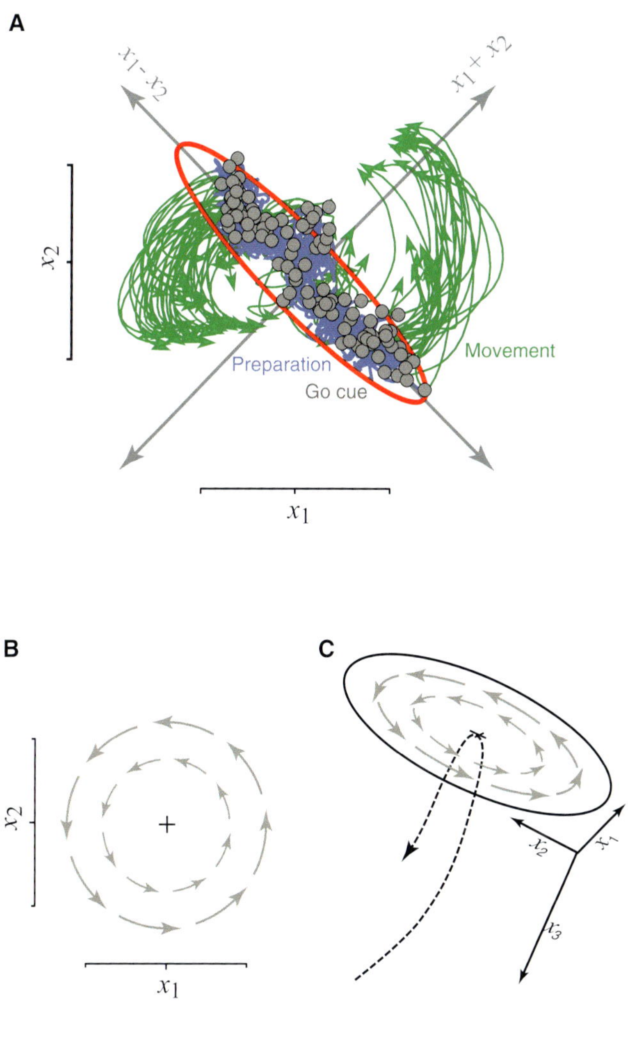

Figure 8. Dynamical hypotheses expressed in state space. (*A*) Illustration, using data, of the "null space hypothesis." The projection is of the population from data set J-array (as in Fig. 5D but with slightly different preprocessing as it was computed for a different study). Preparatory activity is shown in blue; the go cue was given at the time of the gray circles, after which the green traces reveal the trajectory of movement-period activity. The red ellipse shows two standard deviations of the covariance ellipse describing the spread of preparatory activity. (*B*) Hypotheses regarding the interactions between factors can be expressed as flow fields in state space. (*C*) Illustration of how a factor might play a "higher-level" role. In this case, a translation in direction x_3 could bring the neural state into a region of strong oscillatory dynamics, and thus initiate pattern generation. In this case, one dimension (x_3) would be used to recruit pattern generation in a space spanned by other factors (x_1 and x_2). (*A*, Adapted from Kaufman et al. 2014.)

dimensions that map to output muscle activity. This is indeed true of the dynamical model outlined in Figure 3. Preparatory activity is present in the model—no factor is at baseline before the oscillation begins—and plays the role of setting the phase and amplitude of the subsequent oscillation. Yet preparatory activity cancels when the factors are summed to produce the output. Thus, the output is virtually identical across conditions during the preparatory period (Fig. 3C) even though most units have preparatory activity (Fig. 4). Figure 8A illustrates a simple view of this proposal overlaid on a projection of neural data. If preparatory activity (gray dots) is primarily extended in the $x_1 - x_2$ direction then it has little projection onto the $x_1 + x_2$ direction. A downstream target driven by both x_1 and x_2 will mostly ignore preparatory activity. Such an effect is indeed present: Those dimensions that most resemble muscle activity have recently been shown to have the least preparatory activity (Kaufman et al. 2014).

A second outstanding question regards the neural events that trigger movement onset. If activity does not rise past a threshold (as in Fig. 2B), then what triggers movement onset? The model in Figure 2C,E assumes—and the data confirms—the sudden onset of rotational dynamics just before movement onset. We recently hypothesized (Kaufman et al. 2013) that this sudden change in dynamics may be driven by a large overall translation in state space (Fig. 8C). This hypothesis is plausible because local dynamics could be very different in different regions in state space, as suggested by recent modeling (Sussillo et al. 2012). By exploiting such a mechanism, the circuit could control the neural dynamics that drive movement. Hypotheses such as this are readily conceived in a dynamical framework, but difficult to express otherwise.

CONCLUSION

We compared predictions of two idealized models. We considered a traditional model, in which a movement preference appears during the preparatory period and becomes stronger as movement is generated. The basic response features of this model were at odds with the patterns observed in single-neuron responses (Fig. 4). In contrast, a dynamical model produced reasonably realistic single-neuron responses. At the population level, the dynamical model predicted the existence of rotations in state space. Such rotations were also prevalent in the neural data. This prevalence of rotations was not a consequence of the application of the jPCA method: That same method revealed weak rotational tendencies in the population of muscle responses and in simulated population responses of traditional models. More generally, the linear dynamical fit to the neural data was dominated by a rotational component, something not true for the traditional model or the muscle population.

The dynamical model naturally accounts for some seemingly confusing response properties, including the tendency of neurons to display "tuning" that changes with time and with other parameters such as distance and speed (Fig. 6). That said, the dynamical model is currently quite incomplete, and the instantiation shown in Figure 2E and Figure 3A is highly idealized. Future work will have to explain how similar dynamics can be instantiated in neural networks (Schaffer et al. 2006; Sussillo et al. 2012; Hennequin et al. 2014), how dynamics are "turned on" at the right time, the degree to which dynamical motifs other than rotations are important, and the degree to which dynamics are driven by internal versus feedback connections. It also remains an open question why rotations are a dominant motif when other basis sets are possible. One possibility is that neural mechanisms for locomotion were co-opted, on an evolutionary timescale, to produce other kinds of motion (Rokni and Sompolinsky 2012). Alternatively, an oscillatory basis set may be used simply because it is easily generated and provides a robust solution (Sussillo et al. 2012).

In summary, although the dynamical model might initially seem less intuitive, it accounts for the data more successfully than the traditional model. The dynamical model explains a number of features of the data that otherwise seem troublesome or "messy." The model provides a high-level mechanistic explanation for how the system might generate the desired outputs: The output is hypothesized to be the sum of a basis set of patterns that can be generated via internal dynamics. In this view, preparatory activity is truly "preparatory": Its goal is to set the initial state of those dynamics, so that the basis set is recruited appropriately. The subsequent patterns of movement activity may appear complex, but they have a straightforward goal: to provide the building blocks of a temporally structured output.

The above hypothesis is readily extensible. For example, the true basis set is surely more complex than two simple oscillations. Indeed, the data often contain three planes of rotation and significant nonrotational components as well. Thus, rotations may be a dominant motif but they are not the only motif. Furthermore, although an oscillatory basis set will be appropriate for many movements (e.g., those involving multiphasic EMG), other bases will presumably be appropriate for other classes of movement. Indeed, the heart of the dynamical hypothesis is not the particular basis set being used, but the broader strategy at play. This includes three key elements: the presence of more neural dimensions than output dimensions, dynamics that govern the evolution of the neural state within those dimensions, and preparatory activity that plays the role of initializing the dynamics. The dynamical view provides the language and tools for exploring how variations upon this theme may allow motor cortex to generate different classes of movement.

ACKNOWLEDGMENTS

We are very grateful to our colleagues with whom the original research was performed: Krishna Shenoy, Matthew Kaufman, Maneesh Sahani, Stephen Ryu, David Sussillo, Justin Foster, and Paul Nuyujukian. We are deeply grateful to our theoretical colleagues whose work formed much of the inspiration for the dynamical model, in particular Larry Abbott, David Sussillo, and Evan Schaffer. We would also like to acknowledge recent funding sources that have allowed this work to continue: The Grossman Center for the Statistics of Mind, The Burroughs Wellcome Fund, The Searle Scholars Program, The Simons Foundation (SCGB#325171 and SCGB#325233), The Sloan Foundation, The McKnight Foundation, and The National Institutes of Health (NIH) Director's New Innovator Award Program (DP2 NS083037), the Kavli Foundation, and the Gatsby Charitable Trust.

REFERENCES

Ajemian R, Green A, Bullock D, Sergio L, Kalaska J, Grossberg S. 2008. Assessing the function of motor cortex: Single-neuron models of how neural response is modulated by limb biomechanics. *Neuron* **58:** 414–428.

Ashe J, Georgopoulos AP. 1994. Movement parameters and neural activity in motor cortex and area 5. *Cereb Cortex* **4:** 590–600.

Baraduc P, Guigon E, Burnod Y. 2001. Recoding arm position to learn visuomotor transformations. *Cereb Cortex* **11:** 906–917.

Bastian A, Riehle A, Erlhagen W, Schöner G. 1998. Prior information preshapes the population representation of movement direction in motor cortex. *Neuroreport* **9:** 315–319.

Bastian A, Schöner G, Riehle A. 2003. Preshaping and continuous evolution of motor cortical representations during movement preparation. *Eur J Neurosci* **18:** 2047–2058.

Cheney PD, Fetz EE. 1980. Functional classes of primate corticomotoneuronal cells and their relation to active force. *J Neurophysiol* **44:** 773–791.

Churchland MM, Shenoy KV. 2007. Temporal complexity and heterogeneity of single-neuron activity in premotor and motor cortex. *J Neurophysiol* **97:** 4235–4257.

Churchland MM, Santhanam G, Shenoy KV. 2006. Preparatory activity in premotor and motor cortex reflects the speed of the upcoming reach. *J Neurophysiol* **96:** 3130–3146.

Churchland MM, Cunningham JP, Kaufman MT, Ryu SI, Shenoy KV. 2010. Cortical preparatory activity: Representation of movement or first cog in a dynamical machine? *Neuron* **68:** 387–400.

Churchland MM, Cunningham JP, Kaufman MT, Foster JD, Nuyujukian P, Ryu SI, Shenoy KV. 2012. Neural population dynamics during reaching. *Nature* **487:** 51–56.

Cisek P. 2006a. Integrated neural processes for defining potential actions and deciding between them: A computational model. *J Neurosci* **26:** 9761–9770.

Cisek P. 2006b. Preparing for speed. Focus on "Preparatory activity in premotor and motor cortex reflects the speed of the upcoming reach". *J Neurophysiol* **96:** 2842–2843.

Crammond DJ, Kalaska JF. 2000. Prior information in motor and premotor cortex: Activity during the delay period and effect on pre-movement activity. *J Neurophysiol* **84:** 986–1005.

Cunningham JP, Yu BM. 2014. Dimensionality reduction for large-scale neural recordings. *Nat Neurosci* **17:** 1500–1509.

Erlhagen W, Bastian A, Jancke D, Riehle A, Schöner G. 1999. The distribution of neuronal population activation (DPA) as a tool to study interaction and integration in cortical representations. *J Neurosci Methods* **94:** 53–66.

Evarts EV. 1968. Relation of pyramidal tract activity to force exerted during voluntary movement. *J Neurophysiol* **31:** 14–27.

Fetz EE. 1992. Are movement parameters recognizably coded in the activity of single neurons? *Behav Brain Sci* **15:** 679–690.

Fu QG, Flament D, Coltz JD, Ebner TJ. 1995. Temporal encoding of movement kinematics in the discharge of primate primary motor and premotor neurons. *J Neurophysiol* **73:** 836–854.

Georgopoulos AP, Schwartz AB, Kettner RE. 1986. Neuronal population coding of movement direction. *Science* **233:** 1416–1419.

Georgopoulos AP, Crutcher MD, Schwartz AB. 1989a. Cognitive spatial-motor processes. 3. Motor cortical prediction of movement direction during an instructed delay period. *Exp Brain Res* **75:** 183–194.

Georgopoulos AP, Lurito JT, Petrides M, Schwartz AB, Massey JT. 1989b. Mental rotation of the neuronal population vector. *Science* **243:** 234–236.

Gilja V, Nuyujukian P, Chestek CA, Cunningham JP, Yu BM, Fan JM, Churchland MM, Kaufman MT, Kao JC, Ryu SI, et al. 2012. A high-performance neural prosthesis enabled by control algorithm design. *Nat Neurosci* **15:** 1752–1757.

Hanes DP, Schall JD. 1996. Neural control of voluntary movement initiation. *Science* **274:** 427–430.

Hatsopoulos NG, Xu Q, Amit Y. 2007. Encoding of movement fragments in the motor cortex. *J Neurosci* **27:** 5105–5114.

Hennequin G, Vogels TP, Gerstner W. 2014. Optimal control of transient dynamics in balanced networks supports generation of complex movements. *Neuron* **82:** 1394–1406.

Kakei S, Hoffman DS, Strick PL. 2001. Direction of action is represented in the ventral premotor cortex. *Nat Neurosci* **4:** 1020–1025.

Kalaska JF. 2009. From intention to action: Motor cortex and the control of reaching movements. *Adv Exp Med Biol* **629:** 139–178.

Kaufman MT, Churchland MM, Santhanam G, Yu BM, Afshar A, Ryu SI, Shenoy KV. 2010. Roles of monkey premotor neuron classes in movement preparation and execution. *J Neurophysiol* **104:** 799–810.

Kaufman MT, Seely JS, Shenoy KV, Churchland MM. 2013. A large signal in motor cortex that predicts movement onset. In *Neural Control of Movement Annual Meeting*, San Juan, Puerto Rico.

Kaufman MT, Churchland MM, Ryu SI, Shenoy KV. 2014. Cortical activity in the null space: Permitting preparation without movement. *Nat Neurosci* **17:** 440–448.

Lemon RN, Mantel GW, Muir RB. 1986. Corticospinal facilitation of hand muscles during voluntary movement in the conscious monkey. *J Physiol* **381:** 497–527.

Lillicrap TP, Scott SH. 2013. Preference distributions of primary motor cortex neurons reflect control solutions optimized for limb biomechanics. *Neuron* **77:** 168–179.

Moran DW, Schwartz AB. 1999. Motor cortical representation of speed and direction during reaching. *J Neurophysiol* **82:** 2676–2692.

Mussa-Ivaldi FA. 1988. Do neurons in the motor cortex encode movement direction? An alternative hypothesis. *Neurosci Lett* **91:** 106–111.

Orger MB, Kampff AR, Severi KE, Bollmann JH, Engert F. 2008. Control of visually guided behavior by distinct populations of spinal projection neurons. *Nat Neurosci* **11:** 327–333.

Pearce TM, Moran DW. 2012. Strategy-dependent encoding of planned arm movements in the dorsal premotor cortex. *Science* **337:** 984–988.

Pohlmeyer EA, Solla SA, Perreault EJ, Miller LE. 2007. Prediction of upper limb muscle activity from motor cortical discharge during reaching. *J Neural Eng* **4:** 369–379.

Reimer J, Hatsopoulos NG. 2009. The problem of parametric neural coding in the motor system. *Adv Exp Med Biol* **629:** 243–259.

Requin J, Riehle A, Seal J. 1988. Neuronal activity and information processing in motor control: From stages to continuous flow. *Biol Psychol* **26:** 179–198.

Riehle A, Requin J. 1989. Monkey primary motor and premotor cortex: Single-cell activity related to prior information about direction and extent of an intended movement. *J Neurophysiol* **61:** 534–549.

Rokni U, Sompolinsky H. 2012. How the brain generates movement. *Neural Comput* **24:** 289–331.

Sanger TD. 1994. Theoretical considerations for the analysis of population coding in motor cortex. *Neural Comput* **6:** 29–37.

Schaffer ES, Rajan K, Churchland MM, Shenoy KV, Abbott LF. 2006. Generating complex repeatable patterns of activity by gain modulating network neurons. In *Society for Neuroscience Annual Meeting*, Atlanta, GA.

Schieber MH, Rivlis G. 2007. Partial reconstruction of muscle activity from a pruned network of diverse motor cortex neurons. *J Neurophysiol* **97:** 70–82.

Scott SH. 2008. Inconvenient truths about neural processing in primary motor cortex. *J Physiol* **586:** 1217–1224.

Sergio LE, Kalaska JF. 1998. Changes in the temporal pattern of primary motor cortex activity in a directional isometric force versus limb movement task. *J Neurophysiol* **80:** 1577–1583.

Sergio LE, Hamel-Pâquet C, Kalaska JF. 2005. Motor cortex neural correlates of output kinematics and kinetics during isometric-force and arm-reaching tasks. *J Neurophysiol* **94:** 2353–2378.

Shen L, Alexander GE. 1997. Preferential representation of instructed target location versus limb trajectory in dorsal premotor area. *J Neurophysiol* **77:** 1195–1212.

Shenoy KV, Sahani M, Churchland MM. 2013. Cortical control of arm movements: A dynamical systems perspective. *Annu Rev Neurosci* **36:** 337–359.

Sussillo D, Barak O. 2013. Opening the black box: Low-dimensional dynamics in high-dimensional recurrent neural networks. *Neural Comput* **25:** 626–649.

Sussillo D, Churchland MM, Kaufman M, Shenoy KV. 2012. A recurrent neural network that produces EMG from rhythmic dynamics. In *Translational and Computational Motor Control (TCMC)*, New Orleans, LA.

Todorov E. 2000. Direct cortical control of muscle activation in voluntary arm movements: A model. *Nat Neurosci* **3:** 391–398.

Todorov E. 2002. Cosine tuning minimizes motor errors. *Neural Comput* **14:** 1233–1260.

Wang W, Chan SS, Heldman DA, Moran DW. 2007. Motor cortical representation of position and velocity during reaching. *J Neurophysiol* **97:** 4258–4270.

Weinrich M, Wise SP, Mauritz KH. 1984. A neurophysiological study of the premotor cortex in the rhesus monkey. *Brain* **107** (Pt 2)**:** 385–414.

Internal and External Feedback Circuits for Skilled Forelimb Movement

EIMAN AZIM, ANDREW J.P. FINK, AND THOMAS M. JESSELL
Departments of Neuroscience and Biochemistry and Molecular Biophysics, Howard Hughes Medical Institute, Kavli Institute for Brain Science, Mortimer B. Zuckerman Mind Brain Behavior Institute, Columbia University, New York, New York 10032
Correspondence: *tmj1@columbia.edu; ea2471@columbia.edu*

Skilled motor behavior emerges from interactions between efferent neural pathways that induce muscle contraction and feedback systems that report and refine movement. Two broad classes of feedback projections modify motor output, one from the periphery and a second that originates within the central nervous system. The mechanisms through which these pathways influence movement remain poorly understood, however. Here we discuss recent studies that delineate spinal circuitry that binds external and internal feedback pathways to forelimb motor behavior. A spinal presynaptic inhibitory circuit regulates the strength of external feedback, promoting limb stability during goal-directed reaching. A distinct excitatory propriospinal circuit conveys copies of motor commands to the cerebellum, establishing an internal feedback loop that rapidly modulates forelimb motor output. The behavioral consequences of manipulating these two circuits reveal distinct controls on motor performance and provide an initial insight into feedback strategies that underlie skilled forelimb movement.

Goal-directed movements have their basis in neural circuits that channel central commands toward the periphery, translating motor plan into action. The fidelity of motor output requires more than just neural commands, however. Feedback pathways play a critical role in shaping movement, reducing discrepancies between intent and outcome. Defining the organization of such feedback pathways, and how they engage motor circuits, represents a central challenge to understanding the neural basis of movement. In this review, we focus on two classes of feedback pathway that impact motor output: one that originates in the periphery and conveys external sensory information, and another that originates within the central nervous system and mediates internal feedback for rapid motor updating (Fig. 1).

The immediacy of the link between spinal circuitry and muscle contraction has provided an accessible and interpretable experimental system for probing the influence of external and internal feedback pathways on motor output (Baldissera et al. 1981; Jankowska 2001; Pierrot-Deseilligny and Burke 2012; Miri et al. 2013). Among the wide range of mammalian motor behaviors, the control of skilled forelimb movement has come to occupy a central role in studies to define the logic of motor control (Iwaniuk and Whishaw 2000; Shadmehr and Krakauer 2008; Alstermark and Isa 2012; Azim and Alstermark 2015). As such, spinal circuits that govern forelimb movement offer an informative substrate for exploring feedback systems and their influence on skilled motor performance.

PROBLEMS CREATED BY DELAYS IN SENSORY FEEDBACK

Goal-directed reaching movements are characterized by a remarkable regularity. When reaching to a target, limb trajectory and velocity profiles are consistent from trial to trial, despite considerable variation in the patterns of joint movement (Morasso 1981). One potential means of achieving such reproducibility in behavior involves the generation of feed-forward motor commands that take into account the limb's biomechanical response characteristics in driving muscle contraction (Haith and Krakauer 2013). Yet motor command pathways are beset by noise and variability (Harris and Wolpert 1998; Jones et al. 2002; Xu-Wilson et al. 2009; Haith and Krakauer 2013), implying that feedback information is also used to signal error and correct motor output. Proprioception provides one critical source of feedback for motor updating, conveying information about both the state of muscle contraction and the position of the limb in space (Rothwell et al. 1982; Ghez et al. 1995; Gordon et al. 1995). Nevertheless, proprioceptive pathways, and peripheral feedback circuits more generally, take some time to convey signals to relevant sensory recipient centers. These temporal delays arise in part from the mechanics of muscle contraction and spindle activation and the lag incurred in axonal conduction (Wolpert and Miall 1996), and they raise two challenges for motor control (Fig. 1).

The first challenge, documented extensively in the realm of engineering control theory, is that feedback delays destabilize output and results in oscillations and a degradation in performance (Kawato and Gomi 1992; Wolpert and Miall 1996; Ali et al. 1998). One effective means of avoiding motor instability is to limit the impact of sensory feedback on motor output through reduced feedback gain (Stein and Oğuztöreli 1976). At the neuronal level, an inhibitory gain control system could effectively ensure that delays in the provision of peripheral feedback signals do not destabilize the limb during move-

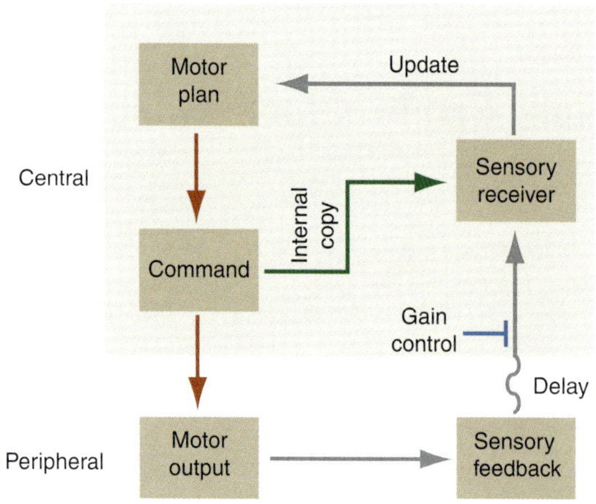

Figure 1. Internal and external feedback motor pathways. During limb movement, motor plans are translated into commands by supraspinal and spinal pathways, eliciting motor output in the form of muscle contraction. Movement generates sensory feedback, including proprioceptive feedback from muscles, which conveys information about motor outcome to central sensory recipient centers. Temporal delays in sensory feedback imply a need for a constraint on sensory feedback gain (blue) to maintain stability and a more rapid mechanism of internal copy feedback (green) for online updating and correction of movement. (Adapted from Azim 2014.)

ment (Fig. 1). The question that emerges then is whether the spinal cord contains a neural mechanism dedicated to the adjustment of proprioceptive sensory gain during the execution of movement, and, if so, what is the functional consequence of its removal?

The second challenge is that delays in feedback limit the rapidity with which motor commands can be refined. During reaching, motor output is continually and rapidly modified as the limb extends to its target (Prablanc and Martin 1992; Messier and Kalaska 1999; Scott 2004), implying the existence of fast feedback pathways that enable online correction of limb movement. One theoretical basis for a rapid source of feedback is the copying of motor commands as they descend to spinal motor centers and the routing of these copies to the cerebellum for prediction of the sensory outcome of movement. In essence, the cerebellum has been argued to use internal copy signals to implement a forward model of the limb (Fig. 1; Wolpert and Miall 1996; Desmurget and Grafton 2000; Shadmehr and Krakauer 2008). In principle, forward model–based predictions offer advantages for motor control. If an incorrect motor command is generated, motor output can be corrected without waiting for sensory information (Shadmehr et al. 2010). Moreover, estimates of the current state of limb position can be made more precise, helping to modify movements in ways beyond the capacity of simple sensorimotor reflex arcs (Todorov and Jordan 2002; Scott 2004; Shadmehr and Krakauer 2008; Azim and Alstermark 2015).

Nevertheless, the neural implementation and behavioral contributions of sensory gain control and internal copy mechanisms have remained a matter of conjecture, in the absence of effective ways of defining their circuitry and evaluating motor performance after manipulating the output of relevant neurons. Traditional experimental methods such as pharmacological silencing and electrical stimulation do not permit selectivity in manipulating spinal circuits for gain control and internal feedback, making it difficult to resolve how their organization confers function. Combining the availability of genetically modified mice with the molecular delineation of neuronal subtype provides a potential strategy for overcoming these limitations, in that they permit the isolation and perturbation of discrete spinal circuits implicated in feedback regulation. Encouragingly, implementation of a detailed three-dimensional quantification of reaching kinematics has revealed the highly stereotyped nature of forelimb movements, even in mice (Whishaw 1996; Azim et al. 2014), providing a sensitive method with which to evaluate the functional consequences of spinal circuit manipulation (Fig. 2). Below, we describe how the use of high-resolution behavioral assays, combined with molecular-genetic manipulations and electrophysiological and modeling approaches, has permitted the functional dissection of some of the spinal circuits implicated in external and internal feedback control (Azim et al. 2014; Fink et al. 2014).

ACHIEVING SMOOTH MOVEMENT THROUGH SENSORY GAIN CONTROL

From first principles, a simple sensorimotor reflex arc subject to high feedback gain has a tendency to oscillate (Stein and Oğuztöreli 1976). Modeling of this sort points to the existence of microcircuits that control the gain of proprioceptive feedback (Fig. 1). The central terminals of primary somatosensory neurons are studded with GABAergic axoaxonic synapses that represent presynaptic inhibitory contacts (Gray 1962; Conradi 1969; Watson and Bazzaz 2001; Betley et al. 2009), and this unusual arrangement has been studied most extensively at synapses between proprioceptive sensory axons and spinal motor neurons (Fig. 3A; Schmidt 1971; Rudomin and Schmidt 1999; Rudomin 2009). A long-recognized feature of presynaptic inhibition is its ability to reduce the excitability of motor neurons without postsynaptic inhibition of the motor neuron itself (Frank and Fuortes 1957; Eccles et al. 1961). The neurons that mediate presynaptic inhibition (which we term GABApre neurons) exert their influence at sensory terminals by reducing the probability of synaptic vesicle release (Kuno 1964; Fink et al. 2014). Moreover, theoretical studies lend support to the idea that presynaptic inhibition reduces gain at sensorimotor synapses (Capaday and Stein 1987, 1989).

How can the behavioral relevance of GABApre neurons best be discerned? Both pre- and postsynaptic inhibitory neurons use GABA as neurotransmitter (Betley et al. 2009), so disentangling the impact of these forms of inhibition through receptor pharmacology has not been possible. Nevertheless, these two inhibitory systems can be distinguished by their molecular provenance and postmitotic identity. GABApre neurons are derived from dorsal dI4 interneurons and express the transcription factor

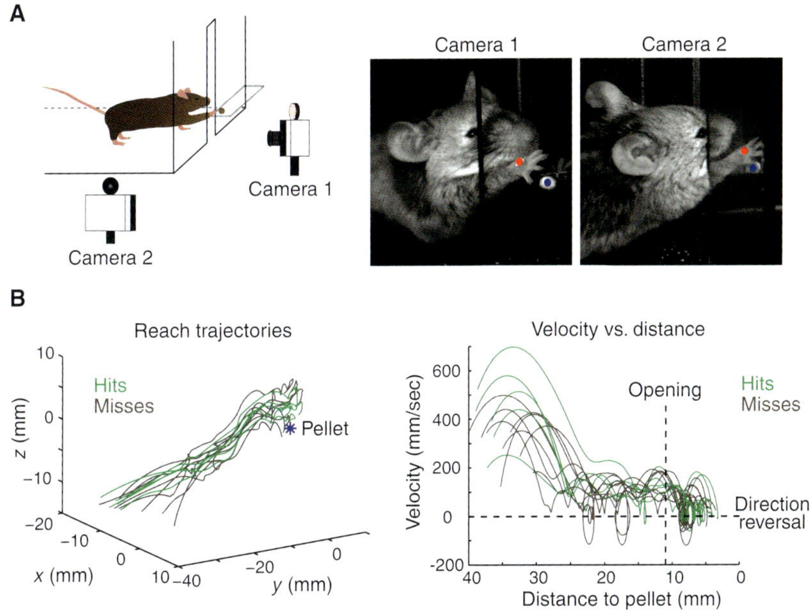

Figure 2. Reaching kinematics in mice. (*A*) Mice were trained to reach for a food pellet through an opening in a transparent box. High-speed, high-resolution video capture and automated tracking of an infrared-reflective marker attached to the right paw were used to reconstruct three-dimensional reaching kinematics. (*B*) Successful (defined as retrieval of the food pellet) and unsuccessful reaches are highly stereotyped in wild-type mice. Trajectories of successful (hits, green traces) and unsuccessful (misses, brown traces) trials from a representative mouse are shown on the *left*. Paw velocity as a function of distance to the pellet is shown on the *right*. The transition from the reach phase to the late grab phase of movement is delineated by the box opening (vertical dashes). Velocity crossings of zero (horizontal dashes) indicate paw direction reversals. (Adapted from Azim et al. 2014.)

Ptf1a, whereas postsynaptic inhibitory (GABApost) neurons, which are derived from V1 and V2b domains, can be defined, respectively, by the transcription factors *En1* and *Gata2/3* (Betley et al. 2009). Both GABApre and GABApost interneurons express the major GABA synthetic enzyme *Gad1*, whereas only GABApre neurons express a second, more specialized, synthetic enzyme *Gad2* (Fig. 3A; Hughes et al. 2005; Betley et al. 2009). The use of *Gad2* as a marker has shown that as they develop, GABApre axons exhibit a stringent specificity for sensory afferent terminals and withdraw from the ventral spinal cord when sensory axons are absent (Betley et al. 2009; Ashrafi et al. 2014). Thus GABApre neurons represent a molecularly defined inhibitory interneuron population whose anatomical organization suggests a potential role in constraining sensory feedback.

Appreciation of the selectivity of *Gad2* expression by GABApre neurons has permitted genetic manipulations that aim to evaluate the contribution of presynaptic inhibition to forelimb movement (Fink et al. 2014). To resolve whether *Gad2*-expressing neurons are functional mediators of presynaptic inhibition, transgenic mice bearing Cre recombinase expression under *Gad2* control (*Gad2-Cre* mice) (Taniguchi et al. 2011) were injected with a virus expressing Cre-dependent channelrhodopsin (ChR2) (Fig. 3B). Photoactivation of $ChR2^{+}$, $Gad2^{+}$ neurons elicited the two canonical features of presynaptic inhibition: a depolarization of primary afferent fibers (primary afferent depolarization, "PAD") (Fig. 3C) and a reduction in the probability of sensory transmitter release that reduces the amplitude of postsynaptic sensory-evoked currents (Fig. 3D). These physiological findings implicate $Gad2^{+}$ neurons as mediators of presynaptic inhibition at spinal sensorimotor synapses.

The behavioral contribution of GABApre neurons to forelimb movement has been explored through the use of a conditional virus that expresses a diphtheria toxin receptor construct under *Gad2* control, and thus selectively in GABApre neurons in adult cervical spinal cord. Administration of diphtheria toxin led to the near-complete absence of GABApre boutons in contact with sensory terminals, without reducing the number of postsynaptic inhibitory boutons (Fig. 4A). Strikingly, analysis of goal-directed reaching behavior revealed that GABApre neuronal ablation produces pronounced limb oscillations as the forepaw approaches the target, as well as a severe degradation in overall reach success (Fig. 4B,C). Moreover, the frequency and decay times of these oscillatory movements were remarkably consistent across reaches and among GABApre-deficient mice (Fig. 4D). A model of the limb as a damped harmonic oscillator driven by feedback at high gain captures many of the core features of this stereotyped motor instability (Fig. 4E), supporting the view that a major role of presynaptic inhibition is to regulate the gain of proprioceptive feedback. Thus, for this set of inhibitory interneurons, a convergence of physiology, behavior, and theory reveal that, during goal-directed reach tasks, GABApre neurons serve as a sensory gain control system for limb stability.

These findings suggest cogent reasons for constructing a specialized presynaptic inhibitory circuit, in addition to the more prevalent postsynaptic inhibitory constraints on

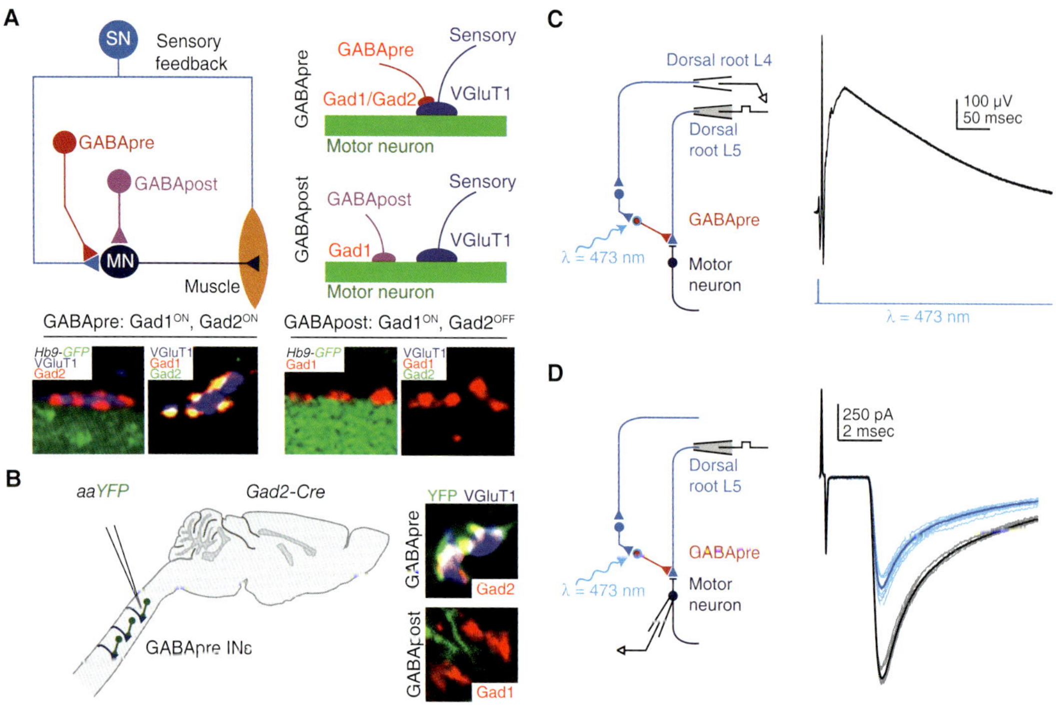

Figure 3. *Gad2*-expressing neurons mediate presynaptic inhibition. (*A*) Proprioceptive sensory neurons (SN) convey sensory feedback signals from muscle to motor neurons (MN). Presynaptic inhibitory (GABApre) neurons contact sensory afferent terminals, whereas postsynaptic inhibitory (GABApost) neurons contact motor neurons directly. GABApre neurons express *Gad2* (*top* schematic and *left* two images; far *left*: red Gad2^{ON} contacts on blue VGluT1$^+$ sensory afferent terminal, adjacent green GFP$^+$ motor neuron labeled in *Hb9-GFP* mice). GABApost neurons express *Gad1* (*bottom* schematic and *right* two images; second from *right*: red Gad1^{ON} contacts adjacent green GFP$^+$ motor neuron). Although GABApre neurons express both *Gad1* and *Gad2* (second image from *left*: yellow Gad1^{ON}/Gad2^{ON} puncta adjacent VGluT1$^+$ sensory terminal), GABApost neurons express *Gad1* alone (far *right* image: red Gad1^{ON}/Gad2^{OFF} puncta). (*B*) Injection of Cre-dependent virus (*AAV-FLEX-ChR2-YFP*) into cervical cord of adult *Gad2-Cre* mice labels GABApre neurons (*top*, YFP$^+$, Gad2$^+$ contact adjacent VGluT1$^+$ sensory terminal) and not GABApost neurons (*bottom*, YFP-negative red puncta). Viral injection marks ~80% of GABApre and ~1% of GABApost boutons. (*C*) After *AAV-FLEX-ChR2-YFP* injection in lumbar spinal cord of neonatal *Gad2-Cre* mice, recordings in isolated spinal cord from sensory afferents (dorsal root L4, extracellular) reveal primary afferent depolarization, and (*D*) recordings from motor neurons (whole cell patch clamp) reveal suppression of monosynaptic sensory-evoked excitatory postsynaptic currents after photostimulation (black, control; blue, 473-nm wavelength (λ) photostimulation). At early ages *Gad2* is also expressed in GABApost boutons; therefore, all behavioral experiments were performed after viral injection in adult mice, when *Gad2* expression is specific for GABApre neurons (*B*) (see Fink et al. 2014 for details). (Adapted from Betley et al. 2009; Fink et al. 2014.)

motor output. One obvious attribute of presynaptic inhibition is its target specificity—this axoaxonic organization provides an anatomical substrate for selective regulation of a single excitatory input to motor neurons, whereas postsynaptic inhibition inevitably exerts a more general impact on the excitability of motor neurons triggered by diverse inputs. In addition, presynaptic inhibition provides a means of scaling the gain of sensory feedback to match the intensity of sensory input. Proprioceptive afferents are known to fire at frequencies spanning three orders of magnitude in the course of a normal motor task (Prochazka 1999), suggesting the need to achieve a compressed dynamic range that can be read out more effectively by central motor circuits (Laughlin 1994; Brenner et al. 2000). One crucial basis for this gain scaling strategy is the fact that proprioceptive afferents also provide excitatory drive to GABApre neurons (Eccles et al. 1961, 1962). Thus enhanced sensory activity will concomitantly increase GABApre-mediated inhibition, providing an effective circuit-based scheme for sensory gain scaling.

The need for presynaptic inhibition to limit the dynamic range of sensory afferent firing also provides a plausible explanation for the selectivity of *Gad2* expression in presynaptic inhibitory neurons, in that Gad2 enzymatic activity within inhibitory boutons has been shown to sustain GABA release at high firing rates (Tian et al. 1999). Together, these anatomical and molecular features would seem to equip GABApre neurons with the ability to rescale sensory input in a dynamic and adaptive manner, an essential constraint given the wide range of sensory firing rates observed in vivo.

Despite these advances, many features of the physiology of spinal presynaptic inhibition still lack molecular or circuit explanations. Work in primates and cats has begun to define the descending circuitry responsible for recruitment of GABApre neurons, but relatively little is known about the physiological triggers for this form of inhibition during limb movement. Studies in monkey have shown that supraspinal systems elicit presynaptic inhibition of cutaneous feedback during voluntary movement of the

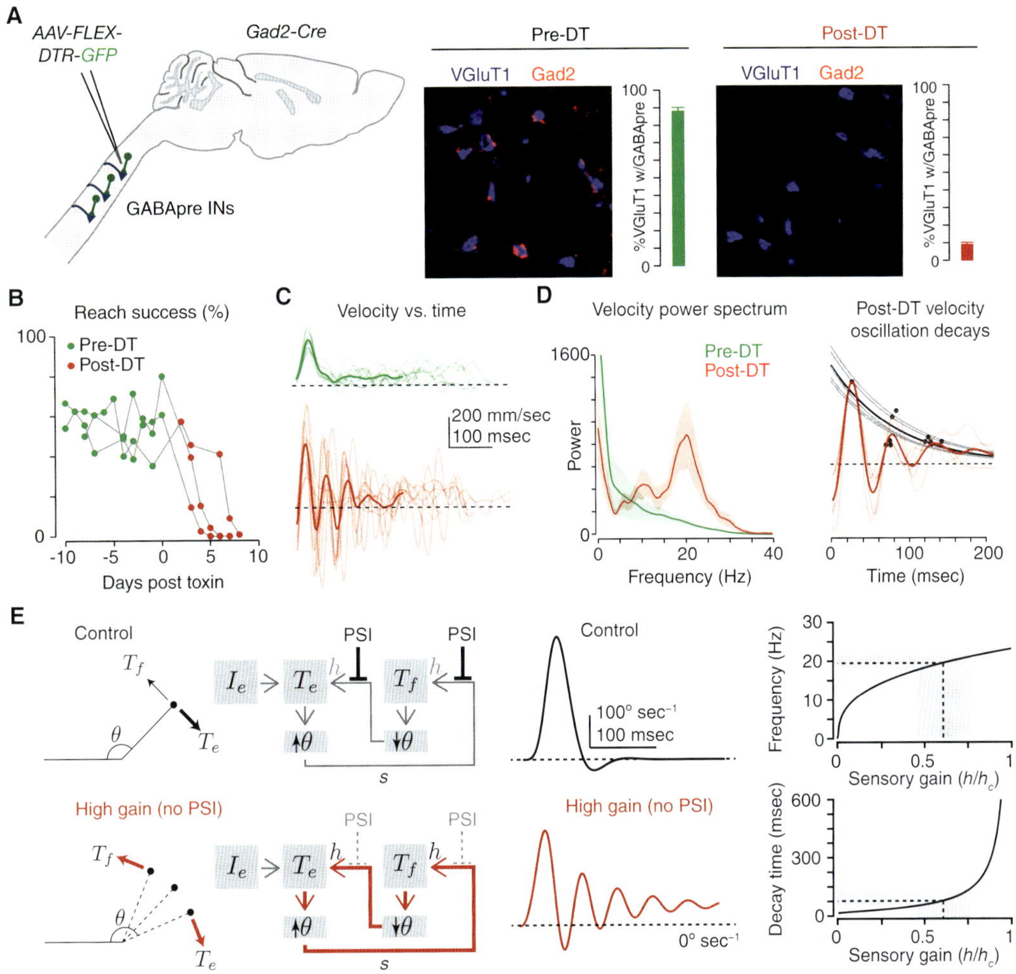

Figure 4. GABApre neuron ablation results in forelimb oscillation during reaching. (*A*) Cervical GABApre interneurons (INs) were targeted for acute ablation via injection of a conditional virus expressing a diphtheria toxin receptor-GFP fusion (*AAV-FLEX-DTR-GFP*) in cervical cord of adult *Gad2-Cre* mice. Before diphtheria toxin (DT) administration (*left*), ~90% of VGluT1$^+$ sensory afferent terminals (blue) are contacted by Gad2$^+$ GABApre boutons (red). Following DT administration (*right*), Gad2$^+$ GABApre boutons contact <10% of sensory afferent terminals. Error bars indicate S.E.M. (*B*) GABApre ablation severely affects reach success; before (green) and after (red) DT administration for three *Gad2-Cre* mice. (*C*) Paw velocity as a function of time for a representative mouse (thin lines, individual reaches; bold lines, mean) before (green) and after (red) DT administration reveals severe oscillations after GABApre ablation. (*D*) Average power spectra (*left*, bold lines, mean across mice; shading, SEM) of pre-DT (green) and post-DT (red) reaches indicate a narrow peak in oscillation frequency focused at ~20 Hz after ablation. Alignment of individual reaches to the point of maximal reach velocity (*right*) reveals consistent oscillatory decay in individual mice (faint red lines) and averaged across mice (bold red line). Fitting single exponential functions to the first three oscillatory peaks (gray dots and faint gray lines, individual mice; black line, mean) reveals a decay time of ~77 msec across mice. (*E*) A simplified model of a joint with sensory feedback controlled by a flexor (T_f) and extensor (T_e) torque and defined by joint angle θ (*left* panels). Sensory feedback (s) depends on joint angle and normally permits smooth joint extension (gray arrows, black control trace) in the presence of external extensor drive (I_e). In the absence of presynaptic inhibition (PSI), simulated by imposing high sensory gain (h), sensory feedback triggers oscillation (red arrows and red high gain trace). Oscillation frequency and decay time change little across a wide range of gain values in the model (*right* plots), providing a possible explanation for why frequency and decay time are so consistent across mice (*D*). Gray shading indicates the range of experimentally observed peak oscillation frequency and decay times in post-DT *Gad2-Cre* mice, as well as the corresponding range of normalized gain values. (Adapted from Fink et al. 2014.)

wrist (Seki et al. 2003), and human studies suggest that descending pathways recruit presynaptic inhibition of proprioceptive afferents on a muscle-by-muscle basis during voluntary contraction (Hultborn et al. 1987). Thus supraspinal systems are likely to engage presynaptic inhibition dynamically, even at the level of an individual motor neuron pool, functional features that imply a highly organized pattern of descending control (Rudomin et al. 1986).

At a local level, presynaptic inhibition shows a finely grained precision in targeting. Anatomical evidence suggests that group Ib but not group Ia sensory boutons in the region of Clarke's column receive axoaxonic contacts (Walmsley et al. 1987). And even at the level of an individual sensory fiber, presynaptic inhibition is likely to act differentially on single axon collaterals (Lomelí et al. 1998). Together, these findings imply a richly endowed organization of presynaptic inhibitory interneurons themselves and of the pathways that recruit them. They also suggest a diversity of GABApre neuronal subtypes and their connections, mirroring the functional heterogeneity of the afferent fibers they control.

INTERNAL FEEDBACK FOR RAPID MOTOR UPDATING

Animals can refine forelimb movements with rapidity, despite incumbent delays in sensory feedback. Internally directed copies of motor commands might provide a faster source of feedback information for motor updating (Fig. 1; Wolpert and Miall 1996; Desmurget and Grafton 2000; Shadmehr and Krakauer 2008; Azim and Alstermark 2015). Yet obtaining experimental evidence that internal feedback does actually influence forelimb movement has been frustratingly hard, in part because of the difficulty of perturbing putative internal copy circuits without simultaneously affecting motor output. Genetic interventions in mice could permit more incisive access to internal copy circuits and their encoded function.

One class of spinal interneurons, cervical propriospinal neurons (PNs), has been implicated in the control of forelimb movement (Alstermark and Isa 2012). Lesion studies in cat suggest a role for PNs in goal-directed reaching and in the rapid updating of limb motor output after sudden changes in target location (Alstermark et al. 1981b; Alstermark and Lundberg 1992). Moreover, recent viral silencing experiments have provided evidence that PNs contribute to both reaching and grasping movements in primates (Kinoshita et al. 2012).

Two anatomical features implicate PNs as mediators of internal copies of forelimb motor commands. They receive input from supraspinal descending pathways (Illert et al. 1978, 1981), and thus serve as potential effectors of motor commands devoted to forelimb movement. In addition, PNs are characterized by a bifurcated output: One axonal branch projects to the cervical motor neurons that control forelimb muscles (Alstermark et al. 1990), while the other projects rostrally to the lateral reticular nucleus (LRN) (Alstermark et al. 1981a), a major precerebellar input (Fig. 5A; Brodal 1949; Arshavsky et al. 1978a; Alstermark and Ekerot 2013). The duality in axonal projection and output constitutes an anatomically direct means of conveying internal copies of premotor signals and raises the question of whether PN internal feedback has any relevance to the control of forelimb movement.

PNs are present in mice, as revealed by in vivo electrophysiology and viral tracing studies, and one prominent set of excitatory PNs expresses the transcription factor *Chx10*, marking them as V2a neurons, one of the cardinal interneuron classes implicated in spinal motor control (Fig. 5B; Azim et al. 2014). The use of a Cre-dependent virus to express diphtheria toxin receptor in *Chx10-Cre* mice has shown that toxin-mediated ablation of cervical V2a interneurons (Fig. 5C) elicits a severe reach-specific dysmetria (Fig. 5D), without major impact on digit abduction or locomotor control. The similarity in behavioral deficit observed after PN disruption in mouse, cat, and monkey (Alstermark and Isa 2012; Kinoshita et al. 2012; Azim et al. 2014) supports the view that PNs have evolved to regulate mammalian reaching behavior.

Nevertheless, elimination of PNs abolishes both motor command and internal copy, and thus does not resolve whether the internal copy pathway alone has any impact on forelimb movement. This issue has been addressed by expressing ChR2 in cervical V2a neurons in *Chx10-Cre* mice, with the aim of gaining selective access to the ascending PN branch (Fig. 6A). Under these conditions, focal photostimulation of PN terminals within the LRN activated the internally directed PN–LRN pathway without affecting PN premotor output. Strikingly, selective PN terminal photostimulation produced a severe disruption in reaching kinematics, eliciting a large increase in the incidence of paw direction reversals (Fig. 6A) without affecting digit movements. The selectivity of these behavioral effects suggests that the PN internal copy pathway is dedicated to the modulation of reaching.

These findings have provided functional evidence that imposed activation of a spinal internal copy pathway perturbs the fidelity of forelimb motor output. Moreover, in vivo recordings revealed that activation of the PN–LRN pathway excites forelimb motor neurons in as little as ~4–5 msec (Fig. 6B). These short latency motor responses are substantially diminished by severing LRN projections to the cerebellum (Fig. 6B), implicating a fast cerebellar–motor loop in modulating forelimb motor neuron activity (Fig. 6C). Viewed in the context of reaching movements that typically last hundreds of milliseconds, the rapidity of these regulatory actions suggests that the PN–LRN circuit can adjust motor output continuously during an individual reach.

To date, PN circuit manipulations have provided functional support for the idea that internal copy pathways directed to the cerebellum contribute to forelimb motor refinement (Alstermark and Lundberg 1992; Alstermark and Isa 2012; Azim et al. 2014). There are, nevertheless, many unresolved issues that impede the understanding of internal copy pathways of use in motor control. One concerns the organization of spinal circuits dedicated to discrete elements of limb movement. LRN-projecting V2a interneurons are restricted to the cervical and upper thoracic spinal cord and absent from more caudal segments (Azim et al. 2014), despite the presence of other classes of neurons in lumbar cord that project to the LRN (Arshavsky et al. 1978a; Ekerot 1990). These observations indicate that the molecular rules of neural circuit organization applicable to the fore- and hindlimb may differ. The concept of stereotypic motor circuits with repeated architecture across spinal segments may therefore be an oversimplification.

From a behavioral perspective, the disruption in reaching upon elimination of cervical V2a interneurons is selective, in that locomotor and digit extension appear unaffected. This selectivity in behavioral perturbation supports the view that forelimb movement is modular in

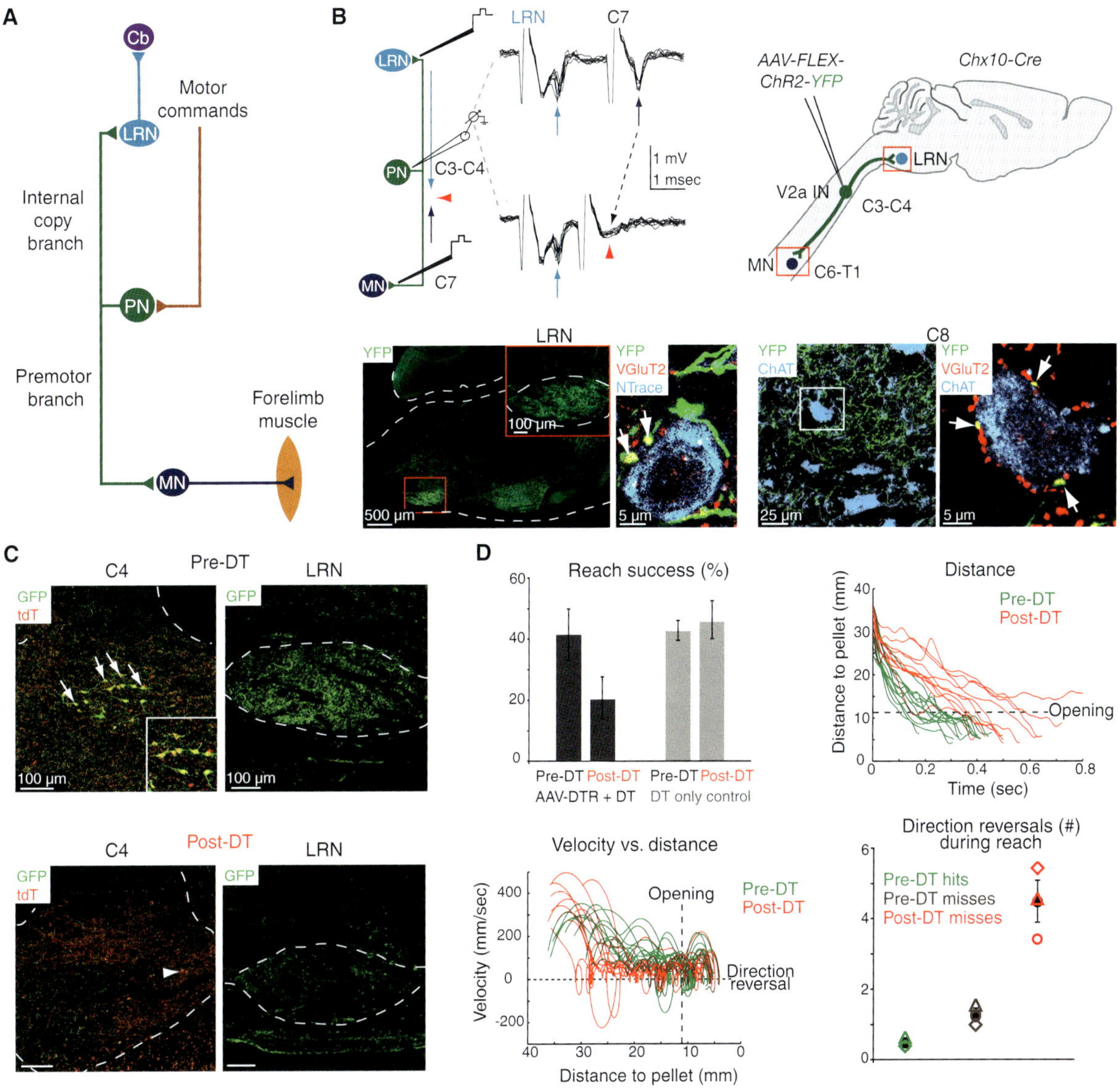

Figure 5. V2a PN ablation disrupts reaching movements. (*A*) PNs receive input from descending motor command pathways. PN axons bifurcate to innervate forelimb motor neurons (MN, premotor branch) and the lateral reticular nucleus (LRN, internal copy branch), which projects to the cerebellum (Cb). (*B*) In vivo extracellular recording in mouse cervical spinal cord while electrically stimulating LRN and C7 ventral horn reveals neurons fired antidromically from both stimulation sites (*top left*, arrows). Spike collision (red arrowheads) shows that the identified neuron projects to both targets, confirming the existence of PNs in mice. Injection of Cre-dependent virus (*AAV-FLEX-ChR2-YFP*) into cervical cord of adult *Chx10-Cre* mice labels V2a interneurons (IN) selectively (*top right*). YFP$^+$ V2a INs project to the LRN (*bottom left*), where they form VGluT2$^+$ excitatory synaptic contacts (red, arrows) onto LRN neurons (blue). Similarly, forelimb motor neurons in C8 (*bottom right*, blue) receive VGluT2$^+$ (red), YFP$^+$ contacts (arrows). (*C*) Cervical V2a INs were targeted for acute ablation via injection of a conditional virus expressing a diphtheria toxin receptor-GFP fusion (*AAV-FLEX-DTR-GFP*) in cervical cord of adult *Chx10-Cre* mice (*top left*, arrows, V2a INs genetically labeled by tdT in red; *top right*, GFP$^+$ projections to the LRN). Diphtheria toxin (DT) administration ablates ~85% of V2a INs (*bottom left*, arrowhead indicates spared neuron), and eliminates GFP$^+$ PN projections to the LRN (*bottom right*). (*D*) V2a IN elimination reduces reaching success (*top left*), and increases movement duration (*top right*) and the number of forelimb direction reversals (*bottom left* and *right*) during the reaching phase of movement (before the box opening). Shapes in *bottom right* plot represent individual mice. Error bars indicate S.E.M. Locomotion and digit abduction were not affected by V2a IN elimination (not shown). (Adapted from Azim et al. 2014).

design and that this modularity emerges at least in part from the differential recruitment of discrete spinal interneuron circuits during distinct behaviors (Bizzi et al. 2008; Alstermark and Isa 2012). Consistent with this notion, genetic inactivation of another set of excitatory premotor interneurons, dI3 interneurons, in mice impacts distal grasp performance (Bui et al. 2013). And as discussed above, the loss of cervical GABApre neurons produces limb oscillations and degrades forelimb movement in a manner quite distinct from the consequences of V2a or dI3 interneuron disruption (Figs. 4C,D, 5D; Azim et al. 2014; Fink et al. 2014). Moreover, the modularity of PN

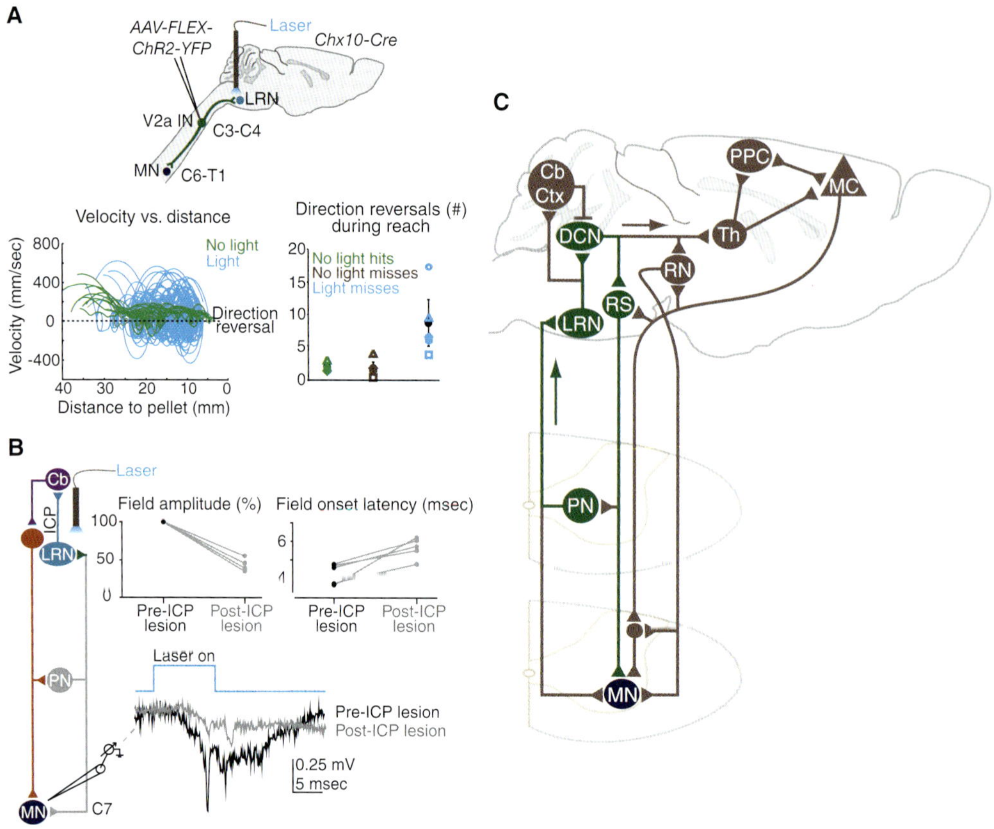

Figure 6. PN internal copy activation modulates forelimb movement through a rapid cerebellar–motor loop. (*A*) ChR2 was expressed in cervical V2a INs via injection of conditional virus (*AAV-FLEX-ChR2-YFP*) in cervical cord of adult *Chx10-Cre* mice. Photostimulation of ChR2$^+$ PN terminals in the LRN activates the PN–LRN pathway selectively, without eliciting antidromic action potentials (not shown). Photostimulation disrupts reaching movements (blue traces), producing a large increase in the incidence of paw direction reversals and variability in velocity and acceleration during reaching, without affecting digit abduction (not shown). Shapes in *right* plot represent individual mice. Error bars indicate S.E.M. (*B*) Motor field potential recordings in cervical spinal segments, as well as intracellular motor neuron recordings (not shown), reveal rapid activation of forelimb motor neurons (earliest activation ~4–5 msec from light onset). Severing the inferior cerebellar peduncles (ICP), which contain the projections from the LRN to the cerebellum, reduces the amplitude of motor field potentials by up to 65% and eliminates the fastest motor field onsets, implicating a rapid cerebellar–motor feedback pathway. (*C*) A putative subcortical PN–LRN feedback loop (green) recruits deep cerebellar nuclei (DCN), which excite reticulospinal (RS) projections to motor neurons (MN). Longer loops (brown) might engage cerebellar cortex (Cb Ctx), rubrospinal neurons (RN), and cortical pathways in posterior parietal cortex (PPC) and motor cortex (MC) via thalamus (Th) (see text). (*A*,*B*, Adapted from Azim et al. 2014.)

circuit organization appears to be malleable since silencing PNs in primates affects both reaching and grasping movements (Kinoshita et al. 2012). Evolutionary modifications in internal copy circuit connectivity and function may therefore underlie aspects of enhanced manual dexterity (Isa et al. 2007).

How does activation of the PN–LRN pathway elicit a rapid modulation of motor neuron activity? Anatomically, the most credible cerebellar feedback loop involves collateral projections from LRN neurons to deep cerebellar nuclei, which in turn excite reticulospinal projections to motor neurons (Carpenter and Nova 1960; Matsushita and Ikeda 1976; Schultz et al. 1979; Wu et al. 1999; Alstermark and Ogawa 2004; Soteropoulos and Baker 2008)—in effect a feedback loop that excites motor neurons three synapses downstream from LRN activation (Fig. 6C). The fact that reticulospinal neurons can activate forelimb motor neurons within ~1 msec supports the plausibility of such a feedback pathway (Azim et al. 2014). One potential advantage of a subcortical positive-feedback loop is that short latency effects could contribute to motor stability, as has been suggested for oculomotor control (Lisberger 2009). This does not exclude the possibility that longer-latency or parallel cerebellar–motor loops supplement the most rapid responses and contribute to a more refined modulation of movement (Fig. 6C). The red nucleus receives direct input from the interposed deep cerebellar nucleus and contains rubrospinal projection neurons that have been implicated in the control of distal forelimb motor neuron pools (Fujito and Aoki 1995). Additionally, a polysynaptic pathway from the dentate deep cerebellar nucleus to the ventral anterior and ventral lateral thalamic nuclei provides inputs to posterior parietal and motor cortical regions thought to integrate feedback, prediction, and reward signals for the flexible updating of movements (Fig. 6C; Andersen et al. 1997; Scott 2004; Shadmehr and Krakauer 2008; Pruszynski et al. 2011). The potential diversity of PN feedback pathways emphasizes the need for selective circuit manipulations when resolving the grain of internal feedback control.

Little is known about the precise structure of copy information conveyed by the PN–LRN pathway, or whether cerebellar circuits use these signals to implement a forward model and predict movement outcome. The LRN–cerebellar system has been speculated to have diverse roles, ranging from relay of spinal pattern generator activity during rhythmic movement (Arshavsky et al. 1978a) to clock-like representation of time (Xu et al. 2013) and the integration of postural and limb motor signals (Alstermark and Ekerot 2013). Resolving the nature of copy signals conveyed by the PN–LRN circuit and how they reflect motor commands should help to pinpoint aspects of feedback relevant for updating forelimb motor acts. Such insight could provide new inroads into cerebellar cortical circuits and their putative contributions to forward model implementation (Desmurget and Grafton 2000; Shadmehr and Krakauer 2008; Bastian 2011).

Finally, a clearer understanding of internal copy pathways in motor control may help to clarify potential nonmotor contributions to cognition and neuropsychiatric disorders (Ito 2008; Strick et al. 2009; Bastian 2011). Internal feedback circuits enabling predictive sensory cancellation during self-initiated behaviors could conceivably be disrupted in patients with schizophrenia (Ford and Mathalon 2012; Shergill et al. 2014). The recent identification in mice of an internal copy pathway involved in inhibiting auditory cortical neurons during self-generated movement (Schneider et al. 2014) provides an experimentally tractable circuit of potential relevance to models of schizophrenia. Thus PNs likely represent an exemplar of other internal copy pathways, the study of which may reveal anatomical and computational features relevant to circuits in motor and sensory systems (Poulet and Hedwig 2007; Azim and Alstermark 2015).

MERGING INTERNAL AND EXTERNAL FEEDBACK

How might presynaptic gain control and internal copy feedback be combined into a more coherent understanding of forelimb motor control? Admittedly, our findings have merely scratched the surface of the ways in which these pathways contribute to forelimb movement. A more complete view of the interaction between external and internal feedback pathways may derive from theoretical considerations of motor circuitry, viewed through the lens of optimal feedback control theory. This scheme describes a formal relationship between motor command, feedback, and cost and provides an explanation of how movement is continually updated and optimized for the task at hand (Todorov and Jordan 2002; Scott 2004). Two salient features of this theory are of particular relevance to PN internal copy and presynaptic inhibitory circuits.

A central proposition of optimal feedback control is that the motor system generates online estimates of the current state of the limb, so as to implement appropriate corrections in command and trajectory (Scott 2004). In this framework, estimates of limb state benefit from a diversity of feedback signals—the more feedback information, the more precise the estimate. Indeed, the accuracy of subjects' proprioceptive assessment of limb location is improved when movements are self-initiated, in contrast to conditions when the limb is moved passively and no commands or copies are conveyed (Fuentes and Bastian 2010; Bhanpuri et al. 2013). These findings are consistent with the idea that internal copy feedback is continuously integrated with delayed sensory information to generate more accurate estimates of the current state of the limb. An extension of this notion is that these estimates will benefit from diverse internal copy pathways.

In the spinal cord, the proposition that internal copies of limb motor signals exist in a variety of anatomical arrangements has received some experimental support. The ventral spinocerebellar tract has long been thought to convey premotor copy information directly to cerebellar circuits (Lundberg 1971; Arshavsky et al. 1978b; Jankowska and Hammar 2013). Moreover, recent studies in mice have revealed several genetically defined excitatory and inhibitory spinal pathways that convey copies of forelimb motor signals to the LRN (Pivetta et al. 2014), providing further evidence that many spinal input systems converge on LRN neurons (Künzle 1973; Clendenin et al. 1974; Ekerot 1990; Alstermark and Lundberg 1992; Alstermark and Ekerot 2013). The repertoire of mammalian forelimb movements is vast, and thus it is not altogether surprising that the diversity of internal copy circuits and the motor behaviors they regulate might be comparably complex.

A second principle of optimal feedback control is that the motor system updates movement by modulating the level of sensory feedback gain. After perturbation of the limb, forelimb movement corrections occur well before the onset of voluntary reaction. Yet in humans, corrective adjustments that occur within ~50–100 msec account for the complexities of limb dynamics and show a biomechanical sophistication beyond the capabilities of simple reflex arcs (Todorov and Jordan 2002; Scott 2004; Pruszynski et al. 2011). Optimal feedback control posits that motor updating is achieved by adjusting dynamically the level of sensory feedback gain and that adjustment occurs throughout the course of movement, permitting high gain at moments beneficial to behavior (Scott 2004; Franklin and Wolpert 2011). In this way, movement can be optimized for task performance by accounting for the changing state of the limb and its environmental context. Indeed, there is substantial experimental evidence that the gain of sensory feedback changes over the course of movement to account for altered conditions—such as a displacement of hand or target location (Franklin and Wolpert 2008; Dimitriou et al. 2012; Crevecoeur and Scott 2013). Neurons in primary motor cortex that integrate feedback to drive movement corrections provide one potential instance of a circuit in which dynamic gain regulation can be translated into appropriate motor commands (Pruszynski et al. 2011).

Presynaptic inhibitory neurons, by virtue of their axo-axonic connections with sensory neurons, have the capacity to modulate specific afferent channels and thus seem well suited for selective and flexible control of feedback gain. Targeted elimination of all presynaptic inhibition in the cervical spinal cord has a profound impact on limb stability (Fink et al. 2014), but discrete subtypes dedicated to modulating distinct afferent pathways with more specialized functions surely exist within this broad population. The selectivity of sensory afferent targeting could permit behaviorally relevant feedback gains in supraspinal centers to remain high, while reducing gains in the spinal cord to maintain local stability (Scott and Crevecoeur 2014). Thus, defining the subtype diversity inherent in presynaptic inhibitory gain control could help to clarify the neural implementation of feedback control theory.

CONCLUSION

Decades of research in primates and cats have identified plausible neural substrates for external and internal feedback pathways and have proposed their influence on movement. Recent advances in the genetic manipulation of circuits in mice hope to build on this foundation, with the goal of linking neuron and circuit to behavior with ever greater precision. The two circuits discussed here illustrate some of the initial insights that can be obtained about the contributions of gain control and internal copy pathways to skilled forelimb movement. The grain of organization of these neural mechanisms, however, is surely far finer than the current state of experimental manipulation. One major challenge for the future will therefore be to obtain greater neuronal subtype resolution in the dissection of motor circuitry. Yet as the number of isolated neural elements and their identified behavioral functions multiply, it remains equally important to design experiments that test and refine theoretical descriptions of motor control and its encoded behaviors.

These thoughts are not entirely new. The French Enlightenment philosopher Denis Diderot wrote, "We are all instruments endowed with feeling and memory. Our senses are so many strings that are struck by surrounding objects and that also frequently strike themselves." Some 250 years later these words still hold relevance in attempts to understand the neural basis of motor behavior. Sensory information is intertwined with behavioral output and thus the comprehension of one is futile without an appreciation of the other. Diderot's words also emphasize a more subtle point—feedback can be triggered by both the environment and the individual. Defining how the brain orchestrates behavior inevitably requires an understanding of the influence of both external and internal sources of feedback information.

ACKNOWLEDGMENTS

We thank our colleagues and collaborators B. Alstermark, J. Jiang, L.F. Abbott, K. Croce, and Z.J. Huang. We are also grateful to J.W. Krakauer, A. Karpova, M. Churchland, and C.E. Schoonover for helpful discussions and feedback. E.A. was supported by a National Institutes of Health (NIH) K99 award (NS088193) and a Helen Hay Whitney Foundation/Howard Hughes Medical Institute Postdoctoral Fellowship. T.M.J. was supported by NIH grant NS033245, the Harold and Leila Y. Mathers Foundation, and Project A.L.S. and is an investigator of the Howard Hughes Medical Institute.

REFERENCES

Ali MS, Hou ZK, Noori MN. 1998. Stability and performance of feedback control systems with time delays. *Comput Struct* **66:** 241–248.

Alstermark B, Ekerot CF. 2013. The lateral reticular nucleus: A precerebellar center providing the cerebellum with overview and integration of motor functions at systems level. A new hypothesis. *J Physiol* **591:** 5453–5458.

Alstermark B, Isa T. 2012. Circuits for skilled reaching and grasping. *Annu Rev Neurosci* **35:** 559–578.

Alstermark B, Lundberg A. 1992. The C3-C4 propriospinal system: Target-reaching and food-taking. In *Muscle afferents and spinal control of movement* (ed. Jami L, Pierrot-Deseilligny E, Zytnicki D), pp. 327–354. Pergamon, Oxford.

Alstermark B, Ogawa J. 2004. In vivo recordings of bulbospinal excitation in adult mouse forelimb motoneurons. *J Neurophysiol* **92:** 1958–1962.

Alstermark B, Lindström S, Lundberg A, Sybirska E. 1981a. Integration in descending motor pathways controlling the forelimb in the cat. 8. Ascending projection to the lateral reticular nucleus from C3–C4 propriospinal also projecting to forelimb motoneurones. *Exp Brain Res* **42:** 282–298.

Alstermark B, Lundberg A, Norrsell U, Sybirska E. 1981b. Integration in descending motor pathways controlling the forelimb in the cat. 9. Differential behavioural defects after spinal cord lesions interrupting defined pathways from higher centres to motoneurones. *Exp Brain Res* **42:** 299–318.

Alstermark B, Kummel H, Pinter MJ, Tantisira B. 1990. Integration in descending motor pathways controlling the forelimb in the cat. 17. Axonal projection and termination of C3-C4 propriospinal neurones in the C6-Th1 segments. *Exp Brain Res* **81:** 447–461.

Andersen RA, Snyder LH, Bradley DC, Xing J. 1997. Multimodal representation of space in the posterior parietal cortex and its use in planning movements. *Annu Rev Neurosci* **20:** 303–330.

Arshavsky YI, Gelfand IM, Orlovsky GN, Pavlova GA. 1978a. Messages conveyed by spinocerebellar pathways during scratching in the cat. I. Activity of neurons of the lateral reticular nucleus. *Brain Res* **151:** 479–491.

Arshavsky YI, Gelfand IM, Orlovsky GN, Pavlova GA. 1978b. Messages conveyed by spinocerebellar pathways during scratching in the cat. II. Activity of neurons of the ventral spinocerebellar tract. *Brain Res* **151:** 493–506.

Ashrafi S, Betley JN, Comer JD, Brenner-Morton S, Bar V, Shimoda Y, Watanabe K, Peles E, Jessell TM, Kaltschmidt JA. 2014. Neuronal Ig/Caspr recognition promotes the formation of axoaxonic synapses in mouse spinal cord. *Neuron* **81:** 120–129.

Azim E. 2014. Neuroscience. Shortcuts and checkpoints on the road to skilled movement. *Science* **346:** 554–555.

Azim E, Alstermark B. 2015. Skilled forelimb movements and internal copy motor circuits. *Curr Opin Neurobiol* **33:** 16–24.

Azim E, Jiang J, Alstermark B, Jessell TM. 2014. Skilled reaching relies on a V2a propriospinal internal copy circuit. *Nature* **508:** 357–363.

Baldissera F, Hultborn H, Illert M. 1981. Integration in spinal neuronal systems. In *Handbook of physiology, Sect. I: The nervous system, Vol. II: Motor control, Part I* (ed. Brookhart JM, Mountcastle VB, Brooks VB), pp. 509–595. American Physiological Society, Bethesda, MD.

Bastian AJ. 2011. Moving, sensing and learning with cerebellar damage. *Curr Opin Neurobiol* **21:** 596–601.

Betley JN, Wright CV, Kawaguchi Y, Erdélyi F, Szabó G, Jessell TM, Kaltschmidt JA. 2009. Stringent specificity in the construction of a GABAergic presynaptic inhibitory circuit. *Cell* **139:** 161–174.

Bhanpuri NH, Okamura AM, Bastian AJ. 2013. Predictive modeling by the cerebellum improves proprioception. *J Neurosci* **33:** 14301–14306.

Bizzi E, Cheung VC, d'Avella A, Saltiel P, Tresch M. 2008. Combining modules for movement. *Brain Res Rev* **57:** 125–133.

Brenner N, Bialek W, de Ruyter van Steveninck R. 2000. Adaptive rescaling maximizes information transmission. *Neuron* **26:** 695–702.

Brodal A. 1949. Spinal afferents to the lateral reticular nucleus of the medulla oblongata in the cat. An experimental study. *J Comp Neurol* **91:** 259–295, incl 252 pl.

Bui TV, Akay T, Loubani O, Hnasko TS, Jessell TM, Brownstone RM. 2013. Circuits for grasping: Spinal dI3 interneurons mediate cutaneous control of motor behavior. *Neuron* **78:** 191–204.

Capaday C, Stein RB. 1987. A method for simulating the reflex output of a motoneuron pool. *J Neurosci Meth* **21:** 91–104.

Capaday C, Stein RB. 1989. The effects of postsynaptic inhibition on the monosynaptic reflex of the cat at different levels of motoneuron pool activity. *Exp Brain Res* **77:** 577–584.

Carpenter MB, Nova HR. 1960. Descending division of the brachium conjunctivum in the cat a cerebello-reticular system. *J Comp Neurol* **114:** 295–305.

Clendenin M, Ekerot CF, Oscarsson O, Rosén I. 1974. Functional organization of two spinocerebellar paths relayed through the lateral reticular nucleus in the cat. *Brain Res* **69:** 140–143.

Conradi S. 1969. Ultrastructure of dorsal root boutons on lumbosacral motoneurons of the adult cat, as revealed by dorsal root section. *Acta Physiol Scand Suppl* **332:** 85–115.

Crevecoeur F, Scott SH. 2013. Priors engaged in long-latency responses to mechanical perturbations suggest a rapid update in state estimation. *PLoS Comput Biol* **9:** e1003177.

Desmurget M, Grafton S. 2000. Forward modeling allows feedback control for fast reaching movements. *Trends Cogn Sci* **4:** 423–431.

Dimitriou M, Franklin DW, Wolpert DM. 2012. Task-dependent coordination of rapid bimanual motor responses. *J Neurophysiol* **107:** 890–901.

Eccles JC, Eccles RM, Magni F. 1961. Central inhibitory action attributable to presynaptic depolarization produced by muscle afferent volleys. *J Physiol* **159:** 147–166.

Eccles JC, Kostyuk PG, Schmidt RF. 1962. Central pathways responsible for depolarization of primary afferent fibres. *J Physiol* **161:** 237–257.

Ekerot CF. 1990. The lateral reticular nucleus in the cat. VIII. Excitatory and inhibitory projection from the bilateral ventral flexor reflex tract (bVFRT). *Exp Brain Res* **79:** 129–137.

Fink AJ, Croce KR, Huang ZJ, Abbott LF, Jessell TM, Azim E. 2014. Presynaptic inhibition of spinal sensory feedback ensures smooth movement. *Nature* **509:** 43–48.

Ford JM, Mathalon DH. 2012. Anticipating the future: Automatic prediction failures in schizophrenia. *Int J Psychophysiol* **83:** 232–239.

Frank K, Fuortes M. 1957. Presynaptic and postsynaptic inhibition of monosynaptic reflexes. *Fed Proc* **16:** 49–50.

Franklin DW, Wolpert DM. 2008. Specificity of reflex adaptation for task-relevant variability. *J Neurosci* **28:** 14165–14175.

Franklin DW, Wolpert DM. 2011. Computational mechanisms of sensorimotor control. *Neuron* **72:** 425–442.

Fuentes CT, Bastian AJ. 2010. Where is your arm? Variations in proprioception across space and tasks. *J Neurophysiol* **103:** 164–171.

Fujito Y, Aoki M. 1995. Monosynaptic rubrospinal projections to distal forelimb motoneurons in the cat. *Exp Brain Res* **105:** 181–190.

Ghez C, Gordon J, Ghilardi MF. 1995. Impairments of reaching movements in patients without proprioception. II. Effects of visual information on accuracy. *J Neurophysiol* **73:** 361–372.

Gordon J, Ghilardi MF, Ghez C. 1995. Impairments of reaching movements in patients without proprioception. I. Spatial errors. *J Neurophysiol* **73:** 347–360.

Gray EG. 1962. A morphological basis for pre-synaptic inhibition? *Nature* **193:** 82–83.

Haith AM, Krakauer JW. 2013. Theoretical models of motor control and motor learning. In *The Routledge handbook of motor control and motor learning* (ed. Gollhofer A, Taube W, Bo Nielsen J), pp. 7–28. Routledge, London.

Harris CM, Wolpert DM. 1998. Signal-dependent noise determines motor planning. *Nature* **394:** 780–784.

Hughes DI, Mackie M, Nagy GG, Riddell JS, Maxwell DJ, Szabó G, Erdélyi F, Veress G, Szucs P, Antal M, et al. 2005. P boutons in lamina IX of the rodent spinal cord express high levels of glutamic acid decarboxylase-65 and originate from cells in deep medial dorsal horn. *Proc Natl Acad Sci* **102:** 9038–9043.

Hultborn H, Meunier S, Pierrot-Deseilligny E, Shindo M. 1987. Changes in presynaptic inhibition of Ia fibres at the onset of voluntary contraction in man. *J Physiol* **389:** 757–772.

Illert M, Lundberg A, Padel Y, Tanaka R. 1978. Integration in descending motor pathways controlling the forelimb in the cat. 5. Properties of and monosynaptic excitatory convergence on C3–C4 propriospinal neurones. *Exp Brain Res* **33:** 101–130.

Illert M, Jankowska E, Lundberg A, Odutola A. 1981. Integration in descending motor pathways controlling the forelimb in the cat. 7. Effects from the reticular formation on C3-C4 propriospinal neurones. *Exp Brain Res* **42:** 269–281.

Isa T, Ohki Y, Alstermark B, Pettersson LG, Sasaki S. 2007. Direct and indirect cortico-motoneuronal pathways and control of hand/arm movements. *Physiology (Bethesda)* **22:** 145–152.

Ito M. 2008. Control of mental activities by internal models in the cerebellum. *Nat Rev Neurosci* **9:** 304–313.

Iwaniuk AN, Whishaw IQ. 2000. On the origin of skilled forelimb movements. *Trends Neurosci* **23:** 372–376.

Jankowska E. 2001. Spinal interneuronal systems: Identification, multifunctional character and reconfigurations in mammals. *J Physiol* **533:** 31–40.

Jankowska E, Hammar I. 2013. Interactions between spinal interneurons and ventral spinocerebellar tract neurons. *J Physiol* **591:** 5445–5451.

Jones KE, Hamilton AF, Wolpert DM. 2002. Sources of signal-dependent noise during isometric force production. *J Neurophysiol* **88:** 1533–1544.

Kawato M, Gomi H. 1992. The cerebellum and VOR/OKR learning models. *Trends Neurosci* **15:** 445–453.

Kinoshita M, Matsui R, Kato S, Hasegawa T, Kasahara H, Isa K, Watakabe A, Yamamori T, Nishimura Y, Alstermark B, et al. 2012. Genetic dissection of the circuit for hand dexterity in primates. *Nature* **487:** 235–238.

Kuno M. 1964. Mechanism of facilitation and depression of the excitatory synaptic potential in spinal motoneurones. *J Physiol* **175:** 100–112.

Künzle H. 1973. The topographic organization of spinal afferents to the lateral reticular nucleus of the cat. *J Comp Neurol* **149:** 103–115.

Laughlin SB. 1994. Matching coding, circuits, cells, and molecules to signals: General principles of retinal design in the fly's eye. *Prog Retin Eye Res* **13:** 165–196.

Lisberger SG. 2009. Internal models of eye movement in the floccular complex of the monkey cerebellum. *Neuroscience* **162:** 763–776.

Lomelí J, Quevedo J, Linares P, Rudomin P. 1998. Local control of information flow in segmental and ascending collaterals of single afferents. *Nature* **395:** 600–604.

Lundberg A. 1971. Function of the ventral spinocerebellar tract. A new hypothesis. *Exp Brain Res* **12:** 317–330.

Matsushita M, Ikeda M. 1976. Projections from the lateral reticular nucleus to the cerebellar cortex and nuclei in the cat. *Exp Brain Res* **24:** 403–421.

Messier J, Kalaska JF. 1999. Comparison of variability of initial kinematics and endpoints of reaching movements. *Exp Brain Res* **125:** 139–152.

Miri A, Azim E, Jessell TM. 2013. Edging toward entelechy in motor control. *Neuron* **80:** 827–834.

Morasso P. 1981. Spatial control of arm movements. *Exp Brain Res* **42:** 223–227.

Pierrot-Deseilligny E, Burke DJ. 2012. *The circuitry of the human spinal cord: Neuroplasticity and corticospinal mechanisms*. Cambridge University Press, Cambridge.

Pivetta C, Esposito MS, Sigrist M, Arber S. 2014. Motor-circuit communication matrix from spinal cord to brainstem neurons revealed by developmental origin. *Cell* **156:** 537–548.

Poulet JF, Hedwig B. 2007. New insights into corollary discharges mediated by identified neural pathways. *Trends Neurosci* **30:** 14–21.

Prablanc C, Martin O. 1992. Automatic control during hand reaching at undetected two-dimensional target displacements. *J Neurophysiol* **67:** 455–469.

Prochazka A. 1999. Quantifying proprioception. *Prog Brain Res* **123:** 133–142.

Pruszynski JA, Kurtzer I, Nashed JY, Omrani M, Brouwer B, Scott SH. 2011. Primary motor cortex underlies multi-joint integration for fast feedback control. *Nature* **478:** 387–390.

Rothwell JC, Traub MM, Day BL, Obeso JA, Thomas PK, Marsden CD. 1982. Manual motor performance in a deafferented man. *Brain* **105(Pt 3):** 515–542.

Rudomin P. 2009. In search of lost presynaptic inhibition. *Exp Brain Res* **196:** 139–151.

Rudomin P, Schmidt RF. 1999. Presynaptic inhibition in the vertebrate spinal cord revisited. *Exp Brain Res* **129:** 1–37.

Rudomin P, Solodkin M, Jiménez I. 1986. PAD and PAH response patterns of group Ia- and Ib-fibers to cutaneous and descending inputs in the cat spinal cord. *J Neurophysiol* **56:** 987–1006.

Schmidt RF. 1971. Presynaptic inhibition in the vertebrate central nervous system. *Ergeb Physiol* **63:** 20–101.

Schneider DM, Nelson A, Mooney R. 2014. A synaptic and circuit basis for corollary discharge in the auditory cortex. *Nature* **513:** 189–194.

Schultz W, Montgomery EB Jr, Marini R. 1979. Proximal limb movements in response to microstimulation of primate dentate and interpositus nuclei mediated by brain-stem structures. *Brain* **102:** 127–146.

Scott SH. 2004. Optimal feedback control and the neural basis of volitional motor control. *Nat Rev Neurosci* **5:** 532–546.

Scott SH, Crevecoeur F. 2014. Neuroscience: Feedback throttled down for smooth moves. *Nature* **509:** 38–39.

Seki K, Perlmutter SI, Fetz EE. 2003. Sensory input to primate spinal cord is presynaptically inhibited during voluntary movement. *Nat Neurosci* **6:** 1309–1316.

Shadmehr R, Krakauer JW. 2008. A computational neuroanatomy for motor control. *Exp Brain Res* **185:** 359–381.

Shadmehr R, Smith MA, Krakauer JW. 2010. Error correction, sensory prediction, and adaptation in motor control. *Annu Rev Neuroscience* **33:** 89–108.

Shergill SS, White TP, Joyce DW, Bays PM, Wolpert DM, Frith CD. 2014. Functional magnetic resonance imaging of impaired sensory prediction in schizophrenia. *JAMA Psychiatry* **71:** 28–35.

Soteropoulos DS, Baker SN. 2008. Bilateral representation in the deep cerebellar nuclei. *J Physiol* **586:** 1117–1136.

Stein RB, Oğuztöreli MN. 1976. Tremor and other oscillations in neuromuscular systems. *Biol Cybern* **22:** 147–157.

Strick PL, Dum RP, Fiez JA. 2009. Cerebellum and nonmotor function. *Annu Rev Neurosci* **32:** 413–434.

Taniguchi H, He M, Wu P, Kim S, Paik R, Sugino K, Kvitsiani D, Fu Y, Lu J, Lin Y, et al. 2011. A resource of Cre driver lines for genetic targeting of GABAergic neurons in cerebral cortex. *Neuron* **71:** 995–1013.

Tian N, Petersen C, Kash S, Baekkeskov S, Copenhagen D, Nicoll R. 1999. The role of the synthetic enzyme GAD65 in the control of neuronal γ-aminobutyric acid release. *Proc Natl Acad Sci* **96:** 12911–12916.

Todorov E, Jordan MI. 2002. Optimal feedback control as a theory of motor coordination. *Nature Neurosci* **5:** 1226–1235.

Walmsley B, Wieniawa-Narkiewicz E, Nicol MJ. 1987. Ultrastructural evidence related to presynaptic inhibition of primary muscle afferents in Clarke's column of the cat. *J Neurosci* **7:** 236–243.

Watson AH, Bazzaz AA. 2001. GABA and glycine-like immunoreactivity at axoaxonic synapses on 1a muscle afferent terminals in the spinal cord of the rat. *J Comp Neurol* **433:** 335–348.

Whishaw IQ. 1996. An endpoint, descriptive, and kinematic comparison of skilled reaching in mice (*Mus musculus*) with rats (*Rattus norvegicus*). *Behav Brain Res* **78:** 101–111.

Wolpert DM, Miall RC. 1996. Forward models for physiological motor control. *Neural Netw* **9:** 1265–1279.

Wu HS, Sugihara I, Shinoda Y. 1999. Projection patterns of single mossy fibers originating from the lateral reticular nucleus in the rat cerebellar cortex and nuclei. *J Comp Neurol* **411:** 97–118.

Xu W, Jones S, Edgley SA. 2013. Event time representation in cerebellar mossy fibres arising from the lateral reticular nucleus. *J Physiol* **591:** 1045–1062.

Xu-Wilson M, Zee DS, Shadmehr R. 2009. The intrinsic value of visual information affects saccade velocities. *Exp Brain Res* **196:** 475–481.

Computations in Sensorimotor Learning

DANIEL M. WOLPERT
Computational and Biological Learning, Department of Engineering,
University of Cambridge, Cambridge CB2 1PZ, United Kingdom
Correspondence: wolpert@eng.cam.ac.uk

Our cognitive abilities can only be expressed on the world through our actions. Here we review the computations underlying the way that the sensorimotor system converts both low-level sensory signals and high-level decisions into action, focusing on the behavioral evidence for the theoretical frameworks. We review recent work that determines how motor memories underlying sensorimotor learning are activated and protected from interference, the role of Bayesian decision theory in sensorimotor control including sources of suboptimality, the role of risk sensitivity in guiding action, and how rapid motor responses may underlie the robustness of the motor system to the vagaries of the world.

Although sensorimotor learning feels to us like a unitary process, its study is often broken down into a number of interacting processes. Here we review several aspects of sensorimotor learning that have seen progress in recent years and focus on computational approaches that have been studied behaviorally in humans. We start by examining how motor memories for different skills are organized, focusing on the rules by which different motor memories are parcellated and accessed. We then focus on the issue of online control and trajectory generation reviewing the theory of optimal feedback control and behavioral tests of this theory. Finally, we focus on the interaction between decision-making and sensorimotor control, examining their bidirectional interaction.

MOTOR MEMORIES

Our ability to learn and store multiple motor skills underlies the extraordinary flexibility with which we can revisit old skills, tools, and environments. A great deal of effort has been expended in trying to understand the rules by which novel skills are parcellated, stored, and protected from interference but yet reactivated and amended as a particular task is reexperienced. The use of motor tasks that are in some sense opposite to each other and therefore interfere has been a powerful tool to study motor memories within the laboratory setting (Brashers-Krug et al. 1996; Krakauer et al. 1999; Karniel and Mussa-Ivaldi 2002; Caithness et al. 2004). For example, robotic interfaces can be used to apply novel dynamics on the arm by having a subject grasp a handle that generates forces on the hand that typically depend on the state of the hand. Using such devices one can, for example, generate velocity-dependent forces that act at right angles to the direction of current motion—the so-called velocity-dependent curl field (Gandolfo et al. 1996). Such a field can be applied in either a clockwise (CW) or a counterclockwise (CCW) direction. When experiencing a single force field, subjects learn to compensate for the dynamics, reverting to the approximately straight-line movements with bell-shaped velocity profiles before the perturbation. However, if CW and CCW fields are intermingled (either randomly or in batches) minimal learning is seen with such opposing force fields leading to strong interference. Even with contextual cues such as overlaying color or a virtual tool whose features fully predict the direction of the force field, learning is at best modest (Howard et al. 2013). This suggests that such opposing skills tend to compete for the same motor memory.

Contextual Rules for Motor Memories

Recently, it has been shown that more dynamic features of a task can allow separate motor memories to form. For example, it is possible to learn opposing force fields in one arm if motion of the other arm (present or absent) indicates the identity of the force field (Nozaki et al. 2006) or if the other arm makes concurrent and different movements for the two force fields (Howard et al. 2010). These results suggest that the coding of motor memories can depend on the conjunction of the movements of the two arms. Other features of movement can also be used to learn opposing perturbations. For example, if similar movements are made that are discrete or rhythmic, opposing perturbations can be learned when each is associated with a different type of movement (Howard et al. 2011). Separating the location of learning either proprioceptively or visually also facilitates learning of opposing force fields (Hwang et al. 2006; Hirashima and Nozaki 2012; Howard et al. 2013). Modeling approaches have suggested that such contextual effects arise from the engagement of separate neural populations (Nozaki and Scott 2009).

Recently, we have examined the effect of the lead-in to a movement. Although there was already evidence for some sequential effects on the ability to learn opposing

 doi: 10.1101/sqb.2014.79.024919

force fields (Wainscott et al. 2005), we examined a task in which we systematically varied the dwell time after a lead-in to a movement that would then be made in a force field (Howard et al. 2012). For example, the lead-in could be made from two different starting locations, followed by a single movement through either a CW or CCW force field. Critically the starting location was fully predictive of the force-field direction that would subsequently be experienced. When subjects were required to dwell for 1000 msec after the lead-in movement before proceeding to the movement through the field, they were unable to learn to compensate for the opposing force fields when randomly interleaved. However, as the dwell time was reduced to <500 msec, learning increased, and with minimal dwell time there was substantial learning. This suggests that the recent history (~ 500 msec) of the state of the arm before experiencing the field affects the currently active motor memory and the more different the history is the greater the ability to separate motor memories. Interestingly, this effect did not require active movement, as it was seen even if the hand was passively moved for the lead-in.

This effect of recent states on separating motor memories led us to explore whether this effect could potentially explain the value of the follow-through in sports. Despite the fact that motion after contact or release with a ball cannot affect the flight of the ball, coaches have learned that the follow-through is important for skilled performance. We hypothesized that future actions that you plan to make might select current motor memories, and therefore different follow-throughs would preactivate distinct motor memories. We showed that it is possible to learn opposing force fields if the follow-through after the movement in the force field is different for each of the fields (Howard et al. 2015). This suggests that the value of making a consistent follow-through is that it activates a single motor memory leading to faster learning. We confirmed this by showing that when learning a single force field, learning is faster if a consistent follow-through is made after each movement through the field compared with when the follow-through is variable across trials. This gives a new interpretation to the value of the follow-through, that rather being specified so that you can optimize planning it may in fact optimize the learning. The neural underpinning of such follow-through effects is not known. However, potential future motor plans will certainly affect neural activity (Cisek and Kalaska 2005) so that different follow-throughs may directly engage separate neural populations that lead to the generation of movement. An alternative, but complementary perspective is that by viewing motor generation as a dynamical system of interacting neurons (Churchland et al. 2012) different follow-throughs will require different initial states of the dynamical system, each of which could be associated with the different force fields.

Structural Learning

Rather than using cues to allow distinct memories to be learned, we have shown that experiencing a variable perturbation increases our ability to learn a new perturbation that comes from the same structure but not from other structures (Braun et al. 2009; Turnham et al. 2012). For example, having experienced visuomotor rotations drawn uniformly from -90° to $+90^\circ$, subjects can then readily learn a fixed $+60^\circ$ rotation and show minimal interference when they transition to a -60° rotation. We proposed that this ability depends on learning the structure, for example, of rotations. Two recent studies have suggested a more mechanistic explanation for the ability to learn from random experience. One suggests that such learning simply leads to increasing error sensitivity globally (Herzfeld et al. 2014), whereas the other suggests that we learn the appropriate weights for primitives (for the set of visuomotor errors we might experience) and that these are activated by prospective (predicted) errors (Takiyama et al. 2015). However, currently there are two issues that these models fail to address, which leaves the question of structural learning open. First, the models already assume the structure such as rotations in the underlying model so half the problem is effectively pre-solved. More troublesome is that the ability to generalize critically depends not on simply experiencing errors that correlate across trials (both models require this) but that the structure be consistent. We found no increased learning of visuomotor rotations when subjects experienced a range of linear visuomotor transformations that includes rotations, shearing, and scalings (Braun et al. 2009). Similarly, for groups of subjects exposed to random shearing or to random scalings, we found facilitation for shearing and scalings, respectively. Several other groups have also provided evidence that the structure of tasks is learned and used to appropriately make feedback and feed-forward adjustments (Yousif and Diedrichsen 2012) and that the degree to which tasks share similar structures determines the ability of people to generalize between them (Ranganathan et al. 2014).

Tool Use

Although learning of force fields and arbitrary visuomotor perturbations has led to a great deal of understanding, it may not reflect most of the learning we perform as adults. That is because although we need to learn the dynamics of our arms or of tools we interact with, similar to force-field learning, this may happen early in life. As an adult we might expect to have rather good structural models of how arms, hammers, and bicycles behave, and what we have to learn is the particular parameters of these such as length, mass, and friction. We have proposed that in naturalistic settings (Ingram and Wolpert 2011), such as when you operate on a tool such as a hammer, using vision you can estimate the structure of the task (assuming you have experience with hammers) but what you lack is precise knowledge of the parameters of the hammer (mass, density distributions, etc.). Consistent with this is that when you operate on a virtual hammer (e.g., rotating it at the handle), the forces that you generate even before any experience of the dynamics is in the correct direction to compensate for the forces induced on the handle by visual motion of the hammer head but for an estimated magnitude (Ingram et al. 2010, 2013). Learning requires adaptation of the magnitude across trials. Such magnitude learning appears to be local to the orientation of the handle in the hand, suggesting that learning of parameters of tools, unlike the structure, is grasp-specific rather than holistic. Consistent with a qualitative difference between these situations is the different time course of learning for force fields compared with more naturalistic learning with tools, with the former an interaction of two processes (a fast and slow) (Smith et al. 2006) and the latter mediated by a single process (Ingram et al. 2011).

ONLINE CONTROL

Our actions are undoubtably variable in that even if we wish to make the same movement repeatedly, we clearly have some variability that cannot be fully controlled. There is currently a debate as to the source of such variability, with different proponents suggesting that substantial variability can arise from sensory noise (Osborne et al. 2005), from suboptimal computation (Beck et al. 2012), or from motor noise (van Beers 2007; Faisal et al. 2008; Chaisanguanthum et al. 2014). However, even if only a proportion of variability arises from motor noise, it still places a premium on planning movement so as to reduce this component of variability, and current models of trajectory control place an emphasis on minimizing costs, which are a mixture of energy and accuracy (for review, see Scott 2012). Such models fall within the framework of optimal feedback control (OFC; Todorov and Jordan 2002; Diedrichsen et al. 2010), which suggests that people do not plan desired trajectories but instead generate time-dependent feedback controllers with the aim of minimizing the movement cost in the presence of potential disturbances. A number of studies have confirmed predictions of OFC in terms of the task-dependent corrections that it predicts (e.g., Diedrichsen 2007; White and Diedrichsen 2010). In addition, the framework of OFC has placed reflexes, or rapid motor responses, at the center of online control, as these reflect the feedback gains and OFC proposes that these should be highly tuned to the task at hand.

Rapid Motor Responses

Several studies have shown elegantly that many rapid motor responses are highly tuned to task demands (Kurtzer et al. 2008; Pruszynski et al. 2008; Yang et al. 2011; Pruszynski and Scott 2012). We have further explored task-dependent reflex tuning and found that both visuomotor and stretch reflexes are highly modifiable by task demands. For example, the visuomotor reflex can be both up-regulated and down-regulated when visuomotor perturbations are task-relevant or task-irrelevant, respectively (Franklin and Wolpert 2008). Moreover, even at the same point within a movement, the visuomotor reflex gain can be regulated to respond differently for leftward or rightward displacements relative to the movement direction (Franklin et al. 2014). Indeed such up-regulation of visuomotor gains is also seen under the initial period of learning dynamic perturbations (Franklin et al. 2012). Such feedback gains can also be examined over the course of a movement and show a time-dependent modulation through the movement that is qualitatively in agreement with predictions of OFC (Dimitriou et al. 2013). Moreover, these gains can adapt within a movement to changes in task demands such as target jumps within 100 msec. Taken together, studies of reflexes within a limb show a level of sophistication that suggests a good match to task demands.

Recently reflexes between limbs have been studied and these show that stretch reflexes between the limbs are also modifiable by task demands (Dimitriou et al. 2012; Omrani et al. 2013). Consider holding a tray with both hands that needs to be maintained level. Vertical perturbations of one hand can be compensated by both reflex responses in the opposite direction to the perturbation in that arm or a response in the same direction in the other arm, both of these will restore the level of the tray. However, if both hands are perturbed vertically in the same direction then there is less need for reflex responses and conversely when they are perturbed in opposite direction reflex responses are needed in both arms. We examined just such a situation in which we could perturb one arm or both arms in the same or opposite direction and examined the rapid motor response (through electromyography [EMG] activity) and found clear task-dependent tuning of reflexes modulated online within a single trial based on the interaction across the arms and the nature of the perturbation (Dimitriou et al. 2012).

Risk Sensitivity

In general, optimal feedback control suggests that what we care about is the expected, or average, cost of the movement. Recently, we have examined the extent to which the variance of the cost is also important to subjects (for review, see Braun et al. 2011). Theories of risk sensitivity posit that the cost is a trade-off between the mean cost and variance, with those who are willing to accept a higher cost to either reduce variance or increase variance being risk-averse or risk-seeking. In a series of sensorimotor tasks we have shown that indeed subjects are often risk-averse or risk-seeking and do not simply optimize the expected reward (Nagengast et al. 2010, 2011a,b).

DECISION-MAKING AND SENSORIMOTOR CONTROL

Both decision-making and motor control require acting in real time on streams of noisy evidence. Thus both rely on inference, termination rules, time constraints, and value/effort costs. Up to now, these two fields have evolved in parallel. Decision-making has focused on complex streams of evidence from the environment but simple, ballistic movements. Conversely, motor control has focused on complex control problems but relatively simple stimuli. Both fields have computational frameworks that underpin them. Decision-making has benefited from the framework of bounded evidence accumulation (e.g., drift-diffusion models) whereby a stream of noisy information is integrated to a threshold, which terminates the decision and establishes the choice outcome (for review, see Shadlen and Kiani 2013). This framework can explain both the timing and the accuracy of behavioral decisions and is supported by a wealth of physiology. Sensorimotor control has benefited from a mathematical framework of optimal feedback control, whereby costs associated with accuracy and effort are combined to produce a time-varying feedback controller. Bayesian decision theory has rapidly become a unifying framework in which to understand how sensorimotor processes deal with uncertainty to make inferences, make decisions, and generate actions.

In general, we tend to perceive our own actions differently to how we perceive actions of others, and most people overrate their ability to drive (most people rate themselves as above average). Recently we have examined whether the prior we have for our own actions are the same as priors for other agent's actions (Wolpe et al. 2014). We found that when performing an action ourselves we tend to have a prior for errors that is narrower than our true error distribution, leading to us perceiving our own performance as better than it really is. In contrast, when we watch another agent perform the same task (in fact our own movements played back to us) the prior now matches our error distribution leading to veridical perception. This suggests that in a simple sensorimotor task overoptimism may arise from our subjective prior being narrower than it should be.

Bayesian Decision Theory

Key to this framework is that humans learn priors (i.e., the statistical distribution) for tasks they need to perform. Many studies have shown that when a prior is experimentally placed upon a task subjects tend to learn the prior and use it in Bayesian way (e.g., Adams et al. 2004; Körding and Wolpert 2004). However, this is true only for simple priors such as Gaussians, and people have much more trouble learning complex priors, at least in a laboratory setting. To examine the sources of such suboptimality we examined whether it was the learning of the prior or the use of the prior that was limited. Subjects performed a Bayesian integration task in which the prior was explicitly presented and therefore did not need to be learned (Acerbi et al. 2014). Our results suggest suboptimality may be primarily due to a problem of acquiring the priors rather than computing with them.

Rather than applying priors it is possible to try to infer the subjects' natural priors, although such examination has until recently been limited to rather simple representations of the prior (Weiss et al. 2002; Turnham et al. 2011). We have recently developed a new method, termed cognitive tomography, which takes simple low-dimensional responses from subjects to calculate high-dimensional priors (Houlsby et al. 2013). As an example, we applied this method to the prior over the physical structure of faces by asking subjects to perform two different tasks. In one task they were presented with computer-generated pairs of faces and had to respond as to which face was more familiar, and in a second task they saw three faces and had to select which face they regarded as the odd one out. Given a generative model of these tasks we could infer each participant's subjective prior over the structure of faces (represented in the first two principal components of facial structure). We found that the priors were complex (not simply Gaussian) and varied widely across participants yet were very similar within each participant across the tasks. Moreover, we could use the prior extracted from one task to predict the performance on the other task. These results provide support for the ideas that there is a single high-dimensional prior for faces that guides behavior in multiple tasks.

Bidirectional Interplay of Decisions and Control

To examine the relation between high-level decision-making and sensorimotor control we used a task in which we could estimate both the evolving decision variable (from drift-diffusion models) and stretch reflex (from OFC) to relate these before a decision was made. The question we addressed is whether decision-making and motor control are a sequential process, accumulation to a bound followed by motor preparation and action, or whether there could be a more continuous flow from the decision process to the motor system. To do this, subjects were required to make a decision of the direction of motion of a random dot kinematogram and indicate the direction by an elbow flexion or extension movement when the display was extinguished (Selen et al. 2012). At the moment the display was extinguished a robotic device rapidly extended the elbow, setting up a stretch reflex whose magnitude could be measured using EMG. By varying the stimulus duration and difficulty (motion coherence) and fitting a model to the behavioral performance we could estimate the decision variable on each trial and relate this to the magnitude of the stretch reflex. We found that the stretch reflex tracked the evolving decision variable from even the smallest amounts of information, suggesting a continuous flow from the decision process (even before a decision has been made) to the motor system, allowing the motor system to pretune its reflexes so as to get to the appropriate target faster.

Such online interactions between the decision and motor systems can also be seen in a change-of-mind phenomenon. We examined subjects performing a simple motion direction discrimination task in which they had to indicate their decision (leftward versus rightward motion) by reaching to one of two choice targets (Resulaj et al. 2009). On initiation of the movement the stimulus was extinguished. Nevertheless, on a proportion of trials subjects appeared to change their mind, reaching initially for one target before deviating to the other. Such changes of mind improved performance even though the stimulus was extinguished. A consideration of the time delays in the sensory, motor, and decision process suggests that even though the stimulus is absent there is further information on the processing pipeline (~350 msec) that could be accumulated during the movement. We accounted for the pattern of the changes of mind with a simple model that assumed that after reaching a bound to make the initial decision, if a change-of-mind bound is then reached (on the opposite side of zero from the original bound but not requiring as much evidence as for the initial decision), then subjects will change their initial decision. Moreover, we and others have shown that the placement of this change-of-mind bound is dependent on the physical effort required to change one's mind (Burk et al. 2014; Moher and Song 2014). For example, more evidence is required to change one's mind if the choice targets are widely separated compared with when they are close together.

CONCLUSION

We have reviewed several aspects of sensorimotor control from motor memories underlying learning, online control, and the interaction of decision-making and control. Several themes have started to emerge within the field. First, the field of sensorimotor control is making links to others fields such as decision-making, working memory (e.g., Ma et al. 2014), and explicit and implicit processes (e.g., Taylor et al. 2014; Trewartha et al. 2014), and such links are likely to become increasingly important. Second, new theories are starting to emerge that can be tested on increasingly large sets of experimental data (e.g., Takiyama et al. 2015), and this is likely to provide an important framework for future theory development. Finally, as the behavioral theories develop, they are able to make increasingly important links to neurophysiological data (e.g., Azim et al. 2014; Zhou et al. 2014).

ACKNOWLEDGMENTS

This work was supported by Wellcome Trust, the Human Frontier Science Program, and a Royal Society Noreen Murray Professorship in Neurobiology to D.M.W.

REFERENCES

Acerbi L, Vijayakumar S, Wolpert DM. 2014. On the origins of suboptimality in human probabilistic inference. *PLoS Comput Biol* **10:** e1003661.

Adams WJ, Graf EW, Ernst MO. 2004. Experience can change the "light-from-above" prior. *Nat Neurosci* **7:** 1057–1058.

Azim E, Jiang J, Alstermark B, Jessell TM. 2014. Skilled reaching relies on a V2a propriospinal internal copy circuit. *Nature* **508:** 357–363.

Beck JM, Ma WJ, Pitkow X, Latham PE, Pouget A. 2012. Not noisy, just wrong: The role of suboptimal inference in behavioral variability. *Neuron* **74:** 30–39.

Brashers-Krug T, Shadmehr R, Bizzi E. 1996. Consolidation in human motor memory. *Nature* **382:** 252–255.

Braun DA, Aertsen A, Wolpert DM, Mehring C. 2009. Motor task variation induces structural learning. *Curr Biol* **19:** 352–357.

Braun DA, Nagengast AJ, Wolpert DM. 2011. Risk-sensitivity in sensorimotor control. *Front Hum Neurosci* **5:** 1.

Burk D, Ingram JN, Franklin DW, Shadlen MN, Wolpert DM. 2014. Motor effort alters changes of mind in sensorimotor decision making. *PLoS ONE* **9:** e92681.

Caithness G, Osu R, Bays P, Chase H, Klassen J, Kawato M, Wolpert DM, Flanagan JR. 2004. Failure to consolidate the consolidation theory of learning for sensorimotor adaptation tasks. *J Neurosci* **24:** 8662–8671.

Chaisanguanthum KS, Shen HH, Sabes PN. 2014. Motor variability arises from a slow random walk in neural state. *J Neurosci* **34:** 12071–12080.

Churchland MM, Cunningham JP, Kaufman MT, Foster JD, Nuyujukian P, Ryu SI, Shenoy KV. 2012. Neural population dynamics during reaching. *Nature* **487:** 51–56.

Cisek P, Kalaska JF. 2005. Neural correlates of reaching decisions in dorsal premotor cortex: Specification of multiple direction choices and final selection of action. *Neuron* **45:** 801–814.

Diedrichsen J. 2007. Optimal task-dependent changes of bimanual feedback control and adaptation. *Curr Biol* **17:** 1675–1679.

Diedrichsen J, Shadmehr R, Ivry RB. 2010. The coordination of movement: Optimal feedback control and beyond. *Trends Cogn Sci* **14:** 31–39.

Dimitriou M, Franklin DW, Wolpert DM. 2012. Task-dependent coordination of rapid bimanual motor responses. *J Neurophysiol* **107:** 890–901.

Dimitriou M, Wolpert DM, Franklin DW. 2013. The temporal evolution of feedback gains rapidly update to task demands. *J Neurosci* **33:** 10898–10909.

Faisal AA, Selen LP, Wolpert DM. 2008. Noise in the nervous system. *Nat Rev Neurosci* **9:** 292–303.

Franklin DW, Wolpert DM. 2008. Specificity of reflex adaptation for task-relevant variability. *J Neurosci* **28:** 14165–14175.

Franklin S, Wolpert DM, Franklin DW. 2012. Visuomotor feedback gains upregulate during the learning of novel dynamics. *J Neurophysiol* **108:** 467–478.

Franklin DW, Franklin S, Wolpert DM. 2014. Fractionation of the visuomotor feedback response to directions of movement and perturbation. *J Neurophysiol* **112:** 2218–2233.

Gandolfo F, Mussa-Ivaldi FA, Bizzi E. 1996. Motor learning by field approximation. *Proc Natl Acad Sci* **93:** 3843–3846.

Herzfeld DJ, Vaswani PA, Marko MK, Shadmehr R. 2014. A memory of errors in sensorimotor learning. *Science* **345:** 1349–1353.

Hirashima M, Nozaki D. 2012. Distinct motor plans form and retrieve distinct motor memories for physically identical movements. *Curr Biol* **22:** 432–436.

Houlsby N, Huszár F, Ghassemi M, Orbán G, Wolpert D, Lengyel M. 2013. Cognitive tomography reveals complex, task-independent mental representations. *Curr Biol* **23:** 2169–2175.

Howard IS, Ingram JN, Wolpert DM. 2010. Context-dependent partitioning of motor learning in bimanual movements. *J Neurophysiol* **104:** 2082–2091.

Howard IS, Ingram JN, Wolpert DM. 2011. Separate representations of dynamics in rhythmic and discrete movements: Evidence from motor learning. *J Neurophysiol* **105:** 1722–1731.

Howard IS, Ingram JN, Franklin DW, Wolpert DM. 2012. Gone in 0.6 seconds: The encoding of motor memories depends on recent sensorimotor states. *J Neurosci* **32:** 12756–12768.

Howard IS, Wolpert DM, Franklin DW. 2013. The effect of contextual cues on the encoding of motor memories. *J Neurophysiol* **109:** 2632–2644.

Howard IS, Wolpert DM, Franklin DW. 2015. The value of the follow-through derives from motor learning depending on future actions. *Curr Biol* **25:** 397–401.

Hwang EJ, Smith MA, Shadmehr R. 2006. Dissociable effects of the implicit and explicit memory systems on learning control of reaching. *Exp Brain Res* **173:** 425–437.

Ingram JN, Wolpert DM. 2011. Naturalistic approaches to sensorimotor control. *Prog Brain Res* **191:** 3–29.

Ingram JN, Howard IS, Flanagan JR, Wolpert DM. 2010. Multiple grasp-specific representations of tool dynamics mediate skillful manipulation. *Curr Biol* **20:** 618–623.
Ingram JN, Howard IS, Flanagan JR, Wolpert DM. 2011. A single-rate context-dependent learning process underlies rapid adaptation to familiar object dynamics. *PLoS Comput Biol* **7:** e1002196.
Ingram JN, Flanagan JR, Wolpert DM. 2013. Context-dependent decay of motor memories during skill acquisition. *Curr Biol* **23:** 1107–1112.
Karniel A, Mussa-Ivaldi FA. 2002. Does the motor control system use multiple models and context switching to cope with a variable environment? *Exp Brain Res* **143:** 520–524.
Körding KP, Wolpert DM. 2004. Bayesian integration in sensorimotor learning. *Nature* **427:** 244–247.
Krakauer JW, Ghilardi MF, Ghez C. 1999. Independent learning of internal models for kinematic and dynamic control of reaching. *Nat Neurosci* **2:** 1026–1031.
Kurtzer IL, Pruszynski JA, Scott SH. 2008. Long-latency reflexes of the human arm reflect an internal model of limb dynamics. *Curr Biol* **18:** 449–453.
Ma WJ, Husain M, Bays PM. 2014. Changing concepts of working memory. *Nat Neurosci* **17:** 347–356.
Moher J, Song JH. 2014. Perceptual decision processes flexibly adapt to avoid change-of-mind motor costs. *J Vis* **14:** 1.
Nagengast AJ, Braun DA, Wolpert DM. 2010. Risk-sensitive optimal feedback control accounts for sensorimotor behavior under uncertainty. *PLoS Comput Biol* **6:** e1000857.
Nagengast AJ, Braun DA, Wolpert DM. 2011a. Risk-sensitivity and the mean-variance trade-off: Decision making in sensorimotor control. *Proc Biol Sci* **278:** 2325–2332.
Nagengast AJ, Braun DA, Wolpert DM. 2011b. Risk sensitivity in a motor task with speed-accuracy trade-off. *J Neurophysiol* **105:** 2668–2674.
Nozaki D, Scott SH. 2009. Multi-compartment model can explain partial transfer of learning within the same limb between unimanual and bimanual reaching. *Exp Brain Res* **194:** 451–463.
Nozaki D, Kurtzer I, Scott SH. 2006. Limited transfer of learning between unimanual and bimanual skills within the same limb. *Nat Neurosci* **9:** 1364–1366.
Omrani M, Diedrichsen J, Scott SH. 2013. Rapid feedback corrections during a bimanual postural task. *J Neurophysiol* **109:** 147–161.
Osborne LC, Lisberger SG, Bialek W. 2005. A sensory source for motor variation. *Nature* **437:** 412–416.
Pruszynski JA, Scott SH. 2012. Optimal feedback control and the long-latency stretch response. *Exp Brain Res* **218:** 341–359.
Pruszynski JA, Kurtzer I, Scott SH. 2008. Rapid motor responses are appropriately tuned to the metrics of a visuospatial task. *J Neurophysiol* **100:** 224–238.
Ranganathan R, Wieser J, Mosier KM, Mussa-Ivaldi FA, Scheidt RA. 2014. Learning redundant motor tasks with and without overlapping dimensions: Facilitation and interference effects. *J Neurosci* **34:** 8289–8299.
Resulaj A, Kiani R, Wolpert DM, Shadlen MN. 2009. Changes of mind in decision-making. *Nature* **461:** 263–266.
Scott SH. 2012. The computational and neural basis of voluntary motor control and planning. *Trends Cogn Sci* **16:** 541–549.
Selen LP, Shadlen MN, Wolpert DM. 2012. Deliberation in the motor system: Reflex gains track evolving evidence leading to a decision. *J Neurosci* **32:** 2276–2286.
Shadlen MN, Kiani R. 2013. Decision making as a window on cognition. *Neuron* **80:** 791–806.
Smith MA, Ghazizadeh A, Shadmehr R. 2006. Interacting adaptive processes with different timescales underlie short-term motor learning. *PLoS Biol* **4:** e179.
Takiyama K, Hirashima M, Nozaki D. 2015. Prospective errors determine motor learning. *Nat Commun* **6:** 5925.
Taylor JA, Krakauer JW, Ivry RB. 2014. Explicit and implicit contributions to learning in a sensorimotor adaptation task. *J Neurosci* **34:** 3023–3032.
Todorov E, Jordan MI. 2002. Optimal feedback control as a theory of motor coordination. *Nat Neurosci* **5:** 1226–1235.
Trewartha KM, Garcia A, Wolpert DM, Flanagan JR. 2014. Fast but fleeting: Adaptive motor learning processes associated with aging and cognitive decline. *J Neurosci* **34:** 13411–13421.
Turnham EJ, Braun DA, Wolpert DM. 2011. Inferring visuomotor priors for sensorimotor learning. *PLoS Comput Biol* **7:** e1001112.
Turnham EJ, Braun DA, Wolpert DM. 2012. Facilitation of learning induced by both random and gradual visuomotor task variation. *J Neurophysiol* **107:** 1111–1122.
van Beers RJ. 2007. The sources of variability in saccadic eye movements. *J Neurosci* **27:** 8757–8770.
Wainscott SK, Donchin O, Shadmehr R. 2005. Internal models and contextual cues: Encoding serial order and direction of movement. *J Neurophysiol* **93:** 786–800.
Weiss Y, Simoncelli EP, Adelson EH. 2002. Motion illusions as optimal percepts. *Nat Neurosci* **5:** 598–604.
White O, Diedrichsen J. 2010. Responsibility assignment in redundant systems. *Curr Biol* **20:** 1290–1295.
Wolpe N, Wolpert DM, Rowe JB. 2014. Seeing what you want to see: Priors for one's own actions represent exaggerated expectations of success. *Front Behav Neurosci* **8:** 232.
Yang L, Michaels JA, Pruszynski JA, Scott SH. 2011. Rapid motor responses quickly integrate visuospatial task constraints. *Exp Brain Res* **211:** 231–242.
Yousif N, Diedrichsen J. 2012. Structural learning in feedforward and feedback control. *J Neurophysiol* **108:** 2373–2382.
Zhou K, Wolpert DM, De Zeeuw CI. 2014. Motor systems: Reaching out and grasping the molecular tools. *Curr Biol* **24:** R269–R271.

Neural Mechanisms Underlying Visual Object Recognition

Arash Afraz, Daniel L.K. Yamins, and James J. DiCarlo

Department of Brain and Cognitive Sciences and McGovern Institute for Brain Research, Massachusetts Institute of Technology, Cambridge, Massachusetts 02139

Correspondence: dicarlo@mit.edu

Invariant visual object recognition and the underlying neural representations are fundamental to higher-level human cognition. To understand these neural underpinnings, we combine human and monkey psychophysics, large-scale neurophysiology, neural perturbation methods, and computational modeling to construct falsifiable, predictive models that aim to fully account for the neural *encoding* and *decoding* processes that underlie visual object recognition. A predictive *encoding* model must minimally describe the transformation of the retinal image to population patterns of neural activity along the entire cortical ventral stream of visual processing and must accurately predict the responses to any retinal image. A predictive *decoding* model must minimally describe the transformation from those population patterns of neural activity to observed object recognition behavior (i.e., subject reports), and, given that population pattern of activity, it must accurately predict behavior for any object recognition task. To date, we have focused on *core object recognition*—a remarkable behavior that is accomplished with image viewing durations of <200 msec. Our work thus far reveals that the neural encoding process is reasonably well explained by a largely feed-forward, highly complex, multistaged nonlinear neural network—the current best neuronal simulation models predict approximately one-half of the relevant neuronal response variance across the highest levels of the ventral stream (areas V4 and IT). Remarkably, however, the decoding process from IT to behavior for all object recognition tasks tested thus far is very accurately predicted by simple direct linear conversion of the inferior temporal neural population state to behavior choice. We have recently examined the behavioral consequences of direct suppression of IT neural activity using pharmacological and optogenetic methods and find them to be well-explained by the same linear decoding model.

Humans rapidly and accurately process visual scenes from their environment, an ability that is critical for everyday functioning. The ease with which this task is accomplished belies the fact that it is an enormous computational feat: The retina presents visual cortex with a barrage of incoming photoreceptor activation patterns (images) from which higher-level structures (e.g., objects) must be extracted, even in the face of dramatic image variation.

The core of the problem is that the information that is explicit in the image (i.e., the luminance of every pixel) is misaligned with the needs of behavior. Relevant behavioral distinctions (e.g., telling the face of a friend from that of a foe) are swamped by low-level variations in pixel-space (e.g., the differences between the same friend's face under variable viewpoints). Thus, the challenge that the brain faces is producing a transform from pixels to a high-level representation that solves this misalignment (DiCarlo et al. 2012).

Extensive research in visual systems neuroscience suggests that this transform is affected by the ventral visual stream, a series of connected cortical areas—V1, V2, V4—situated in the occipital-temporal lobe of humans (Farah 1990; Malach et al. 1995), with homologous areas in nonhuman primates (Desimone et al. 1984; Tanaka 1996; Bell et al. 2009), that culminates in an area called inferior temporal cortex (IT). Somehow, the circuits of each of these areas have constructed a mapping from pixel space to a neuronal representation that supports complex visual behaviors. However, only a decade ago, these mechanisms were mysterious.

In our work on this problem, we find it useful to separate the problem into two basic components (Fig. 1): (1) understanding the transform from pixel stimuli to the neuronal representation in higher visual cortex—that is, the "Image → Neural population state" link—and (2) understanding the mapping from neural representation to visual behaviors—that is, the "Neural population state → Behavioral report" link. Here, what we mean by "understanding," is quite specific: namely, we seek falsifiable, predictive models of each of those links. Thus, we will consider our effort successful when it has produced two specific quantitative models: (1) a model whose inputs are arbitrary images, and whose outputs accurately describe the responses of each neuron in higher visual cortex to these images, and (2) a model whose inputs are neural response patterns, and whose outputs accurately predict behavioral choices across the whole variety of visual tasks supported by the ventral stream. These predictive models should not only be black boxes, but should be mapped to the underlying neural hardware so that they can be explicitly tested (e.g., potentially falsified) and leveraged for future applications such as improved computer vision and visual prosthetics.

In this paper, we discuss our approach and progress to building these two types of models, in the context of what

 doi: 10.1101/sqb.2014.79.024729

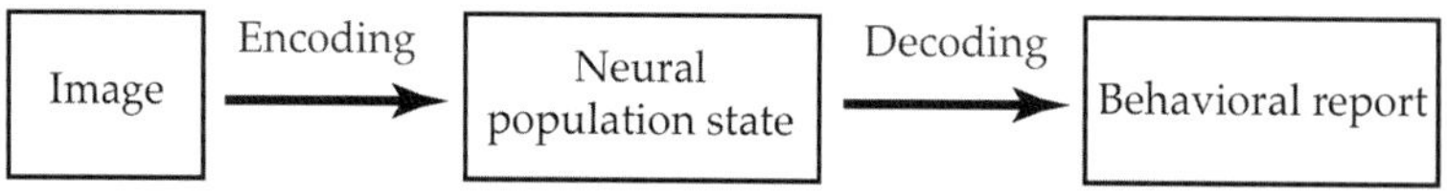

Figure 1. Encoding and decoding processes. We divide our efforts to understand visual information processing in the ventral stream into two goals. Goal one, called "encoding," is to model the transformation of the signal from a pixel-based representation in the retina to a high-level neural population representation that can support the demands of behavior in the face of low-level image variations (pose, location, etc.). Goal two, named "decoding," is to model the mechanisms that transform the population state of the neurons at the top of the ventral cortical stream into behavioral decisions.

we call *core object recognition* behavior. By core object recognition we mean the subdomain of object recognition behavior involving categorization of images containing a single salient object presented briefly (100–200 msec) in the central visual field (within the central 10° of the visual field). In "From neuronal populations activity patterns to behavior" section, we begin by first discussing our work on the neurons-to-behavior (Neural population state → Behavioral report), because it is only by defining this mapping precisely that we will be able to understand which populations of ventral stream neurons we are most interested in modeling, and which properties of these neurons we most care to explain. In other words, a better understanding of the "Neural population state → Behavioral report" link will help us identify those neural events that directly drive behavior, and focus our subsequent modeling efforts on these key events. This section is divided into two subsections: "correlational models" ("Correlative methods to establish a working decoding model" section) and "causal models" ("Causal methods to test our working decoding model" section). In "Correlative methods to establish a working decoding model" section, we bring evidence from monkey electrophysiology and human psychophysics to establish a detailed link between neuronal firing patterns in the inferior temporal cortex and patterns of behavior on challenging object recognition tasks. The work is correlative, but strongly so, in that it uncovers a nearly perfect relationship between neuronal patterns and behavioral patterns over all tested recognition tasks, and it falsifies other possible relationships. Although this approach allows us to infer the form of the neural codes and the underlying mechanisms, it does not directly show that the inferred neural code is the cause of the behavior. Thus, in "Causal methods to test our working decoding model" section, we discuss recent experimental techniques for perturbing brain activity in vivo and present evidence generated from these techniques that directly shows the causal link between the neuronal activity in the inferior temporal cortex and object recognition behavior and lays the foundation for future work along this line.

In "From images to neuronal population activity patterns" section, we discuss our work on modeling the encoding of visual images in higher visual cortical areas. The upshot of that work is that we have recently identified a hierarchical neural networks model who's top output layer is highly predictive of IT spiking responses to complex naturalistic images at both the single site and population levels. The model's intermediate layers are highly predictive of neural responses in the V4 cortex, a midlevel ventral stream visual area that provides the dominant cortical input to IT. Moreover, lower layers are highly predictive of voxel responses from fMRI data in V1 and V2. In addition to substantially advancing our encoding goal (Fig. 1), these results show that our performance-optimization approach—applied in a biologically appropriate model class—can be used to build quantitative predictive models of neural processing.

FROM NEURONAL POPULATION ACTIVITY PATTERNS TO BEHAVIOR

The ventral cortical stream of primates includes a series of hierarchical processing steps that encode image information content (e.g., object category and identity) increasingly explicitly along the ventral stream cortical areas (Gattass et al. 1990; Ungerleider and Haxby 1994; Miyashita and Hayashi 2000; Reddy and Kanwisher 2006). For instance, neurons in the primary visual cortex can be described by Gabor-like filters that extract object contour information (Carandini et al. 2005), although these neurons do not express strong tolerance to geometrical image transformations such as translation and rotation (DiCarlo et al. 2012). In contrast, population patterns of activity in inferior temporal (IT) cortex (highest processing stage of the ventral stream) can directly support, object categorization over a wide range of complex image transformations (Hung et al. 2005; Rust and Dicarlo 2010). Consistently, midlevel ventral stream areas—for example, V4, the main cortical input to IT—express intermediate levels of selectivity for objects under complex image variations (Connor et al. 2007; Rust and Dicarlo 2010). Thus, it is reasonable to hypothesize simple decoding algorithms that drive visual object recognition behavior directly from the neural responses of IT cortex. Although that is a long-standing working paradigm in the field, defining a compact, potentially mechanistic model that explains all object recognition behavior has not yet been accomplished. In what follows, we describe our efforts to accomplish that goal using both strong correlation methods and causal methods.

Correlative Methods to Establish a Working Decoding Model

It has been long known that individual neural responses in higher processing stages of the ventral stream (e.g., IT cortex) encode information about the identity and category of visually presented objects (Tanaka 1996). Such results motivate a potential link between neuronal responses in the ventral stream and behavioral shape judgements; if ventral stream neurons encode shape of the visual stimuli, they may as well be the neural underpinning of behavioral shape judgments. However, this is only a qualitative link, it does not show that ventral stream neural responses support object recognition behavior at high levels of human psychophysical performance, nor does it explain how that behavior can be predicted for any and all object recognitions tasks. To advance understanding, the hypothesis that "object recognition behavior is driven by the neural activity in the ventral stream" must be translated into a potentially mechanistic, predictive *linking model*, and that model must, if correct, consistently predict the relationship between a specific aspect of ventral stream neuronal population activity and behavior over all object recognition tasks.

Although it was not a priori obvious that such a linking model existed, we made it our task to try to find one. That is, we sought a single linking model (a.k.a. decoding model; see Fig. 1) that accurately predicts behavioral difficulty for a wide range of object recognition tasks based on IT neural population activity evoked by each visual image. The first step to answer this question is to operationally and quantitatively define a set of "object recognition" tasks. This is a particularly important step because the choice of psychophysical tasks can potentially change the outcome. We reasoned that the psychophysical tasks that operationally define "object recognition" should be hard enough to push human and monkey observers to the limits of errors, because without such errors it is not possible to evaluate the consistency of the failure patterns of the human observers with that of the ventral stream neurons. More importantly, because we and others assume that the ventral stream is critically important for "invariant object recognition," we wanted the psychophysical tasks to engage that problem. For instance, although the behavioral distinction between a white dove and a black bear may be called "object recognition" in common language, such a behavior can be easily supported by color or simple low-level cues—it does not require tolerance to position, scale, viewpoint, or background clutter and thus does not represent the challenging part of object recognition.

To satisfy these constraints, we operationally defined object recognition by measuring human psychophysical performance in 64 object recognition tests that included thousands of images that explore shape similarity under a range of identity-preserving image variations (Fig. 2A) such as translation, rotation, and variation of the background. Not surprisingly, this showed that human object recognition performance measurements span a wide range—that is, some object recognition tests are more difficult than others (Fig. 2B), and this pattern of difficulty is highly consistent across subjects. Using the same set of visual images and viewing durations, we then measured neuronal responses in V4 and IT of two macaque monkeys in a passive viewing paradigm. We were aware of the numerous potential differences between active psychophysical task performance in human observers and passive, visually evoked neural responses in monkeys who do not have learned behaviors with respect to the stimulus set (e.g., images of cars). However, our hypothesis was that the ventral stream transforms each visual image into a new neuronal population pattern at high levels of the ventral stream that can simply predict the

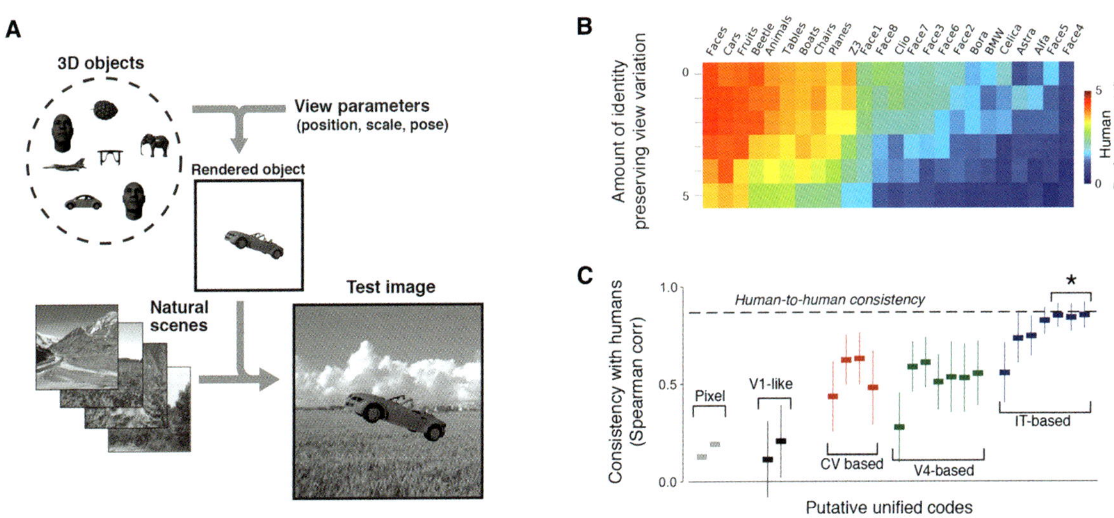

Figure 2. A unified neuronal population code can explain human object recognition behavior. (*A*) We started with 3D models of real world objects (e.g., faces, cars) with multiple exemplars for each category. Given parameters (size, position, pose), we used ray tracing to render the object and place it on an uncorrelated background producing a test image. (*B*) We measured human performance for one eight-way basic-level object categorization (face, cars, planes, animals, tables, chairs, boats, fruits), and two subordinate eight-way identifications (cars: Beetle, BMW z3, Ford, etc.; faces: face1, face2, face3, etc.) and determined d's for 384 tasks (24 labels × 6 levels of identity-preserving view variation). Heat map shows range of performance (high d' = red). (*C*) We find that a single linking hypothesis (aka "unified code" or "decoding model") is sufficient to explain human core object recognition behavior, as showed by its statistically perfect consistency with the pattern of human performance. Only codes marked with asterisks are sufficient. (Adapted from Majaj NJ, Hong H, Solomon EA, DiCarlo JJ, unpubl.)

animal's potential object recognition performance over any object recognition test. More specifically, based on previous work (Miyashita 1993; Freedman et al. 2003) we assumed that that underlying mechanism of learning any new object recognition task (e.g., detecting cars) is implemented by downstream neurons (e.g., prefrontal cortex, perirhinal cortex) modifying the strength of their synaptic inputs received from ventral stream neurons. We tested this linking hypothesis by implementing simple linear classifiers that approximate those downstream learning mechanisms and then explored which aspects of ventral stream neural activity, if any, could predict the pattern of behavioral performance over all object recognition tests.

We refer to those aspects of neural activity as "neural codes," and based on the neural recordings from V4 and IT cortex, we implemented dozens of putative neural codes (results from 14 putative neural codes are shown in Fig. 2). We then compared each code's predicted behavioral performance (neural predicted d's) with the actual (human) pattern of behavioral performance (psychophysical discriminability d') for all of the 64 object recognition tests. The correct neural code and linking hypothesis was expected to predict the entire pattern of human performance across all 64 tests. Remarkably, we found a few neural codes that meet this high bar (Fig. 2C). These few neural codes were all based on simple learned linear sums of mean firing rates over the IT population. Many other IT codes and all V4 neural codes and all simulated neural codes from a V1-like model failed to accurately predict behavior.

These results show that human psychophysical performance on challenging object recognition tasks can be fully predicted by a simple, learned linear sums of IT neural populations firing rates, and strongly suggest that this working model will generalize to a large number (perhaps all) of object recognition tasks. Because this work specifies the parameters of a specific proposed mechanisms for linking neuronal activity to any object recognition tasks, it allows us to make quantitative predictions of how perturbation of IT neural activity population patterns will lead to specific changes in behavioral performance over all object recognition tasks. Testing such predictions requires the validation and application of new tools, which is described in the next section.

Causal Methods to Test Our Working Decoding Model

A visual stimulus such as the image of a face activates a cascade of population neural responses in the visual system at different temporal and spatial scales. At least some parts of this stimulus-driven activity must causally support object recognition behavior, whereas other parts might be epiphenomenal in that they play no role in a particular object recognition task. Indeed, this might even be true for neural activity which has high information for that recognition task and high predictive power for the behavioral pattern of performance (e.g., the working IT-based decoding model outlined above). One must thus use tools that directly interact with neural activity to distinguish among these alternatives. Once the causally relevant neural events are identified, the next critical step is to make quantitative links between them and behavior. For example, imagine a single neuron in IT cortex that can discriminate between male and female faces with 90% accuracy (encoding). Does this neuron play any role in gender discrimination behavior? That is, would inactivation of this neuron take a toll on gender discrimination behavior? If so, how much behavioral deficit would result from inactivation of this neuron? 10%? 1%? 0.1%? 0.01%? And which temporal aspects of the neural activity are most critical to the behavior? Would the inactivation of this neuron also affect other object recognition tasks? If so, by how much? Answering such questions is only possible by directly perturbing the neuronal activity and measuring its effect on behavior.

The potential causal link between neuronal activity and behavior has been studied by multiple research programs for different brain regions. This includes frontal eye fields (Bruce and Goldberg 1985), somatosensory cortex (Romo and Salinas 2001), superior colliculi (Stryker and Schiller 1975), primary visual cortex (Schiller and Tehovnik 2008), and middle temporal cortex (Salzman et al. 1990). However, for the case of IT cortex there are only two neurophysiological studies of the causal relation between the IT neural response properties and object recognition; a study of the effect of microstimulation of IT face-selective cells on face detection behavior (Afraz et al. 2006) and a recent study about the causal role of 3D-structure-selective IT cells in three-dimensional structure categorization (Verhoef et al. 2012). In our work, we seek a comprehensive understanding of the causal role of the IT neural activity in all object recognition tasks. At the current stage of the work, we have selected a specific task—psychophysically demanding face gender discrimination task—to explore the feasibility of this idea.

We trained two macaque monkeys to report the gender of a briefly presented face by looking at one of the two response targets on the screen. Then in separate experiments, we used optogenetic and pharmacological methods to artificially suppress the activity of various subregions of IT cortex while the animals performed the face gender discrimination task. To determine the neural selectivity profile of the targeted IT subregions, we recorded extensively from the IT cortex while the animals simply fixated visually presented images including faces and other objects. We found that (Afraz et al. 2015) both pharmacological and optogenetic inactivation of IT subregions that are traditionally defined as face selective (neural d' for discriminating faces from other objects >1) induced a deficit in face discrimination. The deficit was limited only to the contralateral visual hemifield and it was not observed for other IT subregions where the neural responses were not face selective. Specifically, temporally delimited (200 msec), optogenetic inactivation of face-selective subregions led to a small but significant drop in gender discrimination performance (median = 1.8%, mean = 2.02%, $t_{(16)}$ = 5.99, $P < 0.0001$) for faces presented to the contralateral visual field. Similarly, pharmacological inactivation of the same cortical subregions (induced by microinjection of muscimol) lead to a 5.5% drop in gender discrimination performance only for contralaterally presented stimuli, whereas performance remained unchanged in the ipsilateral hemifield (repeated-measures ANOVA interaction test, $F_{(4,20)} = 7.5, P < 0.001$). To begin to connect these results to our working decoding model derived from correlational methods ("Correlative methods to establish a working decoding model" section) we sought to use that model to predict the causal importance of each subregion of IT to the gender discrimination task. As a first approximation, we computed the predicted causal importance in each IT subregion as the facial gender task performance of a linear classifier that could read only the neural population activity in that subregion ("local linear classification"), based on our recording measurements of about nine (median) sites taken within the subregion. Consistent with our working model, we found that this local linear classification measure was highly predictive of the impact of direct optical inactivation of that subregion on behavioral gender discrimination performance ($r = -0.6, P = 0.01$) (Fig. 3). Linear regression suggests that inactivation of a small IT subregion (~1 mm in diameter) with our highest observed local linear classification for facial gender induces an impact of (2.5%) on psychophysical performance (see Fig. 3).

In summary, these results establish a causal link between the neural activity in IT cortex and at least one type of object discrimination behavior (face gender discrimination). In a hierarchical processing system, it is possible to imagine a situation where inactivation of an upstream area that does not show linear separability for a given feature influences behavioral discrimination of that feature. The correlation between the behavioral impact of inactivation and linear separability (Fig. 3) will, we think, not be found in such an area. That is, we predict that such correlation will be much lower for inactivation of lower-level areas such as V1 or even V4. The fact that we observe a correlation between local neural linear separability of the visual information in each IT cortical subregions and the behavioral impact of inactivation of those subregions suggests that IT representations are relatively close to the read-out mechanisms that drive the behavior, consistent with the working model outlined above. Our next goal is to extend these causal methods to a battery of object recognition tests and to build stronger quantitative links between neural activity and behavior, guided by our working decoding model.

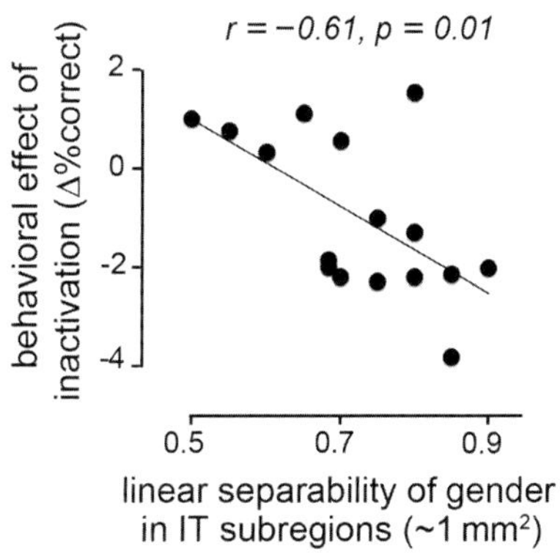

Figure 3. The effect of inactivation of different IT subregions on a gender discrimination task. Behavioral effect of optogenetic inactivation, averaged for small IT cortical subregions. The abscissa shows gender classification performance of a linear classifier trained and tested by multiunit recording data collected during the animals' passive fixation from each of the various subregions (~1 mm) sampled from IT cortex. The ordinate shows the average impact of optogenetic inactivation of those same IT subregions on the animals' active behavior (gender discrimination task). (Adapted from Afraz et al. 2015.)

FROM IMAGES TO NEURONAL POPULATION ACTIVITY PATTERNS

Starting with the seminal ideas of Hubel and Wiesel, work in visual systems neuroscience over the past 50 years has strongly suggested that the ventral visual stream generates invariant object recognition behavior via a hierarchically organized series of cortical areas that encode object identity with increasing selectivity and tolerance (DiCarlo et al. 2012). Our work in visual cortex seeks to go beyond this powerful but broad-stroke understanding to identify concrete predictive models of the underlying neuronal population representation, and then use these models to gain insight inaccessible without large-scale computational precision.

Mathematically, much of our knowledge about the ventral stream can be boiled down to a class of computational architectures known as hierarchical convolutional neural networks (HCNNs), a generalization of Hubel and Wiesel's simple and complex cells (Serre et al. 2007; Pinto et al. 2009). HCNN models are composed of several retinotopic layers combined in series. Each layer is very simple, but together they produce a deep, complex transformation of the input data—in theory, like the transformation produced along the entire depth of the ventral stream. In this formal language, a key step toward understanding would involve identifying a single HCNN model whose internal layers correspond to the known ventral cortical areas and which accurately predict response patterns in those areas (*neural predictivity*). This has proven extremely difficult to do (DiCarlo and Cox 2007; Pinto et al. 2008), in part because subtle parameter changes (e.g., number of layers, local receptive field sizes) can dramatically affect a model's match to neural data (Yamins et al. 2013). Broad stroke understanding is not, by itself, enough.

In high-throughput computational experiments evaluating thousands of HCNN models on both task performance and *neural predictivity* metrics, we found that architectural parameters that performed better on high-level object recognition tasks also tended to have higher *neural predictivity*—that is, they better explained cortical spiking data (Fig. 4A; Yamins et al. 2014). Model param-

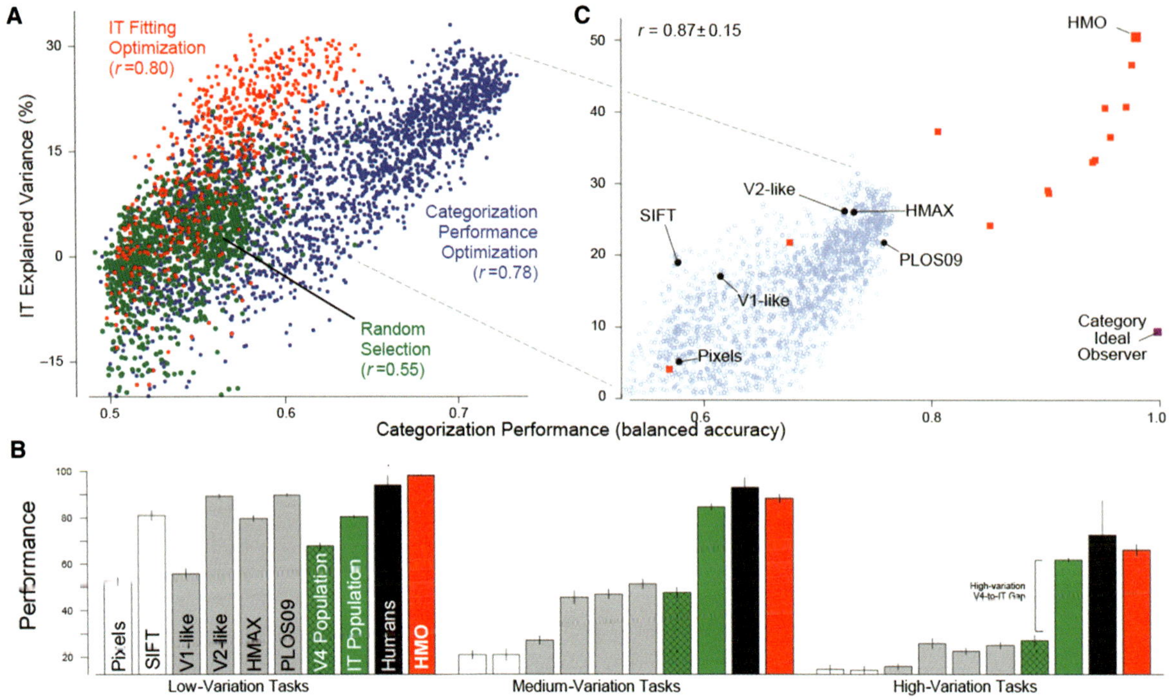

Figure 4. Behavioral optimization yields neurally predictive models. (*A*) Object categorization performance versus IT neural explained variance percentage ("IT Predictivity") for hierarchical convolutional neural network (CNN) models in three independent high-throughput computational experiments (each point is a distinct neural network architecture). The *x*-axis shows performance (balanced accuracy, chance is 0.5) of the model output features on a high-variation categorization task; the *y*-axis shows the median single site IT explained variance percentage ($n = 168$ sites) of that model (i.e., the percentage of explainable response variance where response is defined as the mean IT firing rate 70–170 msec after image onset) computed on held-out test images. Each dot corresponds to a distinct model selected from a large family of convolutional neural network architectures (see text). Models were selected by random draws from parameter space (green dots), object categorization performance-optimization (blue dots), or explicit IT predictivity-optimization (red dots). (*B*) A high-performing neural network was identified that matches human performance on a range of recognition tasks, the hierarchical modular optimization (HMO) model. Object categorization performance results on the test images for eight-way object categorization at three increasing levels of object view variation (*y*-axis units are eight-way categorization percent correct; chance is 12.5%). IT (green bars) and V4 (hatched green bars) neural responses, and computational models (gray and red bars) were collected on the same image set and used to train support vector machine (SVM) linear classifiers from which population performance accuracy was evaluated. Human subject responses on the same tasks were collected via psychophysics experiments (black bars). (*C*) The object categorization performance versus IT neural predictivity correlation extends across a variety of models showing a wide range of performance levels. Black circles include controls and published models; red squares are models produced during the HMO optimization procedure. For reference, light blue circles indicate performance optimized models (blue dots) from *A*. (Modified from Yamins et al. 2014.)

eters were selected for evaluation by one of three rules: (1) random draws from the uniform distribution (Fig. 4A, green dots); (2) optimization for performance (blue dots); and (3) optimization directly for architectures that best fit IT neural data (orange dots). All three conditions show large variation in both performance and IT *neural predictivity* across the parameter space, demonstrating that choices for these architectural parameters have a large effect on a specific model's ability either to perform object recognition effectively or match neural data. Moreover, performance was significantly correlated with *neural predictivity* in all three selection regimes, with models that performed better on the categorization task also being more likely to produce outputs more closely aligned to IT neural responses. Critically, optimizing for *only* task performance significantly increased the correlation with IT *neural predictivity* even though neural data were not used in the optimization. Moreover, the best performing models identified in experiment (2) predicted neural output equally well as the most predictive models identified in experiment (3). Together, these results imply that performance optimization is an efficient means to identify the parameters of HCNN models that best predict the neural population response data—that is, to identify specific encoding models that respond most like the actual ventral visual stream.

Pushing this idea further, we then used Hierarchical Modular Optimization (HMO) to discover a hierarchical neural network model architecture that achieved near-human-level performance on challenging object categorization tasks. The HMO procedure embodies a conceptually simple hypothesis for how high-performing combinations of functionally specialized hierarchical architectures can be efficiently discovered and hierarchically combined. Applying the HMO procedure, we identified a high-performing four-layer HCNN model with 1250 top-level outputs. This model matched the object recognition performance of the IT neural sample, even when

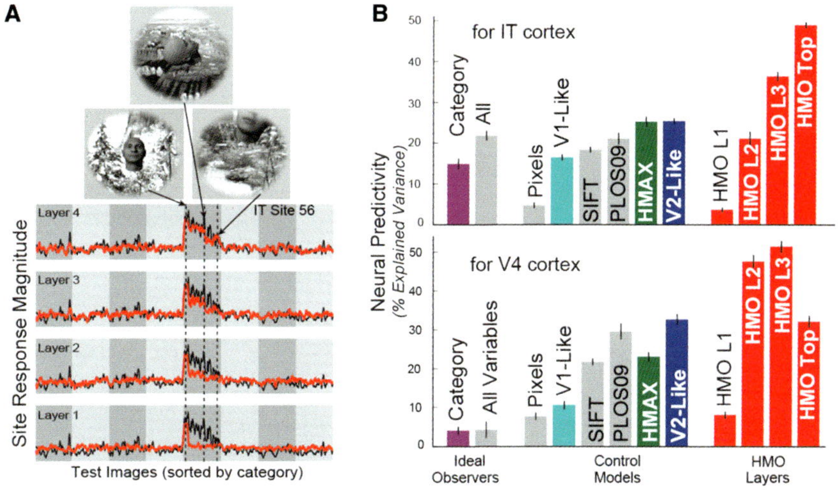

Figure 5. Per-layer. (*A*) Actual neural response (black trace) versus model predictions (colored trace) for a face-selective IT neural site. The *x*-axis in each plot shows 1600 test images sorted first by category identity and then by variation amount, with more drastic image transformations toward the right within each category block. The *y*-axis represents the prediction/response magnitude of the neural site for each test image (those not used to fit the model). The four rows show neural predictions using the visual feature set (i.e., units sampled) from each of the four layers of the HMO model. (*B*) Comparison of IT and V4 neural explained variance percentage for various models. Bar height shows median explained variance, taken over all predicted IT units. (Modified from Yamins et al. 2014.)

faced with large amounts of variation—a hallmark of human object recognition ability (DiCarlo et al. 2012). As a control, we also evaluated a variety of existing models targeting several levels of the ventral hierarchy, including SIFT, a simple baseline computer vision model (Lowe 2004), a V1-like Gabor-based model (Pinto et al. 2008), a V2-like conjunction-of-Gabors model (Freeman and Simoncelli 2011), and HMAX (Serre et al. 2007), a model targeted at explaining higher ventral cortex and that has receptive field sizes similar to those observed in IT. These control models approach IT and human performance levels in low-variation circumstances, but in high-variation conditions all of them except for the HMO model fail to match the performance of human subjects and the IT neural population by a large margin.

Given the HMO model's plausible performance characteristics, we then measured its ability to predict IT neural data, both for the top-level output, as well as for each of the model's three intermediate layers (Figs. 4C, 5, red lines/bars). We found that each successive layer predicted the responses of IT units increasingly well, with category selectivity and tolerance to more drastic image transformations emerging gradually through the multiple layers of the model hierarchy (Fig. 5A). Critically, we found that the top layer of the high-performing HMO model achieves remarkably high IT *neural predictivity*, predicting 48.5 $\pm$ 1.3% of the explainable IT neuronal variance (single site predictivity, median of 168 IT sites; Fig. 5B, top). This represents a nearly 100% improvement over the previous best comparison models. Ideal observer models, with perfect access to all category labels and other image parameters, predict IT units above chance level, but significantly less well than the HMO model, showing that high IT *neural productivity* does not automatically follow from category selectivity. Put another way, IT neurons are not simple categorizers—there is significant noncategorical structure in IT responses that is likely attributable to hierarchical network architecture of the actual ventral visual stream—but the structure in those responses is quite well captured by the HMO network architecture.

Cortical area V4 is the dominant cortical input to IT, and the neural population representation in V4 is known to be even less categorical than that of IT (Rust and Dicarlo 2010). We found that the HMO model's intermediate layers are highly predictive of V4 neural responses (51.7 $\pm$ 2.3% explained V4 variance), providing a significantly better match to V4 than either the model's top or bottom layers (Fig. 5B, bottom). These results suggest that V4 corresponds to an intermediate layer in a hierarchical model whose top layer is an effective model of IT. Unlike the case of IT, ideal observer semantic models explain effectively no response variance in V4, consistent with V4's lack of explicit category selectivity in single neurons. Together these results suggest that performance optimization not only drives top-level output model layers to resemble IT, it also imposes strong biologically consistent constraints that predict the "internal" structure of the ventral stream as well.

In summary, combining two general biological constraints—the behavioral constraint of recognition performance, and the architectural constraint imposed by the HCNN model class—leads to greatly improved models of multiple layers of the visual sensory cascade. Although the models end up being predictive of neural data, they were not explicitly tuned using these data, because model parameters were independently selected to optimize categorization performance on an unrelated image set. Consequently, each layer is a generative model for its corresponding cortical area, from which large numbers of IT- or V4-like units can be sampled. A common assumption in visual neuroscience is that understanding the qualitative structure of tuning curves in lower cortical areas (e.g., Gabor conjunctions in V2 or curvature in V4) is a necessary precursor to explaining higher visual cortex. Our results show that higher-level constraints can yield quantitative models even when bottom-up primitives have not yet been identified.

CONCLUSION

In our work, we have built two basic types of models. One of these models (encoding) maps the retinal input to reasonably accurate predications about neural population responses in V4 and IT cortex. The other one (decoding) maps IT neural population activity to very accurate predictions about behavioral object choice. The encoding model is a hierarchical neural network that computes a complex nonlinear transform of the retinal input. This nonlinearity takes the information that is implicit in these inputs (e.g., the category of an object independent of where and how it appears in the visual field) and renders it explicit and easily accessible to downstream decode mechanisms. Our decoding model is a candidate for exactly this type of downstream read-out mechanism, a simple linear operation that projects a desired behavioral output (e.g., category identity) out of the IT neural population code. Combining these two models provide a working, quantitative end-to-end mechanistic model that begins to capture the neural mechanisms underlying core visual object recognition.

These models are largely feed-forward, suggesting that the core object recognition subdomain of visual behavior (rapid categorization of single centrally presented objects) is mainly supported by a single unidirectional cascade of neural activity along the ventral visual hierarchy. These results also suggest that encoding for core visual object recognition (i.e., the generation of IT population patterns) may be a largely automatic process that is constantly in operation, independent of behavioral state.

Our work is circumscribed by a number of key limitations within the subdomain of core object recognition. Both the encoding and decoding models are imperfect in that they are capturing a significant fraction but not all of the variance in each of these processes. We have explored only a subset of possible behavioral tasks, both in correlational and perturbative studies, leaving open the possibilities that there are images and recognition tasks on which these models would fail. We have not explained the origin of cortical heterogeneity (e.g., spatial clustering of face-selective neurons in multiple interconnected patches) or the functional significance of having multiple cortical layers in each visual area. Although we have provided a quantitative predictive model, we have not yet provided a qualitative characterization of neural tuning functions in intermediate visual areas (e.g., V2 and V4).

Beyond the core object recognition subdomain on which we have established this new foothold, a number of significant advances must be made before neural mechanisms of object recognition can be completely understood. One of the largest remaining gaps is in understanding the role of feedback circuits in complex visual behaviors. These behaviors will likely involve visual memory, visual learning, top-down inference and sensorimotor integration over extended time periods. We believe that a tight loop between quantitative models and large-scale neural and behavioral data collection is the only way to reach that deeper understanding, and that the work described above is an example of the power of that approach.

REFERENCES

Afraz SR, Kiani R, Esteky H. 2006. Microstimulation of inferotemporal cortex influences face categorization. *Nature* **442:** 692–695.

Afraz A, Boyden ES, DiCarlo JJ. 2015. Optogenetic and pharmacological suppression of spatial clusters of "face neurons" reveal their causal role in face discrimination. *Proc Natl Acad Sci* **112:** in press. doi: 10.1073/pnas.1423328112.

Bell AH, Hadj-Bouziane F, Frihauf JB, Tootell RB, Ungerleider LG. 2009. Object representations in the temporal cortex of monkeys and humans as revealed by functional magnetic resonance imaging. *J Neurophysiol* **101:** 688–700.

Bruce CJ, Goldberg ME. 1985. Primate frontal eye fields. I. Single neurons discharging before saccades. *J Neurophysiol* **53:** 603–635.

Carandini M, Demb JB, Mante V, Tolhurst DJ, Dan Y, Olshausen BA, Gallant JL, Rust NC. 2005. Do we know what the early visual system does? *J Neurosci* **25:** 10577–10597.

Connor CE, Brincat SL, Pasupathy A. 2007. Transformation of shape information in the ventral pathway. *Curr Opin Neurobiol* **17:** 140–147.

Desimone R, Albright TD, Gross CG, Bruce C. 1984. Stimulus-selective properties of inferior temporal neurons in the macaque. *J Neurosci* **4:** 2051–2062.

DiCarlo JJ, Cox DD. 2007. Untangling invariant object recognition. *Trends Cogn Sci* **11:** 333–341.

DiCarlo JJ, Zoccolan D, Rust NC. 2012. How does the brain solve visual object recognition? *Neuron* **73:** 415–434.

Farah MJ. 1990. *Visual agnosia: Disorders of object recognition and what they tell us about normal vision.* MIT Press, Cambridge, MA.

Freedman DJ, Riesenhuber M, Poggio T, Miller EK. 2003. A comparison of primate prefrontal and inferior temporal cortices during visual categorization. *J Neurosci* **23:** 5235–5246.

Freeman J, Simoncelli EP. 2011. Metamers of the ventral stream. *Nat Neurosci* **14:** 1195–1201.

Gattass R, Rosa MG, Sousa AP, Pinon MC, Fiorani Junior M, Neuenschwander S. 1990. Cortical streams of visual information processing in primates. *Braz J Med Biol Res* **23:** 375–393.

Hung CP, Kreiman G, Poggio T, DiCarlo JJ. 2005. Fast readout of object identity from macaque inferior temporal cortex. *Science* **310:** 863–866.

Lowe DG. 2004. Distinctive image features from scale-invariant keypoints. *Int J Comput Vis* **60:** 91–110.

Malach R, Reppas JB, Benson RR, Kwong KK, Jiang H, Kennedy WA, Ledden PJ, Brady TJ, Rosen BR, Tootell RB. 1995. Object-related activity revealed by functional magnetic resonance imaging in human occipital cortex. *Proc Natl Acad Sci* **92:** 8135–8139.

Miyashita Y. 1993. Inferior temporal cortex: Where visual perception meets memory. *Annu Rev Neurosci* **16:** 245–263.

Miyashita Y, Hayashi T. 2000. Neural representation of visual objects: Encoding and top-down activation. *Curr Opin Neurobiol* **10:** 187–194.

Pinto N, Cox DD, DiCarlo JJ. 2008. Why is real-world visual object recognition hard? *PLoS Comput Biol* **4:** e27.

Pinto N, Doukhan D, DiCarlo JJ, Cox DD. 2009. A high-throughput screening approach to discovering good forms of

biologically inspired visual representation. *PLoS Comput Biol* **5:** e1000579.
Reddy L, Kanwisher N. 2006. Coding of visual objects in the ventral stream. *Curr Opin Neurobiol* **16:** 408–414.
Romo R, Salinas E. 2001. Touch and go: Decision-making mechanisms in somatosensation. *Annu Rev Neurosci* **24:** 107–137.
Rust NC, Dicarlo JJ. 2010. Selectivity and tolerance ("invariance") both increase as visual information propagates from cortical area V4 to IT. *J Neurosci* **30:** 12978–12995.
Salzman CD, Britten KH, Newsome WT. 1990. Cortical microstimulation influences perceptual judgements of motion direction. *Nature* **346:** 174–177.
Schiller PH, Tehovnik EJ. 2008. Visual prosthesis. *Perception* **37:** 1529–1559.
Serre T, Oliva A, Poggio T. 2007. A feedforward architecture accounts for rapid categorization. *Proc Natl Acad Sci* **104:** 6424–6429.
Stryker MP, Schiller PH. 1975. Eye and head movements evoked by electrical stimulation of monkey superior colliculus. *Exp Brain Res* **23:** 103–112.
Tanaka K. 1996. Inferotemporal cortex and object vision. *Annu Rev Neurosci* **19:** 109–139.
Ungerleider LG, Haxby JV. 1994. 'What' and 'where' in the human brain. *Curr Opin Neurobiol* **4:** 157–165.
Verhoef BE, Vogels R, Janssen P. 2012. Inferotemporal cortex subserves three-dimensional structure categorization. *Neuron* **73:** 171–182.
Yamins D, Hong H, Cadieu C, Dicarlo J. 2013. Hierarchical modular optimization of convolutional networks achieves representations similar to macaque IT and human ventral stream. *Advances in Neural Information Processing Systems*.
Yamins DL, Hong H, Cadieu CF, Solomon EA, Seibert D, DiCarlo JJ. 2014. Performance-optimized hierarchical models predict neural responses in higher visual cortex. *Proc Natl Acad Sci* **111:** 8619–8624.

The Macaque Face Patch System: A Window into Object Representation

DORIS TSAO

Division of Biology and Biological Engineering and Computation and Neural Systems, California Institute of Technology, Pasadena, California 91125

Correspondence: dortsao@caltech.edu

The macaque brain contains a set of regions that show stronger fMRI activation to faces than other classes of object. This "face patch system" has provided a unique opportunity to gain insight into the organizing principles of IT cortex and to dissect the neural mechanisms underlying form perception, because the system is specialized to process one class of complex forms, and because its computational components are spatially segregated. Over the past 5 years, we have set out to exploit this system to clarify the nature of object representation in the brain through a multilevel approach combining electrophysiology, anatomy, and behavior. These experiments reveal (1) a remarkably precise connectivity of face patches to each other, (2) a functional hierarchy for representation of view-invariant identity comprising at least three distinct stages along the face patch system, and (3) the computational mechanisms used by cells in face patches to detect and recognize faces, including measurement of diagnostic local contrast features for detection and measurement of face feature values for recognition.

How does the brain represent objects? This question had its beginnings in philosophy. Our fundamental intuition of the physical world consists of a space containing objects, and philosophers starting from Plato wondered about the basis for the percept of these "pure forms" (e.g., *the tree*) that were clearly different from any real instance. Very early on, the mind could already sense something mysterious about the problem of object perception. Object representation constitutes the basic infrastructure on which the brain operates. We speak in nouns; we remember people, places, and things; and we think in terms of concepts, which can be construed as a generalization of objects. Despite its clear importance, we still understand very little about the neural basis for object perception. In particular, to understand visual object perception, three critical problems need to be solved: (1) How is an object first generated (i.e., how are retinal pixels stitched together into units)? (2) How are these stitched units identified? (3) How are identified units relayed to higher-order brain areas to enable flexible behavior? During the past decade, work in my laboratory has focused largely on the second question, addressing the mechanisms for face processing in macaque inferotemporal (IT) cortex. Here, I describe what we have learned about principles of object identification in the brain from studying a set of regions in the temporal lobe specialized for face processing, the macaque face patch system.

The first stage of visual information processing in the cortex occurs in area V1, where cells extract local stimulus properties like edge orientation, motion, and color contrast. Then, visual information is transmitted through a series of additional stages, V2, V3, V4, which each contain a retinotopic map of space, and must be performing local computations beyond edge detection. The precise nature of these steps remains a mystery. One major transformation appears to be segmentation (i.e., organizing visual information into discrete pieces corresponding to different objects) (Zhou et al. 2000; Bushnell et al. 2011), a highly challenging task owing to partial occlusion and the need to interpolate illusory contours (Fig. 1). Then, visual information proceeds to a large brain region called inferotemporal (IT) cortex, which has been strongly implicated in high-level object recognition (e.g., recognizing a rose, a bird, or a face). A lesion to this part of the brain can create an inability to recognize specific classes of objects such as faces, suggesting this is an important brain area to study if we want to understand object perception.

How are objects represented in IT cortex? Charles Gross and coworkers reported discovery of cells in the temporal lobe that were selective for complex forms such as hands, trees, and faces (Bruce et al. 1981), but the difficulty of finding these cells precluded deeper understanding. In 1997, Nancy Kanwisher, using fMRI in humans, reported the discovery of a face-selective area in the brain (Kanwisher et al. 1997). Remarkably, this area seemed to be in the same place in every subject she scanned, suggesting that face processing occurs in a discrete chunk of cortex. Although this finding was provocative and exciting, it remained a mystery what the cells in these regions might be doing, as the region was found using fMRI, and the relationship between blood flow measured by fMRI and underlying neural activity remains an area of active research (Logothetis 2008; Schummers et al. 2008). Most importantly, fMRI measures activity at a spatial scale of ~ 1 mm^3, whereas neural activity is organized at a much finer scale such that even neighboring cells can have very different tuning properties (Ohki et al. 2006).

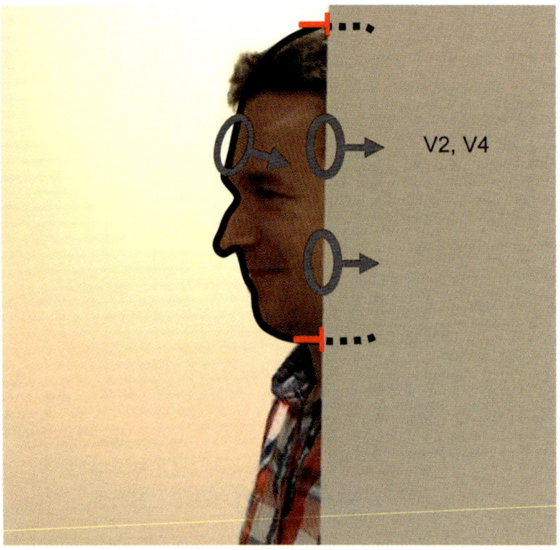

Figure 1. Segmentation processes in extrastriate retinotopic cortex. In area V4, a "boundary curvature" cell tuned for a right angle at south would be suppressed, because the presence of the T junction (red) would signal to the cell that the boundary continues behind the wall instead of making a 90° turn (Bushnell et al. 2011). In area V2, cells tuned for border-ownership (gray) configure their activity to signal the correct ownership of all the contours in the image (Zhou et al. 2000). In this way, a map is generated not just of the location of edges in the image, but which figure owns them, and how they continue behind occluders.

To clarify the link between face cells and fMRI-identified face areas, we performed fMRI experiments in alert monkeys. We found not just one such area, but six of them (Fig. 2A). Moreover, the fact that these six "face patches" were located in the same place across the two hemispheres, and in similar locations across animals, gave the first hint that they constitute a system and not just random islands of face-selective cortex. To study the selectivity of single neurons in these patches, we targeted electrodes to ML/MF, AL, and AM, and asked what the responses of cells in these regions were to the same stimuli that we used in the fMRI localizer experiment. We found that all three regions contained a very high percentage of face-selective cells, with 97% of visually responsive cells in ML/MF giving a mean response to faces at least twice as strong as to other objects (Tsao et al. 2006). This finding was exciting because it meant we now had a system where we could systematically dissect how one visual form is represented.

The macaque face patch system has provided a unique opportunity to gain insight into the organizing principles of IT cortex and to dissect the neural mechanisms underlying form perception, because the system is specialized to process one class of complex forms, and because its computational components are spatially segregated. Over

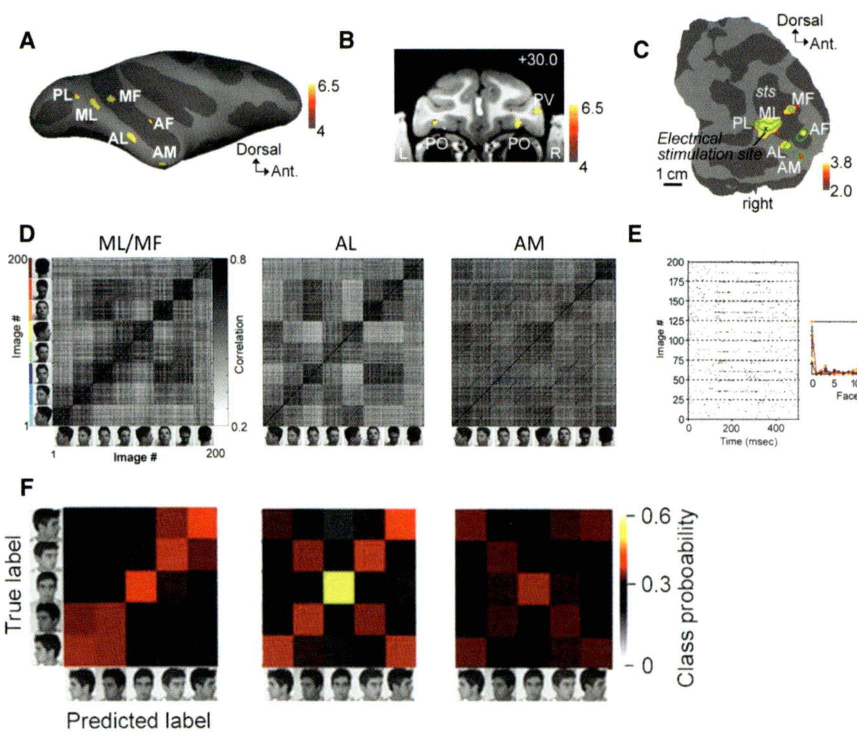

Figure 2. Dissecting face processing in the monkey. (*A*) Six face patches shown on inflated right hemisphere of macaque brain (Tsao et al. 2008a). (*B*) Two prefrontal face-selective patches, PO in the lateral orbital sulcus and PV in the infraprincipal dimple (Tsao et al. 2008b). (*C*) Connectivity of temporal face patches revealed by microstimulation targeted to face patch ML combined with fMRI; areas significantly activated by microstimulation overlaid on a flatmap (Moeller et al. 2008). (*D*) Population similarity matrices in the three face patches. A 200 × 200 matrix of correlation coefficients was computed between responses of all visually responsive cells to a set of 200 stimuli (consisting of 25 different identities each at eight different head orientations) from ML/MF ($N = 121$ cells), AL ($N = 189$ cells), and AM ($N = 158$ cells) (Freiwald and Tsao 2010). (*E*) Mean response time courses of an example sparse, view-invariant identity selective cell from AM to the 200 stimuli. (*Right*) Mean response levels to the 25 individuals at each head orientation (Freiwald and Tsao 2010). (*F*) Decoding of view from fMRI responses in ML/MF, AL, and AM to four identities each at five views (Dubois et al. 2015).

the past 5 years, my laboratory has set out to exploit this system to clarify the nature of object representation in the brain through a multilevel approach combining electrophysiology, anatomy, and behavior, focusing on three questions.

- *Connectivity*: What is the anatomical wiring diagram of the face patches?
- *Functional architecture*: Are the six patches performing different functions?
- *Computational mechanisms*: What are the mechanisms for detection and recognition of faces used by cells in the face patches?

CONNECTIVITY

The existence of six face patches raised obvious questions about anatomical connectivity. Do the patches form a unified system, or is each patch processing faces independently of the others? Does the anatomy reveal any hierarchical relationships? What are the downstream outputs of face patches? To image connectivity of the face patches in vivo, we electrically microstimulated different face patches while the monkey was inside the fMRI scanner (Moeller et al. 2008). Whenever we stimulated one patch, the other patches would light up, but not the surrounding cortex, indicating the patches are strongly connected to each other but not to the other parts of IT cortex (Fig. 2C). In addition, stimulation of face patches activated specific subregions of three subcortical areas: the amygdala, claustrum, and pulvinar. More recently, we have confirmed these results with fMRI-guided anatomical tracer injections (Grimaldi et al. 2012, 2013).

FUNCTIONAL ARCHITECTURE

Because the six face patches span the entire extent of the temporal lobe, it seemed likely that each patch performs a unique function. To discover functional differences between patches, we presented several large sets of face stimuli to animals while recording from multiple patches. In one of these experiments, we presented 25 different identities each at eight different head orientations and discovered that a major functional distinction between the patches concerns how they represent identity across different views (Freiwald and Tsao 2010). Neurons in ML and MF are view-specific; neurons in AL are tuned to identity mirror-symmetrically across views, thus achieving partial view invariance; and neurons in AM, the most anterior face patch, achieve almost full view invariance (Fig. 2D). We further discovered a remarkable cell type in the most anterior face patch AM, which responds extremely sparsely to only a small subset of face identities, invariantly across changes in view (Fig. 2E). Thus, it appears a major goal of the face patches is to build, in stepwise fashion, a representation of individual identity invariant to view direction.

Is there any spatial organization to view and identity tuning? To address this, we presented four identities each at five head orientations in a block-design fMRI experiment (a subset of the stimuli used in Freiwald and Tsao 2010) and found that multivoxel pattern analysis on the fMRI responses from ML/MF, AL, and AM could successfully decode view. Moreover, the view decoding made mirror-symmetric mistakes in AL and AM, just as we had found earlier in the units (Fig. 2F) (Dubois et al. 2015). This suggests that cells tuned to the same view are spatially clustered in each face patch.

Comparison of stimulus selectivity across different patches has revealed other significant differences. There is a clear change in species selectivity going from ML/MF, where most cells respond vigorously to both monkey and human faces, to AM, where many cells are selective for either monkey or human faces (Moeller and Tsao 2011). Experiments in which we presented random face fragments revealed that the effective fragments of a face that trigger firing increase in size and complexity going from posterior to anterior face patches (Cheng et al. 2013). Underscoring this progression in size and complexity, in the most posterior patch PL, Issa and DiCarlo (2012) found that the most effective fragment was the contralateral eye. Overall, our experiments indicate a sparser, more holistic, and more invariant representation as one proceeds anteriorly along the face patch system, consistent with the finding of "Jennifer Aniston cells" one step further in the medial temporal lobe (Quiroga et al. 2005). We do not yet understand the fundamental principle governing why each patch processes faces only up to a certain level of complexity before handing the problem off to the next patch; it seems clear that a deep answer to this question would require not just documentation of phenomenological differences between patches, but a grasp of the fundamental computational architecture.

COMPUTATIONAL MECHANISMS

As a first foray into understanding the computational architecture of the face patches, we have delved into the detailed mechanisms used by single cells to detect and recognize faces, exploiting easily-to-parameterize cartoon faces.

The first step in face processing is face detection (i.e., detecting a face is present somewhere regardless of whose face it is). Faces are robustly detected by computer vision algorithms that search for characteristic coarse contrast features (Viola and Jones 2001; Sinha 2002) (e.g., eyes darker than nose). If one examines the contrast between pairs of regions when a face is illuminated under a large variety of conditions, one finds that for some features, such as upper lip and cheek, there is no consistent contrast relationship. The upper lip is sometimes darker and sometimes brighter than the cheek, depending on the lighting. But for other features, there is a consistent contrast relationship (e.g., the nose is always brighter than the left eye).

Pawan Sinha (2002) suggested that for face detection, the most important features should be ones that are invariant to changes in lighting. To test whether cells in the face patches might be using these illumination-invariant contrast features to detect faces, we constructed an artificial face stimulus consisting of 11 different regions varying in brightness. Individual cells from the middle face patch showed a wide range of responsiveness to these part intensity stimuli, with some stimuli eliciting stronger responses than to a real face, and others eliciting no response at all (Fig. 3A). To determine whether contrast between pairs of parts might be driving this variation, for each pair of parts, we computed the mean response when part A was darker than part B and the mean response when part A was brighter than part B for each of the 55 pairs of parts. Figure 3B shows results for an example unit, and Figure 3C shows the result for the whole population. Remarkably, the cells were completely consistent in their contrast preference (e.g., almost 100 cells preferred the left eye to be darker than the nose, and not a single cell preferred the opposite contrast relationship). Moreover, the preferred features were completely consistent with those predicted from the light-invariance experiments, indicated by the purple arrows. One question often asked about face cells is how do we know these cells are really coding faces and not some other object that we simply have not shown yet, because we obviously cannot show every possible object to a single cell in IT cortex. The consistency of contrast preferences of face cells, with each other and with computational light-invariance experiments, is powerful evidence that these cells are truly coding faces. At the same time, the result shows that these cells are using more primitive mechanisms to detect faces than human observers. Even though both stimuli in Figure 3A appear "face-like" to human observers, they could elicit very different responses in a subset of face cells in ML/MF.

Indeed, the fact that we see both stimuli as faces suggests that contrast cannot be the whole story to face detection. We can readily see faces in line drawings in which there is no contrast. Thus, feature shape must also play an important role. What is the contribution of feature shape to face detection? To address this question, we recorded responses of cells in ML/MF to a cartoon face defined by seven different elementary parts. Responses to the 128 combinations of these seven face parts showed that individual cells are selective for the presence of specific face parts, such as eyes or hair. Figure 4A shows responses of two example cells from the middle face patch to these 128 stimuli, illustrating selectivity for different parts. This result is interesting because it challenges one of the long-standing assumptions about IT cortex—namely, that it is organized into feature columns, like V1, with each column processing various moderately complex shapes that are visually similar (Tanaka 2003). A pair of disks and an upside-down U have nothing visually similar about them. Rather, what they have in common is that they are both defining features of a face, an ethologically meaningful unit. We found neighboring cells within the face patch tuned to such visually dissimilar features, as well as single cells tuned to multiple such features. Thus, the ethological meaning of objects is clearly an important driving

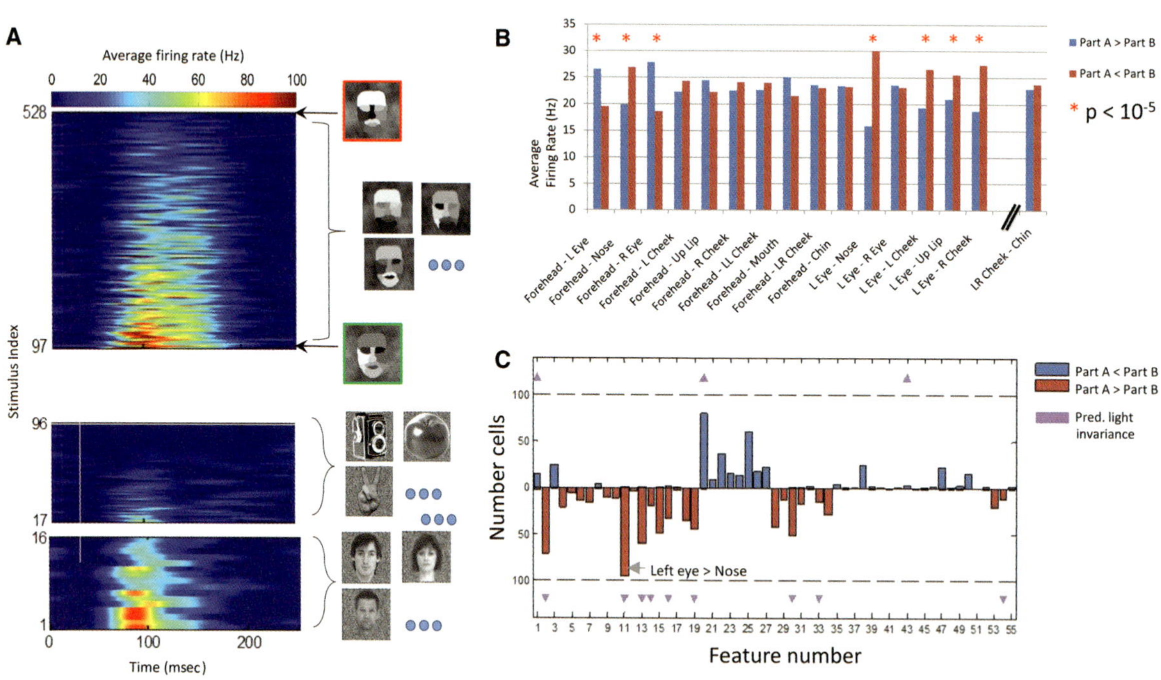

Figure 3. Detecting faces through selectivity for characteristic contrast features (Ohayon et al. 2012). (*A*) Response of an example cell from ML/MF to 16 pictures of real faces (*bottom*), 80 pictures of nonface objects (*middle*), and 432 part intensity stimuli constructed by randomly varying the brightness of 12 face regions. An example ineffective (red outline) and effective (green outline) part intensity stimulus are shown. (*B*) Responses of an example cell from ML/MF to a subset of the 55 feature pairs, showing mean response to both contrast polarities of each pair. Asterisks mark feature pairs for which the cell showed significant contrast selectivity. (*C*) Significant contrast feature histogram. Blue (red) bars indicate the number of cells tuned for intensity in *A* greater (less) than intensity in *B*. Triangles indicate predictions from computational light-invariance experiments.

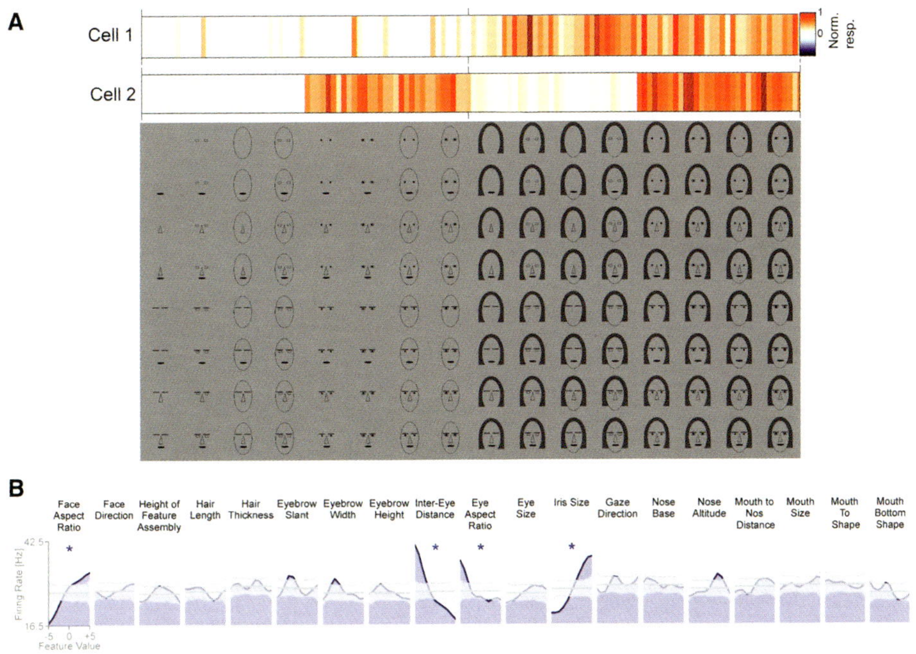

Figure 4. Probing mechanisms for face detection and recognition with cartoon faces. (*A*) Responses of two examples cells from ML/MF to 128 combinations of seven cartoon face parts. Cell 1 was selective for the presence of hair, cell 2 for the presence of irises. (*B*) Tuning of an example cell from ML/MF to 19 cartoon face dimensions. Tuning curve significantly deviating from a shuffle control are indicated by asterisk. This cell was tuned to four parameters: face aspect ratio, inter-eye distance, eye aspect ratio, and iris size.

force in IT organization, above and beyond low-level visual feature similarity.

It is critically important for primates to not only detect other faces, but to recognize them individually. What is the neural mechanism for distinguishing different faces? In general terms, this could be accomplished based on the overall shape of the face (e.g., narrow vs. round), the shape of specific features (e.g., iris size), or the spatial relationship between different features (e.g., inter-eye distance). To distinguish these possibilities, we constructed another set of cartoon faces, this time varied in identity. Each cartoon face was defined by 19 dimensions, and the values of the dimensions were varied randomly and independently; some dimensions described the overall shape of the face, some described the shape of specific features, and some described the spatial relationship between features. We found that individual cells are tuned to subsets of face features. Figure 4B shows tuning curves of an example cell to the 19 feature dimensions; this cell was significantly tuned to four features, face aspect ratio, inter-eye distance, eye aspect ratio, and iris size. Interestingly, all four of the tuning curves are ramp shaped, with a maximum at one extreme and a minimum at the opposite extreme. This was true across the population, suggesting that these cells are acting like rulers, which is consistent with a "face space" representation (Valentine et al. 2015), where cells are measuring deviation from the average face along specific axes rather than encoding specific exemplars. This preference for extreme feature values may explain the power of caricatures, which would be stimulating the population to fire at its maximum dynamic range.

Obviously, one limitation with these cartoon experiments is that it is unclear how the principles we have covered generalize to encoding real faces. For example, if one constructs a realistic face space by performing principle components analysis on a large set of real faces, do cells in the face patches also show ramp-shaped tuning to the realistic face dimensions? And how well can one decode identity of real faces from face patch population activity? If cells truly are encoding specific axes through linear ramps, this suggests that a simple linear regression should be sufficient to decode facial identity. We are currently addressing these questions through ongoing experiments.

SUMMARY AND OUTLOOK

The macaque face patch system is a remarkable gift of nature for understanding the steps of object representation. Even though we are only just beginning to understand the principles underlying the organization of this system, it is already clear that major computational transformations are accomplished between each stage, to generate a code for facial identity in the most anterior face patch AM invariant to transformations such as view, position, and size. Future work will need to clarify whether and how the organization of this system generalizes to other object categories; evidence suggests that systems in IT cortex comprising multiple patches are also used to represent scenes (Kornblith et al. 2013), bodies (Popivanov et al. 2012, 2014), and colored objects (Lafer-Sousa and Conway 2013).

I believe the biggest questions about the face patch system concern how the patches communicate with the rest of the brain, including earlier retinotopic cortex and higher-order brain areas that ultimately drive behavior. The face patches are like a wonderfully lit house in the middle of the woods. What is needed now is to follow the trail of bread crumbs from them, both forward and backward, to gain a deeper level of understanding into (1) how an object first arises as a coherent unit and how this coherent unit is transmitted as such from retinotopic to IT cortex, and (2) how the code for object identity, represented by a distributed population of neurons, is routed to downstream areas to enable flexible, goal-directed behavior. It is clear these processes must involve globally organized interactions that we only have the barest inkling of so far. For example, if two faces are present, how does the brain keep track of the identity, location, and actions of each separately? This "binding problem" is one of the abiding mysteries of systems neuroscience. In his book *Rhythms of the Brain*, Buzsáki (2006) vividly evokes the excitement that greeted the prospect of an imminent solution to the binding problem. It would be exciting if research on face processing, starting from sure knowledge of where the label for facial identity is located in the brain, could bring us closer to that day.

ACKNOWLEDGMENTS

I thank Winrich Freiwald, who started this journey into the face patch system side-by-side with me, and members of my laboratory past and present whose creativity, insight, skill, and hard work are a joy to acknowledge.

REFERENCES

Bruce C, Desimone R, Gross CG. 1981. Visual properties of neurons in a polysensory area in superior temporal sulcus of the macaque. *J Neurophysiol* **46:** 369–384.

Bushnell BN, Harding PJ, Kosai Y, Pasupathy A. 2011. Partial occlusion modulates contour-based shape encoding in primate area V4. *J Neurosci* **31:** 4012–4024.

Buzsáki G. 2006. *Rhythms of the brain*. Oxford University Press, New York.

Cheng X, Crapse T, Tsao DY. 2013. Features that drive face cells: A comparison across face patches. In *Society for Neuroscience Conference*, San Diego, CA.

Dubois J, de Berker AO, Tsao DY. 2015. Single-unit recordings in the macaque face patch system reveal limitations of fMRI MVPA. *J Neurosci* **35:** 2791–2802.

Freiwald WA, Tsao DY. 2010. Functional compartmentalization and viewpoint generalization within the macaque face-processing system. *Science* **330:** 845–851.

Grimaldi P, Saleem KS, Tsao DY. 2012. Anatomical connections of functionally defined anterior face patches in the macaque monkey. In *Society for Neuroscience Conference*, New Orleans, LA.

Grimaldi P, Saleem KS, Tsao DY. 2013. Subcortical connections of the functionally-defined face patches in the macaque monkey. In *Society for Neuroscience Conference*, San Diego, CA.

Issa EB, DiCarlo JJ. 2012. Precedence of the eye region in neural processing of faces. *J Neurosci* **32:** 16666–16682.

Kanwisher N, McDermott J, Chun MM. 1997. The fusiform face area: A module in human extrastriate cortex specialized for face perception. *J Neurosci* **17:** 4302–4311.

Kornblith S, Cheng X, Ohayon S, Tsao DY. 2013. A network for scene processing in the macaque temporal lobe. *Neuron* **79:** 766–781.

Lafer-Sousa R, Conway BR. 2013. Parallel, multi-stage processing of colors, faces and shapes in macaque inferior temporal cortex. *Nat Neurosci* **16:** 1870–1878.

Logothetis NK. 2008. What we can do and what we cannot do with fMRI. *Nature* **453:** 869–878.

Moeller S, Tsao DY. 2011. Representation of face familiarity in AM. In *Society for Neuroscience Conference*, San Diego, CA.

Moeller S, Freiwald WA, Tsao DY. 2008. Patches with links: A unified system for processing faces in the macaque temporal lobe. *Science* **320.** 1353–1359.

Ohayon S, Freiwald WA, Tsao DY. 2012. What makes a cell face selective? The importance of contrast. *Neuron* **74:** 567–581.

Ohki K, Chung S, Kara P, Hubener M, Bonhoeffer T, Reid RC. 2006. Highly ordered arrangement of single neurons in orientation pinwheels. *Nature* **442:** 925–928.

Popivanov ID, Jastorff J, Vanduffel W, Vogels R. 2012. Stimulus representations in body-selective regions of the macaque cortex assessed with event-related fMRI. *NeuroImage* **63:** 723–741.

Popivanov ID, Jastorff J, Vanduffel W, Vogels R. 2014. Heterogeneous single-unit selectivity in an fMRI-defined body-selective patch. *J Neurosci* **34:** 95–111.

Quiroga RQ, Reddy L, Kreiman G, Koch C, Fried I. 2005. Invariant visual representation by single neurons in the human brain. *Nature* **435:** 1102–1107.

Schummers J, Yu H, Sur M. 2008. Tuned responses of astrocytes and their influence on hemodynamic signals in the visual cortex. *Science* **320:** 1638–1643.

Sinha P. 2002. Qualitative representations for recognition. In *Lecture Notes in Computer Science*, pp. 249–262. Springer, New York.

Tanaka K. 2003. Columns for complex visual object features in the inferotemporal cortex: Clustering of cells with similar but slightly different stimulus selectivities. *Cereb Cortex* **13:** 90–99.

Tsao DY, Freiwald WA, Tootell RBH, Livingstone MS. 2006. A cortical region consisting entirely of face-selective cells. *Science* **311:** 670–674.

Tsao DY, Moeller S, Freiwald WA. 2008a. Comparing face patch systems in macaques and humans. *Proc Natl Acad Sci* **105:** 19514–19519.

Tsao DY, Schweers N, Moeller SM, Freiwald WA. 2008b. Patches of face-selective cortex in the macaque frontal lobe. *Nat Neurosci* **11:** 877–879.

Valentine T, Lewis MB, Hills PJ. 2015. Face-space: A unifying concept in face recognition research. *Q J Exp Psychol (Hove)* **2015:** 1–24.

Viola P, Jones M. 2001. Rapid object detection using a boosted cascade of simple features. In *Computer vision and pattern recognition*. IEEE, Piscataway, NJ.

Zhou H, Friedman HS, von der Heydt R. 2000. Coding of border ownership in monkey visual cortex. *J Neurosci* **20:** 6594–6611.

Representation of Naturalistic Image Structure in the Primate Visual Cortex

J. ANTHONY MOVSHON[1] AND EERO P. SIMONCELLI[1,2]
[1]*Center for Neural Science, New York University, New York, New York 10003*
[2]*Howard Hughes Medical Institute, New York University, New York, New York 10003*
Correspondence: movshon@nyu.edu

The perception of complex visual patterns emerges from neuronal activity in a cascade of areas in the primate cerebral cortex. We have probed the early stages of this cascade with "naturalistic" texture stimuli designed to capture key statistical features of natural images. Humans can recognize and classify these synthetic images and are insensitive to distortions that do not alter the local values of these statistics. The responses of neurons in the primary visual cortex, V1, are relatively insensitive to the statistical information in these textures. However, in the area immediately downstream, V2, cells respond more vigorously to these stimuli than to matched control stimuli. Humans show blood-oxygen-level-dependent functional magnetic resonance imaging (BOLD fMRI responses in V1 and V2) that are consistent with the neuronal measurements in macaque. These fMRI measurements, as well as neurophysiological work by others, show that true natural scenes become a more prominent driving feature of cortex downstream from V2. These results suggest a framework for thinking about how information about elementary visual features is transformed into the specific representations of scenes and objects found in areas higher in the visual pathway.

THE FRAMEWORK: CASCADED VISUAL PROCESSING

In primates, the perception of complex visual patterns and objects emerges from neural activity as it is transformed through a cascade of areas in the cerebral cortex. Neurons in the primary visual cortex (V1) are selective for local orientation and spatial scale of visual input (Hubel and Wiesel 1962, 1968; DeValois et al. 1982). Downstream areas contain neurons selective for more complex attributes, and this is presumably achieved by assembling particular combinations of their upstream afferents. But these attributes have proven difficult to discover with either perceptual or physiological measurements.

Given the ubiquity of orientation selectivity in primary visual cortex (Priebe and Ferster 2012), it is intuitively appealing to assume that its computational purpose is to represent the local orientation of edges. Over the past 50 years, the dominant view in both the computational and biological vision communities is that later stages of processing should somehow combine these local edge elements to construct corners, junctions, and more extensive contours, eventually leading to shapes, forms, and objects (Marr 1982; Reisenhuber and Poggio 1999). Until recently, most computational research on object recognition was built around this paradigm, as well as much of the study of midlevel pattern perception, and physiological measurements in areas V2 and V4 of the ventral stream.

The intuitive appeal of the "edge paradigm" is partly due to its constructive nature. We imagine the visual system should analyze a visual scene much the way we would draw a picture of it. But this cartoon reasoning should be viewed with suspicion: The act of recreating a scene with a pencil does not necessarily reveal the processes by which the scene was analyzed by the visual system. In fact, edges and contours make up a very small portion of most naturally occurring visual scenes, and generating realistic drawings relies crucially on the introduction of additional elements such as shading and texture that are less easily described as the assembly of individual strokes. Moreover, 50 years of effort seems not to have brought us much closer to an understanding of form vision.

An alternative (but not exclusive) minority view has coexisted with the edge-based view. In brief, the concept is that the visual system is more concerned with the representation of the "stuff" that lies between the edges, and less concerned with the edges themselves (Adelson and Bergen 1991). To make this more concrete for the present discussion, let us focus on the specific case of visual texture.

VISUAL TEXTURE: MODELS AND HUMAN PERCEPTION

"Visual texture" refers to portions of an image that are filled with repeated elements, often subject to some randomization in their location, size, color, orientation, etc.; for example, an image of leaves, or pebbles, or tree bark (Fig. 1A). Lettvin (1976) offered this insight: "Let us say that to the extent that visible objects are different and far apart, they are *forms*. To the extent that they are similar and congregated they are a *texture*. A man has form; a crowd has man-texture. A leaf has form; an arbor has leaf-

 doi: 10.1101/sqb.2014.79.024844

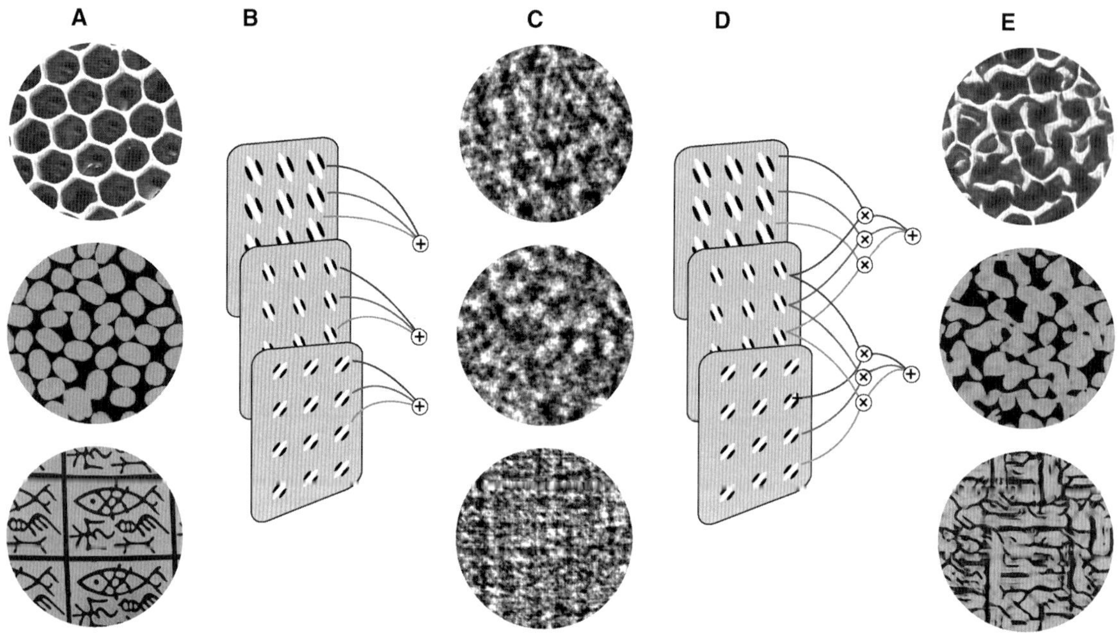

Figure 1. Naturalistic textures and synthetic images with matching statistics. (*A*) Photographs of three different textures. (*B*) Spectral statistics, as captured with a V1-like model. Statistics are computed as spatial averages over responses of arrays of model neurons, whose receptive fields have a particular preferred orientation and scale. (*C*) Images synthesized to have identical spectral statistics to the corresponding original images (column *A*). (*D*) Joint (correlation) statistics of a V1-like model. Statistics are computed as correlations (average of pairwise products) both within and across arrays of model V1 neurons (Portilla and Simoncelli 2000). (*E*) Images synthesized to have identical joint statistics.

texture, and so on." Bela Julesz pioneered the statistical characterization of visual texture by proposing that the statistics of image pixels, measured up to some order, should suffice to partition textures into classes that are indistinguishable to a human observer (Julesz 1962). Implicitly, Julesz was asserting that the human visual system represented visual texture by measuring these statistics, and only these statistics. In this case, the theory predicts that any two images that are identical in this statistical sense must appear identical to a human.

Critical to Julesz' framing of the problem was the desire for a *minimal* set of statistics, as well as the plan to experimentally validate the model by seeking perceptual counterexamples. Julesz et al. (1973) initially thought that pairwise statistics were sufficient, but then disproved this by producing hand-constructed example pairs of textures with identical statistics through second (and even third) order that were easily distinguished by human observers (Caelli and Julesz 1978; Julesz et al. 1978). Given the falsification of this particular instantiation of his statistical theory, Julesz abandoned the approach altogether and began to develop a constructive theory of texture based on randomly placed "texton" features. Like the "edge paradigm," this was appealing for its constructive nature and easily led to stimuli and experiments, but proved far more difficult to interpret in terms of perceptual or physiological representation.

Julesz' statistical-conceptualization was sound, but his definition of the statistical model in terms of the order of pixel statistics was problematic. Increasing statistical order leads to ever more nonlinear interactions of multiple pixels, with no restriction on the spatial extent of those interactions. Responses of early visual neurons, on the other hand, are generally described in terms of particular combinations of *local* image intensity. For example, retinal ganglion responses are commonly described as the rectified response of a center-surround linear filter. V1 simple cell responses are commonly described as rectified (or rectified and squared) responses of oriented linear filters (Heeger 1992). And V1 complex-cell responses are described as an average over these simple cell responses, all having the same orientation preference, but differing in the spatial location of their filters (Hubel and Wiesel 1962; Movshon et al. 1978; Adelson and Bergen 1985).

Measuring statistics in a physiologically consistent format can lead to much more powerful model of texture. Specifically, a model for texture based on the pairwise correlations between model simple- and complex-cell responses, at nearby positions, orientations, and scales (Fig. 1D) can capture many of the salient features of natural texture images (Portilla and Simoncelli 2000). The correlations are computed by averaging the product of pairs of responses over the spatial extent of the image. To show that this model captures the visually relevant attributes of a given texture, the authors followed Julesz' paradigm, synthesizing new images with the same model responses, and then asking whether the resulting images were similar in appearance to the original. In brief, one initializes an image with white noise and then adjusts the pixels according to the gradient of the squared error of the model responses, relative to the desired model responses. This procedure converges after a relatively small number of iterations and produces images that are similar in appearance to the original (Fig. 1E; Portilla and Simoncelli 2000; Balas 2006). It seems remarkable that responses of these simplistic models (cascades of filtering, rectification, averaging), which provide only a crude statistical summary of the image content, are sufficient to capture such sophisticated features. But it is worth noting that a reduced model that uses only the averages of V1 responses, computed for cells at each orientation and size (Fig. 1B) and therefore measuring only spectral information, is insufficient to capture the features of most naturally occurring textures (Fig. 1C).

Given that these synthesized textures capture much of the visual appearance of the textures from which their statistics are derived, it is natural to ask whether (and where) these statistics might be represented in the brain. Although the texture model described in the previous paragraph was not originally intended as a model for post-V1 physiology, two modifications allow it to be interpreted as such, at the abstract level of population representation. First, the statistics can be gathered *locally*, over regions corresponding to receptive fields in a visual area downstream from V1. A natural candidate is V2, which receives strong direct input from V1, and whose neurons have receptive fields that are roughly twice the size of those in V1 at any given eccentricity (Gattass et al. 1981). As in most visual areas, receptive field sizes in V2 increase roughly in proportion to eccentricity. Second, the products of V1-like afferents that are averaged when computing model correlations can be replaced with the squares of summed afferents (Adelson and Bergen 1985). The product is implicitly represented (it is the "cross term"), and two textures with the same correlations will also have the same values in this modified model. More importantly, this modified model assumes the same form as models of V1 complex cells (a local average of squared linear filter responses) and thus instantiates a cascade model of cortical computation in which the same elementary operations are performed in each cortical area, differing only in the inputs they receive.

This physiological version of the texture model, in which statistics are computed over regions roughly the size of V2 receptive fields, can be used to synthesize images whose local statistics over each region are matched to those of an existing photograph. For a human observer whose gaze is directed to the correct location in the image (the sizes of model statistical regions are chosen to match those of V2 receptive fields only for a particular center of gaze), these synthetic images are indistinguishable from the original photograph, despite distortions whose severity increases with eccentricity (Fig. 2; Freeman and Simoncelli 2011). This provides a direct demonstration that the information discarded by the model, which retains only a crude statistical summary of the content within each local spatial region, is also discarded by the human visual system. This also provides a physiological explanation for the visual phenomenon of "crowding" (Bouma 1970), in which humans fail to recognize peripherally presented objects that are surrounded by background clutter (Balas, et al. 2009; Freeman and Simoncelli 2011).

Armed with the hypothesis that V2 represents local statistics of the sort used in this statistical texture model, a natural next step is to examine the physiological responses of V2 neurons to these stimuli.

Figure 2. Example photograph and a random image synthesized to have matching texture statistics over spatial regions the size of V2 receptive fields. V2 receptive fields are roughly twice the diameter of V1 receptive fields and grow approximately linearly with distance from the fovea. When viewed in alternation by human subjects with their gaze fixated on the image center (red dot), the two images are virtually indistinguishable, despite dramatic distortions in the periphery. Thus the statistics of the model, although they provide only a partial summary of the original scene, are sufficient to capture its visual appearance.

MACAQUE CORTICAL UNIT RESPONSES TO NATURALISTIC TEXTURE

It has proved difficult to assign specific visual functions to V2 or at least to deduce them from analysis of V2 neuronal responses to simple parametric stimuli, because these are qualitatively quite similar to responses measured in V1. V2 cells are commonly selective for orientation, direction, spatial frequency, drift rate or speed, color, binocular disparity—all the dimensions for which V1 neurons are selective (Hubel and Wiesel 1965; Hubel and Livingstone 1987; Levitt et al. 1994; Hegde and Van Essen 2007). There are quantitative differences. An idea that has its roots in early work of Hubel and Wiesel (1965) is that cells in V2 are selective for simple extensions of the basic contour responses measured in V1—selectivity has been reported for such elaborated features as curvature, angle, convexity, and illusory borders (von der Heydt and Peterhans 1989; Hegde and Van Essen 2000; Zhou et al. 2000; Ito and Komatsu 2004; Anzai et al. 2007; Willmore et al. 2010). However, although some cells show selectivity for these various attributes, many do not, and, when closely examined, population selectivity does not differ substantially from that seen when comparable measurements are made in V1 (Hegde and Van Essen 2007).

Based on the findings described in the previous section, we hypothesized that cells in V2, but not those in V1, might represent the spatial correlations that distinguish naturalistic texture from spectrally matched control stimuli (Fig. 1). V2 receives a strong direct input from V1, and the size of its receptive fields match human thresholds for discriminating scrambled textures (Fig. 2; Freeman and Simoncelli 2011). We therefore decided to begin with a comparison of response properties of cells in V1 and V2, using a set of textures chosen to span a reasonable range of naturally occurring variation in simple image statistics.

The results of our neurophysiological texture experiments are summarized (Freeman et al. 2013a) in Figure 3. In Figure 3A,B, we show the average response time courses for populations of about 100 units recorded from V1 (Fig. 3A) and V2 (Fig. 3B). In V1, responses to naturalistic textures and spectrally matched noise controls were almost indistinguishable, while in V2 responses to naturalistic textures were robustly larger than responses to noise. The difference is captured in Figure 3C with a "modulation index," which shows strong modulation for the V2 population over the entire time course of response, and only very weak modulation for the V1 population, emerging very late in the response. The average traces do not reveal how consistent

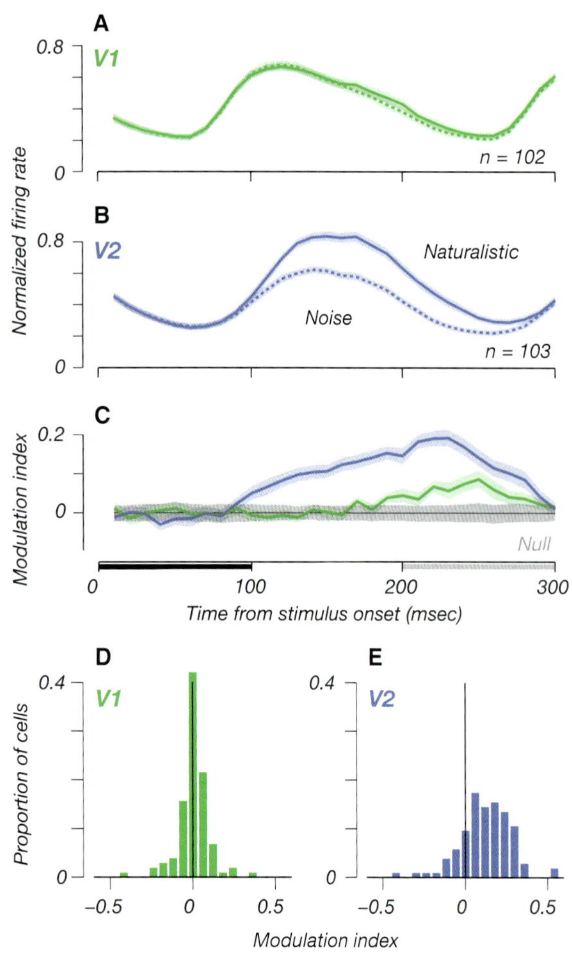

Figure 3. Responses of neurons in macaque V1 and V2 to naturalistic textures and spectrally matched noise controls. We measured the responses of isolated cells in V1 and V2 cells to a randomly interleaved sequence of naturalistic textures and spectrally matched noise (Fig. 1). We presented the textures for 100 msec in a 4° window centered on the receptive field, with a 100-msec blank interval between stimuli. (*A*) Normalized mean responses of 102 cells from V1. Solid line, naturalistic stimuli; dashed line, noise controls. Stimulus time course indicated at bottom. (*B*) Normalized mean responses of 103 cells from V2; conventions as in *A*. (*C*) A modulation index, computed as the difference between naturalistic and noise responses divided by their sum, plotted for V1 and V2 populations. Shaded bands in *A–C* indicate 95% confidence intervals. (*D*) Distribution of the modulation index for 102 cells in V1. (*E*) Distribution of the modulation index for 103 cells in V2. After Freeman et al. (2013a).

the responses were for different neurons, so Figure 3D,E shows the distribution of the modulation index for neuronal populations in V1 and V2. The V1 data show modest dispersion about a mean near 0, whereas the mean for V2 was substantially and significantly shifted to positive values.

HUMAN CORTICAL RESPONSES TO NATURALISTIC TEXTURE AND NATURAL IMAGES

Given the robustly enhanced neuronal responses to naturalistic textures that we observed in unit responses recorded from V2, we decided to look for evidence of similar responses in human visual cortex, using the technique of blood-oxygen-level-dependent functional magnetic resonance imaging (BOLD fMRI) (Freeman et al. 2013b). We measured responses in three subjects to an alternating sequence of spectrally matched control stimuli and naturalistic textures (Fig. 4A), presented in an annular region surrounding the fixation target. We render the responses on a flattened map of the posterior pole of the cortex in a representative subject; the marked borders between visual areas were established in earlier topographic mapping studies using standard techniques (Wandell et al. 2007; Gardner et al. 2008). The alternation of naturalistic and noise stimuli notably failed to modulate the BOLD signal in V1, but in V2 and adjacent areas, the response was robust and reliable. This is entirely consistent with our measurements of neuronal responses in macaque V1 and V2 (Freeman et al. 2013a), as well as the perceptual discrimination of images with statistics matched over regions the size of V2 receptive fields (Freeman and Simoncelli 2011).

The broad view of the brain given by functional imaging allowed us also to explore the way in which image representation begins to move beyond the representation of the kind of naturalistic image statistics that are effective in driving V2; in particular, we have begun to ask whether further elaborations of the cascade of visual processing leads downstream areas to respond differentially not only to images with naturalistic statistics, but also to natural images themselves. Physiological measurements of unit responses in macaque extrastriate areas V4 and IT by Rust and DiCarlo (2010) suggest this possibility, showing that units there respond better to photographs of natural scenes than to scrambled photographs made to match the statistics of the texture model. Using the same block design, we compared fMRI activation in human visual cortex to natural images with those to spectrally matched noise (Fig. 4B), and we also compared activation by natural images and naturalistic textures derived from those images (Fig. 4C) (Freeman et al. 2013b). Although response modulation in V2 and V3 was similar for the exchange of natural images with noise, response modulation in areas anterior to V3 was stronger than when naturalistic textures were exchanged with noise (Fig. 4B). As one might then expect, the exchange of natural images for naturalistic textures evoked little response in early areas (V1–V3), but strongly activated anterior areas (V4 and other downstream areas). In aggregate, these preliminary measurements suggest that areas downstream from V2 and V3 continue the process of elaboration begun in V2, and effect at least part of the job of transforming representations of the stat-

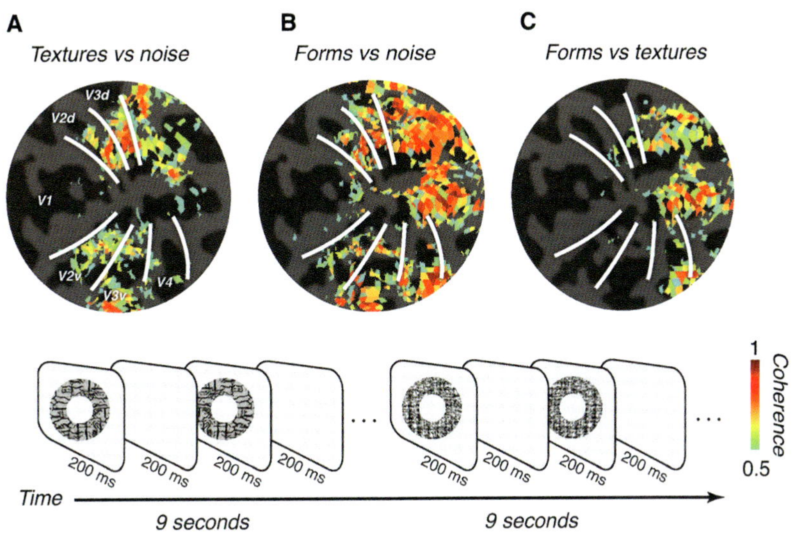

Figure 4. Responses of human visual cortex to natural images, naturalistic textures, and spectrally matched noise controls. Flattened maps of the posterior pole of the right hemisphere of a representative observer showing modulation of BOLD fMRI responses to alternating sequences of stimuli in a block design whose time course is schematically diagrammed at the bottom: briefly presented repeated stimuli (200 msec on, 200 msec off) were presented in 9-sec blocks, alternating between the conditions being compared. Area boundaries are derived from a separate topographic mapping experiment. In each panel, the pseudocolor scale represents the coherence of the BOLD response component synchronized with the time course of the stimulus exchange. (*A*) Alternation between naturalistic textures and spectrally matched noise controls. (*B*) Alternation between natural images and spectrally matched noise controls. (*C*) Alternation between natural images and naturalistic textures. (From Freeman et al. 2013b.)

istical features of images into representations of the particular features of objects and scenes that are found in the visual areas of the inferior temporal lobe (DiCarlo et al. 2012).

CONCLUSION AND OUTLOOK

We have established that V2 reformats the relatively simple visual representation provided by V1 so as to make *explicit* relationships of activity that are *implicit* in the input population. This is akin to the computation of orientation selectivity from nonoriented inputs that is executed by circuits in V1—the orientational structure of the visual image is implicit in the responses of LGN neurons that provide input to V1, but is made explicit in V1 (Hubel and Wiesel 1962; Priebe and Ferster 2012). Another example of such cascaded computation is the selectivity for pattern motion seen in directionally selective neurons in area MT—these neurons respond to complex object motion in a way that is implied, but not directly represented in the inputs from V1 (Movshon et al. 1985; Simoncelli and Heeger 1998; Rust et al. 2006). This recurring motif of cascaded computation is likely the basis of other forms of selectivity in the visual pathway, such as the selective responses to forms and objects observed in V4 and IT (Brincat and Connor 2004; Serre et al. 2007; Rust and DiCarlo 2010). Such elaboration is also likely the basis of our finding that V4 and other areas downstream from V2 seem to respond preferentially to natural images of objects and scenes, compared with statistically matched textures derived from those images (Fig. 4C; Freeman et al. 2013b).

What can we say about the responses of individual V2 neurons on the basis of our analysis of group data? The population of neurons in V1 can represent orientation for purposes, say, of discrimination (Graf et al. 2011). But this is not the same as being able to describe the responses of each V1 cell, for which a particular set of properties and circuits are in play (Priebe and Ferster 2012). The same is true for V2—knowing that the population of V2 neurons can represent statistical structure in images does not uniquely specify the response properties of individual V2 neurons, and a major challenge for ongoing research will be to create such models. Related work in V4 has explored computations that produce some forms of texture specificity (Okazawa et al. 2015), and we have begun to explore models to account for the texture responses of V2 neurons. The structure of the statistical model from which the texture stimuli are generated (Portilla and Simoncelli 2000) provides a starting point. This model combines particular V1-like responses by correlation—an average of pairwise products—a calculation for which a neural implementation has not been shown in mammals (although see Jones and Gabbiani 2012 for an invertebrate example). As shown by Adelson and Bergen (1985), an equivalent model can be constructed by squaring the sums of these pairs, creating an alternative neural circuit implementation that may be more plausible for mammalian cortex. Such models make testable predictions (see Emerson et al. 1992), and fitting and testing V2 models of this form against data from individual neurons is an important next step.

In addition, neural network models that are optimized for performance of visual recognition may provide some constraints (Serre et al. 2007; Khaligh-Razavi and Kriegeskorte 2014; Yamins et al. 2014). In the past few years, hierarchical neural network models, optimized for recognition performance on large image databases, have outstripped previous models working within the classical edge-based paradigm. The success of these networks is still mysterious, and the details of their learned internal representations have not been carefully studied (Khaligh-Razavi and Kriegeskorte 2014; but see Zeiler and Fergus 2014). The design of these networks is often inspired by physiology (Jarrett et al. 2009) and, given the similarity to the construction of the texture model described here, are likely to capture similar texture-like attributes in intermediate stages. Images synthesized from these representations may therefore prove useful as stimuli in perceptual or physiological investigations, and the precise form of internal representation in these networks may provide inspiration for the design of physiological response models.

Our exploration of texture representations in cortex was driven by perceptual phenomena and mechanisms. Now we can begin to ask how to relate cortical neural circuits to those perceptual phenomena. We have found that the average modulation index of V2 neurons for particular texture images is correlated with the ability of human observers to detect the presence of the model statistics in synthetic versions of those images (Freeman et al. 2013a). In the context of a more natural perceptual task, we have obtained preliminary evidence that V2 neurons also provide a better substrate for classifying textures into categories than V1 neurons—as a population, V2 neurons are selective for the properties of particular textures, while being tolerant of image variations that preserve texture identity, such as those that arise during the synthesis procedure (Ziemba et al. 2012). With this regard, V2 responses to texture are analogous to IT responses to objects and forms—selective for particular forms and at the same time robust to transformations that preserve the identity of those forms (Rust and DiCarlo 2010).

A central question is whether the form and object selectivity of these later stages are constructed from texture-selective neurons in area V2, or whether they instead arise through some other form-specific pathway. Lettvin (1976) considers the relationship between form and texture: "One can imagine shapes spatially interfering with each other to comprise texture; or else suppose that texture is primitive and that textures combine to produce forms—just as letters combine to make words." Lettvin himself comes down on the second side, arguing that texture provides the building blocks from which form is constructed. This conjecture is compatible with our observations that mechanisms in V2, relatively early in the visual pathway, respond invariantly to texture. Only later, in areas like the inferotemporal cortex, do responses become selective and invariant for objects and scenes.

ACKNOWLEDGMENTS

The work reviewed in this paper was done in collaboration with Javier Portilla, Jeremy Freeman, Corey Ziemba, and David Heeger, whose many contributions we gratefully acknowledge. This work was supported by a grant from the National Institutes of Health (R01 EY022428) and by the Howard Hughes Medical Institute.

REFERENCES

Adelson EH, Bergen JR. 1985. Spatiotemporal energy models for the perception of motion. *J Opt Soc Am A* **2:** 284–299.

Adelson EH, Bergen JR. 1991. The plenoptic function and the elements of early vision. In *Computational models of visual processing* (eds Landy MS, Movshon JA), pp. 3–20. MIT Press, Cambridge, MA.

Anzai A, Peng X, Van Essen DC. 2007. Neurons in monkey visual area V2 encode combinations of orientations. *Nat Neurosci* **10:** 1313–1321.

Balas B. 2006. Texture synthesis and perception: using computational models to study texture representations in the human visual system. *Vision Res* **46:** 299–309.

Balas B, Nakano L, Rosenholtz R. 2009. A summary-statistic representation in peripheral vision explains visual crowding. *J Vis* **9:** 13–18.

Bouma H. 1970. Interaction effects in parafoveal letter recognition. *Nature* **226:** 177–178.

Brincat SL, Connor CE. 2004. Underlying principles of visual shape selectivity in posterior inferotemporal cortex. *Nat Neurosci* **7:** 880–886.

Caelli T, Julesz B. 1978. On perceptual analyzers underlying visual texture discrimination: Part I. *Biol Cybern* **28:** 167–175.

De Valois RL, Albrecht DG, Thorell LG. 1982. Spatial frequency selectivity of cells in macaque visual cortex. *Vis Res* **22:** 545–559.

DiCarlo JJ, Zoccolan D, Rust NC. 2012. How does the brain solve visual object recognition? *Neuron* **73:** 415–434.

Emerson RC, Bergen JR, Adelson EH. 1992. Directionally selective complex cells and the computation of motion energy in cat visual cortex. *Vision Res* **32:** 203–218.

Freeman J, Simoncelli EP. 2011. Metamers of the ventral stream. *Nat Neurosci* **14:** 1195–1201.

Freeman J, Ziemba CM, Heeger DJ, Simoncelli EP, Movshon JA. 2013a. A functional and perceptual signature of the second visual area in primates. *Nat Neurosci* **16:** 974–981.

Freeman J, Ziemba CM, Simoncelli EP, Movshon JA. 2013b. Functionally partitioning the ventral stream with controlled natural stimuli. *Society for Neuroscience Abstracts*, 406.01. Online.

Gardner JL, Merriam EP, Movshon JA, Heeger DJ. 2008. Maps of visual space in human occipital cortex are retinotopic, not spatiotopic. *J Neurosci* **28:** 3988–3999.

Gattass R, Gross CG, Sandell JH. 1981. Visual topography of V2 in the macaque. *J Comp Neurol* **201:** 519–539.

Graf AB, Kohn A, Jazayeri M, Movshon JA. 2011. Decoding the activity of neuronal populations in macaque primary visual cortex. *Nat Neurosci* **14:** 239–245.

Heeger DJ. 1992. Half-squaring in responses of cat striate cells. *Vis Neurosci* **9:** 427–443.

Hegdé J, Van Essen DC. 2000. Selectivity for complex shapes in primate visual area V2. *J Neurosci* **20:** RC61.

Hegdé J, Van Essen DC. 2007. A comparative study of shape representation in macaque visual areas v2 and v4. *Cereb Cortex* **17:** 1100–1116.

Hubel DH, Livingstone MS. 1987. Segregation of form, color, and stereopsis in primate area 18. *J Neurosci* **7:** 3378–3415.

Hubel DH, Wiesel TN. 1962. Receptive fields, binocular interaction and functional architecture in the cat's visual cortex. *J Physiol* **160:** 106–154.

Hubel DH, Wiesel TN. 1965. Receptive fields and functional architecture in two nonstriate visual areas (18 and 19) of the cat. *J Neurophysiol* **28:** 229–289.

Hubel DH, Wiesel TN. 1968. Receptive fields and functional architecture of monkey striate cortex. *J Physiol* **195:** 215–243.

Ito M, Komatsu H. 2004. Representation of angles embedded within contour stimuli in area V2 of macaque monkeys. *J Neurosci* **24:** 3313–3324.

Jarrett K, Kavukcuoglu K, Ranzato M, LeCun Y. 2009. What is the best multi-stage architecture for object recognition? In *Computer Vision, 2009 IEEE 12th International Conference*, IEEE, pp. 2146–2153.

Jones PW, Gabbiani F. 2012. Impact of neural noise on a sensory-motor pathway signaling impending collision. *J Neurophysiol* **107:** 1067–1079.

Julesz B. 1962. Visual pattern discrimination. *IRE Trans Inf Theory* **IT-8:** 84–92.

Julesz B, Gilbert EN, Shepp LA, Frisch HL. 1973. Inability of humans to discriminate between visual textures that agree in second-order statistics-revisited. *Perception* **2:** 391–405.

Julesz B, Gilbert EN, Victor JD. 1978. Visual discrimination of textures with identical third-order statistics. *Biol Cybern* **31:** 137–140.

Khaligh-Razavi SM, Kriegeskorte N. 2014. Deep supervised, but not unsupervised, models may explain IT cortical representation. *PLoS Comput Biol* **10:** e1003915.

Lettvin JY. 1976. On seeing sidelong. *Sciences* **16:** 10–20.

Levitt JB, Kiper DC, Movshon JA. 1994. Receptive fields and functional architecture of macaque V2. *J Neurophysiol* **71:** 2517–2542.

Marr D. 1982. *Vision*. W.H. Freeman, San Francisco.

Movshon JA, Thompson ID, Tolhurst DJ. 1978. Receptive field organization of complex cells in the cat's striate cortex. *J Physiol* **283:** 79–99.

Movshon JA, Adelson EH, Gizzi MS, Newsome WT. 1985. The analysis of moving visual patterns. *Pattern Recognition Mechanisms* **54:** 117–151.

Okazawa G, Tajima S, Komatsu H. 2015. Image statistics underlying natural texture selectivity of neurons in macaque V4. *Proc Natl Acad Sci* **112:** E351–E360.

Portilla J, Simoncelli EP. 2000. A parametric texture model based on joint statistics of complex wavelet coefficients. *Int J Comput Vis* **40:** 49–70.

Priebe NJ, Ferster D. 2012. Mechanisms of neuronal computation in mammalian visual cortex. *Neuron* **75:** 194–208.

Reisenhuber M, Poggio T. 1999. Hierarchical models of object recognition in cortex. *Nat Neurosci* **2:** 1019–1025.

Rust NC, DiCarlo JJ. 2010. Selectivity and tolerance ("invariance") both increase as visual information propagates from cortical area V4 to IT. *J Neurosci* **30:** 12978–12995.

Rust NC, Mante V, Simoncelli EP, Movshon JA. 2006. How MT cells analyze the motion of visual patterns. *Nat Neurosci* **9:** 1421–1431.

Serre T, Oliva A, Poggio T. 2007. A feedforward architecture accounts for rapid categorization. *Proc Natl Acad Sci* **104:** 6424–6429.

Simoncelli EP, Heeger DJ. 1998. A model of neuronal responses in visual area MT. *Vision Res* **38:** 743–761.

von der Heydt R, Peterhans E. 1989. Mechanisms of contour perception in monkey visual cortex. I. Lines of pattern discontinuity. *J Neurosci* **9:** 1731–1748.

Wandell BA, Dumoulin SO, Brewer AA. 2007. Visual field maps in human cortex. *Neuron* **56:** 366–383.
Willmore BD, Prenger RJ, Gallant JL. 2010. Neural representation of natural images in visual area V2. *J Neurosci* **30:** 2102–2114.
Yamins DL, Hong H, Cadieu CF, Solomon EA, Seibert D, DiCarlo JJ. 2014. Performance-optimized hierarchical models predict neural responses in higher visual cortex. *Proc Natl Acad Sci* **111:** 8619–8624.
Zeiler ND, Fergus R. 2014. Visualizing and understanding convolutional networks. *Computer Vision—ECCV 2014*, 816–833.
Zhou H, Friedman HS, Von Der Heydt R. 2000. Coding of border ownership in monkey visual cortex. *J Neurosci* **20:** 6594–6611.
Ziemba CM, Freeman J, Movshon JA, Simoncelli EP. 2012. Selectivity and invariance are greater in macaque V2 than V1. CoSyNe Abstracts II-91. Online.

The Influence of Gaze Control on Visual Perception: Eye Movements and Visual Stability

REBECCA M. KROCK[1] AND TIRIN MOORE[1,2]

[1]*Department of Neurobiology, Stanford University School of Medicine, Stanford, California 94305*

[2]*Howard Hughes Medical Institute, Stanford University School of Medicine, Stanford, California 94305*

Correspondence: tirin@stanford.edu

Primates make several saccadic eye movements each second, and yet the retinal motion these movements generate goes unnoticed. Saccadic suppression is a profound loss of visual sensitivity occurring around the time of eye movements, and it is thought to contribute to visual stability by blunting the perception of self-generated motion. Neurophysiological studies have produced evidence that neurons throughout the visual system, including both the dorsal and ventral streams of extrastriate visual cortex, show a reduction in visual responses or sensitivity around the time of saccades. However, the source of this suppression remains unknown. We review evidence that oculomotor regions such as the superior colliculus and frontal eye field may play a role, as well as anatomical data that place constraints on possible mechanisms of suppression.

In spite of the constant shifting of our gaze from one stimulus to another, we perceive the visual world to be a stable and continuous whole (Fig. 1A,B). By comparison, a quick displacement of an eye with one's finger results in a marked perception of visual motion. This difference suggests the existence of some form of active mechanism that contributes to perceived stability during centrally generated eye movements. Although substantial progress has been made in understanding the neural mechanisms of visual processing during passive fixation, saccadic eye movements (saccades) punctuate these fixations approximately two to four times each second during free viewing. The mechanisms by which the brain compensates for the visual disruption that eye movements introduce remain largely unknown. A recent body of evidence implicates the saccadic system in the modulation of visual cortical representations and of perception (e.g., Moore 2006; Squire et al. 2013), but large gaps remain in our understanding of the neural circuitry that could specifically underlie the illusion of stability during eye movements.

One psychophysical phenomenon thought to reflect an active mechanism in perceived stability during eye movements is saccadic suppression. Saccadic suppression is a profound loss of visual sensitivity that takes place around the time of saccadic eye movements (Fig. 1C). For example, when human observers are asked to report the presence or absence of a visual probe appearing just before, during, or after a saccade, their contrast detection thresholds show an approximately fivefold elevation above those measured when the probe occurs during fixation (Latour 1962; Zuber and Stark 1966; Bridgeman et al. 1975; Burr et al. 1982, 1994; Shioiri and Cavanagh 1989; Diamond et al. 2000; Knöll et al. 2011). Saccadic suppression has also been reported in nonhuman primates, both during the fast (saccadic) phase of optokinetic nystagmus (Mohler and Cechner 1975), as well as during microsaccades (Hass and Horwitz 2011).

Given the frequency at which primates make saccadic eye movements, these findings imply that primate vision is impaired much of the time. The benefit conferred by a reduction of self-generated visual motion during saccades evidently outweighs the disadvantages of frequent visual impairment. Consistent with this, some patients with cerebellar or brainstem lesions experience visual motion in conjunction with abnormal spontaneous eye movements that they report to be disabling and distressing (Corkill and Vijayan 1976; Tilikete and Vighetto 2011). Suppression of vision during saccades therefore appears to play a critical role in perceptual stability across saccades. Here, we review (1) psychophysical studies exploring possible mechanisms of saccadic suppression, (2) neural correlates in the primate visual system, (3) anatomical data that constrain possible mechanisms of saccadic suppression, and (4) functional evidence that the saccadic system plays a role in modulating perisaccadic visual representations in the visual system.

PSYCHOPHYSICAL EVIDENCE FOR A CENTRAL SOURCE OF SUPPRESSION

Psychophysical studies have examined two basic classes of potential mechanisms for saccadic suppression. Active mechanisms for suppression would likely incorporate some form of a centrally generated (extraretinal) signal coupled to the motor command to move the eyes, known as corollary discharge or efference copy. It has

 doi: 10.1101/sqb.2014.79.024836

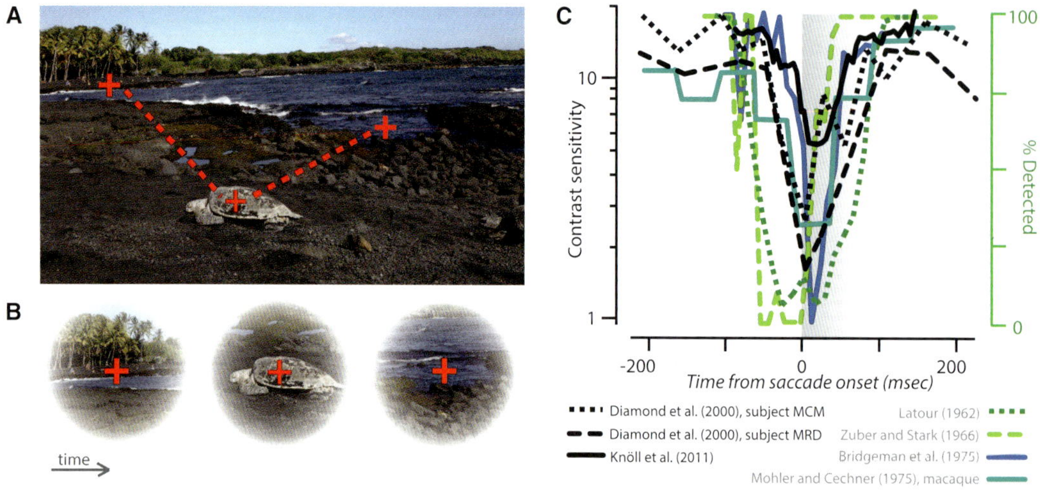

Figure 1. (*A*) Primates make frequent saccades to inspect features of the visual world (crosses, points of fixation; dashed lines, saccade vectors). (*B*) Between successive fixations (red crosses), no visual motion is perceived. Instead, visual perception consists of a series of stable snapshots. (*C*) A selection of psychophysical studies demonstrating that around the time of eye movements, contrast sensitivity (black *y*-axis) and the likelihood of detecting a visual stimulus (green *y*-axis) decline sharply.

long been argued that an active mechanism must be at play (Holt 1903; von Helmholtz 1925; Sperry 1950; von Holst and Mittelstaedt 1950). Passive mechanisms are those that would depend solely on visual feedback (reafference) generated by the saccade. These could include an inability of neurons early in the visual system to resolve high-speed visual stimuli and/or visual masking, in which the ability to perceive one visual stimulus is compromised by the appearance of another stimulus close by in space and time. Active and passive mechanisms are not mutually exclusive, and support has been found for both.

If passive mechanisms are responsible for saccadic suppression, then moving a visual stimulus across the eye at the high speeds achieved by saccades should reduce its visibility compared with stable stimulus. (Saccades shift the direction of gaze >100° of visual angle per second; 1° of visual angle is approximately the width of one's thumb held at arm's length.) Consistent with this, Dorr and Bex (2013) found that moving naturalistic stimuli across the eyes generated a saccadic suppression-like effect. Other studies, however, have found variable effects depending on the image features. Burr and Ross (1982) observed that although humans can resolve low-spatial-frequency sine-wave gratings moving at saccadic speeds, psychophysical contrast detection thresholds for moving, high-spatial-frequency gratings are in fact elevated, especially at lower illumination levels. These data suggest that passive mechanisms may specifically play a role in reducing the visibility of high-frequency, low-luminance stimuli during saccades. Yet sensitivity to low spatial frequencies moving at saccadic speeds is maintained and even enhanced for the lowest spatial frequencies, and it is at low spatial frequencies (below ~0.3 cycles/dva) that saccadic suppression is strongest (Burr et al. 1982). Thus, an active mechanism seems necessary to explain saccadic suppression of at least the low-spatial-frequency content of images.

Diamond et al. (2000) addressed the relative roles of passive and active mechanisms by comparing visual sensitivity when subjects made saccades over an image (the "active" condition) and when the image was moved over their eyes during fixation (the "passive" condition). When this image included high-contrast noise, they found that passive visual motion was sufficient to generate a suppressive effect equal to that observed in the active condition. However, when the image consisted solely of a horizontal sinusoidal grating that was swept horizontally across the eyes, such that there was little visual motion, suppression was observed in the "active" condition but not in the "passive" condition. Thus, saccadic suppression persisted even when passive mechanisms were excluded. This evidence is consistent with the idea that passive mechanisms may be sufficient to explain saccadic suppression under some visual conditions, but an active mechanism or mechanisms are also at play.

SACCADIC SUPPRESSION OF NEURONAL RESPONSES TO VISUAL STIMULI

Neural correlates of saccadic suppression have been observed in the responses of neurons throughout the primate visual system, including the lateral geniculate nucleus (LGN), the superior colliculus (SC), the pulvinar nucleus, primary visual cortex (V1), and extrastriate areas V2, V3, V4, MT, LIP, and VIP (Table 1; Fig. 2). Many of these studies have used strategies to minimize visual reafference and isolate any active mechanisms of saccadic suppression, such as presenting stimuli just before or after eye movements while the eye is still stable (e.g., Ibbotson et al. 2008; Bremmer et al. 2009; Han et al. 2009; Berman and Wurtz 2011), using spatially uniform stimuli (e.g., Mohler and Cechner 1975; Reppas et al. 2002; Sylvester et al. 2005), and/or comparing responses to active and

Table 1. Summary of past studies of neural correlates of saccadic suppression in humans and nonhuman primates

Year	Authors	Region	Saccadic suppression	% Suppression (max across time)	% Cells suppressed (if applicable)	Posts accadic enhancement
1975	Mohler and Cechner	V1	✓	[]	[]	
1976	Bartlett, Doty, Lee, and Sakakura	Geniculostriate tract	✓	43% (Magnocellular); 17% (parvocellular)	✓	
1976	Robinson and Wurtz	SC	✓	[]	61%	
1991	Robinson, McClurkin, Kertzman, and Petersen	SC	✓	[]	75%	
		Pi/Pl	✓	[]	45%	
2001	Ramcharan, Gnadt, and Sherman	LGN			✓	
2011	Hass and Horwitz	V1	✓	17%	[]	✓
2002	Reppas, Usrey, and Reid	LGN	✓	20%	90% (Magnocellular); 21% (parvocellular)	✓
2002	Thiele, Henning, Kubischik, and Hoffman	MST	✓	[]	68.1%	
		MT	✓	[]	66%	
2004	Kleiser, Seitz, and Krekelberg	V1/V2 (Human)	✓	24%		
		V3 (Human)				
		V4 (Human)	✓	50%		
		hMT+ (Human)	✓	34%		
		V7 (Human)	✓	46%		
		IPS (Human)	✓	14%		
2005	Sylvester, Haynes, and Rees	LGN (Human)	✓	[]		
		V1 (Human)	✓	[]		
		V2 (Human)				
		V3 (Human)				
		MT (Human)				
2006	Sylvester and Rees	LGN (Human)	✓	[]		
		V1 (Human)	✓	[]		
		V2 (Human)				
		V3 (Human)	✓	[]		
2007	Ibbotson, Price, Crowder, Ono, and Mustari	MT/MST	✓	[]	83%	✓
2008	Ibbotson, Crow der, Cloherty, Price, and Mustari	MSTd	✓	81%	[]	✓
2008	Kagan, Gur, and Snodderly	V1	✓	[]	[]	✓
2009	Bremmer, Kubischik, Hoffmann, and Krekelberg	MT	✓	26%	[]	✓
		MST	✓	19%	[]	✓
		VIP	✓	20%	[]	✓
		LIP	✓	22%	[]	×
2009	Han, Xian, and Moore	V4	✓	43% (decrease in contrast sensitivity)	42.6%	
2010	Cloherty, Mustari, Rosa, and Ibbotson	MSTd	✓	80%	79%	✓
2010	Hafed and Krauzlis	SC	✓	30%	[]	×
2011	Berman and Wurtz	P	✓	[]	24%	

Although paradigmatic differences make precise comparison between studies impossible, the strength of suppression is estimated here by the percent reduction in visual responses across all cells, at the time of deepest suppression with respect to the saccade (column five) and/or the percent of cells suppressed (column six). A number of studies report postsaccadic enhancement, usually following perisaccadic suppression, in the LGN, V1, MT, MST, and VIP (column seven).

[], exact value not reported; entry left blank if not applicable.

passive visual motion (e.g., Robinson and Wurtz 1976; Reppas et al. 2002; Thiele et al. 2002; Ibbotson et al. 2007). Correlates of saccadic suppression are consistently observed under all of these conditions, supporting a role for an active, extraretinal mechanism of suppression.

In addition to these neural correlates, saccadic suppression of perception has also been observed in the macaque. Mohler and Cechner (1975) showed that macaques' ability to detect briefly flashed stimuli is impaired by 0.5–0.8 log units during the fast (saccadic) phase of optokinetic nystagmus. Similarly to human saccadic suppression, this impairment stretches from ~50 msec before to 50 msec after eye movement initiation and is accompanied by a loss of visual responses in striate cortex. More recently, Hafed and Krauzlis (2010) observed perimicrosaccadic suppression of visual bursts in the SC accompanied by elevation of saccadic reaction times. Finally, Hass and Horwitz (2011) found that the occurrence of microsaccades during stimulus presentation impaired detection of achromatic stimuli.

Saccadic suppression occurs for luminance-modulated stimuli, but not equiluminant chromatic stimuli (Burr

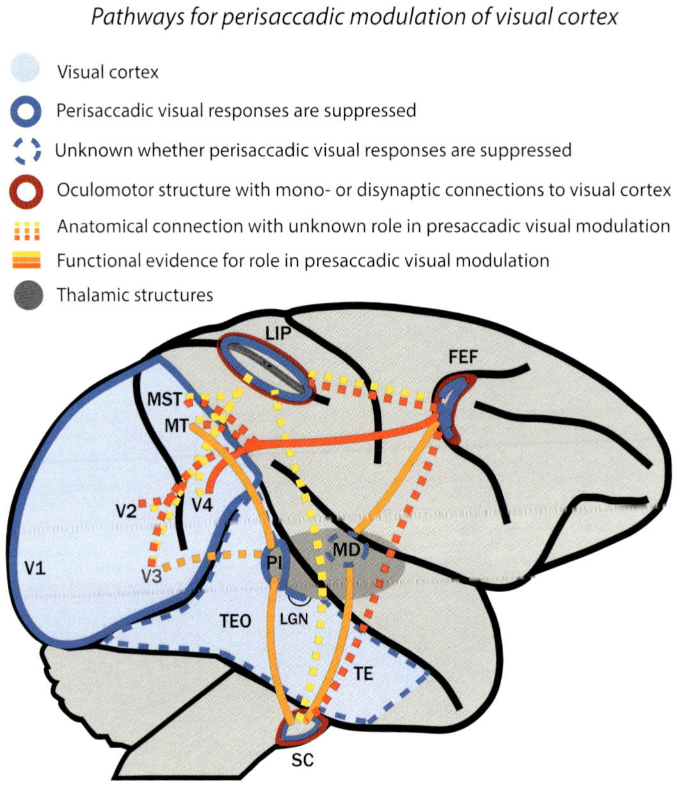

Figure 2. Summary of mono- and disynaptic pathways between oculomotor areas and visual cortex. Saccadic suppression (blue outlines) is observed in both the dorsal and ventral streams of visual cortex (Kleiser et al. 2004; Bremmer et al. 2009; Han et al. 2009) as well as in oculomotor structures including LIP (Bremmer et al. 2009), the superior colliculus (Robinson and Wurtz 1976; Hafed and Krauzlis 2010), and the frontal eye field (Krock and Moore, abstract in Society for Neuroscience 2014). The superior colliculus sends disynaptic projections (medium orange lines) to dorsal stream but not ventral stream visual areas (Lyon et al. 2010), whereas the FEF (red-orange lines) and LIP (yellow lines) are reciprocally connected with both ventral and dorsal visual cortex (Baizer et al. 1991; Shipp and Zeki 1995; Stanton et al. 1995; Ungerleider et al. 2008; Anderson et al. 2011). Extensive connectivity also exists between these oculomotor regions, including a corollary discharge pathway from the SC to FEF (Sommer and Wurtz 2000, 2002, 2006; Ferraina et al. 2002; Berman et al. 2009). FEF, frontal eye field; LIP, lateral intraparietal area; SC, superior colliculus; PI, inferior pulvinar; LGN, lateral geniculate nucleus; MD, mediodorsal thalamus; MST, medial superior temporal; MT, middle temporal.

et al. 1994). Evidence for maintenance of chromatic contrast sensitivity during saccades was also found in macaques (Hass and Horwitz 2011). This observation has prompted a number of experiments exploring the hypothesis that saccadic suppression first occurs in the color-insensitive magnocellular pathway of the LGN, sparing representations of color in the parvocellular and koniocellular layers. In support of this argument, suppression may be more common in magnocellular neurons than parvocellular neurons. However, identical patterns of suppression beginning ~100 msec before saccades are present in magnocellular, parvocellular, and koniocellular LGN cells (Reppas et al. 2002; Royal et al. 2006). Furthermore, neuronal firing rate responses and blood-oxygen-level-dependent (BOLD) responses to both chromatic and achromatic stimuli appear similarly suppressed in human LGN and V1 (Kleiser et al. 2004; Sylvester and Rees 2006; Hass and Horwitz 2011). These studies also show that saccadic suppression in the LGN is followed by much stronger postsaccadic enhancement, a pattern that is not observed in psychophysical measurements (Burr et al. 1994; Diamond et al. 2000; Knöll et al. 2011) or in downstream visual areas (Bremmer et al. 2009).

Another potential explanation for the luminance selectivity of saccadic suppression could be that suppression is biased toward the relatively color-insensitive dorsal stream, though neurophysiological evidence does not unequivocally support this view. Visual responses are suppressed in dorsal areas V3, MT, and MST (Thiele et al. 2002; Kleiser et al. 2004; Ibbotson et al. 2007; Bremmer et al. 2009; Cloherty et al. 2010). However, strong luminance-selective suppression is present in the ventral stream as well (Kleiser et al. 2004). Han et al. (2009) measured V4 neuronal contrast sensitivity using a staircasing procedure to determine the threshold contrast that elicited a significant response from each neuron (Fig. 3). They found a 0.5 log-unit decrease in sensitivity to brief, luminance-contrast modulated presaccadic visual probes. In contrast to the LGN and V1, perisaccadic sensitivity to red-green modulated equiluminant stimuli was less reduced in V4, paralleling human perception. Although the luminance specificity of saccadic suppression may yet be an important clue to its neural implementa-

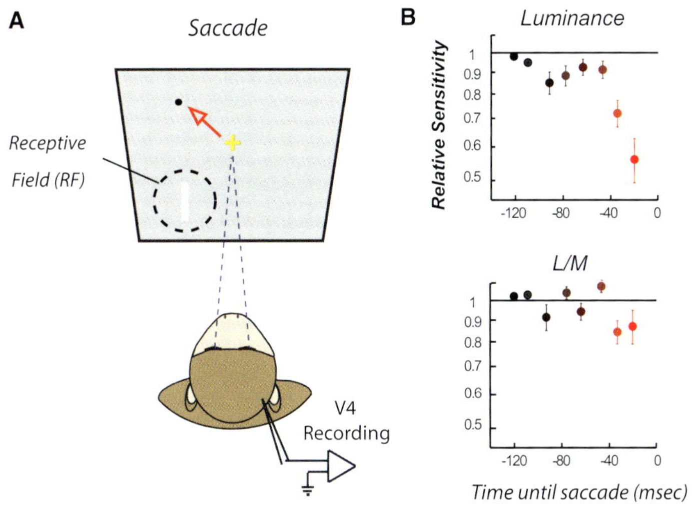

Figure 3. Saccadic suppression of luminance contrast sensitivity of V4 neurons. (*A*) The monkey was cued to make a visually guided saccade (red arrow, planned saccade vector; yellow cross, eye position.) A brief visual probe consisting of an oriented bar was presented inside the V4 receptive field at varying times with respect to the saccade onset. Luminance or chromatic contrast of the probe was varied across trials in a staircase procedure to determine the threshold contrast that elicited a neuronal response. (*B*) For luminance-contrast modulated probes, neuronal sensitivity declined by 50% just before movement onset. There was minimal change to neuronal sensitivity to red-green equiluminant modulated stimuli. (Modified from Han et al. 2009.)

tion, the existing data on the subject leave many questions unanswered.

ANATOMIC CONSTRAINTS ON POSSIBLE MECHANISMS OF SUPPRESSION

Although psychophysical and neurophysiological studies support a role for an active suppressive mechanism in saccadic suppression, exactly what form this mechanism might take is unclear. One type of mechanism that is frequently postulated to play a role is corollary discharge, defined as a copy of a motor signal that is used to alter sensory representations. A structure that could produce a corollary discharge would need to have information about impending saccades, implicating particular oculomotor structures. To date, the clearest evidence of a saccadic corollary discharge pathway involves the indirect pathway from the SC to the FEF. Sommer and Wurtz (2008) identified a pathway from the SC to the frontal eye field via the mediodorsal thalamus that appears necessary both for accurate behavioral updating of eye position (Sommer and Wurtz 2002) and for modulating eye movement signals within the FEF (Sommer and Wurtz 2006).

In addition, one might expect corollary discharge signals to directly convey information about impending saccades to the visual system to alter visual representations (Wurtz et al. 2011). Indeed, a simple implementation of a corollary discharge might consist of an inhibitory signal that originates in an oculomotor structure and suppresses visual representations. Yet because long-range cortical connections are overwhelmingly excitatory, it is unlikely that any saccade command signal directly inhibits neurons in visual cortex. For example, ~90% of the projections of FEF neurons to extrastriate area V4 terminate directly onto the dendritic spines of pyramidal cells, precluding the possibility that FEF neurons directly inhibit neurons in this area (Anderson et al. 2011). Another way that oculomotor structures could contribute to perisaccadic insensitivity of visual cortex is via a decrease in excitatory modulation. Berman and Wurtz (2010) used orthodromic and antidromic stimulation to identify a disynaptic relay from the superior colliculus to area MT via the inferior pulvinar. These pulvinar relay cells showed a robust suppression of activity around the time of saccades (Berman and Wurtz 2011). These data are consistent with the idea that the SC contributes to saccadic suppression of visual representations in MT via a reduction of excitation. In addition, recent evidence indicates that SC projection through the pulvinar and into extrastriate visual cortex is exclusively projected to areas of the dorsal visual stream (e.g., MT) and not to ventral stream areas (e.g., V4) (Berman and Wurtz 2010; Lyon et al. 2010). This suggests a possible basis for the relative sparing of chromatic and high spatial frequency vision during saccadic suppression (Burr et al. 1994; Han et al. 2009).

If the SC indeed transmits suppressed visual responses to visual cortex during saccades, where might suppression within the superior colliculus originate? The avian optic tectum (OT), a homolog of the primate SC, is part of a circuit that generates global competitive suppression and sharpens selection of saccade target representations (Mysore and Knudsen 2013; Goddard et al. 2014). The intermediate and deep layer neurons of the OT send topographic projections to the isthmi pars magnocellularis (Imc), the avian homolog of the lateral tegmental nucleus, which in turn sends global inhibitory projections back to the OT. The global inhibition mediated by the Imc is powerful even at great distances from the chosen target representation. This network could generate global suppression of visual responses within the OT around the time of eye movements. In the future, it will be of interest to determine whether activity in the intermediate and deep layers of the primate SC is necessary for saccadic suppression of visual responses in the superficial and intermediate SC, in visual cortex, and for saccadic suppression measured behaviorally.

MODULATION OF VISUAL CORTICAL SIGNALS BY THE FEF

It is now evident that the visually driven responses of neurons within posterior visual cortex are significantly influenced by the activity of neurons within the FEF (for review, see Squire et al. 2013). Of particular importance are the observations that changes in FEF neuronal activity are sufficient to alter the gain of stimulus-driven responses in extrastriate visual cortex (Moore and Armstrong 2003; Armstrong et al. 2006; Ekstrom et al. 2009; Noudoost and Moore 2011a). The FEF is directly connected to neurons within extrastriate visual cortex, including extrastriate areas MT and V4 (Fig. 2; Stanton et al. 1995). The influence of FEF neurons on posterior visual cortex appears to be accomplished, at least in part, via direct excitatory connections from FEF neurons onto pyramidal neurons within retinotopically corresponding columns within extrastriate cortex (e.g., area V4) (Anderson et al. 2011). One current view posits that saccade-related activity within the FEF neurons participates in a recurrent excitatory circuit that can simultaneously amplify visual responses at prospective saccadic target locations and further specify the precise saccade vector needed to foveate particular visual targets (Moore et al. 2003; Schafer and Moore 2007; Hamker et al. 2008; Noudoost and Moore 2011b). In the proposed circuit, the recurrent excitatory interactions between the FEF and extrastriate cortex also facilitates competitive, local inhibitory interactions between cortical columns representing different locations in space, resulting in progressively greater gain disparities between target and nontarget locations (Hamker et al. 2008; Noudoost and Moore 2011b). Such a circuit is consistent with the observation that stimulation of FEF sites both facilitates the visually driven responses at retinotopically corresponding locations and suppresses the responses at noncorresponding locations (Fig. 4; Moore and Armstrong 2003). It is also consistent with the observation that visual receptive fields within visual cortex (Tolias et al. 2001) and within the FEF (Zirnsak et al. 2014) appear to converge toward the location of impending saccadic targets (Hamker et al. 2008). In addition, the proposed circuit also seems consistent with basic phenomenon of saccadic suppression in that it predicts a reduction in visual processing and sensitivity at all locations outside of the target location during saccadic eye movements. Indeed, both the psychophysical and the physiological data to date indicate that around the time of saccades, visual sensitivity at the target location is increased in exchange for large decrements in sensitivity elsewhere (Zirnsak and Moore 2014).

CONCLUDING REMARKS

We have reviewed evidence that although numerous psychophysical and neurophysiological studies have showed a role for an active suppressive mechanism in

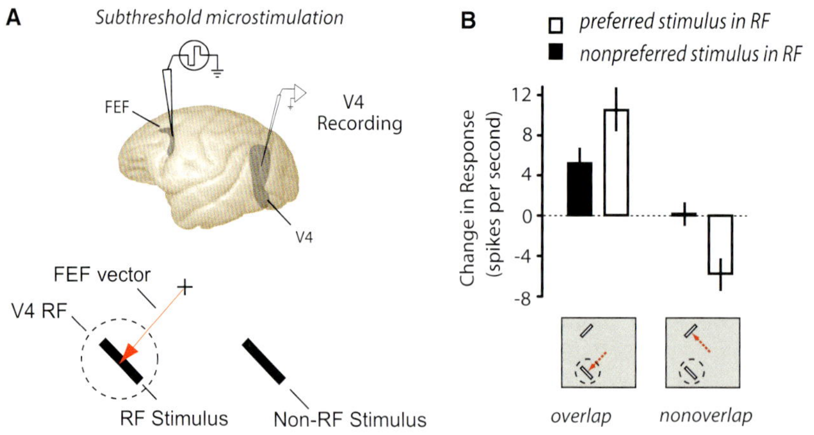

Figure 4. Modulation of visually driven responses of area V4 neurons with microstimulation of the FEF. (*A*) Sites within the FEF were electrically stimulated while recording from neurons in area V4. Cartoon shows a side view of the macaque brain. (*Top*) The locations of the FEF in the anterior bank of the arcuate sulcus and of area V4 in the prelunate gyrus and below the inferior occipital sulcus are shown (both shaded). Monkeys performed a fixation task while oriented bar stimuli were presented inside the recorded V4 neuron's receptive field (RF) (dotted circle) and at another location outside the RF. The stimulation and recording sites in the FEF and area V4, respectively, could be chosen such that the FEF saccade vector (arrow) and the area V4 neuron's RF overlapped spatially. (*B*) The stimulation and recording sites in the FEF and area V4, respectively, could be chosen for each experiment such that the FEF saccade vector (dotted red arrow) and the area V4 neuron's RF (dotted circle) were either spatially overlapping (*left*) or nonoverlapping (*right*). In the overlapping configuration, FEF stimulation enhanced the V4 neuron's response to visual stimuli appearing in the RF, and these enhancements mirrored the modulations observed during covert spatial attention in that more enhancement was seen for preferred than nonpreferred stimuli (*left*). In contrast, in the nonoverlapping configuration no response enhancement was produced.

saccadic suppression, many questions remain about the precise form this mechanism might take. Neural correlates of saccadic suppression have been observed throughout the visual and saccadic systems, but these data leave open the question of what the source(s) of this suppression could be. Although corollary discharge seems to be a likely mechanism, direct inhibition of visual cortex by oculomotor structures is unlikely; instead, structures such as the FEF and SC could contribute to saccadic suppression of visual cortical representations through a loss of excitatory feedback. Future studies are needed to determine whether the SC, the FEF, or other structures in the saccadic system play a causal role in the generation of saccadic suppression of neural visual representations and psychophysical contrast sensitivity.

REFERENCES

Anderson JC, Kennedy H, Martin KA. 2011. Pathways of attention: Synaptic relationships of frontal eye field to V4, lateral intraparietal cortex, and area 46 in macaque monkey. *J Neurosci* **31:** 10872–10881.

Armstrong KM, Fitzgerald JK, Moore T. 2006. Changes in visual receptive fields with microstimulation of frontal cortex. *Neuron* **50:** 791–798.

Baizer JS, Ungerleider LG, Desimone R. 1991. Organization of visual inputs to the inferior temporal and posterior parietal cortex in macaques. *J Neurosci* **11:** 168–190.

Berman RA, Wurtz RH. 2010. Functional identification of a pulvinar path from superior colliculus to cortical area MT. *J Neurosci* **30:** 6342–6354.

Berman RA, Wurtz RH. 2011. Signals conveyed in the pulvinar pathway from superior colliculus to cortical area MT. *J Neurosci* **31:** 373–384.

Berman RA, Joiner WM, Cavanaugh J, Wurtz RH. 2009. Modulation of presaccadic activity in the frontal eye field by the superior colliculus. *J Neurophysiol* **101:** 2934–2942.

Bremmer F, Kubischik M, Hoffman KP, Krekelberg B. 2009. Neural dynamics of saccadic suppression. *J Neurosci* **29:** 12374–12383.

Bridgeman B, Hendry D, Stark L. 1975. Failure to detect displacement of the visual world during saccadic eye movements. *Vision Res* **15:** 719–722.

Burr DC, Ross J. 1982. Contrast sensitivity at high velocities. *Vision Res* **22:** 479–484.

Burr DC, Holt J, Johnston JR, Ross J. 1982. Selective depression of motion sensitivity during saccades. *J Physiol* **333:** 1–15.

Burr DC, Morrone MC, Ross J. 1994. Selective suppression of the magnocellular visual pathway during saccadic eye movements. *Nature* **371:** 511–513.

Cloherty SL, Mustari MJ, Rosa MG, Ibbotson MR. 2010. Effects of saccades on visual processing in primate MSTd. *Vision Res* **50:** 2683–2691.

Corkill G, Vijayan N. 1976. Saccadic visual suppression and down-beat nystagmus: A case of Arnold-Chiari malformation. *Surg Neurol* **5:** 95–96.

Diamond MR, Ross J, Morrone MC. 2000. Extraretinal control of saccadic suppression. *J Neurosci* **20:** 3449–3455.

Dorr M, Bex PJ. 2013. Peri-saccadic natural vision. *J Neurosci* **33:** 1211–1217.

Ekstrom LB, Roelfsema PR, Arsenault JT, Kolster H, Vanduffel W. 2009. Modulation of the contrast response function by electrical microstimulation of the macaque frontal eye field. *J Neurosci* **29:** 10683–10694.

Ferraina S, Paré M, Wurtz RH. 2002. Comparison of cortico-cortical and cortico-collicular signals for the generation of saccadic eye movements. *J Neurophysiol* **87:** 845–858.

Goddard CA, Mysore SP, Bryant AS, Huguenard JR, Knudsen EI. 2014. Spatially reciprocal inhibition of inhibition within a stimulus selection network in the avian midbrain. *PLoS One* **9:** e85865.

Hafed ZM, Krauzlis RJ. 2010. Microsaccadic suppression of visual bursts in the primate superior colliculus. *J Neurosci* **30:** 9542–9547.

Hamker FH, Zirnsak M, Calow D, Lappe M. 2008. The perisaccadic perception of objects and space. *PLoS Comput Biol* **4:** e31.

Han X, Xian SX, Moore T. 2009. Dynamic sensitivity of area V4 neurons during saccade preparation. *Proc Natl Acad Sci* **106:** 13046–13051.

Hass CA, Horwitz GD. 2011. Effects of microsaccades on contrast detection and V1 responses in macaques. *J Vis* **11:** 1–17.

Holt E. 1903. Eye movement and central anaesthesia. *Harvard Psychol Stud* **1:** 3–45.

Ibbotson MR, Price NS, Crowder NA, Ono S, Mustari MJ. 2007. Enhanced motion sensitivity follows saccadic suppression in the superior temporal sulcus of the macaque cortex. *Cereb Cortex* **17:** 1129–1138.

Ibbotson MR, Crowder NA, Cloherty SL, Price NS, Mustari MJ. 2008. Saccadic modulation of neural responses: Possible roles in saccadic suppression, enhancement, and time compression. *J Neurosci* **28:** 10952–10960.

Kleiser R, Seitz RJ, Krekelberg B. 2004. Neural correlates of saccadic suppression in humans. *Curr Biol* **14:** 386–390.

Knöll J, Binda P, Morrone MC, Bremmer F. 2011. Spatiotemporal profile of peri-saccadic contrast sensitivity. *J Vis* **11:** 1–12.

Latour PL. 1962. Visual threshold during eye movements. *Vision Res* **2:** 261–262.

Lyon DC, Nassi JJ, Callaway EM. 2010. A disynaptic relay from superior colliculus to dorsal stream visual cortex in macaque monkey. *Neuron* **65:** 270–279.

Mohler CW, Cechner R. 1975. Saccadic suppression in the monkey. *Vision Res* **15:** 1157–1160.

Moore T. 2006. The neurobiology of visual attention: Finding sources. *Curr Opin Neurobiol* **16:** 159–165.

Moore T, Armstrong KM. 2003. Selective gating of visual signals by microstimulation of frontal cortex. *Nature* **421:** 370–373.

Moore T, Armstrong KM, Fallah M. 2003. Visuomotor origins of covert spatial attention. *Neuron* **40:** 671–683.

Mysore SP, Knudsen EI. 2013. A shared inhibitory circuit for both exogenous and endogenous control of stimulus selection. *Nat Neurosci* **16:** 473–478.

Noudoost B, Moore T. 2011a. Control of visual cortical signals by prefrontal dopamine. *Nature* **474:** 372–375.

Noudoost B, Moore T. 2011b. The role of neuromodulators in selective attention. *Trends Cogn Sci* **15:** 585–591.

Ramcharan EJ, Gnadt JW, Sherman SM. 2001. The effects of saccadic eye movements on the activity of geniculate relay neurons in the monkey. *Vis Neurosci* **18:** 253–258.

Reppas JB, Usrey WM, Reid RC. 2002. Saccadic eye movements modulate visual responses in the lateral geniculate nucleus. *Neuron* **35:** 961–974.

Robinson DL, Wurtz RH. 1976. Use of an extraretinal signal by monkey superior colliculus neurons to distinguish real from self-induced stimulus movement. *J Neurophysiol* **39:** 852–870.

Robinson DL, McClurkin JW, Kertzman C, Petersen SE. 1991. Visual responses of pulvinar and collicular neurons during eye movements of awake, trained macaques. *J Neurophysiol* **66:** 485–496.

Royal DW, Sáry G, Schall JD, Casagrande VA. 2006. Correlates of motor planning and postsaccadic fixation in the macaque monkey lateral geniculate nucleus. *Exp Brain Res* **168:** 62–75.

Schafer RJ, Moore T. 2007. Attention governs action in the primate frontal eye field. *Neuron* **56:** 541–551.

Shioiri S, Cavanagh P. 1989. Saccadic suppression of low-level motion. *Vision Res* **29:** 915–928.

Shipp S, Zeki S. 1995. Segregation and convergence of specialized pathways in macaque monkey visual cortex. *J Anat* **187:** 547–562.

Sommer MA, Wurtz RH. 2000. Composition and topographic organization of signals sent from the frontal eye field to the superior colliculus. *J Neurophysiol* **83:** 1979–2001.

Sommer MA, Wurtz RH. 2002. A pathway in primate brain for internal monitoring of movements. *Science* **296:** 1480–1482.

Sommer MA, Wurtz RH. 2006. Influence of the thalamus on spatial visual processing in frontal cortex. *Nature* **444:** 374–377.

Sommer MA, Wurtz RH. 2008. Brain circuits for the internal monitoring of movements. *Annu Rev Neurosci* **31:** 317–338.

Sperry RW. 1950. Neural basis of the spontaneous optokinetic response produced by visual inversion. *J Comp Physiol Psychol* **43:** 482–489.

Squire RF, Noudoost B, Schafer RJ, Moore T. 2013. Prefrontal contributions to visual selective attention. *Annu Rev Neurosci* **36:** 451–466.

Stanton GB, Bruce CJ, Goldberg ME. 1995. Topography of projections to posterior cortical areas from the macaque frontal eye fields. *J Comp Neurol* **353:** 291–305.

Sylvester R, Rees G. 2006. Extraretinal saccadic signals in human LGN and early retinotopic cortex. *NeuroImage* **30:** 214–219.

Sylvester R, Haynes JD, Rees G. 2005. Saccades differentially modulate human LGN and V1 responses in the presence and absence of visual stimulation. *Curr Biol* **15:** 37–41.

Thiele A, Henning P, Kubischik M, Hoffman KP. 2002. Neural mechanisms of saccadic suppression. *Science* **295:** 2460–2462.

Tilikete C, Vighetto A. 2011. Oscillopsia: Causes and management. *Curr Opin Neurol* **24:** 38–43.

Tolias AS, Moore T, Smirnakis SM, Tehovnik EJ, Siapas AG, Schiller PH. 2001. Eye movements modulate visual receptive fields of V4 neurons. *Neuron* **29:** 757–767.

Ungerleider LG, Galkin TW, Desimone R, Gattass R. 2008. Cortical connections of area V4 in the macaque. *Cereb Cortex* **18:** 477–499.

von Helmholtz H. 1925. *Physiological optics* (ed. Southall J), Vol. III. Banta, Menasha, WI.

von Holst E, Mittelstaedt H. 1950. The reafference principle. Interaction between the central nervous system and the periphery. In *Selected papers of Erich von Holst: The behavioural physiology of animals and man*, Vol. 1:1, pp. 39–73. Methuen, London.

Wurtz RH, Joiner WM, Berman RA. 2011. Neuronal mechanisms for visual stability: Progress and problems. *Philos Trans R Soc Lond B Biol Sci* **366:** 492–503.

Zirnsak M, Moore T. 2014. Saccades and shifting receptive fields: Anticipating consequences or selecting targets? *Trends Cogn Sci* **18:** 621–628.

Zirnsak M, Steinmetz NA, Noudoost B, Xu KZ, Moore T. 2014. Visual space is compressed in prefrontal cortex before eye movements. *Nature* **507:** 504–507.

Zuber BL, Stark L. 1966. Saccadic suppression: Elevation of visual threshold associated with saccadic eye movements. *Exp Neurol* **16:** 65–79.

Neural Circuits for Motion Vision in the Fly

Alexander Borst

Max-Planck-Institute of Neurobiology, 82152 Martinsried, Germany

Correspondence: borst@neuro.mpg.de

Seeing the direction of motion is fundamental for visual navigation, predator avoidance, and prey capture. However, the direction of motion is not explicitly represented at the level of individual photoreceptors. Rather, directional motion information needs to be extracted from the photoreceptor array by comparing the signals of neighboring photoreceptors over time. The exact nature of this process as implemented in the *Drosophila* visual system is currently being studied in great detail, and much progress has recently been made in determining the neural circuits giving rise to directional motion information in this species.

THE CENTRAL PHENOMENON

Image motion can have two different causes: when the observer is standing still and something else is moving in the environment, or when the observer itself is moving. In the first case, image motion may indicate the presence of a predator, a prey, or a conspecific; in the second case, image motion is indicative for the exact way the observer is moving, whether it is rotating around or translating along a particular body axis. Also, during self-motion, nearby objects will move relative to the background. Thus, image motion can also be used for segmenting the visual scenery. In any case, motion information is of prime importance for the appropriate action the animal should take and therefore is indispensable for survival. However, the direction in which something is moving is not explicitly encoded at the level of single photoreceptors. As depicted in Figure 1, if a bright bar is moved from left to right and back again in front of a fly's eye, the electrical response of a single photoreceptor looks the same. Thus, from the responses of an individual photoreceptor, one cannot tell the direction of local image motion. The fly's optic lobe consists of four different neuropiles called lamina, medulla, lobula, and lobula plate. All these neuropiles are built from repetitive columns, arranged in a retinotopic way. In the lobula plate, large so-called "lobula plate tangential cells" cover many hundreds of different columns with their dendrites. If the response from one of these neurons is recorded during the same stimulation protocol as described previously, a very different behavior is found: During bar motion to the right (their "preferred direction"), the cells depolarize; during bar motion to the left (their "null direction"), they hyperpolarize. In contrast to the photoreceptor signal, these cells apparently encode the direction of image motion in their membrane potential. Obviously, somewhere between the photoreceptor terminals and the dendrites of the lobula plate tangential cells, the direction of motion has been computed. This process is called "elementary motion detection" and is the central topic of interest.

THE HASSENSTEIN–REICHARDT DETECTOR

Analyzing the turning tendency of the beetle *Chlorophanus viridis* walking on a spherical Y-maze, Hassenstein and Reichardt (1956) proposed a specific model of elementary motion detection that could account for their observations in a quantitative way. The idea was that the beetle's nervous system contains many hundreds of such elementary units that, collectively, cover the whole visual field, each extracting locally the direction of image motion. The algorithmic model for such an elementary motion detector consists of two subunits, which are mirror-symmetrical to each other (Fig. 2; Hassenstein and Reichardt 1956; Reichardt 1961, 1987; Borst and Egelhaaf 1989). Each subunit reads the luminance values measured in two adjacent facets. These values pass an input filter before they are fed through temporal filters with different characteristics (low-pass filter, high-pass filter). Next, they are multiplied (M). The output values of both subunits are finally subtracted to result in a directionally selective signal. This model is referred to in the literature as Hassenstein–Reichardt detector, correlation detector, or Reichardt detector. The model makes predictions which are both quantitative and counterintuitive. These predictions have been tested and confirmed in many studies using either the optomotor turning response or the electrical signals of the lobula plate tangential cells as a readout (for the blow fly *Calliphora*, see, e.g., Egelhaaf and Borst 1989; Brenner et al. 2000; Borst et al. 2003, 2005; Reisenman et al. 2003; Haag et al. 2004; for the house fly *Musca*, see, e.g., Fermi and Reichardt 1963; Eckert 1973; Borst and Bahde 1986; for the fruit fly *Drosophila*, see, e.g., Goetz 1964, 1965; Buchner 1976; Joesch et al. 2008; Schnell et al. 2010).

doi: 10.1101/sqb.2014.79.024695

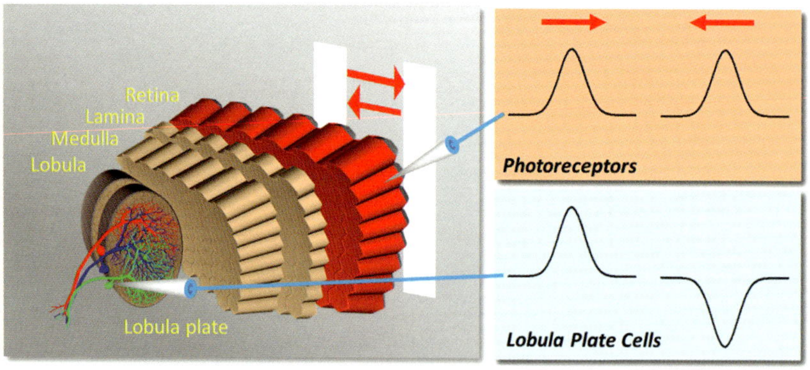

Figure 1. From nondirectional to directional responses. (*Left*) Schematic of the optic lobe of *Drosophila*. The retina (red) built from repetitive facets is followed by four layers of neuropile, called lamina, medulla, lobula, and lobula plate. Within the lobula plate, the group of three horizontal system (HS) cells is shown as reconstructed from stained tissue. (*Right*) Schematic signals recorded from a single photoreceptor (*top*) and from a tangential cell (*bottom*) during motion of a bar. Although the photoreceptor signal is nondirectional, the signal of the tangential cell is strongly directional.

NEURAL IMPLEMENTATION OF THE HASSENSTEIN–REICHARDT DETECTOR

Given the evidence that has been accumulated for Hassenstein–Reichardt detector in the optic lobes of different fly species, the question naturally arises: Which neurons form the Hassenstein–Reichardt detector? To answer this question, one first of all needs to know which neurons exist between the photoreceptors and the lobula plate tangential cells. Here, an almost complete catalog has been worked out: Starting with the work of Cajal and Sanchez (1915), the columnar cell types of the lamina, medulla, lobula, and lobula plate have all been identified and described on the basis of Golgi impregnations in the housefly (Strausfeld 1976) as well as in *Drosophila* (Fischbach and Dittrich 1989). Each lamina column (or cartridge, as it is usually referred to) contains a set of 12 cell types, connected to the photoreceptors either directly or indirectly (Fig. 3A; Meinertzhagen and O'Neil 1991). These lamina neurons connect the photoreceptors to specific layers of the medulla (Takemura et al. 2008, 2013). In the medulla, a single column houses more than 60 different cell types. They can be roughly grouped into four groups: medulla intrinsic (Mi) neurons connect different layers of the medulla to each other (Fig. 3B), transmedulla (Tm) neurons connect specific layers of the medulla to various layers in the lobula (Fig. 3C), transmedulla Y (TmY) neurons connect specific layers of the medulla to various layers in the lobula and lobula plate (Fig. 3D), and bushy T4 cells connect medulla layer 10 to the four layers of the lobula plate (Fig. 3E). In addition, a similar group of bushy T5 cells connect the posterior layer of the lobula to the four layers of the lobula plate (Fig. 3E). However, the small size of all these neurons prohibited electrophysiological recordings in most cases. Therefore, although a rather complete map of all columnar neurons was at hands for long, the visual response properties of most of them were completely unknown until recently, and so were their contributions to motion computation.

Only recently, the advent of sophisticated neurogenetics methods in *Drosophila* allowed for elucidating the circuits for elementary motion detection. These techniques are all based on a two-component expression system in which a so-called "driver line" defining the neurons where a certain effector gene is expressed is crossed with another line, the so-called "reporter line,"

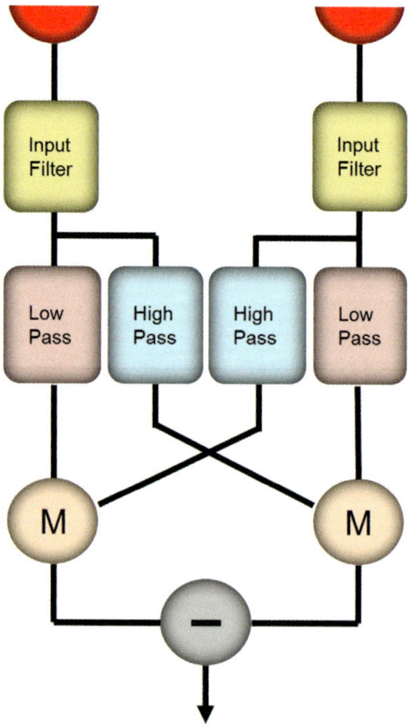

Figure 2. The Reichardt model of elementary motion detection. The model consists of two mirror-symmetrical subunits sharing the same two inputs after passing a filter. Within each subunit, the signal from one input is processed by a temporal low-pass filter and subsequently multiplied (M) with the high-pass-filtered signal derived from the neighboring input. The signals from both subunits are subtracted (−), giving rise to a directionally selective output.

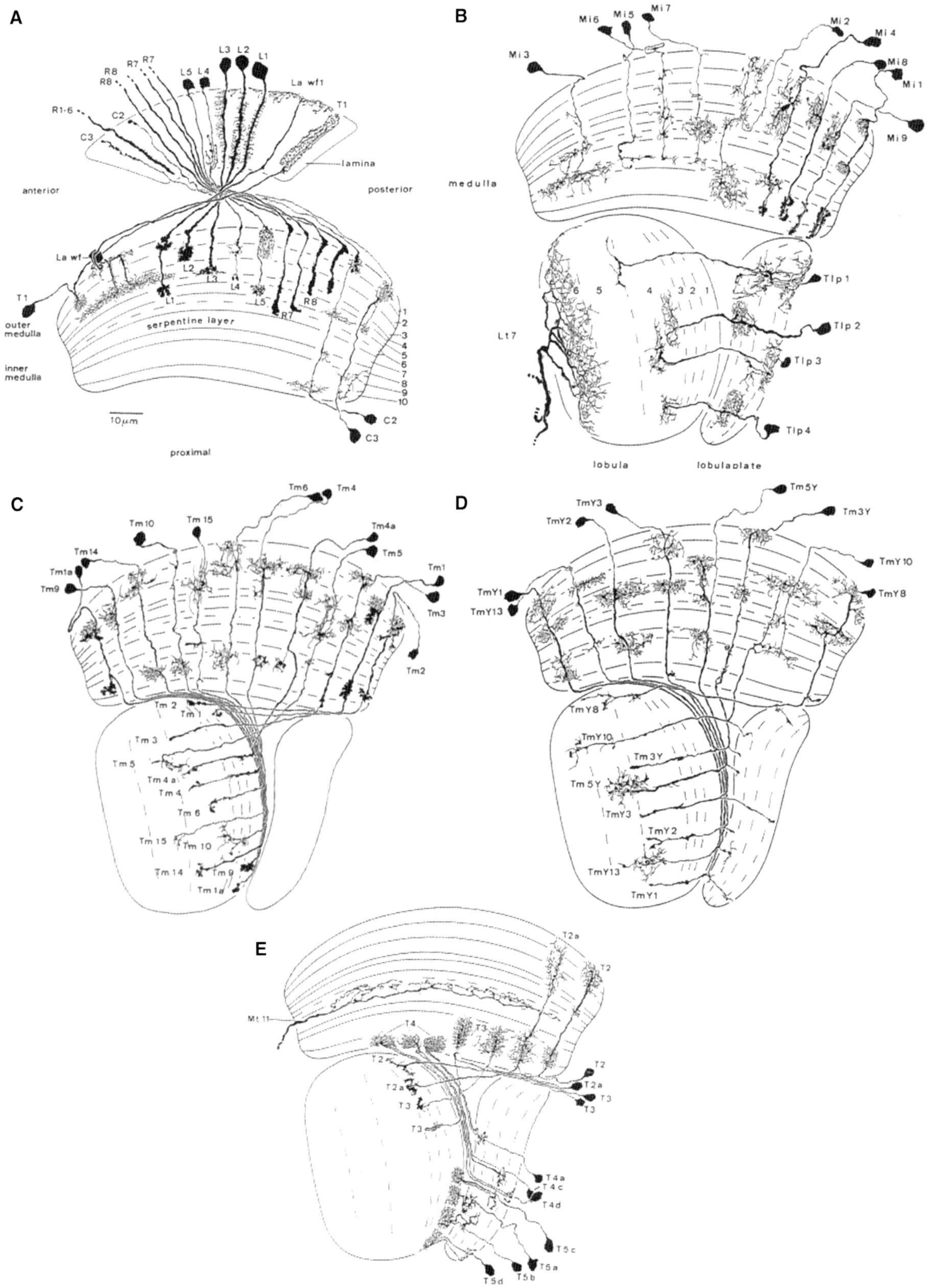

Figure 3. The Golgi catalog of columnar neurons of the *Drosophila* optic lobe. (Adapted from Fischbach and Dittrich 1989.)

defining what gene is expressed (Brand and Perrimon 1993). Today, thousands of different driver lines are available, many of which reveal a high degree of specificity for expression in individual cell types of the optic lobe (Pfeiffer et al. 2008; Tuthill et al. 2013). Furthermore, many reporter lines have been developed, allowing for blocking the synaptic output of neurons, activating neurons optogenetically, and recording from neurons via genetically encoded calcium indicators (for review, see Borst 2009; Venken et al. 2011). Applying these techniques to the problem of local motion detection revealed that (a) luminance information from fly photoreceptors R1–R6 is split into two parallel motion circuits, specialized to detect the motion of luminance increments (ON channel) and decrements (OFF channel) separately, (b) lamina neurons L1 and L2 are the primary input neurons to these circuits (L1→ON channel, L2→OFF channel), and (c) T4 and T5 cells carry their output signals (ON→T4, OFF→T5) (Joesch et al. 2010, 2013; Eichner et al. 2011; Maisak et al. 2013).

INPUT ELEMENTS

A single facet of the fly eye comprises two sets of photoreceptors: the peripheral photoreceptors R1–R6 and the central R7,8. Early work on different mutants where either the central or the peripheral ones were defective indicated that motion vision relies primarily on R1–R6, and not on R7 and R8 (Heisenberg and Buchner 1977). The two most prominent target neurons of the peripheral photoreceptors R1–R6 are the large monopolar cells L1 and L2 of the lamina. First evidence for the contribution of L1 and L2 in motion vision was provided by experiments where either behavioral responses (Rister et al. 2007) or electrophysiological recordings from lobula plate tangential cells were used as a readout (Joesch et al. 2010): Blocking both L1 and L2 led to a complete abolishment of the responses to gratings moving along any direction. The individual contributions of L1 and L2 were discovered in flies where either L1 or L2 was blocked, and, instead of gratings, edges of different contrast polarity were moved to probe the responses of lobula plate tangential cells (Joesch et al. 2010). Although control flies responded to moving brightness increments (ON edges) and decrements (OFF edges) with about the same amplitude, flies with lamina neurons L1 blocked completely lost their response to moving ON edges, whereas their response to moving OFF edges was largely intact. The opposite was found for flies with lamina neuron L2 blocked: Here, the response to moving ON edges was roughly as large as the one of control flies, whereas the response to moving OFF edges was severely reduced. Together, these experiments suggest that the photoreceptor input from R1–R6 is split into parallel channels depending on the contrast polarity of the incoming signal: Although lamina neurons L1 specifically transmit information about brightness increments to downstream motion circuits, lamina neurons L2 only allow information about brightness decrements to be passed onto further motion processing units. Technically, such signal processing is referred to as "half-wave rectification." Using a genetically encoded calcium indicator, Reiff et al. (2010) indeed showed that in the OFF pathway, such half-wave rectification starts already in the axon terminal of L2 neurons in layer 2 of the medulla.

TWO PARALLEL MOTION PATHWAYS

The findings about the specific contribution of L1 and L2 to motion processing of brightness increments and decrements immediately suggests that in *Drosophila*, motion is detected in several separate circuits. Theoretically, there could be four such channels comprising all possible combinations between ON and OFF signals. However, apparent motion experiments in blow flies indicated that only two such channels exist, one for the interaction of ON signals and one for the interaction of OFF signals between neighboring image points (Riehle and Franceschini 1984). Recordings from lobula plate tangential cells in *Drosophila* arrived at the same conclusion: Only sequences of light pulses of the same contrast polarity (ON–ON and OFF–OFF) elicited significant responses, whereas light pulses of differing contrast polarity (ON–OFF and OFF–ON) failed to evoke any responses in lobula plate tangential cells (Eichner et al. 2011). Interestingly, using brightness steps instead of brightness pulses led to different results: Although again, ON–ON and OFF–OFF sequences along the preferred direction of the cell led to positive responses, ON–OFF and OFF–ON sequences elicited negative responses (Egelhaaf and Borst 1992; Eichner et al. 2011). Although this so-called "reverse phi" phenomenon (Anstis and Rogers 1975) was usually taken as evidence for an interaction between ON and OFF stimuli, as in the original Hassenstein–Reichardt detector, careful simulations of such circuits as well as experiments with flies with either lamina neurons L1 or L2 blocked revealed that such an inversion of the response is not necessarily indicative of an interaction between ON and OFF channels. Rather, it can be explained by the residual sustained response component of the lamina neurons taking into account the contribution of motion detectors at the edge of the stimulated areas in the visual field of the fly (Eichner et al. 2011; Joesch et al. 2013).

In summary, the Hassenstein–Reichardt detector is implemented in the visual system of the fly in two parallel circuits, one ON channel and one OFF channel (Fig. 4). What could be the advantage of such a doubling of wiring expenditure? With just a single motion detector dealing with brightness going positive as well as negative, a multiplicative interaction—according to the sign rule of multiplication—would imply the output being positive when both input signals go positive as well as when both go negative. Such a processing rule seems hard to implement biophysically and indeed is unheard of in cellular neuroscience. If, however, the input signals are split according to their polarity with the negative ones being sign-invert-

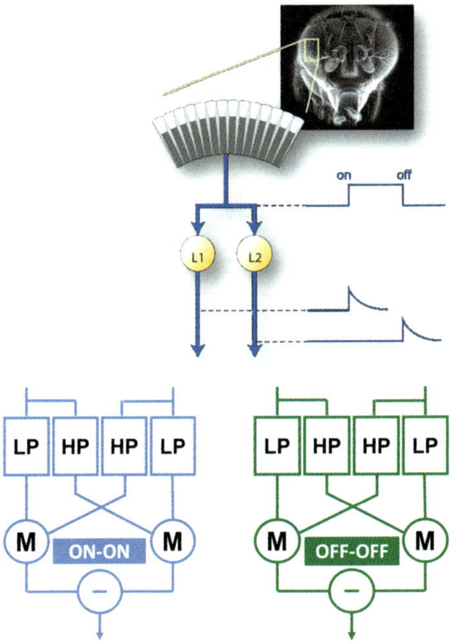

Figure 4. Input organization and general layout of the Reichardt detector as implemented in the *Drosophila* nervous system. At the level of the lamina, the photoreceptor input is split and half-wave rectified into an ON (L1) and an OFF (L2) signal. Both signals feed into separate motion detection circuits.

ed, subsequent circuits only deal with positive input signals reducing the problem of multiplication to a supralinear input–output characteristic as can be implemented by, for example, a voltage-gated ion channel. Interestingly, analog electronics faced the same problem of multiplying positive and negative signals, and the solution in these days was what is called a "four-quadrant multiplier." There, the input signals, before multiplication, were half-wave rectified, split into parallel ON and OFF channels carrying positive signals only, and subsequently multiplied in four separate multipliers. Their output signals were then added with the appropriate sign (Hassenstein and Reichardt 1956). This looks similar to what we find in the fly, except that the fly seems to discard two of the four units, reducing a true multiplication to a "two-quadrant multiplier." In the context of motion detection, however, this does not have any functional consequences: Because motion of rigid objects gives rise to spatial correlations between time-delayed signals at subsequent image pixels, an ON event at one location will always correlate with an ON event at a later point in time at a neighboring location, and the same holds true for OFF events. Consequently, a comparison of a four- and a two-quadrant detector reveals no significant differences in response to realistic visual motion (Eichner et al. 2011).

LOCAL DIRECTIONALLY SELECTIVE NEURONS

Having identified lamina neurons L1 and L2 as the decisive input elements to two parallel motion detector circuits allowed anatomy to guide the next steps. Indeed, two parallel processing streams had previously been postulated in the fly visual system (Bausenwein and Fischbach 1992; Bausenwein et al. 1992). These studies indicated that an L1 pathway should indirectly lead to T4 cells and an L2 pathway to T5 cells, with both T4 and T5 cells projecting into the lobula plate. Given that four different subtypes of T4 and T5 cells exist terminating in four different layers of the lobula plate (Fischbach and Dittrich 1989), and deoxyglucose labeling showed activity in one of the four layers according to the direction of the stimulus (Buchner et al. 1984), T4 and T5 cells were prime candidates for local elementary motion detectors. Their small size, however, prohibited electrophysiological recordings and let their visual response properties be unknown for long. A recent study (Maisak et al. 2013) circumvented this problem by using driver lines specific for T4 and T5 cells and combining them with a genetically encoded calcium indicator (Akerboom et al. 2012). Using two-photon calcium imaging and stimulating the flies with grating motion in four cardinal directions (front-to-back, back-to-front, upward, downward), Maisak et al. recorded directionally selective activity from T4 and T5 cells in each one of the four lobula plate layers depending on the direction in which the grating was moving (Fig. 5). To assess the particular contribution of T4 and T5 cells to the signals observed in the above experiments, driver lines specific for either T4 or T5 cells were used. Applying the same stimulus protocol as described before, identical results were obtained for both the T4- and the T5-specific driver line. Further experiments revealed similar response properties for T4 and T5 cells with respect to their orientation and velocity tuning (Maisak et al. 2013). If, however, instead of gratings, moving edges with either positive or negative contrast polarity were used as visual stimuli, T4 cells were found to strongly and selectively respond to moving ON edges, with little or no responses to moving OFF edges, whereas T5 cells selectively responded to moving OFF edges and mostly failed to respond to moving ON edges.

To investigate the specific contributions of T4 and T5 cells to postsynaptic lobula plate cells and visually driven behavior, T4 and T5 cells were genetically blocked and flies subsequently tested. Blocking both T4 and T5 cells led to a complete loss of the motion response in lobula plate tangential cells (Schnell et al. 2012) and of the optomotor response of tethered walking flies (Bahl et al. 2013). Blocking T4 cells specifically led to selective loss of the responses to moving ON edges, in the electrical signal of tangential cells as well as in optomotor behavior. Conversely, blocking T5 cells led to a loss of the responses to moving OFF edges, again both in the electrical signal of tangential cells as well as in optomotor behavior. In summary, the selective defects of T4 block flies for ON edges and of T5 block flies for OFF edges not only corroborate the above findings about the selective preference of T4 and T5 cells for different contrast polarities, but also show that the signals of T4 and T5 cells are indeed the major, if not exclusive, inputs to downstream circuits and motion-driven behaviors.

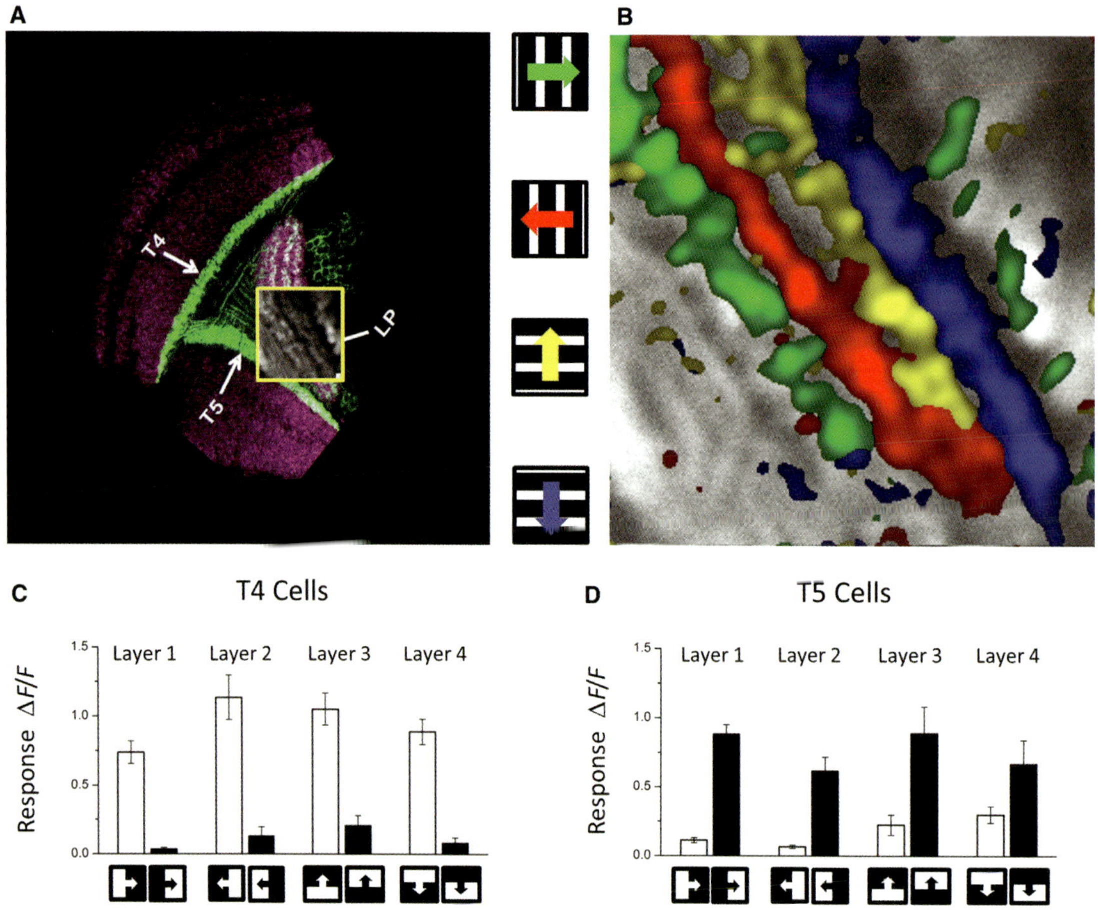

Figure 5. T4 and T5 cells represent the elementary motion detector outputs. (*A*) Confocal image of a driver line giving rise to expression in both T4 and T5 cells, shown as a horizontal cross section. (Modified from Schnell et al. 2012.) Neurons are marked in green (Kir2.1-EGFP), revealing clear labeling (in green) in the medulla (T4 cell dendrites), in the lobula (T5 cell dendrites), and in four distinct layers of the lobula plate, representing the terminal arborizations of the four subpopulations of both T4 and T5 cells. The *inset* is from a two-photon microscope and contains the area shown enlarged to the *right*. (*B*) Two-photon calcium imaging in the lobula plate reveals directionally selective signals. Following stimulation with grating motion along four cardinal directions, activity is confined to mostly one of the four layers. A particular color has been assigned to each pixel depending on the stimulus direction during which it responded most strongly. (Modified from Maisak et al. 2013.) (*C,D*) Responses of T4 (*C*) and T5 (*D*) cells to ON (in white) and OFF (in black) edges moving along all four cardinal directions. Although T4 cells respond most strongly to moving ON edges, T5 cells almost exclusively respond to moving OFF edges. (Adapted from Maisak et al. 2013.)

IMPLEMENTING MOTION OPPONENCY

The final processing step of the Hassenstein–Reichardt detector consists of subtracting the signals of motion-sensitive units with opposite preferred direction. This can readily account for the depolarization of lobula plate tangential cells during preferred direction motion, and their hyperpolarization during null direction motion. But what might be the cellular basis of this subtraction stage? In both blow flies and fruit flies, evidence exists that the subtraction is realized by a synaptic push–pull organization with local motion detectors of opposite preferred direction providing excitatory and inhibitory input onto the dendrites of lobula plate tangential cells (Borst and Egelhaaf 1990; Borst et al. 1995; Joesch et al. 2008). However, at present it is not known whether and how individual T4 and T5 cells can provide, at the same time, excitation to one and inhibition to the adjacent lobula plate layer. One possibility could be that there exists a separate set of local, inhibitory neurons, like inhibitory twins of the T4/T5 cells. This, however, seems unlikely because blocking of T4 and T5 cells abolishes both the preferred and the null direction response of lobula plate tangential cells in *Drosophila* (Schnell et al. 2012). As an alternative explanation, a recent study suggests that inhibition is transferred from excitatory T4/T5 signals within one layer of the lobula plate to the adjacent one via local inhibitory interneurons: When T4/T5 cells are optogenetically activated, a fast EPSP (excitatory postsynaptic potential) is recorded in tangential cells, followed by a delayed IPSP (inhibitory postsynaptic potential) (Mauss et al. 2014). Current experiments aim to identify such local inhibitory interneurons and to test whether they indeed provide the inhibition to tangential cells during null direction motion.

CONCLUSION AND OUTLOOK

Local motion detection in *Drosophila* has long been shown to follow the predictions of the Hassenstein–Reichardt detector. However, only the last few years have witnessed a major breakthrough in our understanding of its neural implementation. Genetically targeting individual cell types led to the following picture (Fig. 6): Starting with L1 and L2, the visual input is split into separate ON and OFF pathways, and motion along all four cardinal directions is computed separately within each pathway. The output signals of these eight different motion-detectors are then sorted such that ON (T4) and OFF (T5) motion detectors with the same directional tuning converge in the same layer of the lobula plate, jointly providing the input to downstream circuits and motion-driven behaviors. These findings are in striking similarity to motion processing in the mammalian retina, where the photoreceptor output is split into ON and OFF pathways as well, and ON–OFF direction-selective ganglion cells are found in four variants with each group responding preferentially to one of the cardinal directions (for review, see Borst and Euler 2011).

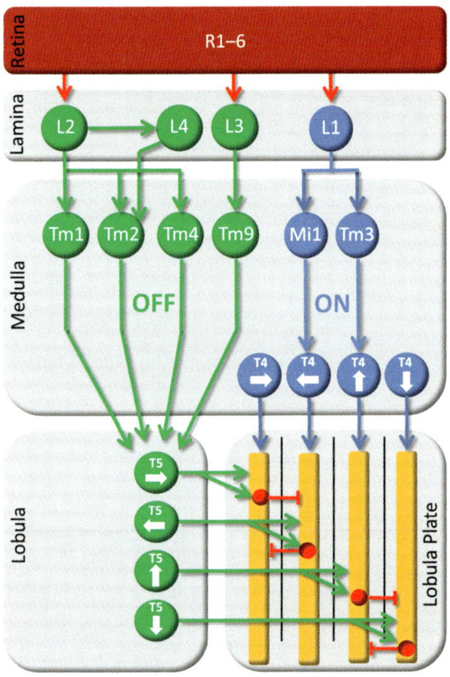

Figure 6. Circuit diagram of the neurons involved in the fly elementary motion detection. Visual input from photoreceptors R1–6 is split into parallel pathways at the level of the lamina. The ON pathway (to the *right*) is shown to involve lamina neuron L1 and two postsynaptic cells, Mi1 and Tm3, in the medulla. These cells contact the dendrites of T4 cells. The OFF pathway (to the *left*) involves more neurons. Here lamina cells L2 and L4 synapse onto medulla neurons Tm1, Tm2, and Tm4. In addition, lamina cell L3 synapses onto Tm9 (for involvement of L3, see Silies et al. 2013). All four medulla neurons contact the dendrites of T5 cells. Directionally selective signals are carried via T4 and T5 cells to the four layers of the lobula plate where T4 and T5 cells with the same preferred direction converge again on the dendrites of the tangential cells (in yellow). Inhibition is conveyed via hypothetical, local interneurons from one layer to the adjacent one (in red).

As an immediate hypothesis derived from the above results, the dendrites of T4 and T5 cells could be the place where the signals from neighboring facets converge, giving rise to T4 and T5 cell direction selectivity. Using large-scale serial sectioning transmission electron microscopy and computer-aided reconstruction, Takemura et al. (2013) found two neurons, Mi1 and Tm3, to connect the output terminals of L1 neurons to the dendrites of T4 cells. In a similar way, four different cells were found to convey the signals from L2 and L3 cells to the dendrites of T5 cells (i.e., Tm1, Tm2, Tm4, and Tm9 [Shinomiya et al. 2014]). Some of these neurons have meanwhile been characterized physiologically (Behnia et al. 2014; Meier et al. 2014; Strother et al. 2014) confirming their selectivity for either brightness increases or decreases, depending on whether they participate in the ON or the OFF pathway. In line with the above hypothesis, none of these cells showed any sign of direction selectivity. Thus, the major cellular components have been identified and physiologically characterized, and their connectivity established. Nevertheless, we do not know yet how the circuit works. In this context, a recent study made an interesting observation (Behnia et al. 2014): As revealed by whole-cell patch recording from Mi1, Tm1, Tm2, and Tm3 neurons, the dynamics of these four cells show an overall low-pass characteristic with slight differences in the time constants between Mi1 and Tm3, and Tm1 and Tm2, respectively, of 18 msec on average. The authors propose that Mi1 and Tm3 could represent the delayed and nondelayed input channel, respectively, to T4 cells, whereas Tm1 and Tm2 play analogous roles for T5 cells (Behnia et al. 2014). With the measured differences of the time constants, however, the model reproduces a temporal tuning curve consistent with experimental data from T4 and T5 cells (Maisak et al. 2013) only after subtraction of mirror-symmetrical subunits, a process which is thought to be implemented only on postsynaptic tangential cells (Mauss et al. 2014). Furthermore, T5 cells receive input not just from Tm1 and Tm2 cells but from two additional types of interneurons. Hence, for these cells, no immediate correspondence to a simple motion-detection scheme exists. As an alternative, the delay might be implemented in the dendritic membrane receptors of T4 and T5 cells: As shown by RNA profiling, T4 and T5 cells express both nicotinic and muscarinic acetylcholine receptors, which could mediate a fast ionotropic and a slow metabotropic signal (Shinomiya et al. 2014). It is therefore left to future studies to determine which cell corresponds to which element of the Hassenstein–Reichardt detector, where and how the delay is implemented, and what the biophysical nature of the nonlinear signal interaction is.

REFERENCES

Akerboom J, Chen TW, Wardill TJ, Tian L, Marvin JS, Mutlu S, Calderón NC, Esposti F, Borghuis BG, Sun XR, et al. 2012. Optimization of a GCaMP calcium indicator for neural activity imaging. *J Neurosci* **32:** 13819–13840.

Anstis SM, Rogers BJ. 1975. Illusory reversal of visual depth and movement during changes of contrast. *Vision Res* **15:** 957–961.

Bahl A, Ammer G, Schilling T, Borst A. 2013. Object tracking in motion-blind flies. *Nat Neurosci* **16:** 730–738.

Bausenwein B, Fischbach KF. 1992. Activity labeling patterns in the medulla of *Drosophila melanogaster* caused by motion stimuli. *Cell Tissue Res* **270:** 25–35.

Bausenwein B, Dittrich AP, Fischbach KF. 1992. The optic lobe of *Drosophila melanogaster*. II. Sorting of retinotopic pathways in the medulla. *Cell Tissue Res* **267:** 17–28.

Behnia R, Clark DA, Carter AG, Clandinin TR, Desplan C. 2014. Processing properties of ON and OFF pathways for *Drosophila* motion detection. *Nature* **512:** 427–430.

Borst A. 2009. *Drosophila*'s view on insect vision. *Curr Biol* **19:** R36–R47.

Borst A, Bahde S. 1986. What kind of movement detector is triggering the landing response of the housefly? *Biol Cybern* **55:** 59–69.

Borst A, Egelhaaf M. 1989. Principles of visual motion detection. *Trends Neurosci* **12:** 297–306.

Borst A, Egelhaaf M. 1990. Direction selectivity of blowfly motion-sensitive neurons is computed in a two-stage process. *Proc Natl Acad Sci* **87:** 9363–9367.

Borst A, Euler T. 2011. Seeing things in motion: Models, circuits, and mechanisms. *Neuron* **71:** 974–994.

Borst A, Egelhaaf M, Haag J. 1995. Mechanisms of dendritic integration underlying gain control in fly motion-sensitive interneurons. *J Comput Neurosci* **2:** 5–18.

Borst A, Reisenman C, Haag J. 2003. Adaptation of response transients in fly motion vision. II: Model studies. *Vision Res* **43:** 1309–1322.

Borst A, Flanagin VL, Sompolinsky H. 2005. Adaptation without parameter change: Dynamic gain control in motion detection. *Proc Natl Acad Sci* **102:** 6172–6176.

Brand AH, Perrimon N. 1993. Targeted gene expression as a means of altering cell fates and generating dominant phenotypes. *Development* **118:** 401–415.

Brenner N, Bialek W, de Ruyter van Steveninck R. 2000. Adaptive rescaling maximizes information transmission. *Neuron* **26:** 695–702.

Buchner E. 1976. Elementary movement detectors in an insect visual system. *Biol Cybern* **24:** 86–101.

Buchner E, Buchner S, Bülthoff I. 1984. Deoxyglucose mapping of nervous activity induced in *Drosophila* brain by visual movement. *J Comp Physiol A* **155:** 471–483.

Cajal SR, Sanchez D. 1915. *Contribución al conocimiento de los centros nerviosos de los insectos*. Imprenta de Hijos de Nicholas Moja, Madrid.

Eckert H. 1973. Optomotorische Untersuchungen am visuellen System der Stubenfliege *Musca domestica* L. *Kybernetik* **14:** 1–23.

Egelhaaf M, Borst A. 1989. Transient and steady-state response properties of movement detectors. *J Opt Soc Am A* **6:** 116–127.

Egelhaaf M, Borst A. 1992. Are there separate on- and off-channels in fly motion vision? *Vis Neurosci* **8:** 151–164.

Eichner H, Joesch M, Schnell B, Reiff DF, Borst A. 2011. Internal structure of the fly elementary motion detector. *Neuron* **70:** 1155–1164.

Fermi G, Reichardt W. 1963. Optomotorische Reaktionen der Fliege *Musca domestica*. *Kybernetik* **2:** 15–28.

Fischbach KF, Dittrich APM. 1989. The optic lobe of *Drosophila melanogaster*. I. A Golgi analysis of wild-type structure. *Cell Tissue Res* **258:** 441–475.

Goetz KG. 1964. Optomotorische Untersuchung des visuellen Systems einiger Augenmutanten der Fruchtfliege *Drosophila*. *Kybernetik* **2:** 77–92.

Goetz KG. 1965. Die optischen Übertragungseigenschaften der Komplexaugen von *Drosophila*. *Kybernetik* **2:** 215–221.

Haag J, Denk W, Borst A. 2004. Fly motion vision is based on Reichardt detectors regardless of the signal-to-noise ratio. *Proc Natl Acad Sci* **101:** 16333–16338.

Hassenstein B, Reichardt W. 1956. Systemtheoretische Analyse der Zeit-, Reihenfolgen- und Vorzeichenauswertung bei der Bewegungsperzeption des Rüsselkäfers *Chlorophanus*. *Z Naturforsch* **11b:** 513–524.

Heisenberg M, Buchner E. 1977. The role of retinula cell types in visual behavior of *Drosophila melanogaster*. *J Comp Physiol A* **117:** 127–162.

Joesch M, Plett J, Borst A, Reiff DF. 2008. Response properties of motion-sensitive visual interneurons in the lobula plate of *Drosophila melanogaster*. *Curr Biol* **18:** 368–374.

Joesch M, Schnell B, Raghu S, Reiff DF, Borst A. 2010. ON and OFF pathways in *Drosophila* motion vision. *Nature* **468:** 300–304.

Joesch M, Weber F, Eichner H, Borst A. 2013. Functional specialization of parallel motion detection circuits in the fly. *J Neurosci* **33:** 902–905.

Maisak MS, Haag J, Ammer G, Serbe E, Meier M, Leonhardt A, Schilling T, Bahl A, Rubin GM, Nern A, et al. 2013. A directional tuning map of *Drosophila* elementary motion detectors. *Nature* **500:** 212–216.

Mauss A, Meier M, Serbe E, Borst A. 2014. Optogenetic and pharmacologic dissection of feedforward inhibition in *Drosophila* motion vision. *J Neurosci* **34:** 2254–2263.

Meier M, Serbe E, Maisak MS, Haag J, Dickson BJ, Borst A. 2014. Neural circuit components of the *Drosophila* OFF motion vision pathway. *Curr Biol* **24:** 385–392.

Meinertzhagen IA, O'Neil SD. 1991. Synaptic organization of columnar elements in the lamina of the wild type in *Drosophila melanogaster*. *J Comp Neurol* **305:** 232–263.

Pfeiffer BD, Jenett A, Hammonds AS, Ngo TB, Misra S, Murphy C, Scully A, Carlson JW, Wan KH, Laverty TR, et al. 2008. Tools for neuroanatomy and neurogenetics in *Drosophila*. *Proc Natl Acad Sci* **105:** 9715–9720.

Reichardt W. 1961. Autocorrelation, a principle for the evaluation of sensory information by the central nervous system. In *Sensory communication* (ed. Rosenblith WA), pp. 303–317. The MIT Press and John Wiley & Sons, New York.

Reichardt W. 1987. Evaluation of optical motion information by movement detectors. *J Comp Physiol A* **161:** 533–547.

Reiff DF, Plett J, Mank M, Griesbeck O, Borst A. 2010. Visualizing retinotopic half-wave rectified input to the motion detection circuitry of *Drosophila*. *Nat Neurosci* **13:** 973–978.

Reisenman C, Haag J, Borst A. 2003. Adaptation of response transients in fly motion vision. I: Experiments. *Vision Res* **43:** 1291–1307.

Riehle A, Franceschini N. 1984. Motion detection flies: Parametric control over ON-OFF pathways. *Exp Brain Res* **54:** 390–394.

Rister J, Pauls D, Schnell B, Ting CY, Lee CH, Sinakevitch I, Morante J, Strausfeld NJ, Ito K, Heisenberg M. 2007. Dissection of the peripheral motion channel in the visual system of *Drosophila melanogaster*. *Neuron* **56:** 155–170.

Schnell B, Joesch M, Forstner F, Raghu S, Otsuna H, Ito K, Borst A, Reiff DF. 2010. Processing of horizontal optic flow in three visual interneurons of the *Drosophila* brain. *J Neurophysiol* **103:** 1646–1657.

Schnell B, Raghu S, Nern A, Borst A. 2012. Columnar cells necessary for motion responses of wide-field visual interneurons in *Drosophila*. *J Comp Physiol A* **198:** 389–395.

Shinomiya K, Karuppudurai T, Lin T-Y, Lu Z, Lee C-H, Meinertzhagen IA. 2014. Candidate neural substrates of Off-edge motion detection in *Drosophila*. *Curr Biol* **24:** 1062–1070.

Silies M, Gohl DM, Fisher YE, Freifeld L, Clark DA, Clandinin TR. 2013. Modular use of peripheral input channels tunes motion-detecting circuitry. *Neuron* **79:** 111–127.

Strausfeld NJ. 1976. *Atlas of an insect brain*. Springer, Berlin.

Strother JA, Nern A, Reiser MB. 2014. Direct observation of ON and OFF pathways in the *Drosophila* visual system. *Curr Biol* **24:** 976–983.

Takemura SY, Lu Z, Meinertzhagen IA. 2008. Synaptic circuits of the *Drosophila* optic lobe: The input terminals to the medulla. *J Comp Neurol* **509:** 493–513.

Takemura SY, Bharioke A, Zhiyuan L, Nern A, Vitaladevuni S, Rivlin PK, Katz WT, Olbris DJ, Plaza SM, Winston P, et al. 2013. A visual motion detection circuit suggested by *Drosophila* connectomics. *Nature* **500:** 175–181.

Tuthill JC, Nern A, Holtz SL, Rubin GM, Reiser MB. 2013. Contributions of the 12 neuron classes in the fly lamina to motion vision. *Neuron* **79:** 128–140.

Venken KJT, Simpson JH, Bellen HJ. 2011. Genetic manipulation of genes and cells in the nervous system of the fruit fly. *Neuron* **72:** 202–230.

How Optic Flow and Inertial Cues Improve Motion Perception

Dora E. Angelaki
Department of Neuroscience, Baylor College of Medicine, Houston, Texas 77030
Correspondence: angelaki@bcm.edu

Estimating our egocentric heading direction is an important component of navigation. Recent studies have explored how inertial cues from the vestibular system and optic flow signals from the visual system interact to improve perceptual precision and accuracy. Heading precision is improved through multisensory integration, whereas heading accuracy is maintained through multisensory calibration mechanisms. Neural correlates of these behaviors are found in a large interconnected cortical network, although the specific contributions of each area remain to be explored. Whether and how this multisensory self-motion cortical circuit contributes to navigation and how egocentric heading signals interact with the allocentric representations for foraging and exploration also remain challenges for the future.

Navigation is a fundamental behavior in most animals. During navigation and foraging, information about spatial relationships between environmental landmarks is integrated with self-motion (idiothetic) cues (Mittelstaedt and Mittelstaedt 1980) into a cognitive map of the environment (Stuchlik and Bures 2002). Navigating by self-motion involves path integration, a process that allows the animal to return fairly directly from any point along its path back to its starting position (Etienne et al. 1986). For example, to navigate from point A to point B using idiothetic cues, multisensory signals must encode three distinct components of self-motion (Fig. 1): the linear speed, the angular speed, and egocentric heading direction. During movement between A and B, these egocentric signals allow path integration processes to update spatial information about current allocentric head direction and location. While using path integration to navigate, a subject must align their idiothetic sense of position by referencing their location relative to known stable landmarks, which may be either substrate-based local cues or more distant orientation beacons.

Identifying the sensory origins of the signals contributing to spatial orientation and navigation has long posed a challenge for researchers due to the multisensory contributions of motor, vestibular, visual, proprioceptive, and other cues that are difficult to isolate. The field of rodent navigation has focused on investigating how linear and angular speed are used to convert self-motion cues into allocentric head direction, place, and grid signals (for recent reviews, see Knierim and Zhang 2012; Hartley et al. 2013; Moser et al. 2013; Barry and Burgess 2014; Geva-Sagiv et al. 2015). In contrast, the neural basis of egocentric heading direction has been primarily investigated in macaques. This review provides a summary of what is currently known about how visual and vestibular cues influence and control our perceived egocentric heading direction (see also Britten 2008; Angelaki et al. 2009, 2011; Fetsch et al. 2010, 2011; DeAngelis and Angelaki 2012).

MULTISENSORY CUES FOR HEADING PERCEPTION

Patterns of image motion across the retina ("optic flow") can be strong cues of self-motion, as evidenced by the fact that optic flow alone can elicit the illusion of self-motion (Mach 1875). Multiple studies have since characterized the behavioral observation that large-field optic flow stimulation induces self-motion perception (Brandt et al. 1972; Berthoz et al. 1975; Dichgans and Brandt 1978). In parallel, many visual psychophysical and theoretical studies have shown that optic flow also provides powerful cues of heading direction (Gibson 1950) and have examined how heading can be computed from optic flow (for review, see Warren 2003). During translation, the resulting retinal pattern consists of a radial flow field from which the point of zero velocity can be used to estimate heading (Tanaka et al. 1986; Warren et al. 1988; Duffy and Wurtz 1995; Britten 2008). However, retinal image motion is often confounded by eye/head and object movement, such that heading perception can often be ambiguous. Thus, visual cues alone might not always be appropriate for accurate heading estimation.

Independent information about the motion of our head/body in space can arise from the vestibular system (Angelaki and Cullen 2008). Psychophysical studies of heading discrimination using two-alternative forced-choice tasks have reported vestibular heading discrimination thresholds in darkness that are as small as a few degrees (Gu et al. 2007; Fetsch et al. 2009; Butler et al. 2010, 2015; de Winkel et al. 2010; Drugowitsch et al. 2014). Such threshold values are comparable (although larger) with those described in visual heading discrimi-

 doi: 10.1101/sqb.2014.79.024638

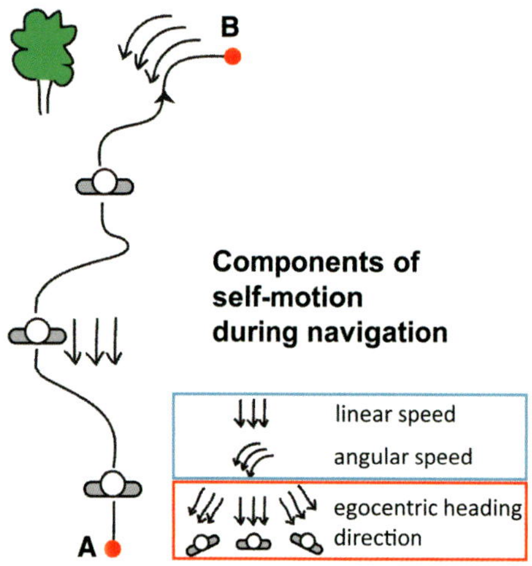

Figure 1. Schematic illustrating the three components of self-motion cues important for spatial navigation. This review focuses specifically on egocentric heading direction.

nation tasks (Warren and Hannon 1990; Royden et al. 1992; van den Berg and Brenner 1994; Stone and Perrone 1997). Heading thresholds in darkness increase by more than 10-fold after bilateral labyrinthectomy, suggesting that vestibular cues are critical for heading discrimination during inertial motion (Gu et al. 2007).

Thus, both visual and vestibular cues work together to estimate heading, and many recent studies have investigated both the perceptual benefits of multisensory integration and the neural correlates of these interactions. Next, we summarize the established properties of multisensory heading perception by first defining the concepts of precision and accuracy.

DEFINITION OF PRECISION AND ACCURACY

When throwing darts (Fig. 2A), we want to be both accurate (unbiased, darts centered on target) and precise (low variability of hit distribution). A skilled observer would hit the center bullseye each time; thus, dart-throwing performance would be accurate (Fig. 2A, left panels). Alternatively, the observer may always miss the bullseye and consistently throw darts at a different location, resulting in low accuracy in hitting the bullseye (Fig. 2A, right panels). On the other hand, the performance of an experienced observer would also show low variability (Fig. 2A, top panels), whereas the behavior of a novice observer may exhibit high variability (Fig. 2A, bottom panels). Thus, in Figure 2A, a small bias indicates high accuracy, whereas a low variability indicates high precision. Accuracy, represented by the bias, and precision, represented by the variability, together characterize the quality of motor performance.

Similarly, proficiency of sensory perception can be assessed by two independent and complementary properties: *Precision* is the sensitivity or threshold (also referred to as "reliability," computed as the "sigma" of a cumulative Gaussian fit of a psychometric function; Fig. 2C,D), whereas *accuracy* is the bias of the percept in relation to the actual stimulus (mean of psychometric function; Fig. 2D). An ideal percept is both precise (minimum threshold) and accurate (unbiased), although the two properties are independent, such that some precise cues could also be inaccurate (Fig. 2D).

MULTISENSORY INTEGRATION IMPROVES PERCEPTUAL PRECISION

Numerous studies have demonstrated that multisensory integration increases the precision of perceptual performance (e.g., Ernst and Banks 2002; Alais and Burr 2004; Knill and Pouget 2004). In particular, as illustrated by the psychometric functions in Figure 2C, the precision of perception in a heading discrimination task (Fig. 2B) increases when both vestibular (i.e., inertial) and visual (i.e., optic flow) signals indicate self-motion as compared with when either cue is present alone (Gu et al. 2008; Fetsch et al. 2009, 2011, 2013; Butler et al. 2010, 2015; de Winkel et al. 2013).

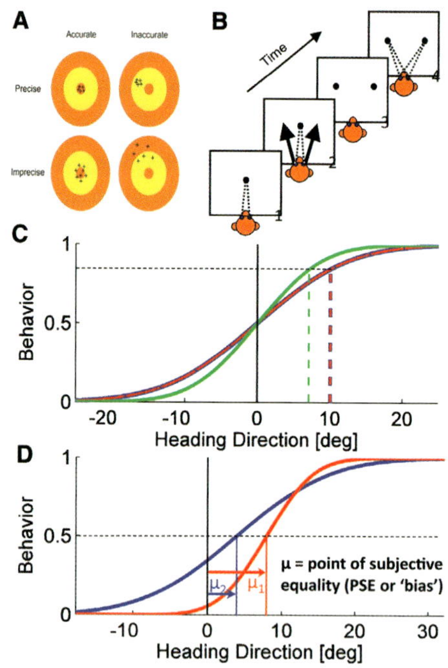

Figure 2. Schematics of how multisensory cue integration and calibration control precision and accuracy. (*A*) Definition of precision and accuracy using the example of darts. (*B*) Heading discrimination task. (*C,D*) Definition of precision and accuracy in psychometric functions. *Precision* is the sensitivity or threshold (also called "reliability," computed as the "sigma" of the cumulative Gaussian fit of a psychometric function) (*C*, vertical dashed lines). *Accuracy* is the bias of the percept in relation to the actual stimulus (mean of cumulative Gaussian fit of psychometric function; μ in *D*). An ideal percept is both reliable (minimum threshold) and accurate (unbiased), although some reliable cues might be inaccurate (*D*). In this case, the combined estimate would also be inaccurate according to Equation 2. Red/blue, single-cue psychometric functions. Green, combined cue psychometric function.

Statistically optimal cue integration predicts that threshold in the combined condition σ_{com} can be calculated from the single-cue vestibular and visual thresholds according to the equation

$$\sigma_{com} = \sqrt{\frac{\sigma_{vest}^2 \sigma_{vis}^2}{\sigma_{vest}^2 + \sigma_{vis}^2}}. \quad (1)$$

This is because an optimal estimator (in terms of minimizing the variance of the final estimate) should combine sensory information using a rule that weights the single-cue estimates in proportion to their reliability, or the inverse of their variance (Knill and Saunders 2003; Hillis et al. 2004):

$$\begin{aligned} \hat{S}_{com} &= w_{vest}\hat{S}_{vest} + w_{vis}\hat{S}_{vis}, \\ w_{vest} &= (1/\sigma_{vest}^2)/(1/\sigma_{vest}^2 + 1/\sigma_{vis}^2), \quad (2) \\ w_{vis} &= (1/\sigma_{vis}^2)/(1/\sigma_{vest}^2 + 1/\sigma_{vis}^2). \end{aligned}$$

As a result, more precise cues should have a higher weighting and less precise cues should have a lower weighting in the combined estimate. Indeed, empirically measured heading discrimination thresholds during combined vestibular–visual heading perception were close to the predictions produced by a statistically optimal cue combination rule (Gu et al. 2008; Fetsch et al. 2009, 2011, 2013; Butler et al. 2010, 2015; de Winkel et al. 2013).

MULTISENSORY CALIBRATION IMPROVES PERCEPTUAL ACCURACY

Whereas many studies have shown that multisensory cue integration can lead to a small improvement in precision (according to Eq. 1), it is important to appreciate that multisensory integration can lead to compromised accuracy, if the most reliable cue is inaccurate. Because of the combination rule for cue-weighting (Eq. 2), the combined estimate would lie close to the most reliable cue, regardless of whether or not that cue is accurate—often compromising the accuracy of the combined perception.

Such a solution would be problematic for everyday life, unless robust calibration mechanisms existed to maintain long-term accuracy. Robust plasticity mechanisms maintain the accuracy of sensory perception during development (Gori et al. 2008, 2010; Nardini et al. 2008; Stein and Rowland 2011) and in adulthood (Burge et al. 2010; Ernst and Di Luca 2011; Zaidel et al. 2011, 2013). In fact, there is growing evidence that the ability to optimally integrate cues (in the form of improved thresholds) develops late in adolescence (Gori et al. 2008; Nardini et al. 2008). It is thought that the reason for this developmental delay is that maintaining perceptual accuracy through calibration is more important and more fundamental than the small increase ($\leq\sqrt{2}$) in precision brought about through multisensory integration (Eqs. 1, 2). Thus, until robust calibration mechanisms are set in place during development to ensure perceptual accuracy, optimal integration, in the form of threshold improvement, does not occur (Gori et al. 2010, 2011a,b, 2012a,b,c).

Given the demonstrated importance of maintaining accuracy (through multisensory *calibration*) versus improving precision (through multisensory *integration*) during development, it is logical to assume that the brain maintains the ability to calibrate our senses relative to each other and relative to the environment, even in adulthood. Indeed, recent macaque and human psychophysics experiments have revealed the existence of two adult calibration mechanisms (Zaidel et al. 2011, 2013): In the absence of external feedback, an implicit, "unsupervised," calibration mechanism acts alone to reduce or eliminate a discrepancy between the cues by shifting single-cue percepts toward each other (Zaidel et al. 2011). In the presence of external feedback, which provides information regarding cue accuracy, the unsupervised mechanism works in parallel with a second "supervised" calibration mechanism, which relies only on the combined (multisensory) estimate. The goal of the *unsupervised calibration* is "internal consistency" and is achieved by simultaneously comparing the cues with one another and calibrating them individually. In contrast, the goal of the *supervised calibration* is "external accuracy" and is achieved by comparing the combined cue estimate with feedback from the environment (Zaidel et al. 2013). Although *unsupervised calibration* is more implicit and thus likely to represent a perceptual shift, the *supervised* component seems to use a more explicit mechanism, possibly targeting the mapping between perception and action. In combination, these two calibration mechanisms can ultimately achieve both internal consistency and external accuracy. A similar distinction between implicit and explicit components is found in studies of sensorimotor plasticity (Mazzoni and Krakauer 2006; Simani et al. 2007; Haith et al. 2008; Taylor and Ivry 2011).

BIASES IN HEADING PERCEPTION IN THE PRESENCE OF MOVING OBJECTS

The behavioral studies summarized above have examined heading perception in a rigid environment comprising only world-stationary objects. However, this is not an accurate representation of our natural surroundings as the world around us comprises both world-stationary objects and independently moving objects. For effective navigation in such a complex environment, we should be able to accurately judge our heading direction robustly. However, object motion is known to induce biases and reduce the accuracy of heading perception in the visual system (Warren and Saunders 1995; Saunders and Warren 1996). The movement of objects through space disrupts the pattern of optic flow due to the motion of the observer, making it difficult to distinguish between self-motion and object motion (Warren and Saunders 1995; Royden and Hildreth 1996).

Most previous studies that considered nonstationary environments have examined the influence of object mo-

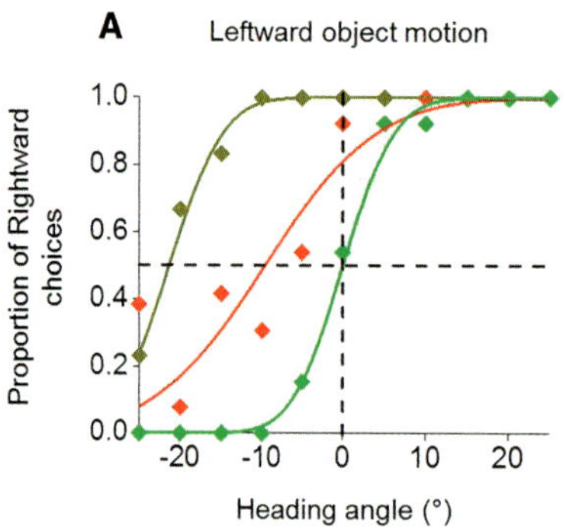

Figure 3. Heading perception in the presence of moving object. (*A*,*B*) Example psychometric functions from a monkey in the presence of an object moving leftward (*A*) or rightward (*B*). Heading perception is heavily biased in vestibular (dark yellow) and visual (red) single cue, but not combined (green), conditions.

tion only when self-motion was specified exclusively by visual signals (i.e., optic flow simulating observer movement in the world). In contrast, self-motion is usually always accompanied by inertial motion wherein vestibular signals provide independent information about head/body movement in space. It is thus plausible that vestibular signals would play a role in improving perceptual performance when visual signals alone cannot resolve this perceptual ambiguity completely. Indeed, recent studies have shown that vestibular signals play an important role in allocentric judgments of object trajectory in moving observers (Fajen and Matthis 2013; Dokka et al. 2015). We therefore hypothesized that the addition of vestibular cues to optic flow would improve the accuracy of heading perception in the presence of independently moving objects.

As illustrated in Figure 3, combined visual and vestibular heading stimuli improve the accuracy of heading discrimination in a nonrigid environment containing a large, independently moving object. A rhesus monkey was trained to perform a heading discrimination task with vestibular, visual, and combined vestibular–visual headings in both the absence and presence of an independently moving object. Remarkably, a moving object significantly reduced the accuracy of heading discrimination (i.e., there were large biases in perceived heading that depended on the direction of object motion) when self-motion was specified by visual or vestibular cues individually (Fig. 3, red or dark yellow). Importantly, the accuracy of heading discrimination improved significantly in the presence of multisensory vestibular–visual signals (Fig. 3, green).

BIASES IN HEADING PERCEPTION IN THE PRESENCE OF EYE AND HEAD MOVEMENTS

But it is not only moving objects that cause biases in heading perception. In addition, eye-in-head, head-in-body, and body-in-world rotations alter the optic flow patterns encoded in the retina, such that strong perceptual biases can result (Royden et al. 1992, 1994; Crowell et al. 1998). Although psychophysical studies have provided evidence that visual cues may play a role in reducing perceptual biases and accurately estimating heading direction in the presence of rotations (Grigo and Lappe 1999; Li and Warren 2000, 2002, 2004; Crowell and Andersen 2001; Royden et al. 2006), significant perceptual biases persist (Royden et al. 1992; Royden 1994; Banks et al. 1996). It has been well established that motor-related movement information from extra-retinal efference copy signals help reduce these biases (for a recent literature review, see Sunkara et al. 2015). Whether the multisensory influence of vestibular cues also reduces rotation-induced biases in heading perception is possible, but has not yet been tested.

NEURAL CORRELATES OF MULTISENSORY HEADING PERCEPTION

In recent years, the neural correlates of multisensory heading perception have also been systematically explored. Figure 4A summarizes a cortical circuit including areas with neurons tuned to both optic flow and inertial heading cues (varying shades of red and dark yellow, respectively). Based on these findings, heading perception likely involves several multisensory cortical regions, such as the dorsal medial superior temporal (MSTd) area (Bremmer et al. 1997; Duffy 1998; Page and Duffy 2003; Gu et al. 2006; Maciokas and Britten 2010), the ventral intraparietal (VIP) area (Bremmer et al. 1999, 2002; Schlack et al. 2002; Zhang et al. 2004; Maciokas and Britten 2010; Chen et al. 2011a,b, 2013a,b), the visual posterior sylvian (VPS) area (Chen et al. 2011c), and the pursuit area of the frontal eye fields (FEF; Y Gu, GC DeAngelis, DE Angelaki, unpubl.).

Whereas MSTd, VIP, VPS, and FEF are all examples of cortical areas shown to process both optic flow and vestibular heading, other cortical areas, such as the parieto-insular vestibular cortex (PIVC), are tuned to vestibular heading cues only (Chen et al. 2010). Furthermore, other cortical areas, such as the medial temporal area (MT) (Chowdhury et al. 2009) and visual area V6 (R Fan, GC DeAngelis, DE Angelaki, unpubl.), are tuned only to optic flow.

Figure 4B summarizes the mean heading tuning strength of each area, using a direction discrimination in-

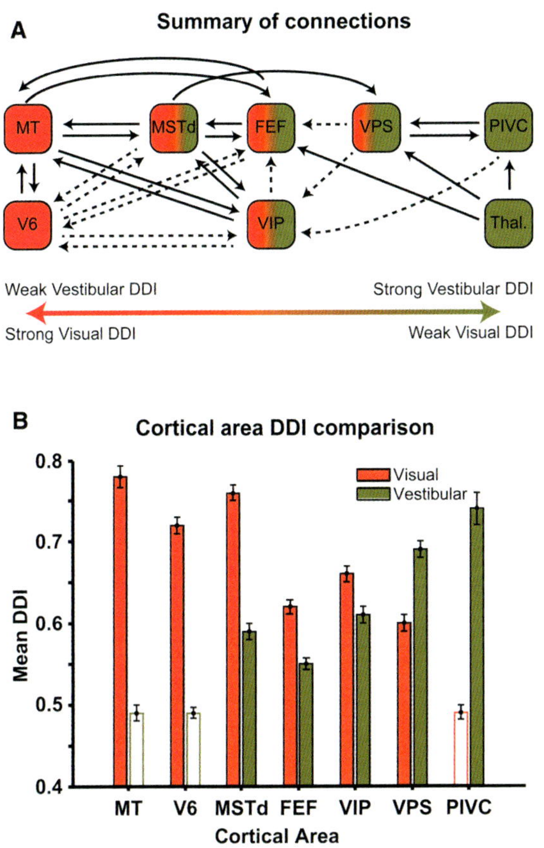

Figure 4. Cortical network of visual/vestibular heading areas. (*A*) Schematic of area connectivity, with each area color-coded according to visual/vestibular tuning strength. (*B*) Distributions of mean visual (red) and vestibular (green) DDI (Direction Discrimination Index) ± SE across cortical areas. Open bars are used for areas without significant tuning.

dex (DDI), which takes on values between 0 (no tuning) and 1 (strong tuning). Because the same experimental protocols and analyses were used to measure how well all areas are tuned to heading, the mean DDI_{vis} and mean DDI_{vest} values are directly comparable (Fig. 4B). Of particular note is the fact that there appears to be a progression of tuning strength to visual and vestibular cues across the four multisensory areas, MSTd, VIP, FEF, and VPS.

Of these multisensory areas, MSTd and VIP have been studied the most. Although some differences in response properties between VIP and MSTd have been noted (e.g., Chen et al. 2011a; Yang et al. 2011), MSTd and VIP share many similarities in their tuning to visual and vestibular stimuli and also show strong correlations between neural activity and perceptual choice (Gu et al. 2008; Maciokas and Britten 2010; Chen et al. 2011a,b, 2013a,b,c). As illustrated in Figure 5, where firing rate is plotted as a function of heading direction for the vestibular (dark yellow) and visual (red) conditions, multisensory MSTd/VIP cells fall into one of two groups, which were identified by constructing spatial tuning curves as heading direction was varied: (1) "Congruent" neurons had similar visual/vestibular preferred stimuli and thus signaled the same heading direction under both single-cue stimulus conditions. For example, the congruent MSTd cell preferred leftward (negative) headings for both stimuli (note that heading directions are referenced to either the real or simulated self-motion; thus, similar tuning in the visual and vestibular conditions defines a congruent cell). In contrast, the congruent VIP cell example preferred rightward (positive) heading directions. (2) "Opposite" neurons preferred opposite directions of heading under the visual and vestibular stimulus conditions, as shown with the MSTd and VIP examples in Figure 5. The extent of

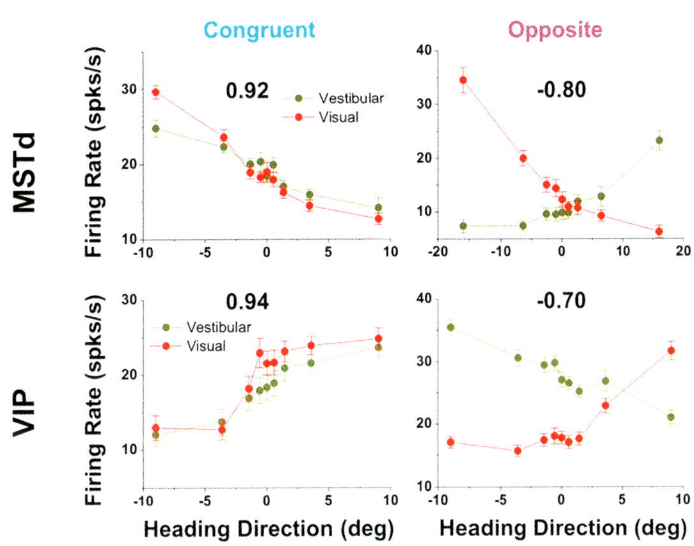

Figure 5. Two types of multisensory neurons in MSTd (*top*) and VIP (*bottom*). Heading tuning curves, for example, congruent and opposite cells (dark yellow: vestibular; red: visual). 0° heading denotes straight forward translation, whereas positive/negative numbers indicate rightward/leftward directions, respectively. Data (from Gu et al. 2008 and Chen et al. 2013a) illustrate firing rate means and SE. The numbers on each panel illustrate the "Congruency Index" for each cell.

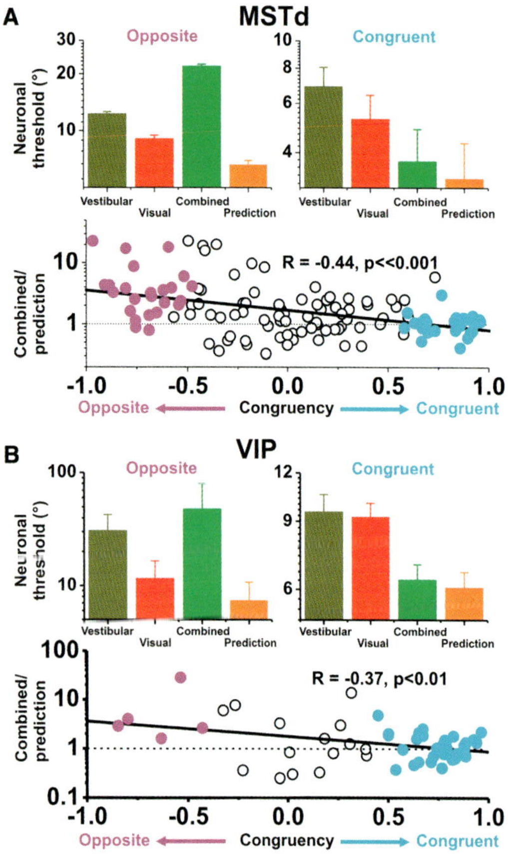

Figure 6. Improvement in neuronal threshold under cue combination depends on congruency of visual and vestibular tuning for (*A*) MSTd (from Gu et al. 2008) and (*B*) VIP (from Chen et al. 2013a). Data are shown as scatter plots of the ratio of the threshold measured in the combined condition relative to the prediction for optimal cue integration (Eq. 1) versus the Congruency Index. Cyan symbols are used for "congruent" neurons; magenta symbols are used for "opposite" neurons; open black symbols mark "intermediate" neurons. Dashed horizontal lines indicate that threshold in the combined condition is equal to the prediction. Solid lines show type II linear regressions. Neuronal thresholds are also plotted as means ± SE for opposite and congruent neurons (*top* panels).

"congruency" was quantified using a "Congruency" index that varied from +1 for cells with congruent multisensory integration to −1 for cells with opposite multisensory integration.

Gu et al. (2008) (see also Gu et al. 2014) showed that congruent and opposite MSTd neurons play different roles in heading perception. As illustrated in Figure 6A, the neuronal threshold of congruent MSTd neurons was lowest in the combined condition and remarkably similar to the prediction from optimal cue integration models (Eq. 1; Fig. 6A, top; compare green and orange bars for the congruent cell panel), confirming that these cells increased their ability to discriminate small variations in heading direction when both cues were present. This finding suggests that congruent cells in MSTd satisfy one major prediction of optimal cue integration theory (Eq. 1). Fetsch et al. (2011) further showed that congruent MSTd neurons also satisfy the second prediction of optimal cue integration theory (Eq. 2). In contrast, opposite cells became less sensitive during combined stimulation (Fig. 6A, top; green bar for the opposite cell panel). In fact, there was a significant correlation between how well multisensory thresholds of MSTd neurons followed the predictions of optimal cue integration and the congruency index for those neurons (Fig. 6A, bottom panel). Remarkably, very similar results also characterize multisensory neurons in VIP (Fig. 6B; Chen et al. 2013a). The properties of FEF and VPS multisensory neurons during heading discrimination have not yet been examined.

CONCLUSION

Much progress has been made in recent years to understand the multisensory properties of heading discrimination in the parietal cortex of macaques, although these responses appear to remain egocentric (Fetsch et al. 2007; Chen et al. 2013b,c). The focus of this work has been primarily to derive a mechanistic understanding of how neural circuits give rise to multisensory perception in the context of probabilistic, statistical decision theory and the analysis of ideal observers. As summarized by Fetsch et al. (2013), this strategy has been based on linking neural responses with psychophysical performance in behaving animals, combined with computational modeling that has allowed comparisons with normative predictions.

Much work remains to be done to determine whether and how this multisensory circuit contributes to navigation and how egocentric heading signals interact with the allocentric representations of head direction, grid, and place cells in the medial temporal hippocampal networks. Evidence is rapidly mounting that multisensory interactions are more fundamental to these allocentric neural representations than previously thought (Ravassard et al. 2013). Hence, the quest to untangle and exploit the normative rules and principles of multisensory processing will be vital for understanding decision-making in the intersection of sensation, action, and experience.

ACKNOWLEDGMENTS

Experiments were supported by NIH EY017866 and DC007620. We thank Kalpana Dokka, Aihua Chen, Kaushik Lakshminarasimhan, and Reuben Fan for their assistance with figures and Michael Shinder for assistance with references.

REFERENCES

Alais D, Burr D. 2004. The ventriloquist effect results from near-optimal bimodal integration. *Curr Biol* **14:** 257–262.

Angelaki DE, Cullen KE. 2008. Vestibular system: The many facets of a multimodal sense. *Annu Rev Neurosci* **31:** 125–150.

Angelaki DE, Gu Y, DeAngelis GC. 2009. Multisensory integration: Psychophysics, neurophysiology, and computation. *Curr Opin Neurobiol* **19:** 452–458.

Angelaki DE, Gu Y, Deangelis GC. 2011. Visual and vestibular cue integration for heading perception in extrastriate visual cortex. *J Physiol* **589:** 825–833.

Banks MS, Ehrlich SM, Backus BT, Crowell JA. 1996. Estimating heading during real and simulated eye movements. *Vision Res* **36:** 431–443.

Barry C, Burgess N. 2014. Neural mechanisms of self-location. *Curr Biol* **24:** R330–R339.

Berthoz A, Pavard B, Young LR. 1975. Perception of linear horizontal self-motion induced by peripheral vision (linearvection) basic characteristics and visual-vestibular interactions. *Exp Brain Res* **23:** 471–489.

Brandt T, Dichgans J, Koenig E. 1972. Perception of self-rotation (circular vection) induced by optokinetic stimuli. *Pflugers Arch* **332** (Suppl 332): R398.

Bremmer F, Ilg UJ, Thiele A, Distler C, Hoffmann KP. 1997. Eye position effects in monkey cortex. I. Visual and pursuit-related activity in extrastriate areas MT and MST. *J Neurophysiol* **77:** 944–961.

Bremmer F, Kubischik M, Pekel M, Lappe M, Hoffmann KP. 1999. Linear vestibular self-motion signals in monkey medial superior temporal area. *Ann NY Acad Sci* **871:** 272–281.

Bremmer F, Duhamel JR, Ben Hamed S, Graf W. 2002. Heading encoding in the macaque ventral intraparietal area (VIP). *Eur J Neurosci* **16:** 1554–1568.

Britten KH. 2008. Mechanisms of self-motion perception. *Annu Rev Neurosci* **31:** 389–410.

Burge J, Girshick AR, Banks MS. 2010. Visual-haptic adaptation is determined by relative reliability. *J Neurosci* **30:** 7714–7721.

Butler JS, Smith ST, Campos JL, Bülthoff HH. 2010. Bayesian integration of visual and vestibular signals for heading. *J Vis* **10:** 23.

Butler JS, Campos JL, Bülthoff HH. 2015. Optimal visual-vestibular integration under conditions of conflicting intersensory motion profiles. *Exp Brain Res* **233:** 587–597.

Chen A, DeAngelis GC, Angelaki DE. 2010. Macaque parieto-insular vestibular cortex: Responses to self-motion and optic flow. *J Neurosci* **30:** 3022–3042.

Chen A, DeAngelis GC, Angelaki DE. 2011a. A comparison of vestibular spatiotemporal tuning in macaque parietoinsular vestibular cortex, ventral intraparietal area, and medial superior temporal area. *J Neurosci* **31:** 3082–3094.

Chen A, DeAngelis GC, Angelaki DE. 2011b. Representation of vestibular and visual cues to self-motion in ventral intraparietal cortex. *J Neurosci* **31:** 12036–12052.

Chen A, DeAngelis GC, Angelaki DE. 2011c. Convergence of vestibular and visual self-motion signals in an area of the posterior sylvian fissure. *J Neurosci* **31:** 11617–11627.

Chen A, Deangelis GC, Angelaki DE. 2013a. Functional specializations of the ventral intraparietal area for multisensory heading discrimination. *J Neurosci* **33:** 3567–3581.

Chen X, Deangelis GC, Angelaki DE. 2013b. Diverse spatial reference frames of vestibular signals in parietal cortex. *Neuron* **80:** 1310–1321.

Chen X, DeAngelis GC, Angelaki DE. 2013c. Eye-centered representation of optic flow tuning in the ventral intraparietal area. *J Neurosci* **33:** 18574–18582.

Chowdhury SA, Takahashi K, DeAngelis GC, Angelaki DE. 2009. Does the middle temporal area carry vestibular signals related to self-motion? *J Neurosci* **29:** 12020–12030.

Crowell JA, Andersen RA. 2001. Pursuit compensation during self-motion. *Perception* **30:** 1465–1488.

Crowell JA, Banks MS, Shenoy KV, Andersen RA. 1998. Visual self-motion perception during head turns. *Nat Neurosci* **1:** 732–737.

DeAngelis GC, Angelaki DE. 2012. Visual–vestibular integration for self-motion perception. In *The neural bases of multisensory processes* (ed. Murray MM, Wallace MT), Chapter 31. CRC Press, Boca Raton, FL.

de Winkel KN, Weesie J, Werkhoven PJ, Groen EL. 2010. Integration of visual and inertial cues in perceived heading of self-motion. *J Vis* **10:** 1.

de Winkel KN, Soyka F, Barnett-Cowan M, Bülthoff HH, Groen EL, Werkhoven PJ. 2013. Integration of visual and inertial cues in the perception of angular self-motion. *Exp Brain Res* **231:** 209–218.

Dichgans J, Brandt T. 1978. Visual-vestibular interaction: Effects on self-motion perception and postural control. In *Handbook of sensory physiology* (ed. Held R, Leibowitz HW, Teuber HL), pp. 756–804. Springer, Berlin.

Dokka K, MacNeilage PR, DeAngelis GC, Angelaki DE. 2015. Multisensory self-motion compensation during object trajectory judgments. *Cereb Cortex* **25:** 619–630.

Drugowitsch J, DeAngelis GC, Klier EM, Angelaki DE, Pouget A. 2014. Optimal multisensory decision-making in a reaction-time task. *Elife* **3:** doi: 10.7554/eLife.03005.

Duffy CJ. 1998. MST neurons respond to optic flow and translational movement. *J Neurophysiol* **80:** 1816–1827.

Duffy CJ, Wurtz RH. 1995. Response of monkey MST neurons to optic flow stimuli with shifted centers of motion. *J Neurosci* **15:** 5192–5208.

Ernst MO, Banks MS. 2002. Humans integrate visual and haptic information in a statistically optimal fashion. *Nature* **415:** 429–433.

Ernst MO, Di Luca M. 2011. Multisensory perception: From integration to remapping. In *Sensory cue integration* (ed. Trommershauser J, Kording K, Landy MS), pp. 224–250. Oxford University Press, Oxford.

Etienne AS, Maurer R, Saucy F, Teroni E. 1986. Short-distance homing in the golden hamster after a passive outward journey. *Anim Behav* **34:** 696–715.

Fajen BR, Matthis JS. 2013. Visual and non-visual contributions to the perception of object motion during self-motion. *PLoS One* **8:** e55446.

Fetsch CR, Wang S, Gu Y, Deangelis GC, Angelaki DE. 2007. Spatial reference frames of visual, vestibular, and multimodal heading signals in the dorsal subdivision of the medial superior temporal area. *J Neurosci* **27:** 700–712.

Fetsch CR, Turner AH, DeAngelis GC, Angelaki DE. 2009. Dynamic reweighting of visual and vestibular cues during self-motion perception. *J Neurosci* **29:** 15601–15612.

Fetsch CR, DeAngelis GC, Angelaki DE. 2010. Visual-vestibular cue integration for heading perception: Applications of optimal cue integration theory. *Eur J Neurosci* **31:** 1721–1729.

Fetsch CR, Pouget A, DeAngelis GC, Angelaki DE. 2011. Neural correlates of reliability-based cue weighting during multisensory integration. *Nat Neurosci* **15:** 146–154.

Fetsch CR, DeAngelis GC, Angelaki DE. 2013. Bridging the gap between theories of sensory cue integration and the physiology of multisensory neurons. *Nat Rev Neurosci* **14:** 429–442.

Geva-Sagiv M, Las L, Yovel Y, Ulanovsky N. 2015. Spatial cognition in bats and rats: From sensory acquisition to multiscale maps and navigation. *Nat Rev Neurosci* **16:** 94–108.

Gibson JJ. 1950. *The perception of the visual world*. Houghton-Mifflin, Boston, MA.

Gori M, Del Viva M, Sandini G, Burr DC. 2008. Young children do not integrate visual and haptic form information. *Curr Biol* **18:** 694–698.

Gori M, Sandini G, Martinoli C, Burr D. 2010. Poor haptic orientation discrimination in nonsighted children may reflect disruption of cross-sensory calibration. *Curr Biol* **20:** 223–225.

Gori M, Mazzilli G, Sandini G, Burr D. 2011a. Cross-sensory facilitation reveals neural interactions between visual and tactile motion in humans. *Front Psychol* **2:** 55.

Gori M, Sciutti A, Burr D, Sandini G. 2011b. Direct and indirect haptic calibration of visual size judgments. *PLoS One* **6:** e25599.

Gori M, Giuliana L, Sandini G, Burr D. 2012a. Visual size perception and haptic calibration during development. *Dev Sci* **15:** 854–862.

Gori M, Sandini G, Burr D. 2012b. Development of visuo-auditory integration in space and time. *Front Integr Neurosci* **6:** 77.

Gori M, Squeri V, Sciutti A, Masia L, Sandini G, Konczak J. 2012c. Motor commands in children interfere with their haptic perception of objects. *Exp Brain Res* **223:** 149–157.

Grigo A, Lappe M. 1999. Dynamical use of different sources of information in heading judgments from retinal flow. *J Opt Soc Am A Opt Image Sci Vis* **16:** 2079–2091.
Gu Y, Watkins PV, Angelaki DE, DeAngelis GC. 2006. Visual and nonvisual contributions to three-dimensional heading selectivity in the medial superior temporal area. *J Neurosci* **26:** 73–85.
Gu Y, DeAngelis GC, Angelaki DE. 2007. A functional link between area MSTd and heading perception based on vestibular signals. *Nat Neurosci* **10:** 1038–1047.
Gu Y, Angelaki DE, DeAngelis GC. 2008. Neural correlates of multisensory cue integration in macaque MSTd. *Nat Neurosci* **11:** 1201–1210.
Gu Y, Angelaki DE, DeAngelis GC. 2014. Contribution of correlated noise and selective decoding to choice probability measurements in extrastriate visual cortex. *Elife* **3:** doi: 10.7554/eLife.02670.
Haith A, Jackson C, Miall C, Vijayakumar S. 2008. Interactions between sensory and motor components of adaptation predicted by a Bayesian model. In *Workshop on advances in computational motor control (ACMC 2008)*.
Hartley T, Lever C, Burgess N, O'Keefe J. 2013. Space in the brain: How the hippocampal formation supports spatial cognition. *Philos Trans R Soc Lond B Biol Sci* **369:** 20120510.
Hillis JM, Watt SJ, Landy MS, Banks MS. 2004. Slant from texture and disparity cues: Optimal cue combination. *J Vis* **4:** 967–992.
Knierim JJ, Zhang K. 2012. Attractor dynamics of spatially correlated neural activity in the limbic system. *Annu Rev Neurosci* **35:** 267–285.
Knill DC, Pouget A. 2004. The Bayesian brain: The role of uncertainty in neural coding and computation. *Trends Neurosci* **27:** 712–719.
Knill DC, Saunders JA. 2003. Do humans optimally integrate stereo and texture information for judgments of surface slant? *Vision Res* **43:** 2539–2558.
Li L, Warren WH Jr. 2000. Perception of heading during rotation: Sufficiency of dense motion parallax and reference objects. *Vision Res* **40:** 3873–3894.
Li L, Warren WH Jr. 2002. Retinal flow is sufficient for steering during observer rotation. *Psychol Sci* **13:** 485–491.
Li L, Warren WH Jr. 2004. Path perception during rotation: Influence of instructions, depth range, and dot density. *Vision Res* **44:** 1879–1889.
Mach E. 1875. *Grundlinien der Lehre von den Bewegungsempfindungen*. W. Engelmann, Leipzig.
Maciokas JB, Britten KH. 2010. Extrastriate area MST and parietal area VIP similarly represent forward headings. *J Neurophysiol* **104:** 239–247.
Mazzoni P, Krakauer JW. 2006. An implicit plan overrides an explicit strategy during visuomotor adaptation. *J Neurosci* **26:** 3642–3645.
Mittelstaedt M-L, Mittelstaedt H. 1980. Homing by path integration in a mammal. *Naturwissenschaften* **67:** 566–567.
Moser EI, Moser MB, Roudi Y. 2013. Network mechanisms of grid cells. *Philos Trans R Soc Lond B Biol Sci* **369:** 20120511.
Nardini M, Jones P, Bedford R, Braddick O. 2008. Development of cue integration in human navigation. *Curr Biol* **18:** 689–693.
Page WK, Duffy CJ. 2003. Heading representation in MST: Sensory interactions and population encoding. *J Neurophysiol* **89:** 1994–2013.
Ravassard P, Kees A, Willers B, Ho D, Aharoni D, Cushman J, Aghajan ZM, Mehta MR. 2013. Multi-sensory control of hippocampal spatiotemporal selectivity. *Science* **340:** 1342–1346.
Royden CS. 1994. Analysis of misperceived observer motion during simulated eye rotations. *Vision Res* **34:** 3215–3222.
Royden CS, Hildreth EC. 1996. Human heading judgments in the presence of moving objects. *Percept Psychophys* **58:** 836–856.
Royden CS, Banks MS, Crowell JA. 1992. The perception of heading during eye movements. *Nature* **360:** 583–585.
Royden CS, Crowell JA, Banks MS. 1994. Estimating heading during eye movements. *Vision Res* **34:** 3197–3214.
Royden CS, Cahill JM, Conti DM. 2006. Factors affecting curved versus straight path heading perception. *Percept Psychophys* **68:** 184–193.
Saunders JA, Warren WH. 1996. Perceived heading biased by a moving object: Effects of disparity and object position. *Invest Ophthalmol Vis Sci* **37:** S454.
Schlack A, Hoffmann KP, Bremmer F. 2002. Interaction of linear vestibular and visual stimulation in the macaque ventral intraparietal area (VIP). *Eur J Neurosci* **16:** 1877–1886.
Simani MC, McGuire LM, Sabes PN. 2007. Visual-shift adaptation is composed of separable sensory and task-dependent effects. *J Neurophysiol* **98:** 2827–2841.
Stein BE, Rowland BA. 2011. Organization and plasticity in multisensory integration: Early and late experience affects its governing principles. *Prog Brain Res* **191:** 145–163.
Stone LS, Perrone JA. 1997. Human heading estimation during visually simulated curvilinear motion. *Vision Res* **37:** 573–590.
Stuchlik A, Bures J. 2002. Relative contribution of allothetic and idiothetic navigation to place avoidance on stable and rotating arenas in darkness. *Behav Brain Res* **128:** 179–188.
Sunkara A, DeAngelis GC, Angelaki DE. 2015. Role of visual and non-visual cues in constructing a rotation-invariant representation of heading in parietal cortex. *Elife* **4:** doi: 10.7554/eLife.04693.
Tanaka M, Weber H, Creutzfeldt OD. 1986. Visual properties and spatial distribution of neurones in the visual association area on the prelunate gyrus of the awake monkey. *Exp Brain Res* **65:** 11–37.
Taylor JA, Ivry RB. 2011. Flexible cognitive strategies during motor learning. *PLoS Comput Biol* **7:** e1001096.
van den Berg AV, Brenner E. 1994. Humans combine the optic flow with static depth cues for robust perception of heading. *Vision Res* **34:** 2153–2167.
Warren WH Jr. 2003. Optic flow. In *The visual neurosciences* (ed. Chalupa LM, Werner JS), pp. 1247–1259. MIT Press, Cambridge, MA.
Warren WH Jr, Hannon DJ. 1990. Eye movements and optical flow. *J Opt Soc Am A* **7:** 160–169.
Warren WH Jr, Saunders JA. 1995. Perceiving heading in the presence of moving objects. *Perception* **24:** 315–331.
Warren WH Jr, Morris MW, Kalish M. 1988. Perception of translational heading from optical flow. *J Exp Psychol Hum Percept Perform* **14:** 646–660.
Yang Y, Liu S, Chowdhury SA, DeAngelis GC, Angelaki DE. 2011. Binocular disparity tuning and visual-vestibular congruency of multisensory neurons in macaque parietal cortex. *J Neurosci* **31:** 17905–17916.
Zaidel A, Turner AH, Angelaki DE. 2011. Multisensory calibration is independent of cue reliability. *J Neurosci* **31:** 13949–13962.
Zaidel A, Ma WJ, Angelaki DE. 2013. Supervised calibration relies on the multisensory percept. *Neuron* **80:** 1544–1557.
Zhang T, Heuer HW, Britten KH. 2004. Parietal area VIP neuronal responses to heading stimuli are encoded in head-centered coordinates. *Neuron* **42:** 993–1001.

How Attention Affects Spatial Resolution

Marisa Carrasco[1,2] and Antoine Barbot[1,3]
[1]*Department of Psychology, New York University, New York, New York 10003*
[2]*Center for Neural Science, New York University, New York, New York 10003*
Correspondence: marisa.carrasco@nyu.edu

We summarize and discuss a series of psychophysical studies on the effects of spatial covert attention on spatial resolution, our ability to discriminate fine patterns. Heightened resolution is beneficial in most, but not all, visual tasks. We show how endogenous attention (voluntary, goal driven) and exogenous attention (involuntary, stimulus driven) affect performance on a variety of tasks mediated by spatial resolution, such as visual search, crowding, acuity, and texture segmentation. Exogenous attention is an automatic mechanism that increases resolution regardless of whether it helps or hinders performance. In contrast, endogenous attention flexibly adjusts resolution to optimize performance according to task demands. We illustrate how psychophysical studies can reveal the underlying mechanisms of these effects and allow us to draw linking hypotheses with known neurophysiological effects of attention.

ATTENTION IS A SELECTIVE PROCESS

Each time we open our eyes we are confronted with an overwhelming amount of information. Yet, we seemingly understand our visual world effortlessly. To make sense of a scene, we need to detect, localize, and identify relevant information. By focusing on a certain location or aspect of the visual scene, attention allows us to selectively process information, prioritizing some aspects of information while ignoring others. Attention lies at the crossroads between perception and cognition, bringing together scientists using psychophysics, neurophysiology, neuroimaging, and computational neuroscience techniques. Significant advances in visual attention have been facilitated by fruitful cross talk among these fields and levels of analyses. The interest in visual attention has exponentially grown; a PubMed search yields more than 3500 articles dealing with visual attention since 1970 ("visual attention" in title or abstract), with half of them published since 2008 (Fig. 1). Changes in an observer's attentional state while keeping the retinal image constant can affect perceptual performance and appearance, as well as the activity of "sensory" neurons throughout visual cortex.

Selective attention arises from the brain's limited capacity to process information. The fixed amount of overall energy available to the brain and the high bioenergetic cost of the neuronal activity involved in cortical computation require the use of efficient representational codes that rely on a sparse collection of active neurons, as well as the flexible allocation of metabolic resources according to task demands. These energy limitations allow only a small fraction of the machinery to be engaged concurrently, and provide a neurophysiological basis for selective attention (Lennie 2003; Carrasco 2011). The notion that stimuli compete for limited resources has been long recognized (Broadbent 1958; Neisser 1967; Kinchla 1980, 1992) and supported by electrophysiological, neuroimaging, and behavioral studies (for reviews, see Desimone and Duncan 1995; Reynolds and Chelazzi 2004; Carrasco 2011, 2014; Beck and Kastner 2014; Posner 2014).

Attention optimizes the use of the system's limited resources by enhancing representations of the relevant locations or features of our environment while diminishing the representations of less relevant locations or features. Attentional trade-offs emerge across different tasks and displays, including noncluttered displays, in which only two simple stimuli are competing for processing; the benefit at the attended location has a concomitant cost at unattended locations (Luck et al. 1994; Lu and Dosher 1998; Pestilli and Carrasco 2005; Anton-Erxleben et al. 2007; Pestilli et al. 2007; Giordano et al. 2009; Montagna et al. 2009; Herrmann et al. 2010; Yeshurun and Rashal 2010; Barbot et al. 2011, 2012a).

SPATIAL COVERT ATTENTION: ENDOGENOUS AND EXOGENOUS

Knowledge and assumptions about the world, the behavioral state of the organism, and the sudden appearance of possibly relevant information, facilitate the processing of sensory input. Attention can be allocated overtly, by moving one's eyes toward a location, and covertly, by attending to a given location without directing one's gaze toward it. Covert attention aids us monitor our crowded environment and inform subsequent eye movements to locations where relevant information is likely. We deploy covert attention in many everyday situations: searching for objects, walking, driving, dancing—as well as in social situations, to conceal intentions eye movements would reveal (e.g., in competitive sports). Attention

[3]Present address: Flaum Eye Institute and Center for Visual Science, University of Rochester Medical Center, Rochester, New York 14642.

 doi: 10.1101/sqb.2014.79.024687

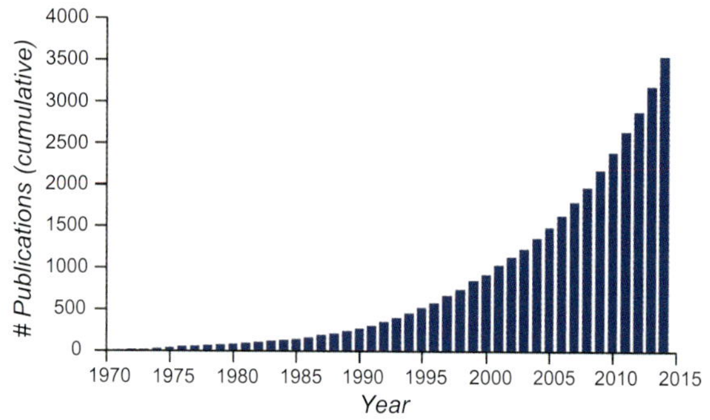

Figure 1. Cumulative numbers of publications reported by PubMed since 1970 containing the key word "visual attention" in either the title or the abstract.

affects performance and appearance in many tasks mediated by early visual dimensions—contrast sensitivity and spatial resolution—once considered to be preattentive (for reviews, see Carrasco and Yeshurun 2009; Carrasco 2011, 2014; Anton-Erxleben and Carrasco 2013).

Two types of covert attention facilitate selective processing of information: endogenous attention, which enables us to voluntarily monitor information at a given location, and exogenous attention, which allows us to involuntarily orient to a location where sudden stimulation has occurred. Observers deploy voluntary attention in ~300 msec and can sustain it as needed; the involuntary deployment of attention is transient, peaks at ~100–120 msec and decays quickly (Müller and Rabbitt 1989; Nakayama and Mackeben 1989; Remington et al. 1992; Ling and Carrasco 2006a,b; Liu et al. 2007; Koenig-Robert and Vanrullen 2011). Whereas observers can allocate resources according to cue validity when deploying endogenous attention, they cannot do so when deploying exogenous attention (Sperling and Melchner 1978; Kinchla 1980; Mangun and Hillyard 1990; Yantis and Jonides 1996; Giordano et al. 2009). Exogenous cues cannot be ignored: Involuntary transient shifts of attention occur even when the cues are known to be uninformative and irrelevant (Pestilli and Carrasco 2005; Pestilli et al. 2007; Montagna et al. 2009; Herrmann et al. 2010; Yeshurun and Rashal 2010; Barbot et al. 2011) and even when they impair performance (Yeshurun and Carrasco 1998, 2000, 2008; Talgar and Carrasco 2002; Yeshurun and Levy 2003; Yeshurun 2004; Carrasco et al. 2006b; Hein et al. 2006; Bocanegra and Zeelenberg 2011). The different temporal characteristics and degrees of automaticity suggest that exogenous attention may be phylogenetically older than endogenous attention, allowing us to respond automatically and quickly to stimuli that may provide behaviorally relevant information.

SPATIAL RESOLUTION

Vision is limited by many factors, including spatial resolution—our ability to discriminate fine patterns. The visual system's spatial resolution is best at the fovea (the focus of our gaze) and decays toward the periphery. The systematic resolution differences across eccentricity are due to several factors, starting with the spacing of photoreceptors in the retina. The visual system is retinotopic and arranged hierarchically from the retina to early extrastriate areas. In the visual cortex, the neurons that receive projections from the fovea have the smallest RFs, and RF sizes increase both with eccentricity and along the visual pathway (Fig. 2A). Additionally, the number and density of RFs decreases and the spacing among them increases with eccentricity (Levi et al. 1985; Martin 1986; DeValois and DeValois 1988; Zhaoping 2014). Thus, as eccentricity increases, information is pooled over a larger area, diminishing sensitivity to fine patterns. The central part of the visual field is overrepresented. A greater proportion of cortex is devoted to processing input from the central part than the periphery (cortical magnification). In area V1, ~25% of cortex is devoted to processing the central 2.5° of visual angle (DeValois and DeValois 1988). Population receptive field (pRF) models provide a tool for quantitatively linking the signal evoked by a visual stimulus to the fMRI response at a point on the cortical surface. The pRF model accounts for both the most effective point in space that produces an fMRI response at a cortical site (pRF center) and the spatial extent of the visual field that a cortical site is sensitive to (pRF size). The pRF size is an important measure of visual encoding that varies systematically with eccentricity within a visual field map and between visual field maps (Fig. 2B) (Dumoulin and Wan-

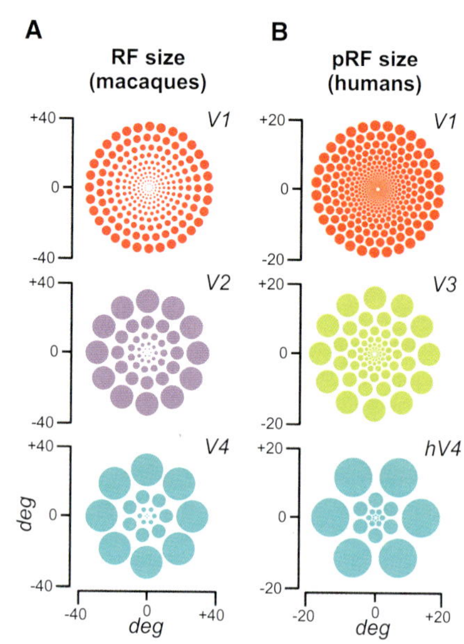

Figure 2. Schematic depiction of (*A*) receptive field (RF) size and (*B*) population receptive field (pRF) size as a function of eccentricity based on physiological measurements in macaque areas V1, V2, and V4, and fMRI measurements in human areas V1, V3, and hV4. The center of each array corresponds to the fovea. The size of each circle is proportional to its eccentricity, based on the corresponding scaling parameters. At a given eccentricity, a larger scaling parameter implies larger receptive fields. (*A*, Reprinted by permission from Macmillan Publishers Ltd. from Freeman and Simoncelli 2011; *B*, reproduced with the permission of Jonathan Winawer and Hiroshi Horiguchi; https://archive.nyu.edu/handle/2451/33887.)

dell 2008; Wandell and Winawer 2015). Note the similarity between the two sets of measures.

RF size is correlated with spatial frequency tuning, which measures a neuron's sensitivity to different spatial scales of variation in contrast. Thus, selectivity for spatial frequency also varies across the visual field. The visual system is most sensitive to higher spatial frequencies near the fovea, and its peak sensitivity shifts toward lower spatial frequencies with eccentricity (Rovamo et al. 1978; Wright and Johnston 1983; Kitterle 1986; Strasburger et al. 2011). Consequently, we can clearly see an object in the center of the field but cannot discriminate its fine details when it is in the peripheral field. Signals from the field's center are processed more accurately and faster (Rijsdijk et al. 1980; Cannon 1985; Carrasco et al. 1995). In many tasks, these performance differences are eliminated when stimulus size is enlarged according to the cortical magnification factor, which equates the size of the cortical representation for stimuli at different eccentricities (Rovamo and Virsu 1979; Virsu and Rovamo 1979; Kitterle 1986; Carrasco and Frieder 1997; Carrasco et al. 1998a; Harvey and Dumoulin 2011), or when thresholds are expressed in distance along the V1 surface rather than in degrees of visual angle (Duncan and Boynton 2003; Pelli and Tillman 2008). However, compensating for cortical magnification does not eliminate qualitative differences in neural processing between central and peripheral vision (Kitterle 1986; Strasburger et al. 2011). Moreover, at isoeccentric locations performance is better along the horizontal than the vertical meridian, and in the lower than the upper region of the vertical meridian (Rovamo et al. 1982; Talgar and Carrasco 2002; Fuller et al. 2008; Montaser-Kouhsari and Carrasco 2009).

COVERT ATTENTION ALTERS SPATIAL RESOLUTION

To investigate covert attention, it is necessary to keep both the task and stimuli constant across conditions while manipulating attention. Psychophysical studies have shown that we can differentially engage endogenous and exogenous attention by using different spatial cues. For endogenous attention, a cue—typically a line or a symbolic cue (number or color) at fixation—indicates the most likely location of the upcoming target. For exogenous attention, a brief peripheral cue is presented adjacent to the upcoming target or distractor.

Research in peripheral vision has historically neglected the role of spatial attention (Strasburger et al. 2011). When spatial attention is directed to a given location, performance improves in tasks mediated by spatial resolution—visual search, crowding, acuity, and texture segmentation (unless resolution is already too high for the task; see below). These psychophysical studies provide evidence for the "resolution hypothesis" (Yeshurun and Carrasco 1998), which states that attention can enhance spatial resolution, and that the magnitude of such an effect increases with eccentricity. Moreover, they illustrate that systematic psychophysics studies can reveal the underlying attention mechanisms. In this section, we discuss studies that have contributed to our understanding of the effects of covert attention on performance in spatial resolution tasks. Figure 3 summarizes the findings reviewed below.

Visual Search

To identify objects in a scene, the visual system first needs to isolate them. In search, we need to separate a target stimulus from distracting stimuli in the scene (e.g., when we are trying to find a friend in a crowd or a book on the bookshelf). The area over which the visual system integrates information limits performance in search tasks: The larger the integration area, the more likely it includes irrelevant information that interferes with target identification. In a search task, observers are typically required to detect the presence of a predefined target appearing among other irrelevant items; for instance, in a feature search task, a purple tilted line appearing among purple horizontal lines. In most real-life situations, however, we search for targets that differ from the distractors by a conjunction of different features—that is, the target shares a feature with a subset of distractors (e.g., its color) and another feature with another subset of distractors (e.g., its orientation), but the unique combination of the two features defines the target (e.g., a purple tilted line among purple horizontal and orange tilted lines; Treisman and Gelade 1980; Treisman 1985). Typically, for feature searches the number of distractors does not affect performance, but for conjunction searches performance decreases (longer reaction times and higher errors) with the number of distractors—the set-size effect. There are, however, feature searches yielding a set-size effect (Carrasco and Yeshurun 1998; Carrasco and McElree 2001; Carrasco et al. 2004, 2006a; Dosher et al. 2004) and conjunction searches that do not (Nakayama and Silverman 1986; McLeod et al. 1988; Duncan and Humphreys 1989; Carrasco et al. 1998b).

The set-size effect had traditionally been explained by a serial shift of attention to each item in a display (Treisman and Gelade 1980; Treisman 1985; Wolfe et al. 1989; Eckstein 2011). With more distractors, the target is more likely to appear at more eccentric positions. Consequently, the set-size effect can also be explained by the decrease of spatial resolution with eccentricity, as more eccentric RFs integrate over a larger area and include more distractors. Consistent with this spatial explanation, performance deteriorates with target eccentricity (Carrasco and Chang 1995; Carrasco et al. 1995, 1998a; Carrasco and Frieder 1997; Carrasco and Yeshurun 1998). Indeed, adjusting the stimulus size to its eccentricity according to the cortical magnification factor (to equate cortical surface activated by different stimuli) eliminates the eccentricity effect (Fig. 4A), indicating that resolution is a limiting factor in visual search (Carrasco and Frieder 1997; Carrasco et al. 1998a). Interestingly, the eccentricity effect is also practically eliminated for both features and conjunctions when, instead of magnifying the stimulus size, we manipulated exogenous attention (Fig. 4B) (Carrasco and Yes-

	Exogenous attention	**Endogenous attention**
Visual search		
attended	improves Carrasco and Yeshurun (1998); Morgan et al. (1998) Baldassi and Burr (2000); Carrasco and McElree (2001) Carrasco et al. (2004; 2006); Giordano et al. (2009)	improves Carrasco and McElree (2001) Giordano et al. (2009)
unattended	impairs Giordano et al. (2009)	impairs Giordano et al. (2009)
Crowding		
attended	improves Yeshurun and Rashal (2010); Grubb et al. (2013)	improves Montaser-Kouhsari and Rajimehr (2005)
unattended	impairs Yeshurun and Rashal (2010)	*impairs?*
Acuity		
attended	improves Shiu and Pashler (1995); Yeshurun and Carrasco (1999) Carrasco et al. (2002); Golla et al. (2004); Montagna et al. (2009)	improves Montagna et al. (2009)
unattended	impairs Montagna et al. (2009)	impairs Montagna et al. (2009)
Texture segmentation	central locations / periphery	central locations / periphery
attended	impairs / improves Yeshurun and Carrasco (1998; 2000; 2008); Talgar and Carrasco (2002) Carrasco et al. (2006b); Yeshurun et al. (2008) Barbot et al. (2013)	improves / improves Yeshurun et al. (2008) Barbot et al. (2012b)
unattended	improves / impairs Barbot et al. (2013)	*impairs?* / *impairs?*
Spatial resolution	**automatically increased** **by ↗ sensitivity of high-SF filters** Carrasco et al. (2006b)	**flexibly adjusted to task demands** **by ↗ or ↘ sensitivity og high-SF filters** Barbot et al. (2012b)

Figure 3. Summary of the effects of exogenous and endogenous attention on spatial resolution tasks described in this paper.

hurun 1998). These studies show that, similar to the effects of stimulus magnification, attention reduces the eccentricity effect by enhancing resolution, which in turn reduces the set-size effect.

Search studies assessing orientation thresholds also support the idea that attention enhances resolution. Orientation discrimination thresholds worsen with increasing number of distractors, but peripherally cueing the target location reduces orientation thresholds to the same level found when the target is alone (Morgan et al. 1998; Baldassi and Burr 2000). Together with these studies, findings that in discrimination, detection and localization tasks, both exogenous and endogenous attention not only reduce the set-size effect but also boost performance when the target is presented alone, support distractor exclusion and signal enhancement as mechanisms underlying attentional selection (Carrasco and McElree 2001; Cameron et al. 2004; Giordano et al. 2009).

Crowding

Another consequence of spatial resolution limits is crowding—a stimulus that is easily recognized when presented in isolation in peripheral vision can be unidentifiable when presented among nearby distractors. For example, you can identify a book on your shelf without looking at it directly, but it is a lot more difficult to recognize it when other books surround it. One of the dominant explanations for crowding posits that when we identify a target we pool information across several feature detectors over the "integration field" area (Pelli et al. 2004). Integration fields expand with eccentricity, increasing the probability that the integration process will include flankers into the target signal (Parkes et al. 2001). The critical distance in crowding increases with eccentricity, consistent with the increase of RF size. Accordingly, the crowding effect depends on a critical distance (center-to-center spacing between the target and the distractors) and its ratio with eccentricity (Levi 2008; Pelli and Tillman 2008). Exogenous attention affects the critical distance between the target and distractors: Valid cues decrease the critical distance whereas invalid cues increase it (Freeman and Pelli 2007; Yeshurun and Rashal 2010; Grubb et al. 2013). Similarly, endogenous attention decreases crowding (Montaser-Kouhsari and Rajimehr 2005). These results further support the idea that attention enhances spatial resolution.

Acuity

Investigating attention's role on the processing of stimuli used to test visual acuity further supports the hypothesis that attention enhances resolution. The retinal mosaic limits detection of a small gap in a Landolt square, whereas cortical processes limit offset discrimination with Vernier targets, which is hyperacute (Levi et al.

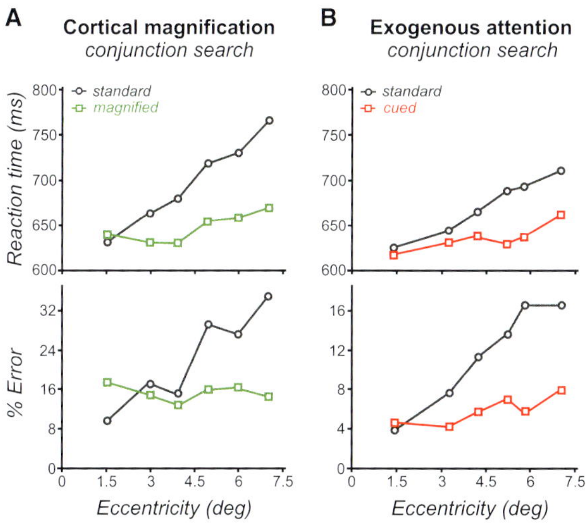

Figure 4. Effects of attention in visual search tasks. (*A*) Performance in conjunction visual search decreases with eccentricity, as depicted by slower reaction times and higher error rates. Adjusting stimulus size according to the cortical magnification factor at a given eccentricity eliminates the eccentricity effect in visual search, indicating that spatial resolution constrains performance. (*B*) Manipulating exogenous attention has a similar effect to cortical magnification, strongly reducing the eccentricity effect. These results support the idea that attention increases resolution at the attended location, restoring visual search performance in the periphery where low resolution is a limiting factor. (*A*, Adapted from Carrasco and Frieder 1997; *B*, adapted from Carrasco and Yeshurun 1998.)

1985; Olzak and Thomas 1986). Performance in acuity tasks decreases with eccentricity, consistent with decreased resolution in the periphery. Directing exogenous attention to the target location improves performance in both acuity (Yeshurun and Carrasco 1999; Carrasco et al. 2002; Golla et al. 2004; Montagna et al. 2009; Bocanegra and Zeelenberg 2011) and hyperacuity (Shiu and Pashler 1995; Yeshurun and Carrasco 1999) tasks, and the benefits intensify with eccentricity. These results were obtained even when the display contained no added sources of external noise (local masks, global masks, or distractors). Furthermore, effects of covert attention on Landolt acuity are very similar in humans and nonhuman primates (Fig. 5A) (Golla et al. 2004). Moreover, for both endogenous and exogenous attention, there is a resolution trade-off; increased spatial acuity at the attended location is coupled with decreased acuity at unattended locations, gap-size discrimination thresholds are lower in the cued and higher in the uncued condition, compared to the neutral baseline condition (Fig. 5B) (Montagna et al. 2009).

Given that attention cannot operate at the level of the retina, these findings imply that cortical or subcortical attentional processes can influence acuity by altering the spatial frequency tuning of the relevant spatial filters. Alternative hypotheses regarding attentional mechanisms, such as shifts in the decisional criterion and reduction of location uncertainty or reduction of external noise, could not account for these attention effects, because (1) the peripheral cue conveyed neither information regarding the target location; (2) observers could not rely on the cue for correct discrimination; (3) indistinguishable results emerge with and without a local postmask; and (4) one or two suprathreshold stimuli at fixed locations cannot be confused with the blanks at other locations. Thus, improved performance in acuity and hyperacuity tasks can only be accounted for by enhanced resolution.

Texture Segmentation

An attentional mechanism that increases spatial resolution by default can be useful because heightened resolution is advantageous in most everyday tasks, such as localizing and discriminating objects, searching for small objects, identifying fine details, and reading. In certain situations, however, resolution enhancement is not beneficial; for example, when navigating under poor atmospheric conditions (e.g., fog or haze) or when a global assessment of a scene is required (e.g., when seeing a whole tree rather than its individual leaves or viewing paintings using pointillism or photomosaic/pixellated [Fig. 6A] techniques).

The detrimental effect of heightened resolution enabled a crucial test of the resolution hypothesis: If attention enhanced resolution, performance at the attended location would be impaired rather than improved (Yeshurun and Carrasco 1998). The visual system performs a texture segmentation task whenever we process a scene, to assign foreground and background and to parse different objects. The area over which information is integrated, hence, also limits segmentation. We used a basic texture segmentation task in which the to-be-detected texture target is em-

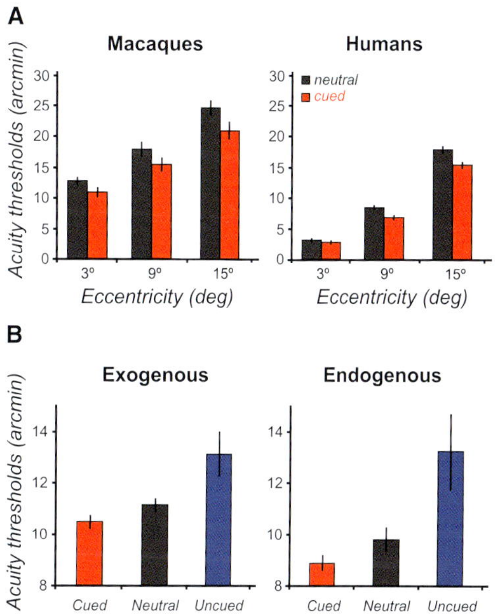

Figure 5. Effects of attention in acuity tasks. (*A*) Attention improves performance in acuity tasks in both humans and nonhuman primates (macaques), resulting in lower acuity thresholds with attention. (*B*) For human observers, both exogenous and endogenous attention trade-off acuity, increasing acuity at the attended location at the cost of decreased acuity at unattended locations. (*A*, Adapted from Golla et al. 2004; *B*, adapted from Montagna et al. 2009.)

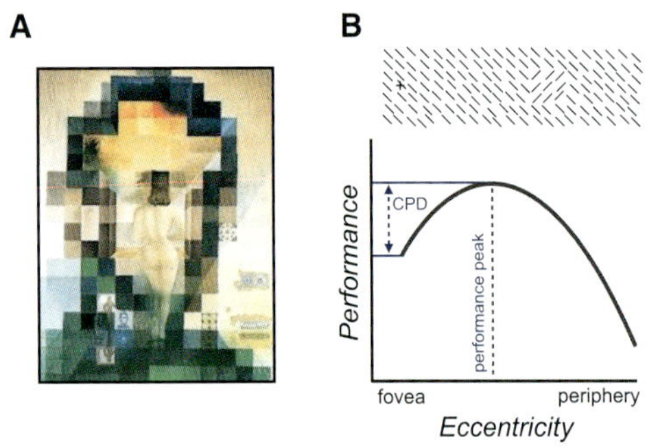

Figure 6. (*A*) *Lincoln in Dalivision* (1977) by Salvador Dalí. Depending on the dominance of high- or low-spatial frequency content of the image, the observer will perceive Gala's body or Lincoln's face (note small inserts on the *lower left*). (*B*) Texture segmentation task used in the studies described. Note that performance peaks at perifoveal locations and decreases at central locations (central performance drop; CPD), where resolution is too high for the task, as well as at peripheral locations, where resolution is too low.

bedded in a background of an orthogonal orientation presented along the horizontal meridian (Fig. 6B). In this task, observers' performance peaks at mid-peripheral locations, and drops when the target appears at more central or farther peripheral locations. This "central performance drop" (CPD) is attributed to the average size of spatial filters at the fovea being too small and spatial resolution being too high for the scale of the target patch. The filters' average size increases gradually with eccentricity and is optimal around the performance peak. At farther locations, the filters are too big and the resolution is too low for the task and performance drops. Accordingly, enlarging or decreasing the texture scale shifts the performance peak to farther or more central locations, respectively (Joffe and Scialfa 1995; Gurnsey et al. 1996; Kehrer 1997; Yeshurun and Carrasco 1998; Kehrer and Meinecke 2003; Yeshurun et al. 2008).

Exogenous attention improves performance at the periphery where the resolution is too low, but impairs performance near the fovea where the resolution is already too high for the task, in a smaller or larger range of eccentricities depending on the texture scale (Fig. 7A) (Yeshurun and Carrasco 1998). Thus, the texture scale and the filters' average size at a given eccentricity determine whether exogenous attention helps or hinders performance. Along the vertical meridian (VM), performance peaks at farther eccentricities in the lower than the upper VM (Fig. 8A), consistent with the higher resolution in the former than the latter (Rovamo et al. 1982; Talgar and Carrasco 2002; Montaser-Kouhsari and Carrasco 2009), but the attention benefit and cost are the same in both regions of the VM in relation to their crossover eccentricity (Fig. 8B) (Talgar and Carrasco 2002). These findings provide further evidence that the VM asymmetry is limited by visual rather than attentional factors (Carrasco et al. 2001) and that attention enhances spatial resolution. Note that the attentional impairment at central locations cannot be explained by shifts in the decisional criterion, reduction of location uncertainty, or reduction of external noise, which would predict a benefit throughout all eccentricities.

We investigated the adaptability of exogenous attention by examining whether the cue size modulates the attention effect on the resolution at the attended location. Employing endogenous cues of different sizes or dual tasks has revealed that the larger the attended region, the lower the resolution (Hock et al. 1998; Goto et al. 2001; Muller et al. 2003; Greenwood and Parasuraman 2004). We tested whether this would also be the case with exogenous attention, considered to be less flexible than endogenous attention. Were the gradual increase in cue size to result in a gradual resolution decrement, performance should gradually improve at central locations and deteriorate at peripheral locations, and the performance peak's eccentricity and the CPD should continuously decrease. We replicated the attention enhancement of resolution with small cues, but there was no significant effect for larger cues. Thus, we found no evidence that exogenous attention can flexibly lower resolution when it is attracted to a broader spatial region by large cues, it either increases resolution or has no effect (Yeshurun and Carrasco 2008).

The effect of exogenous attention on texture segmentation also reveals an automatic resolution trade-off. Cap-

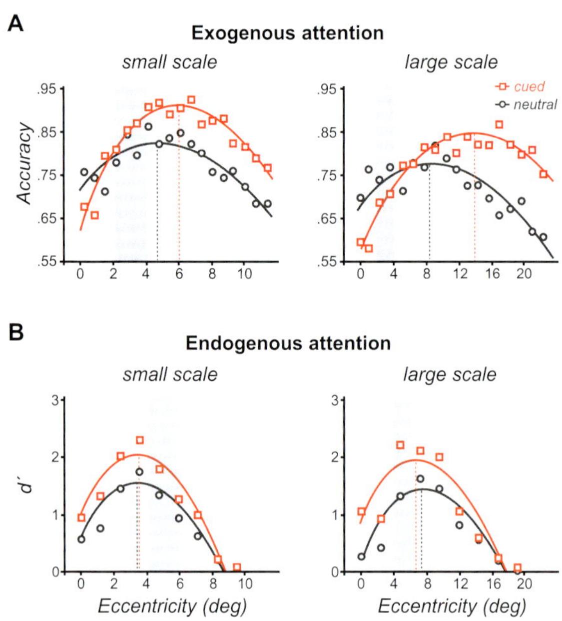

Figure 7. Effects of attention in texture segmentation tasks. (*A*) Exogenous attention automatically increases spatial resolution, improving texture segmentation performance in the periphery where the resolution is too low and impairing performance at central locations where the resolution is already too high, for the scale of the texture. The vertical dashed lines indicate the eccentricity of the performance peak. The blue overlay shows a range of eccentricity in which exogenous attention has opposite effects improving performance (*left*) or impairing performance (*right*), depending on the scale of the texture. (*B*) Endogenous attention benefits performance across eccentricities, regardless of whether performance is limited by the resolution being too low or too high. The vertical dashed lines indicate the eccentricity of the performance peak. (*A*, Adapted from Yeshurun and Carrasco 1998; *B*, adapted from Yeshurun et al. 2008.)

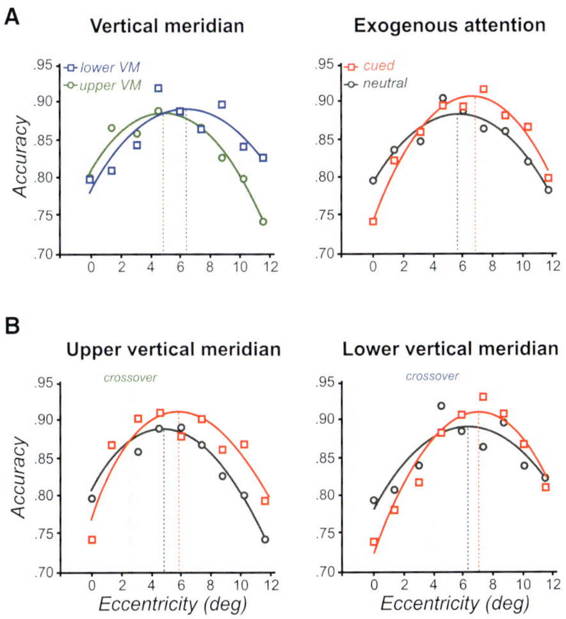

Figure 8. Changes in resolution with attention along the vertical meridian (VM). (*A*) Changes in texture segmentation along the vertical meridian and with exogenous attention. Spatial resolution is higher in the lower VM than the upper VM, resulting in higher performance in the periphery where resolution is too low and worse performance at central locations where resolution is too high for the scale of the texture. Similarly, by enhancing resolution, exogenous attention impairs and improves performance at central and peripheral locations, respectively. (*B*) Effects of exogenous attention on texture segmentation along the upper and lower VM. Exogenous attention impairs and improves performance across eccentricity consistent with enhanced resolution. Note that the attentional crossover (indicated by the colored area) occurs closer to the fovea along the upper (lower resolution) than the lower (higher resolution) VM, consistent with the idea that increasing resolution impairs or improves performance according to resolution constraints. (Adapted from Talgar and Carrasco 2002, with kind permission from Springer Science+Business Media.)

italizing on the fact that increased resolution can hinder performance in texture segmentation, we uncovered a seemingly counterintuitive situation in which decreasing resolution benefits performance at unattended locations. When the peripheral cues are invalid and divert attention away from the target location, performance decreases at peripheral locations but improves at central locations. The impairment at unattended peripheral locations where resolution is too low and the improvement at unattended central locations where resolution is too high can only be explained by decreased resolution at the unattended location (Barbot et al. 2013). All alternative hypotheses would predict a cost on performance at unattended locations throughout all eccentricities.

To assess the level of processing at which exogenous attention affects resolution, we used textures composed of narrow-band stimuli to ensure that first- or second-order filters of various specific scales would be differentially stimulated. Theoretically, texture segmentation involves passage of visual input through two layers of spatial linear filters, separated by a point-wise nonlinearity. The first-order linear filters perform a more local analysis of spatial frequency and orientation, and reflect the activity of simple cortical cells in area V1. The second-order linear filters are of a larger scale, perform a more global analysis on the output of the first-order filters plus the intermediate nonlinearity, and reflect the activity of cortical cells in striate and extrastriate areas (for reviews, see Landy and Graham 2004; Landy 2013). Accuracy is higher for cued than neutral trials at peripheral eccentricities, but is lower at central locations, for both first-order low- and high-frequency conditions. In contrast, for the second-order spatial frequency content attention impairs performance in a larger eccentricity range for the low- than the high-frequency condition, and an attention benefit emerges only for the high-frequency condition. These findings are consistent with the idea that attention operates by reducing the size of the second-order filters in texture segmentation tasks (Yeshurun and Carrasco 2000). Note, however, that a similar performance pattern with eccentricity has been observed using first-order, luminance-defined target patches, indicating that the CPD is not exclusive to the second-order stage of processing (Kehrer and Meinecke 2006; Meinecke and Kehrer 2007).

To test the hypothesis that exogenous attention enhances spatial resolution by increasing sensitivity to high-spatial frequencies, we used a selective adaptation procedure in conjunction with peripheral cueing. This procedure is used to assess the visual system's spatiotemporal properties (Blakemore and Campbell 1969; Movshon and Lennie 1979; Graham 1989). While keeping the stimulus content identical, the availability of spatial frequency information is manipulated by reducing observers' sensitivity to a range of frequencies. Selective adaptation to high-SF filters should reduce observers' sensitivity to high-SF, shift sensitivity toward lower frequencies, displace the performance peak toward the fovea, and reduce the CPD. Conversely, adapting to low-SF should reduce observers' sensitivity to low-SF, shift sensitivity toward higher frequencies, displace the performance peak toward the periphery, and exacerbate the CPD (Fig. 9A). Furthermore, adapting to high-spatial frequencies should eliminate the attentional impairment observed with exogenous attention at central locations (Fig. 9B). Experimental results show that the CPD is primarily due to the predominance of high-spatial frequencies at central locations, and that exogenous attention automatically enhances resolution by increasing sensitivity to higher spatial frequencies (Carrasco et al. 2006b). Consistent with this finding, a classification image study showed that with exogenous attention, the perceptual templates become sharper and are characterized by stronger high-spatial frequency components (Megna et al. 2012). In addition, both exogenous (Gobell and Carrasco 2005) and endogenous (Abrams et al. 2010) attention increased perceived spatial frequency at the attended area.

The texture segmentation studies above described show that exogenous attention increases resolution even when it is detrimental to the task. Given that endogenous attention is allocated more flexibly according to task demands, we compared its effects with those of exogenous

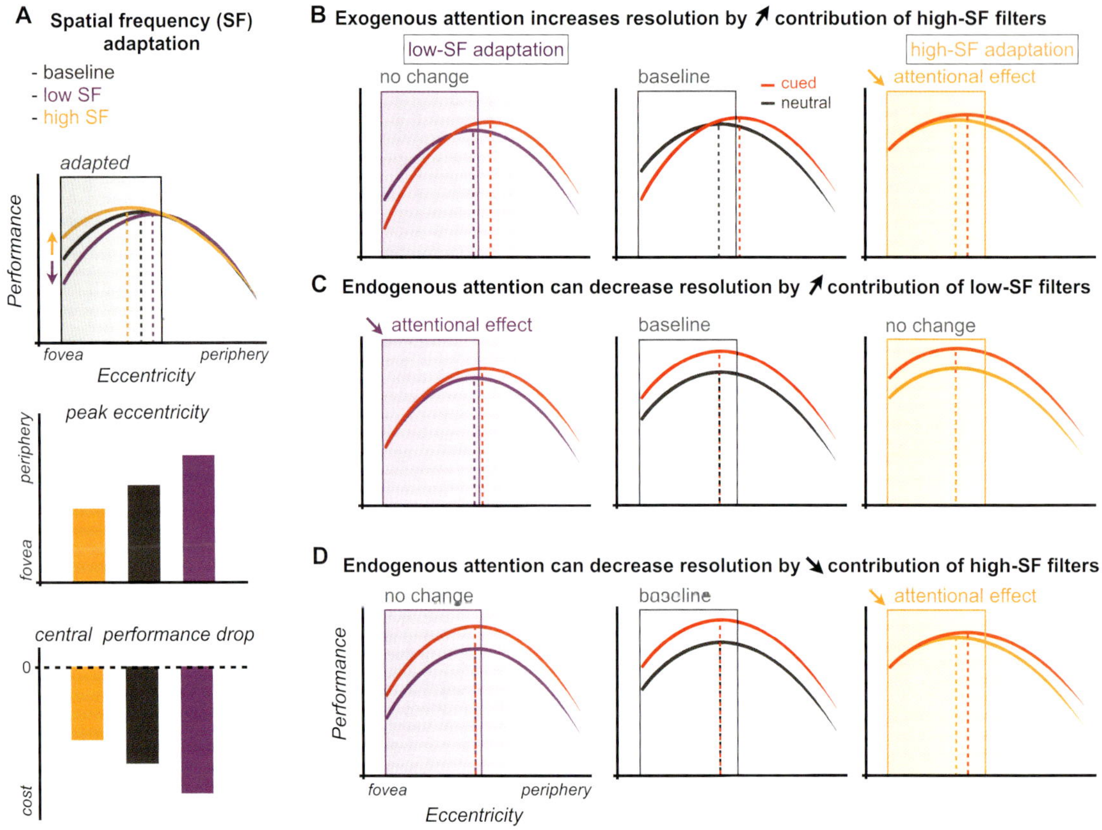

Figure 9. Schematic depiction of the effects of spatial frequency adaptation on texture segmentation performance and visual attention effects. (*A*) Effects of selective adaptation to spatial frequencies (SF) at central locations on texture segmentation. Adapting to high-SF shifts sensitivity toward lower SF, reducing the CPD and shifting the peak closer to the fovea, similar to a decrease in resolution. Conversely, adapting to low-SF shifts sensitivity toward higher-SF, increasing the CPD and shifting the peak toward the periphery similar to an increase in resolution. (*B*) Adapting to high SF, but not low SF, reduces the effects of exogenous attention at central locations, suggesting that exogenous attention enhances resolution by increasing the contribution of the high-SF filters. (*C,D*) Endogenous attention may benefit performance at central locations either by (*C*) enhancing activity in the low-SF filters or (*D*) suppressing activity in the high-SF filters. Adapting to high SF, but not to low SF, reduced the attentional benefits supporting a mechanism that can either boost or suppress the high SF to optimize resolution to the task demands.

attention. Conversely to exogenous attention, for two different texture scales, endogenous attention improves performance at all eccentricities (Fig. 7B) (Yeshurun et al. 2008). Furthermore, comparing the effects of precues and postcues revealed that the precue benefit goes well beyond the mere effect of location uncertainty at the decisional stage. The improvement at all eccentricities supports the idea that endogenous attention is more flexible than exogenous attention, and suggests that endogenous attention may increase resolution at peripheral locations but *decrease* resolution at central locations to match task demands (Yeshurun et al. 2008).

To characterize how endogenous attention affects resolution at central locations, we used a selective adaptation paradigm to modulate spatial frequency sensitivity at central locations (CPD). We hypothesized two attentional mechanisms that can decrease spatial resolution to benefit performance at central locations: by increasing the sensitivity of the low-spatial frequency filters or by decreasing the sensitivity of the high-spatial frequency filters. Accordingly, selectively adapting to either the low-spatial or high-spatial frequency filters at central locations should decrease their availability and suppress the attention effect (Fig. 9C,D). Experimental results show that adapting to high-spatial frequencies diminishes the benefit of endogenous attention at central locations, pointing to high-spatial frequency filters as responsible for the benefit of endogenous attention at central locations (Fig. 9D; Barbot et al. 2012b). Therefore, high-spatial frequency filters mediate both endogenous and exogenous attention effects on resolution, but whereas exogenous attention reflexively increases their contribution, endogenous can flexibly modulate their contribution to modify resolution according to task demands.

Linking Neurophysiology and Behavior

Physiologically, attention modulates neuronal responses and alters the profile and position of receptive fields near the attended location. Changes in RF size and position can qualitatively account for the aforementioned behavioral effects. Anton-Erxleben and Carrasco (2013) proposed a linking hypothesis based on two neural mechanisms. First, by concentrating processing resources at the attentional focus, RF shift toward the focus of attention (Connor et al. 1996, 1997; Ben Hamed et al. 2002; Womelsdorf et al. 2006; Anton-Erxleben et al. 2009) could improve performance in search, acuity, hyperacuity, and texture segmentation tasks. Second, RF shrinkage (Womelsdorf et al. 2006; Anton-Erxleben et al. 2009) could improve performance by reduction of filter size and integration area. The combination of RF shift and RF shrinkage leads to more and smaller RFs at the attended location, and thus to better resolution. The ability of smaller RFs to resolve finer details is correlated with a reduction of the area over which a single RF integrates information. Thus, attention could improve performance in search and crowding tasks, in which the integration area is critical to isolate a target from nearby distractors.

In texture segmentation tasks, the increase in spatial resolution can be explained by RF shrinkage, or by stronger weighting of small, high-spatial frequency-selective RFs (Yeshurun and Carrasco 2000; Carrasco et al. 2006b). The improved performance at central locations with endogenous attention seems to be mediated by decreased resolution (Figs. 7B and 9D), achieved by enlarging RFs or by decreased weighting of high-pass relative to low-pass filters (Yeshurun et al. 2008; Barbot et al. 2012b). RF expansion has been reported when attention is directed to a stimulus next to the RF (Anton-Erxleben et al. 2009) and in attentive tracking tasks (Niebergall et al. 2011). Additionally, attention's modulation of center-surround interactions (Anton-Erxleben et al. 2009; Sundberg et al. 2009; Schwartz and Coen-Cagli 2013) could result in finer or coarser perceptual analysis.

Attention also affects resolution at the population level in human fMRI studies. Directing attention to a particular location decreases the spatial overlap for adjacent locations in BOLD responses, indicating a narrowing of the population's integration area and increased resolution (Fischer and Whitney 2009). Conversely, withdrawing attention from the periphery results in larger pRFs and blurrier representations (de Haas et al. 2014), consistent with decreased resolution at unattended locations (Montagna et al. 2009; Barbot et al. 2013). Furthermore, attention attracts pRFs toward the focus of attention across the visual field and throughout the visual system (Klein et al. 2014). Inspired by psychophysical and neurophysiological findings, several computational models have implemented ways in which spatial attention can increase resolution (Lee et al. 1999; Deco and Zihl 2001; Cutzu and Tsotsos 2003; Compte and Wang 2006; Womelsdorf et al. 2008; Miconi and VanRullen 2011; Baruch and Yeshurun 2014). For example, some have proposed that the attentional effects on RFs can best be explained by a combination of attentional modulation of feed-forward connections with reciprocal modulatory feedback and local inhibition (Miconi and VanRullen 2011). But others have proposed a feed-forward model in which attention changes the gain of inputs to the RF, and a multiplicative interaction between the baseline RF and the attentional modulation results in a Gaussian profile that is narrower and shifted toward the attentional focus (Womelsdorf et al. 2008).

CONCLUSION

Covert attention enables us to better resolve fine details at attended parts of the visual scene, thus overcoming limitations in processing and partially restoring performance in the periphery. When attending to a particular location, observers' performance improves in a variety of tasks mediated by spatial resolution, such as search, crowding, acuity, and texture segmentation. Endogenous attention is a flexible mechanism that adjusts its operation on resolution to meet task demands and optimize performance. Conversely, exogenous attention is automatic and increases resolution even when detrimental for the task. Attention enhances the visual system's effective resolution by concentrating neuronal resources and reducing the area of spatial integration at the attended locations. Consistent with a selective representation of the world, there are processing trade-offs in resolution at the attended and unattended locations. As a selective process that optimizes relevant details, attention provides an organism with a heightened representation of the sensory input.

ACKNOWLEDGMENTS

This publication is supported by the U.S. National Institutes of Health (NIH) grant NIH-R01-EY019693 and NIH-R01-EY016200 (to M.C.)

REFERENCES

Abrams J, Barbot A, Carrasco M. 2010. Voluntary attention increases perceived spatial frequency. *Atten Percept Psychophys* **72:** 1510–1521.

Anton-Erxleben K, Carrasco M. 2013. Attentional enhancement of spatial resolution: Linking behavioural and neurophysiological evidence. *Nat Rev Neurosci* **14:** 188–200.

Anton-Erxleben K, Henrich C, Treue S. 2007. Attention changes perceived size of moving visual patterns. *J Vis* **7:** 5.1–5.9.

Anton-Erxleben K, Stephan VM, Treue S. 2009. Attention reshapes center-surround receptive field structure in macaque cortical area MT. *Cereb Cortex* **19:** 2466–2478.

Baldassi S, Burr DC. 2000. Feature-based integration of orientation signals in visual search. *Vision Res* **40:** 1293–1300.

Barbot A, Landy MS, Carrasco M. 2011. Exogenous attention enhances 2nd-order contrast sensitivity. *Vision Res* **51:** 1086–1098.

Barbot A, Landy MS, Carrasco M. 2012a. Differential effects of endogenous and exogenous attention on second-order texture contrast sensitivity. *J Vis* **12:** 1–15.

Barbot A, Montagna B, Carrasco M. 2012b. Endogenous attention optimizes performance by adjusting spatial resolution: Evidence from selective adaptation. *J Vis* **12:** art.387.

Barbot A, Bustamante L, Carrasco M. 2013. Diverting attention impairs or improves performance by decreasing spatial resolution. *Perception* **42:** 90.

Baruch O, Yeshurun Y. 2014. Attentional attraction of receptive fields can explain spatial and temporal effects of attention. *Vis Cogn* **22:** 704–736.

Beck DM, Kastner S. 2014. Neural systems for spatial attention in the human brain: Evidence from neuroimaging in the framework of biased competition. In *The Oxford handbook of attention* (ed. Nobre AC, Kastner S). Oxford University Press, New York.

Ben Hamed S, Duhamel JR, Bremmer F, Graf W. 2002. Visual receptive field modulation in the lateral intraparietal area during attentive fixation and free gaze. *Cereb Cortex* **12:** 234–245.

Blakemore C, Campbell FW. 1969. Adaptation to spatial stimuli. *J Physiol* **200:** 11P–13P.

Bocanegra BR, Zeelenberg R. 2011. Emotional cues enhance the attentional effects on spatial and temporal resolution. *Psychon Bull Rev* **18:** 1071–1076.

Broadbent D. 1958. *Perception and communication*. Pergamon Press, London.

Cameron EL, Tai JC, Eckstein MP, Carrasco M. 2004. Signal detection theory applied to three visual search tasks—Identification, yes/no detection and localization. *Spat Vis* **17:** 295–325.

Cannon MW Jr. 1985. Perceived contrast in the fovea and periphery. *J Opt Soc Am A* **2:** 1760–1768.

Carrasco M. 2011. Visual attention: The past 25 years. *Vis Res* **51:** 1484–1525.

Carrasco M. 2014. Spatial attention: Perceptual modulation. In *The Oxford handbook of attention* (ed. Nobre AC, Kastner S), pp. 183–230. Oxford University Press, New York.

Carrasco M, Chang I. 1995. The interaction of objective and subjective organizations in a localization search task. *Percept Psychophys* **57:** 1134–1150.

Carrasco M, Frieder KS. 1997. Cortical magnification neutralizes the eccentricity effect in visual search. *Vis Res* **37:** 63–82.

Carrasco M, McElree B. 2001. Covert attention accelerates the rate of visual information processing. *Proc Natl Acad Sci* **98:** 5363–5367.

Carrasco M, Yeshurun Y. 1998. The contribution of covert attention to the set-size and eccentricity effects in visual search. *J Exp Psychol Hum Percept Perform* **24:** 673–692.

Carrasco M, Yeshurun Y. 2009. Covert attention effects on spatial resolution. *Prog Brain Res* **176:** 65–86.

Carrasco M, Evert DL, Chang I, Katz SM. 1995. The eccentricity effect: Target eccentricity affects performance on conjunction searches. *Percept Psychophys* **57:** 1241–1261.

Carrasco M, McLean TL, Katz SM, Frieder KS. 1998a. Feature asymmetries in visual search: Effects of display duration, target eccentricity, orientation and spatial frequency. *Vis Res* **38:** 347–374.

Carrasco M, Ponte D, Rechea C, Sampedro MJ. 1998b. "Transient structures": The effects of practice and distractor grouping on within-dimension conjunction searches. *Percept Psychophys* **60:** 1243–1258.

Carrasco M, Talgar CP, Cameron EL. 2001. Characterizing visual performance fields: Effects of transient covert attention, spatial frequency, eccentricity, task and set size. *Spat Vis* **15:** 61–75.

Carrasco M, Williams PE, Yeshurun Y. 2002. Covert attention increases spatial resolution with or without masks: Support for signal enhancement. *J Vis* **2:** 467–479.

Carrasco M, Giordano AM, McElree B. 2004. Temporal performance fields: Visual and attentional factors. *Vis Res* **44:** 1351–1365.

Carrasco M, Giordano AM, McElree B. 2006a. Attention speeds processing across eccentricity: Feature and conjunction searches. *Vis Res* **46:** 2028–2040.

Carrasco M, Loula F, Ho YX. 2006b. How attention enhances spatial resolution: Evidence from selective adaptation to spatial frequency. *Percept Psychophys* **68:** 1004–1012.

Compte A, Wang XJ. 2006. Tuning curve shift by attention modulation in cortical neurons: A computational study of its mechanisms. *Cereb Cortex* **16:** 761–778.

Connor CE, Gallant JL, Preddie DC, Van Essen DC. 1996. Responses in area V4 depend on the spatial relationship between stimulus and attention. *J Neurophysiol* **75:** 1306–1308.

Connor CE, Preddie DC, Gallant JL, Van Essen DC. 1997. Spatial attention effects in macaque area V4. *J Neurosci* **17:** 3201–3214.

Cutzu F, Tsotsos JK. 2003. The selective tuning model of attention: Psychophysical evidence for a suppressive annulus around an attended item. *Vis Res* **43:** 205–219.

Deco G, Zihl J. 2001. A neurodynamical model of visual attention: Feedback enhancement of spatial resolution in a hierarchical system. *J Comput Neurosci* **10:** 231–253.

de Haas B, Schwarzkopf DS, Anderson EJ, Rees G. 2014. Perceptual load affects spatial tuning of neuronal populations in human early visual cortex. *Curr Biol* **24:** R66–R67.

Desimone R, Duncan J. 1995. Neural mechanisms of selective visual attention. *Annu Rev Neurosci* **18:** 193–222.

DeValois RL, DeValois KK. 1988. *Spatial vision*. Oxford University Press, New York.

Dosher BA, Han S, Lu ZL. 2004. Parallel processing in visual search asymmetry. *J Exp Psychol Hum Percept Perform* **30:** 3–27.

Dumoulin SO, Wandell BA. 2008. Population receptive field estimates in human visual cortex. *Neuroimage* **39:** 647–660.

Duncan RO, Boynton GM. 2003. Cortical magnification within human primary visual cortex correlates with acuity thresholds. *Neuron* **38:** 659–671.

Duncan J, Humphreys GW. 1989. Visual search and stimulus similarity. *Psychol Rev* **96:** 433–458.

Eckstein MP. 2011. Visual search: A retrospective. *J Vis* **11**.

Fischer J, Whitney D. 2009. Attention narrows position tuning of population responses in V1. *Curr Biol* **19:** 1356–1361.

Freeman J, Pelli DG. 2007. An escape from crowding. *J Vis* **7:** 22.1–22.14.

Freeman J, Simoncelli EP. 2011. Metamers of the ventral stream. *Nat Neurosci* **14:** 1195–1201.

Fuller S, Rodriguez RZ, Carrasco M. 2008. Apparent contrast differs across the vertical meridian: Visual and attentional factors. *J Vis* **8:** 16.11–16.16.

Giordano AM, McElree B, Carrasco M. 2009. On the automaticity and flexibility of covert attention: A speed-accuracy trade-off analysis. *J Vis* **9:** 30.1–30.10.

Gobell J, Carrasco M. 2005. Attention alters the appearance of spatial frequency and gap size. *Psychol Sci* **16:** 644–651.

Golla H, Ignashchenkova A, Haarmeier T, Thier P. 2004. Improvement of visual acuity by spatial cueing: A comparative study in human and non-human primates. *Vis Res* **44:** 1589–1600.

Goto M, Toriu T, Tanahashi JI. 2001. Effect of size of attended area on contrast sensitivity function. *Vis Res* **41:** 1483–1487.

Graham N. 1989. *Visual pattern analyzers*. Oxford University Press, New York.

Greenwood PM, Parasuraman R. 2004. The scaling of spatial attention in visual search and its modification in healthy aging. *Percept Psychophys* **66:** 3–22.

Grubb MA, Behrmann M, Egan R, Minshew NJ, Heeger DJ, Carrasco M. 2013. Exogenous spatial attention: Evidence for intact functioning in adults with autism spectrum disorder. *J Vis* **13:** 9.

Gurnsey R, Pearson P, Day D. 1996. Texture segmentation along the horizontal meridian: Nonmonotonic changes in performance with eccentricity. *J Exp Psychol Hum Percept Perform* **22:** 738–757.

Harvey BM, Dumoulin SO. 2011. The relationship between cortical magnification factor and population receptive field size in human visual cortex: Constancies in cortical architecture. *J Neurosci* **31:** 13604–13612.

Hein E, Rolke B, Ulrich R. 2006. Visual attention and temporal discrimination: Differential effects of automatic and voluntary cueing. *Vis Cogn* **13:** 29–50.

Herrmann K, Montaser-Kouhsari L, Carrasco M, Heeger DJ. 2010. When size matters: Attention affects performance by contrast or response gain. *Nat Neurosci* **13:** 1554–1559.

Hock HS, Balz GW, Smollon W. 1998. Attentional control of spatial scale: Effects on self-organized motion patterns. *Vis Res* **38:** 3743–3758.

Joffe KM, Scialfa CT. 1995. Texture segmentation as a function of eccentricity, spatial frequency and target size. *Spat Vis* **9:** 325–342.

Kehrer L. 1997. The central performance drop in texture segmentation: A simulation based on a spatial filter model. *Biol Cybern* **77:** 297–305.

Kehrer L, Meinecke C. 2003. A space-variant filter model of texture segregation: Parameter adjustment guided by psychophysical data. *Biol Cybern* **88:** 183–200.

Kehrer L, Meinecke C. 2006. A 'first stage' central performance drop in a Gabor luminance-modulation detection task. *Spat Vis* **19:** 427–437.
Kinchla RA. 1980. The measurement of attention. in *Attention and performance VIII* (ed. Nickerson R), pp. 213–238. Psychology Press, Princeton, NJ.
Kinchla RA. 1992. Attention. *Annu Rev Psychol* **43:** 711–742.
Kitterle FL. 1986. Psychophysics of lateral tachistoscopic presentation. *Brain Cogn* **5:** 131–162.
Klein BP, Harvey BM, Dumoulin SO. 2014. Attraction of position preference by spatial attention throughout human visual cortex. *Neuron* **84:** 227–237.
Koenig-Robert R, VanRullen R. 2011. Spatiotemporal mapping of visual attention. *J Vis* **11:** 1–16.
Landy MS. 2013. Texture analysis and perception. In *The new visual neurosciences* (ed. Werner JS, Chalupa LM), pp. 639–652. MIT Press, Cambridge, MA.
Landy MS, Graham N. 2004. Visual perception of texture. (ed. Chalupa LM, Werner JS), pp. 1106–1118. MIT Press, Cambridge, MA.
Lee DK, Itti L, Koch C, Braun J. 1999. Attention activates winner-take-all competition among visual filters. *Nat Neurosci* **2:** 375–381.
Lennie P. 2003. The cost of cortical computation. *Curr Biol* **13:** 493–497.
Levi DM. 2008. Crowding—An essential bottleneck for object recognition: A mini-review. *Vis Res* **48:** 635–654.
Levi DM, Klein SA, Aitsebaomo AP. 1985. Vernier acuity, crowding and cortical magnification. *Vis Res* **25:** 963–977.
Ling S, Carrasco M. 2006a. Sustained and transient covert attention enhance the signal via different contrast response functions. *Vis Res* **46:** 1210–1220.
Ling S, Carrasco M. 2006b. When sustained attention impairs perception. *Nat Neurosci* **9:** 1243–1245.
Liu T, Stevens ST, Carrasco M. 2007. Comparing the time course and efficacy of spatial and feature-based attention. *Vis Res* **47:** 108–113.
Lu ZL, Dosher BA. 1998. External noise distinguishes attention mechanisms. *Vis Res* **38:** 1183–1198.
Luck SJ, Hillyard SA, Mouloua M, Woldorff MG, Clark VP, Hawkins HL. 1994. Effects of spatial cuing on luminance detectability: Psychophysical and electrophysiological evidence for early selection. *J Exp Psychol Hum Percept Perform* **20:** 887–904.
Mangun GR, Hillyard SA. 1990. Allocation of visual attention to spatial locations: Tradeoff functions for event-related brain potentials and detection performance. *Percept Psychophys* **47:** 532–550.
Martin G. 1986. Psychophysics. Limits of visual resolution. *Nature* **319:** 540.
McLeod P, Driver J, Crisp J. 1988. Visual search for a conjunction of movement and form is parallel. *Nature* **332:** 154–155.
Megna N, Rocchi F, Baldassi S. 2012. Spatio-temporal templates of transient attention revealed by classification images. *Vis Res* **54:** 39–48.
Meinecke C, Kehrer L. 2007. A central performance drop with luminance stimuli requiring spatial integration. *Percept Psychophys* **69:** 923–929.
Miconi T, VanRullen R. 2011. A feedback model of attentional effects in the visual cortex. In *Proceedings of the 2011 IEEE CIMSIVP*, pp. 106–113. IEEE, Piscataway, NJ.
Montagna B, Pestilli F, Carrasco M. 2009. Attention trades off spatial acuity. *Vis Res* **49:** 735–745.
Montaser-Kouhsari L, Carrasco M. 2009. Perceptual asymmetries are preserved in short-term memory tasks. *Atten Percept Psychophys* **71:** 1782–1792.
Montaser-Kouhsari L, Rajimehr R. 2005. Subliminal attentional modulation in crowding condition. *Vis Res* **45:** 839–844.
Morgan MJ, Ward RM, Castet E. 1998. Visual search for a tilted target: Tests of spatial uncertainty models. *Q J Exp Psychol A* **51:** 347–370.
Movshon JA, Lennie P. 1979. Pattern-selective adaptation in visual cortical neurones. *Nature* **278:** 850–852.
Müller HJ, Rabbitt PM. 1989. Reflexive and voluntary orienting of visual attention: Time course of activation and resistance to interruption. *J Exp Psychol Hum Percept Perform* **15:** 315–330.
Muller NG, Donner TH, Bartelt OA, Brandt SA, Villringer A, Kleinschmidt A. 2003. The functional neuroanatomy of visual conjunction search: A parametric fMRI study. *Neuroimage* **20:** 1578–1590.
Nakayama K, Mackeben M. 1989. Sustained and transient components of focal visual attention. *Vis Res* **29:** 1631–1647.
Nakayama K, Silverman GH. 1986. Serial and parallel processing of visual feature conjunctions. *Nature* **320:** 264–265.
Neisser U. 1967. *Cognitive psychology*. Appleton-Century-Crofts, New-York.
Niebergall R, Khayat PS, Treue S, Martinez-Trujillo JC. 2011. Expansion of MT neurons excitatory receptive fields during covert attentive tracking. *J Neurosci* **31:** 15499–15510.
Olzak LA, Thomas JP. 1986. *Seeing spatial patterns*. Wiley-Interscience, New York.
Parkes L, Lund J, Angelucci A, Solomon JA, Morgan M. 2001. Compulsory averaging of crowded orientation signals in human vision. *Nat Neurosci* **4:** 739–744.
Pelli DG, Tillman KA. 2008. The uncrowded window of object recognition. *Nat Neurosci* **11:** 1129–1135.
Pelli DG, Palomares M, Majaj NJ. 2004. Crowding is unlike ordinary masking: Distinguishing feature integration from detection. *J Vis* **4:** 1136–1169.
Pestilli F, Carrasco M. 2005. Attention enhances contrast sensitivity at cued and impairs it at uncued locations. *Vis Res* **45:** 1867–1875.
Pestilli F, Viera G, Carrasco M. 2007. How do attention and adaptation affect contrast sensitivity? *J Vis* **7:** 9.1–9.12.
Posner MI. 2014. Orienting of attention: Then and now. *Q J Exp Psychol*: 1–12.
Remington RW, Johnston JC, Yantis S. 1992. Involuntary attentional capture by abrupt onsets. *Percept Psychophys* **51:** 279–290.
Reynolds JH, Chelazzi L. 2004. Attentional modulation of visual processing. *Annu Rev Neurosci* **27:** 611–647.
Rijsdijk JP, Kroon JN, van der Wildt GJ. 1980. Contrast sensitivity as a function of position on the retina. *Vis Res* **20:** 235–241.
Rovamo J, Virsu V. 1979. An estimation and application of the human cortical magnification factor. *Exp Brain Res* **37:** 495–510.
Rovamo J, Virsu V, Nasanen R. 1978. Cortical magnification factor predicts the photopic contrast sensitivity of peripheral vision. *Nature* **271:** 54–56.
Rovamo J, Virsu V, Laurinen P, Hyvarinen L. 1982. Resolution of gratings oriented along and across meridians in peripheral vision. *Invest Ophthalmol Vis Sci* **23:** 666–670.
Schwartz O, Coen-Cagli R. 2013. Visual attention and flexible normalization pools. *J Vis* **13:** 1–24.
Shiu LP, Pashler H. 1995. Spatial attention and vernier acuity. *Vis Res* **35:** 337–343.
Sperling G, Melchner MJ. 1978. The attention operating characteristic: Examples from visual search. *Science* **202:** 315–318.
Strasburger H, Rentschler I, Juttner M. 2011. Peripheral vision and pattern recognition: A review. *J Vis* **11:** 13.
Sundberg KA, Mitchell JF, Reynolds JH. 2009. Spatial attention modulates center-surround interactions in macaque visual area v4. *Neuron* **61:** 952–963.
Talgar CP, Carrasco M. 2002. Vertical meridian asymmetry in spatial resolution: Visual and attentional factors. *Psychon Bull Rev* **9:** 714–722.
Treisman AM. 1985. Preattentive processing in vision. *Comput Vis Graph Image Process* **31:** 157–177.

Treisman AM, Gelade G. 1980. A feature-integration theory of attention. *Cogn Psychol* **12:** 97–136.
Virsu V, Rovamo J. 1979. Visual resolution, contrast sensitivity, and the cortical magnification factor. *Exp Brain Res* **37:** 475–494.
Wandell BA, Winawer J. 2015. Computational neuroimaging and population receptive fields. *Trends Cogn Sci* **19:** in press.
Wolfe JM, Cave KR, Franzel SL. 1989. Guided search: An alternative to the feature integration model for visual search. *J Exp Psychol Hum Percept Perform* **15:** 419–433.
Womelsdorf T, Anton-Erxleben K, Pieper F, Treue S. 2006. Dynamic shifts of visual receptive fields in cortical area MT by spatial attention. *Nat Neurosci* **9:** 1156–1160.
Womelsdorf T, Anton-Erxleben K, Treue S. 2008. Receptive field shift and shrinkage in macaque middle temporal area through attentional gain modulation. *J Neurosci* **28:** 8934–8944.
Wright MJ, Johnston A. 1983. Spatiotemporal contrast sensitivity and visual field locus. *Vis Res* **23:** 983–989.
Yantis S, Jonides J. 1996. Attentional capture by abrupt onsets: New perceptual objects or visual masking? *J Exp Psychol Hum Percept Perform* **22:** 1505–1513.
Yeshurun Y. 2004. Isoluminant stimuli and red background attenuate the effects of transient spatial attention on temporal resolution. *Vis Res* **44:** 1375–1387.
Yeshurun Y, Carrasco M. 1998. Attention improves or impairs visual performance by enhancing spatial resolution. *Nature* **396:** 72–75.
Yeshurun Y, Carrasco M. 1999. Spatial attention improves performance in spatial resolution tasks. *Vis Res* **39:** 293–306.
Yeshurun Y, Carrasco M. 2000. The locus of attentional effects in texture segmentation. *Nat Neurosci* **3:** 622–627.
Yeshurun Y, Carrasco M. 2008. The effects of transient attention on spatial resolution and the size of the attentional cue. *Percept Psychophys* **70:** 104–113.
Yeshurun Y, Levy L. 2003. Transient spatial attention degrades temporal resolution. *Psychol Sci* **14:** 225–231.
Yeshurun Y, Rashal E. 2010. Precueing attention to the target location diminishes crowding and reduces the critical distance. *J Vis* **10:** 16.1–16.12.
Yeshurun Y, Montagna B, Carrasco M. 2008. On the flexibility of sustained attention and its effects on a texture segmentation task. *Vis Res* **48:** 80–95.
Zhaoping L. 2014. *Understanding vision: Theory, models, and data*. Oxford University Press, New York.

Causal Model Comparison Shows That Human Representation Learning Is Not Bayesian

ANDRA GEANA[1] AND YAEL NIV[1,2]
[1]*Psychology Department, Princeton University, Princeton, New Jersey 08540*
[2]*Princeton Neuroscience Institute, Princeton University, Princeton, New Jersey 08540*
Correspondence: ageana@princeton.edu; yael@princeton.edu

How do we learn what features of our multidimensional environment are relevant in a given task? To study the computational process underlying this type of "representation learning," we propose a novel method of *causal model comparison*. Participants played a probabilistic learning task that required them to identify one relevant feature among several irrelevant ones. To compare between two models of this learning process, we ran each model alongside the participant during task performance, making predictions regarding the values underlying the participant's choices in real time. To test the validity of each model's predictions, we used the predicted values to try to perturb the participant's learning process: We crafted stimuli to either facilitate or hinder comparison between the most highly valued features. A model whose predictions coincide with the learned values in the participant's mind is expected to be effective in perturbing learning in this way, whereas a model whose predictions stray from the true learning process should not. Indeed, we show that in our task a reinforcement-learning model could help or hurt participants' learning, whereas a Bayesian ideal observer model could not. Beyond informing us about the notably suboptimal (but computationally more tractable) substrates of human representation learning, our manipulation suggests a sensitive method for model comparison, which allows us to change the course of people's learning in real time.

We live in a rich, complex environment, in which we are constantly bombarded with a wide variety of sensory input. Even an action as simple as walking down the street carries with it a large volume of low-quality information in the form of people we see, places we walk by, cars, colors, voices, noises, emotional content, etc. Intuitively, one would imagine that given sufficient resources, it is best to always represent every aspect of the environment so that any detail can potentially be acted upon. However, the "curse of dimensionality" (Bellman 1957) posits that task representations that involve unnecessary stimulus dimensions will not afford efficient learning and decision-making, where efficiency is measured in the number of examples needed to learn the task. In particular, an increase in the number of dimensions of the problem (in our case, the dimensions of the environment that the brain may represent) implies that the learner needs to collect exponentially larger quantities of data to learn to solve the problem. If we want learning to be feasible it is therefore both computationally optimal and a practical imperative to represent tasks with as compact a representation as possible.

What are the computational strategies that humans use to learn a representation for a given task? To address this question, we tested participants on a multidimensional trial-and-error choice task, in which only one dimension was relevant to predicting reward (Wilson and Niv 2012; Niv et al. 2015). To test the explanatory power of different models of learning dynamics, we developed a method that compares two models against each other in terms of their causal effects on behavior. Specifically, we used each model to manipulate participants' learning in real time, and asked which model was more effective in changing behavior. This is at the same time a novel, intuitive measure of how well a model captures participants' strategies, and it constitutes evidence that it is possible to use model predictions to impact learning in real time, by manipulating the stimuli that are presented to the participant.

It goes without saying that trial-and-error learning depends on what information is available to the learner. Indeed, work in machine learning and information theory has established how information in any given task might be optimally selected so as to maximally discriminate between competing hypotheses and accelerate learning (optimal experimental design; Sebastiani and Wynn 2000). Although human learning does not always mirror these optimal strategies, judicious choice of information has been shown to improve learning, for instance of category boundaries (Gureckis and Markant 2012) or speech motor learning (Knock et al. 2000). Moreover, the order in which information is presented is relevant to determining what is learned (Ritter 2007). We thus set forth to use our candidate models to manipulate the timing and availability of information in such a way as to aid or hinder participants' learning trajectory.

This kind of effort to manipulate learning, however, is heavily dependent on having a good model of how participants structure and update their representations of the environment. How to compare and select a "best"

 doi: 10.1101/sqb.2014.79.024851

model for a complex cognitive process is not trivial (Cutting et al. 1992; Pitt et al. 2002): Models that fit the data better on some common goodness-of-fit measures may not fit better on other such measures; models that posit very different processes may perform similarly in terms of average fit (Townsend 1990; Rust et al. 1995); and a model that seems to describe behavior better might do so because of a more flexible function form or different numbers of parameters, and not necessarily because it better captures the underlying cognitive processes (Busemeyer and Wang 2000). We therefore developed an *interventional* method for model comparison.

We used our candidate models to predict in real time what hypotheses a participant might be testing and to design stimuli that will make it easier or harder to distinguish between the competing hypotheses. We reasoned that a model that does not capture the participants' beliefs about the available stimuli would not be effective at such a manipulation of learning. In contrast, a model whose predictions are well-matched to the cognitive processes underlying participants' behavior should allow us to manipulate learning in real time. In particular, while participants played our multidimensional probabilistic learning task, we inferred their value representations in real-time using either a Bayesian or a reinforcement-learning model. We used those inferred representations to present participants with information that would either help discriminate their competing hypotheses about which dimension of the stimuli is relevant for reward, or specifically hurt such a discrimination.

Despite much work suggesting that the human brain is Bayes-optimal (Körding and Wolpert 2004; Beierholm et al. 2009), and in line with our previous findings (Niv et al. 2015), our manipulation was only effective when we based it on predictions of the reinforcement-learning model. Our ability to manipulate the learning process both precisely and in real time consists of a proof of concept for the new proposed model-testing tool, and is a step in the right direction in terms of development of individualized tools to improve learning in general.

METHODS

Participants

Twenty-five participants (16 females) recruited from the Princeton University undergraduate community gave informed consent and were compensated $12 an hour plus a performance bonus of up to $5 depending on their final score in the task. The average pay was $15. Study materials and procedures were approved by the Princeton University Institutional Review Board.

Task

Participants played a probabilistic learning task. Each trial involved choosing one of three compound stimuli displayed on the screen (see Fig. 1A). Each stimulus was comprised of three features defined on three dimensions: a color (red, yellow, or green), a shape (triangle, square, or circle), and a texture (dots, plaid, or waves). No two stimuli could share the same feature (i.e., there was only one red stimulus, one triangle, etc., per trial). Choosing a stimulus resulted in immediate feedback in the form of either one or zero points. Participants were instructed to try to obtain as many points as possible.

The task was designed so that, of the three dimensions of a stimulus, only one dimension was relevant to determining reward. Within that dimension, one feature was the target feature–choosing the stimulus that contained this feature led to one point 75% of the time and zero points 25% of the time. Choosing any other stimulus resulted in one point only 25% of the time. To maximize their score, participants therefore had to aggregate over previous choices and outcomes to learn which feature is the target feature. Participants were explicitly instructed about these aspects of the task structure in advance.

The task consisted of 52 "games." Participants were informed that the target feature would not change within a game, but would change between games. Ends of games were explicitly signaled on-screen. The first 12 games (referred to as the baseline phase, described below) included 30 trials each, whereas the remaining games (the manipulation phase) consisted of 36 trials each, for a total of 1800 trials. After each game, participants were asked how difficult they found the game (on a scale from very easy [1] to very difficult [9]; Fig. 1B) and to identify the target feature in that game. They could select any of the nine features, as well as an "I do not know" option. If they did select a feature, they were also asked to rate how confident they were about their choice.

After the baseline phase, participants took a 1-min break, during which we used their baseline phase data to fit the free parameters of the two candidate models we would later test in the manipulation phase. The remaining 40 games of the task comprised of the manipulation phase, in which we manipulated stimuli according to predictions from each model, to either help or hurt participants' learning (see below).

Modeling

The two models we compared represent two different ways of thinking about human representation learning. The first is a Bayesian model that assumes statistically optimal updating of the posterior probabilities of each feature being the target feature, and the second model uses reinforcement-learning principles to update values via trial and error. Both models compute the value of a compound stimulus by estimating values of individual features and combining them: The Bayesian model estimates, for each feature, the posterior probability that it is the target feature, whereas the reinforcement-learning (RL) model learns the values of each feature based on prediction errors. In both models, current values of stimuli depend on the history of choices and rewards.

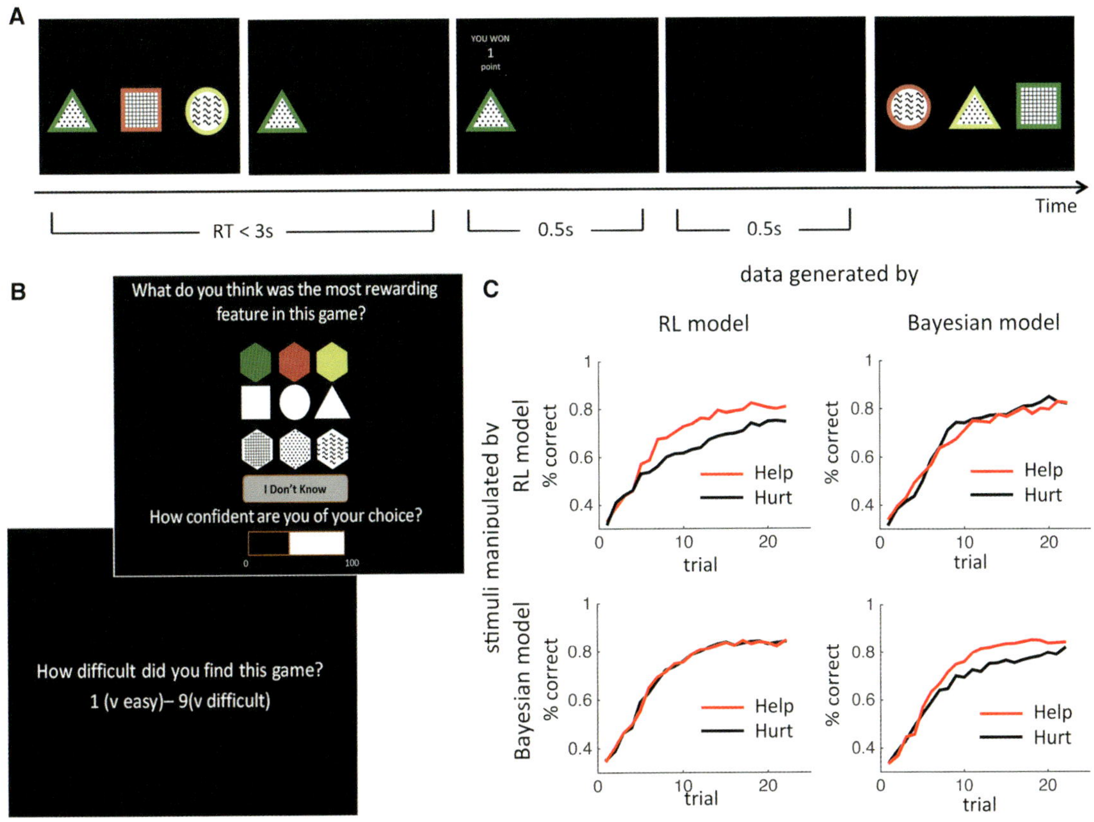

Figure 1. The dimensions task. (*A*) Example trial—stimulus presentation, choice, positive feedback, new stimulus presentation. (*B*) Query screens—difficulty ratings, identifying correct feature, confidence ratings. (*C*) Simulations of model-based manipulation. The manipulation was effective ("Help" improves performance as compared with "Hurt") only when stimuli were manipulated according to predictions from the same model that generated the choice data (*top left*, *bottom right* panels). The task was designed so that, of the three dimensions of a stimulus, only one dimension was relevant to determining reward. Within that dimension, one feature was the target feature—choosing the stimulus that contained this feature led to one point 75% of the time and zero points 25% of the time. Choosing any other stimulus resulted in one point only 25% of the time. To maximize their score, participants therefore had to aggregate over previous choices and outcomes to learn which feature is the target feature. Participants were explicitly instructed about these aspects of the task structure in advance.

Bayesian model. The Bayesian model tracks the posterior probability that each feature f is the target feature f^*. At the end of each trial, the posterior is updated by combining the prior (i.e., the posterior from the previous trial) and the likelihood of the observed data if f were the target feature. The prior depends on the history of choices C and rewards R, $D_{1,t-1} = \{C_{1:t-1}; R_{1:t-1}\}$, from the beginning of the game and up until the current trial (not inclusive). The likelihood depends on the reward probabilities imposed by the experimenter; for instance, the likelihood of a win if the chosen stimulus contains the target feature is 0.75.

At the beginning of the game, the prior is initialized at 1/9 (all features are equally likely to be the target feature). After each trial, the posterior is updated according to

$$P(f = f^*|D_{1:t}) \propto P(R_t|f = f^*, C_t)P(f = f^*|D_{1:t-1}). \tag{1}$$

The value of a stimulus S is then calculated as the probability of obtaining a 1 point reward for choosing that stimulus on the current trial t,

$$\begin{aligned} V(S) &= P(R = 1|S, D_{1,t-1}) \\ &= \sum\nolimits_{f \in S} [P(R = 1|f = f^*)P(f = f^*|D_{1,t-1})] \\ &\quad + P(R = 1|f^* \notin S) \\ &\quad \times \left(1 - \sum\nolimits_{f \in S} P(f = f^*|D_{1,t-1})\right), \end{aligned} \tag{2}$$

where $P(R = 1|f = f^*) = 0.75$ for all features contained in S, and $P(R = 1|f^* \notin S) = 0.25$. The model can be considered an ideal observer because it maintains a full probability distribution over the identity of f^* and updates this distribution in a statistically optimal way.

To afford this model, the same temporal locality as the reinforcement model (described below), we also allowed some degree of decay for all feature posteriors toward a uniform value of 1/9,

$$\tilde{P}(f = f^*) = (1 - d)P(f = f^*) + d*1/9, \quad (3)$$

and used $\tilde{P}$ instead of P in Equations 1 and 2 above. Although suboptimal, the decay component has been shown to significantly improve the models' fit to behavioral choices in our task (Niv et al. 2015).

Finally, we assumed that the probability of choosing stimulus S_i on each trial is proportional to the value of the stimulus, according to the softmax probability choice function:

$$P(\text{choose } S_i) = \frac{e^{\beta V(S_i)}}{\sum_{j=1}^{3} e^{\beta V(S_j)}}. \quad (4)$$

The positive-valued inverse-temperature parameter β sets the level of noise in the decision process, with large β resulting in near-deterministic choice of the highest value option, whereas small β results in high decision noise and more random decisions. In all, this model has two free parameters, β and the decay rate d.

Reinforcement-learning model. This model takes advantage of the fact that, in our task, features determine reward independently, and uses reinforcement learning to learn values for each of the nine features. The values of all features were initialized at zero at the beginning of each game; on each trial, values of the three features of the stimulus that was chosen were then updated according to

$$V_t(f) = V_{t-1}(f) + \eta(R_t - V_{t-1}(S_{\text{chosen}})), \quad \forall f \in S_{\text{chosen}}, \quad (5)$$

where η represents the learning rate, and $[R_t - V_{t-1}(S_{\text{chosen}})]$ is a prediction error—the discrepancy between the actual reward on the current trial and the reward that was expected based on choosing this stimulus.

Based on our previous findings (Niv et al. 2015), we also included in this model a decay of the values of the unchosen features to zero:

$$V_t(f) = (1 - d)V_{t-1}(f), \quad \forall f \notin S_{\text{chosen}}, \quad (6)$$

with d a free parameter determining the decay rate. As in the Bayesian model, action probabilities were determined by a softmax probability function on stimulus values (Equation 4). The RL model thus had three free parameters, the learning rate η, the softmax inverse temperature β, and the decay rate d.

Model fitting. At the end of the baseline phase (described above), participants were given a 1-min break while the computer fit both models to their data. Free parameters of both models were set for each participant separately, and were selected so as to maximize the likelihood of the data from the baseline phase (12 games and a total of 360 trials), by using the Matlab routine fmincon and fitting the data five times, initializing at different random starting points. Parameter values from the run that obtained the best likelihood for the data were then used to fully specify the model for the manipulation phase of the experiment, in which we used each model to track feature values in real time.

Real-Time Manipulation

In the remaining 40 games of our task, we aimed to help or hurt learning by manipulating stimulus presentation. Specifically, we manipulated how the available features (colors, shapes, and textures) were combined into three different stimuli, so as to either make available or obscure information about which feature is more likely to be the target feature. This manipulation relies on the idea—common to both models we tested—that while playing the task, participants update values for each feature, with the goal of ultimately learning which is the most rewarding feature. These feature values carry predictions regarding the reward associated with each feature and thus can be seen as "hypotheses" as to which is the target feature.

For every manipulated trial, to help learning, the highest-valued feature in each dimension was selected, and each of the three stimuli presented to the participant was designed to include only one of these three highest-valued features. This manipulation facilitates hypothesis testing, allowing participants to test one high-value feature independently of the other two. Conversely, to hurt learning, the three highest-valued features (one in each dimension) were combined into a single stimulus, thus preventing the participant from assigning credit for the feedback to just one of the three competing features. Both types of manipulation can potentially impact the rate of learning in the game, but only as long as the inferred values are close to the participants' actual values (Fig. 1C).

In each manipulated game, we manipulated every other trial from the fourth to the thirtieth trial for a total of 14 manipulated trials, using one of the two models and either helping or hurting learning throughout the game. Because our manipulation affects not only learning, but the likelihood to make the correct choice on the current trial, to measure learning we analyzed choices only in nonmanipulated (neutral) trials, in which stimuli were constructed so as to specifically not include all three highest-valued features in the same stimulus, nor separate them into three different stimuli. (Therefore, in nonmanipulated trials, one stimulus always contained exactly two of the highest-valued features—and these trials did not overlap with either the helping or the hurting manipulation.)

Each of the 40 games in the manipulation phase consisted of 36 trials, with 14 manipulated and 22 neutral trials. The last six trials in each game were not manipu-

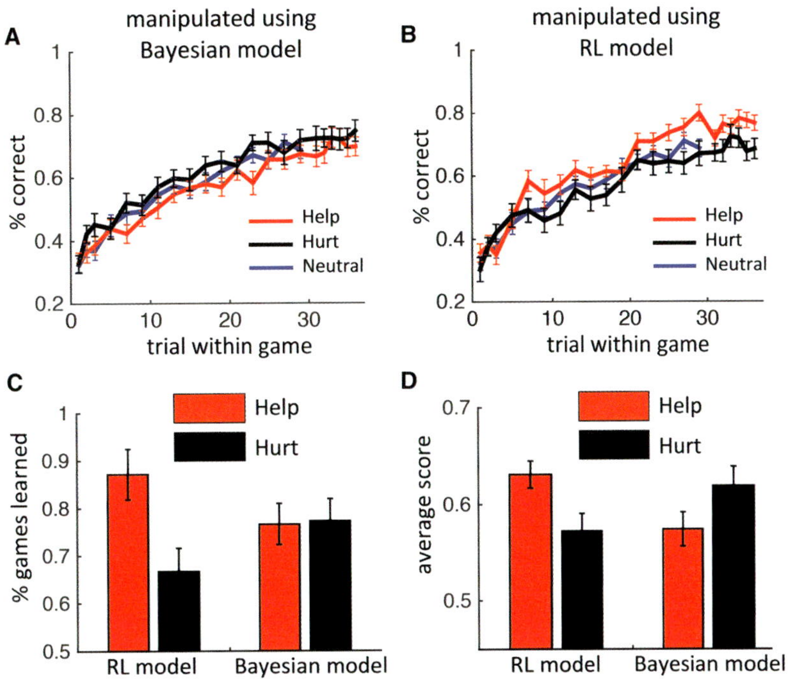

Figure 2. Model-based manipulation of stimuli affects learning only when using predictions from the RL model, not from the Bayesian model. (*A*) Learning curves for Help (red) and Hurt (black) conditions overlap when the manipulation used predictions from Bayesian model. (*B*) When the manipulation used predictions from the RL mode, learning curves for the Help condition show significantly better learning at the end of the game as compared with the Hurt condition. (Blue line) Data from the baseline phase. Similar effects of the manipulation are also seen in *C*, the overall number of learned games, and *D*, the average score across games. Error bars: S.E.M.

lated so as to allow measurement of steady-state learning at the end of the game.

RESULTS

To understand the dynamic process of learning a compact and sufficient task representation in a multidimensional environment, we tested human participants on a probabilistic three-dimensional choice task. While they played the task, we used the real-time trial-by-trial predictions from two competing models to manipulate the presented stimuli so as to help or hurt learning. Success of the causal manipulation would attest to congruence between the model and participants' learning strategies.

As shown in Figure 2, the manipulation had a significant effect on learning only in those games in which we manipulated stimuli using the RL model (Fig. 2A), and did not alter learning in games in which we manipulated stimuli using the Bayesian model (Fig. 2B). For RL-manipulated games, average performance on the last six trials of games (all nonmanipulated) in the Help condition (red line) was significantly better than in the Hurt condition (black line). A two-way Performance × Manipulation repeated-measures ANOVA yielded a significant interaction ($F_{(1,48)} = 5.22$, $P = 0.02$) with no main effects, and the average performance at the end of the game was significantly higher in the Help than the Hurt condition for RL-manipulated games (one-sided paired *t*-test, $t_{(24)} = 2.25$, $P = 0.01$), but did not differ between conditions for the Bayesian-manipulated games ($t_{(24)} = -0.96$, $P = 0.82$).

Participants did not learn all the games, that is, for some games, they could not identify the correct feature when probed at the end of the game (Fig. 1B). The real-time manipulation had an impact on the total number of games learned (Fig. 2C): Here too a two-way Model × Manipulation repeated-measures ANOVA showed a significant interaction ($F_{(1,48)} = 7.23$, $P = 0.009$) with no main effects. When the RL model was used to manipulate stimuli, the number of learned games differed significantly depending on whether the game was designed to help or hurt learning ($F_{(1,24)} = 12.64$, $P < 0.01$). On average, the Help condition resulted in an ~30% increase in the number of learned games. Conversely, when the Bayesian model was used to manipulate stimuli, there was no difference between the help and hurt conditions ($F_{(1,24)} = 0.77$, $P = 0.73$).

A similar pattern was observed in the average score per game (Fig. 2D). For games manipulated using the RL model, the average score (number of correct choices) was higher in the Help condition than in the Hurt condition (paired *t*-test; $t_{(24)} = 2.72$, $P = 0.011$). In contrast, scores for the Help and Hurt conditions were similar when the Bayesian model was used to manipulate the stimuli ($t_{(24)} = -1.91$, $P = 0.07$).

The impact of the stimulus manipulation was also evident in participants' difficulty (Fig. 3A) and confidence (Fig. 3B) ratings. For RL-manipulated games, participants rated "help" games as easier compared with the "hurt"

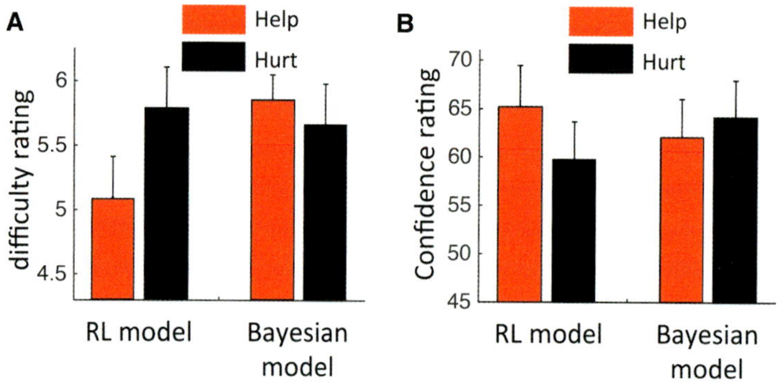

Figure 3. Difficulty (*A*) and confidence (*B*) ratings reflect the fact that the manipulation was only effective in RL-manipulated games, not in games manipulated using the Bayesian model. Error bars: S.E.M. Three subjects were excluded from this analysis, two because they consistently did not rate difficulty/confidence, and one participant who reported reversing the difficulty scale for most of the experiment.

games ($t_{(21)} = -2.23, P = 0.03$, paired t-test). This effect was not present in the games manipulated using the Bayes model ($t_{(21)} = 0.734$, $P = 0.47$, paired t-test). A similar pattern was seen in the confidence ratings (Fig. 3B), where confidence in RL-manipulated "help" games was rated as significantly higher than confidence in the "hurt" games ($t_{(21)} = 2.57, P = 0.01$), but ratings in games manipulated using the Bayesian model were not significantly different across conditions ($t_{(21)} = -1.26, P = 0.22$).

DISCUSSION

Using a multidimensional choice task, we investigated the computational strategies by which humans determine what dimensions of the environment are relevant for obtaining reward, and which can be safely ignored. The assumption underlying our work is that naturalistic tasks require such a representation learning process: In any given scenario, only a subset of information is relevant to the task at hand, and, moreover, the particular environmental dimensions that are relevant to one task are not necessarily relevant for another. For instance, the color of cars is irrelevant for crossing the street, but relevant for hailing a taxi, and the identity of the shops across the street is irrelevant to both tasks (but of course not for the task of navigating to the coffee shop).

To compare between qualitatively different accounts of how humans may learn what dimension of the environment is relevant for the current task, we showcased a novel method that compares two learning models by attempting to use each model in a causal, real-time manipulation of participants' learning. That is, we used each model to predict what hypotheses participants might be testing at each point in time, and manipulated stimuli to either help or hinder comparison of these hypotheses. This model-based manipulation can affect learning only to the extent that a model indeed captures participants' underlying learning process. We found that when stimuli were manipulated based on a RL model, games designed to help learning resulted in faster and more complete learning than games designed to hurt the learning process. In contrast, manipulating games using a Bayesian model had no significant effect on learning. Our method thus provides a stringent measure of how well each model captures people's strategies, and at the same time, our results provide evidence that it is indeed possible to impact representation learning in real time, by manipulating the stimuli that people have access to.

Our method is related to the framework of "optimal experimental design" in which experiments are designed so as to optimally elicit information about the process under investigation (Sebastiani and Wynn 2000; Atkinson et al. 2007). Normative statistical principles from Bayesian inference can, in some cases, be used to select an experimental design that will best resolve the details of participants' underlying cognitive processes (e.g., set the free parameters of a model of the process under scrutiny; Rafferty et al. 2012). One way to optimize our manipulation would be to choose, on each trial, the stimulus configuration that would allow participants to glean the maximum (or minimum) amount of information regarding the identity of the target stimulus, assuming participants' underlying cognitive processes matched each of the candidate models. This manipulation, although more normative than the one we designed, is more computationally intensive and, importantly, relies on further assumptions regarding the optimality of participants' actions. In particular, if participants are not selecting stimuli in an effort to maximize information (e.g., because they are also maximizing reward), this manipulation may not be more effective than ours. It is due to this interaction between information acquisition and reward acquisition that we assessed performance only in nonmanipulated trials—our "help" manipulation, although affording better information, made it more difficult to obtain reward on manipulated trials than did our "hurt" manipulation.

Rather than assume that the highest-valued features correspond to the hypotheses that the participant is comparing, another alternative is to infer these hypotheses using Bayesian inference. We have previously used such a method in association with the dimensions task and shown that recent choice history can effectively identify tested hypotheses (Wilson and Niv 2012). However, in that work, inference was only optimal if we assume that participants test one hypothesis at a time—an assumption that is incompatible with our current results. If participants were indeed focusing only on one hypothesis (feature) at any point in time, then neither of our manipulations would have affected learning.

Our findings join others (Eckstein et al. 2004) in suggesting that human learning is not always Bayes-optimal and, in particular, that humans do not solve the difficult task of representation learning in a Bayes-optimal way (see also Wilson and Niv 2012; Niv et al. 2015). These findings stand in contrast to multiple demonstrations of Bayes-optimality (Doya 2007) in perceptual decision-making (Gold and Shadlen 2002; Knill and Pouget 2004), motor control (Trommershäuser et al. 2003, 2005), multimodal integration (Körding and Wolpert 2004), reasoning (Oaksford and Chater 1994), and even setting metalearning parameters for reinforcement learning (Behrens et al. 2007; Yu 2007). However, whereas Bayesian inference may be computationally feasible (and indeed, simple) in scenarios that can be reduced to a several-alternative forced-choice decision (Gold and Shadlen 2002) or a choice between lotteries (Wu et al. 2009), representation learning in natural environments places much heavier computational demands on the learning system. In particular, the dimensionality of the environment is essentially unbounded (given that dimensions such as previous actions and events can be, and frequently are, relevant for task performance), and whereas feedback is often available for one's actions, the environment does not provide any supervision regarding one's representations.

To overcome these difficulties, our results suggested that humans might use a suboptimal but computationally more tractable strategy based on reinforcement learning. However, we note that we only compared two very different models, partly as a proof-of-concept for our novel model-comparison method. It is entirely possible—in fact, likely—that the feature-level RL model that we suggested also falls short of fully capturing participants' learning strategies. Future applications of this method will hopefully delineate more precisely the workings of representation learning in the human brain.

Finally, our "dimensions task" lends itself easily to manipulation of presented information. Work on "active learning" (Cohn et al. 1996) in categorization and perceptual estimation tasks has used a related manipulation, effectively allowing participants to design their experiment optimally (Kruschke 2008; Castro et al. 2009; Gureckis and Markant 2009; Juni et al. 2011; Markant and Gureckis 2014). Some adjustments will likely be needed to apply this model-comparison method to other task structures, though we are optimistic as to the method's wider applicability (Nelson et al. 2010).

In sum, we have described a real-time manipulation of information presented to participants, and have suggested that basing this manipulation on predictions of different models can allow for a new, sensitive and *causal* means of model comparison. Using this method and a RL model, we have shown that human representation learning can be improved or hampered. Beyond the implications for effective, individual-difference-sensitive model selection, such "access" to participants' mental strategies suggests exciting applications, particularly in the domain of education and tailoring the flow of information toward individual learning.

ACKNOWLEDGMENTS

This work was funded in part by the National Institutes of Health grant R01MH098861 and Army Research Office award W911NF-14-1-0101. The information herein reflects the opinion of the authors and does not necessarily reflect the opinion or policy of the federal government.

REFERENCES

Atkinson A, Doney A, Tobias R. 2007. *Optimum experimental designs, with SAS*. (ed. Atkinson RJCA, Hand DJ, Pierce DA, Schervish MJ, Titterington DM). Oxford University Press, Oxford.

Behrens TE, Woolrich MW, Walton ME, Rushworth MF. 2007. Learning the value of information in an uncertain world. *Nat Neurosci* **10:** 1214–1221.

Beierholm UR, Quartz SR, Shams L. 2009. Bayesian priors are encoded independently from likelihoods in human multisensory perception. *J Vis* **9:** 23.

Bellman R. 1957. *Dynamic programming*. Princeton University Press, Princeton.

Busemeyer JR, Wang YM. 2000. Model comparisons and model selections based on generalization criterion methodology. *J Math Psychol* **44:** 171–189.

Castro RM, Kalish C, Nowak R, Qian R, Rogers T, Zhu X. 2009. Human active learning. In *Advances in neural information processing systems*, pp. 241–248.

Cohn DA, Ghahramani Z, Jordan MI. 1996. Active learning with statistical models. *J Artif Intell Res* **4:** 129–145.

Cutting JE, Bruno N, Brady NP, Moore C. 1992. Selectivity, scope, and simplicity of models: A lesson from fitting judgments of perceived depth. *J Exp Psychol Gen* **121:** 364.

Doya K. (Ed.). 2007. *Bayesian brain: Probabilistic approaches to neural coding*. MIT Press, Cambridge, MA.

Eckstein MP, Abbey CK, Pham BT, Shimozaki SS. 2004. Perceptual learning through optimization of attentional weighting: Human versus optimal Bayesian learner. *J Vis* **4:** 3.

Gold JI, Shadlen MN. 2002. Banburismus and the brain: Decoding the relationship between sensory stimuli, decisions, and reward. *Neuron* **36:** 299–308.

Gureckis T, Markant D. 2009. Active learning strategies in a spatial concept learning game. In *Proceedings of the Thirty-First Annual Conference of the Cognitive Science Society*, pp. 3145–3150.

Gureckis TM, Markant DB. 2012. Self-directed learning: A cognitive and computational perspective. *Perspect Psychol Sci* **7:** 464–481.

Juni MZ, Gureckis TM, Maloney LT. 2011. Don't stop 'til you get enough: Adaptive information sampling in a visuomotor estimation task. In *Proceedings of the Thirty-Third Annual Conference of the Cognitive Science Society* (ed. Carlson L, Hölscher C, Shipley T). Cognitive Science Society, Austin, TX.

Knill DC, Pouget A. 2004. The Bayesian brain: The role of uncertainty in neural coding and computation. *Trends Neurosci* **27:** 712–719.

Knock TR, Ballard KJ, Robin DA, Schmidt RA. 2000. Influence of order of stimulus presentation on speech motor learning: A principled approach to treatment for apraxia of speech. *Aphasiology* **14:** 653–668.

Körding KP, Wolpert DM. 2004. Bayesian integration in sensorimotor learning. *Nature* **427:** 244–247.

Kruschke JK. 2008. Bayesian approaches to associative learning: From passive to active learning. *Learn Behav* **36:** 210–226.

Markant DB, Gureckis TM. 2014. Is it better to select or to receive? Learning via active and passive hypothesis testing. *J Exp Psychol Gen* **143:** 94.

Nelson JD, McKenzie CR, Cottrell GW, Sejnowski TJ. 2010. Experience matters: Information acquisition optimizes probability gain. *Psychol Sci* **21:** 960–969.

Niv Y, Daniel R, Geana A, Gershman S, Leong YC, Radulescu A, Wilson RC. 2015. Reinforcement learning in multidimensional environments relies on attention mechanisms. *J Neurosci* **35:** in press.

Oaksford M, Chater N. 1994. A rational analysis of the selection task as optimal data selection. *Psychol Rev* **101:** 608.

Pitt MA, Myung IJ, Zhang S. 2002. Toward a method of selecting among computational models of cognition. *Psychol Rev* **109:** 472.

Rafferty AN, Zaharia M, Griffiths TL. 2012. Optimally designing games for cognitive science research. In *Proceedings of the 34th Annual Conference of the Cognitive Science Society*, pp. 893–898.

Ritter FE. 2007. *In order to learn: How the sequence of topics influences learning*. Oxford University Press, New York.

Rust RT, Simester D, Brodie RJ, Nilikant V. 1995. Model selection criteria: An investigation of relative accuracy, posterior probabilities, and combinations of criteria. *Manag Sci* **41:** 322–333.

Sebastiani P, Wynn HP. 2000. Maximum entropy sampling and optimal Bayesian experimental design. *J R Stat Soc Series B Stat Methodol* **62:** 145–157.

Townsend JT. 1990. Serial vs. parallel processing: Sometimes they look like Tweedledum and Tweedledee but they can (and should) be distinguished. *Psychol Sci* **1:** 46–54.

Trommershäuser J, Maloney LT, Landy MS. 2003. Statistical decision theory and the selection of rapid, goal-directed movements. *J Opt Soc Am A Opt Image Sci Vis* **20:** 1419–1433.

Trommershäuser J, Gepshtein S, Maloney LT, Landy MS, Banks MS. 2005. Optimal compensation for changes in task-relevant movement variability. *J Neurosci* **25:** 7169–7178.

Wilson R, Niv Y. 2012. Inferring relevance in a changing world. *Front Hum Neurosci* **5:** 189.

Wu SW, Delgado MR, Maloney LT. 2009. Economic decision-making compared with an equivalent motor task. *Proc Natl Acad Sci* **106:** 6088–6093.

Yu AJ. 2007. Adaptive behavior: Humans act as Bayesian learners. *Curr Biol* **17:** R977–R980.

Understanding the Hows and Whys of Decision-Making: From Expected Utility to Divisive Normalization

Paul Glimcher
Institute for the Interdisciplinary Study of Decision Making, New York University, New York, New York 10013
Correspondence: *pg3@nyu.edu*

Over the course of the last century, economists and ethologists have built detailed models from first principles of how humans and animals *should* make decisions. Over the course of the last few decades, psychologists and behavioral economists have gathered a wealth of data at variance with the predictions of these economic models. This has led to the development of highly descriptive models that can often predict *what* choices people or animals will make but without offering any insight into *why* people make the choices that they do—especially when those choices reduce a decision-maker's well-being. Over the course of the last two decades, neurobiologists working with economists and psychologists have begun to use our growing understanding of how the nervous system works to develop new models of *how* the nervous system makes decisions. The result, a growing revolution at the interdisciplinary border of neuroscience, psychology, and economics, is a new field called *Neuroeconomics*. Emerging neuroeconomic models stand to revolutionize our understanding of human and animal choice behavior by combining fundamental properties of neurobiological representation with decision-theoretic analyses. In this overview, one class of these models, based on the widely observed neural computation known as *divisive normalization*, is presented in detail. The work demonstrates not only that a discrete class of computation widely observed in the nervous system is fundamentally ubiquitous, but how that computation shapes behaviors ranging from visual perception to financial decision-making. It also offers the hope of reconciling economic analysis of what choices we should make with psychological observations of the choices we actually do make.

The postenlightenment study of human decision-making has its roots in the work of mathematicians and economists like Blaise Pascal, Daniel Bernoulli, Adam Smith, and David Ricardo. These scholars sought to explain regularities in human behavior from first principles. For these early theorists, that meant defining algorithms that maximized individual gain, profit, or well-being and that thus gave rise to observable market level regularities. These scholars took as their focus the study of "why"—why it was that people made the choices they did from a mathematical and theoretical point of view. While revising many of the key assumptions of these early scholars, economists of the early 1900s like Paul Samuelson (Samuelson 1947) and John von Neumann (Neumann and Morgenstern 1944) made huge strides in this direction by defining how decision-making ought to work in humans. They developed theories that fundamentally described how "optimal" decision-makers should behave.

In the 1960s, however, it started to become clear that humans did not always behave in exactly the ways predicted by these theories. Work by economists like Maurice Allais (Allais 1953) and psychologists like Amos Tversky and Daniel Kahneman (Kahneman and Tversky 1979) began to reveal systematic patterns of behavior that violated the predictions of the economic models. These were behaviors that appeared to reduce the well-being of individuals and even of whole economies. The psychologist Herbert Simon (Simon 1955) responded to this observation by suggesting that the limitations of our cognitive or neurobiological architecture might be the cause of these inefficiencies. The majority of those cataloging these inefficiencies, however, devoted their energies to developing descriptive models. These were models that could predict *what* people would choose in any situation, even if the models offered little or no insight into *why* biological decision-makers produced these apparently self-destructive behaviors.

In the early 2000s, a group of neurobiologists, working with economists and psychologists, sought to reconcile this tension by studying *how* the nervous system actually makes decisions. These interdisciplinary scholars, working with like-minded economists and psychologists, rapidly gave birth to the interdisciplinary field of *Neuroeconomics*. Within Neuroeconomics, a number of models of how the nervous system encodes and processes decision-related information began to emerge in the 2000s and early 2010s. One of these models, the focus of this review, has its roots in the *Efficient Coding Hypothesis* of Horace Barlow (Barlow 1961). Barlow had hypothesized that because both building internal representations of the outside world and performing computations on those representations is metabolically costly, the representational frameworks and computational processes used by organ-

 doi: 10.1101/sqb.2014.79.024778

isms should naturally trade off completeness against metabolic cost. He hypothesized that an animal nervous system would be "efficient" in the sense that it would devote energy only to valuable computations and goals, thus minimizing costly redundancies in its internal representations.

In the mid-2000s, my colleagues and I began to see some parallels between these statements and Herbert Simon's hypothesis that limitations imposed by our cognitive or neurobiological architecture might be the cause of the apparent "inefficiencies" in human choice behavior identified by psychologists. It might be, we hypothesized, that decision-making systems were trading off representational costs at the neurobiological level against benefits at the level of behavior—and that this trade-off might account for the apparent gap between the *whys* of economists and the *whats* of psychologists. If that was true, then it might even be the case that human choice behavior was efficient, once one took into account the neurobiological costs of representation and computation.

EFFICIENT CODING AND THE THEORY OF CORTICAL REPRESENTATION

The theory of efficient coding emerged, in part, from Keffer Hartline's observation (Ratliff and Hartline 1959) that the activity of sensory neurons in the retina encodes not just the photons falling directly on the *centers* of their receptors but is also influenced by the structure of *surrounding* visual stimuli. Barlow noted that the precise relationship between center and surround described by Hartline would have the effect of creating more neural activity when the center and surround differed and very little activity when center and surround were homogeneous. Drawing on Claude Shannons' *Theory of Information* (Shannon and Weaver 1949), Barlow proposed that to be efficient, neurons must devote action potentials to sensory stimuli in proportion to the degree to which those stimuli are *surprising*, and he proposed that it was the differences, rather than the homogeneities, that carried surprising information.

By the early 1990s, it had become clear that the relationship between center and surround in visual cortical neurons was more complex than even Barlow had suspected. Cortical area V1 neurons sensitive to vertically oriented bars, for example, responded more strongly when horizontally oriented bars were arranged outside their "classical" receptive fields and were suppressed in a nonlinear way when vertically oriented bars were placed at those same locations. David Heeger and his colleagues (Heeger 1992; Heeger et al. 1996) responded to that observation by developing what came to be known as the *Heeger normalization model* (Fig. 1A). In that model, the firing rate of a given neuron (FR_1) is described by

$$FR_1 = \frac{Stim_1}{\sigma^2 + \sum_i Stim_i}. \quad (1)$$

Heeger's idea was that the activity of a V1 neuron reflected a direct excitatory input from its receptive field ($Stim_1$) that was modulated by the summed activity of all of its neighbors (the set of all other $Stim_i$). This modulation was achieved by dividing the input activity by the summed activity of the neighbors, a model that decreased the FR_1 when the surround was similar. The model also included a constant in the denominator: σ^2. Note that the magnitude of σ^2 controls the rate of rise of the firing rate function, as shown in Figure 1B. It should be noted that Heeger's model also included a number of other features glossed over here.

In the late 1990s, Eero Simoncelli and others (Olshausen and Field 1997; Simoncelli 2003) began to test the hypothesis that models of this kind formally achieve Barlow's goal: They maximize the information carried by each action potential and by each neuron. To understand their approach, consider a neural system that encodes the luminance of a single square pixel surrounded by eight other pixels (Fig. 2A). Simoncelli and his colleagues recognized that in the real visual world, the luminance of the central pixel is not independent (or, formally, *uncorrelated*) with the luminance of the surrounding pixels. In the real world, if we knew that the eight outer pixels were black, we could be almost certain that the central pixel

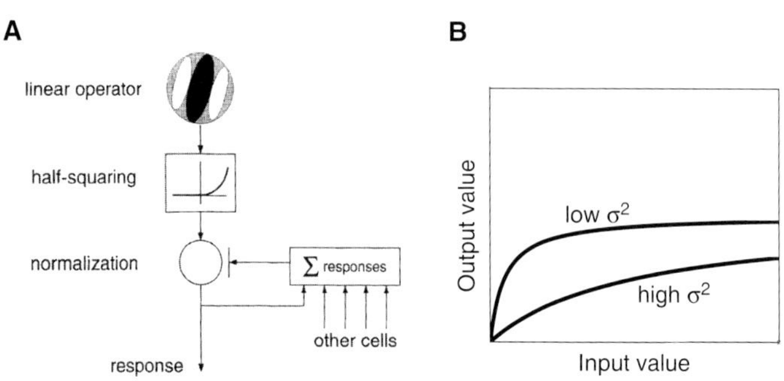

Figure 1. (*A*) The original divisive normalization model of Heeger and colleagues (Heeger 1992; Heeger et al. 1996). Visually sensitive neurons receive a direct excitatory input that is rectified (negative responses are removed) by half-squaring. This rectified output is then divided by the sum of the activity of similarly sensitive neighbors plus a constant, usually called σ^2. (*B*) Firing rate curves for a simulated normalized neuron are shown for a high and a low value of σ^2. A low σ^2 focuses the dynamic range of the function at low input values. A high σ^2 distributes the firing rate function over a larger range. (*A*, Courtesy of David Heeger.)

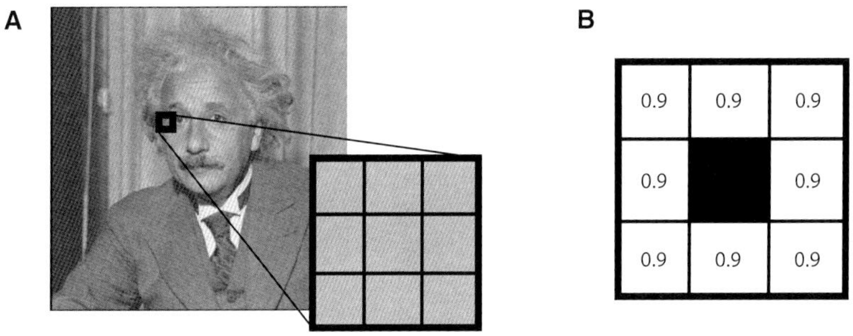

Figure 2. (*A*) An array of local pixels from a photograph. (*B*) The correlation between a central pixel and eight surrounding pixels is quite high in natural images.

was also black. Conditional on the outer pixels being black, there is very little information gained by learning that the central pixel is also black. An efficient neuron would know that and would waste very few action potentials to communicate this unsurprising fact.

Simoncelli and his colleagues (Schwartz and Simoncelli 2001; Wainwright et al. 2002) thus took two approaches to revising the Heeger equation. At a purely mathematical level, they set out to prove that there exists some variant of the Heeger model that maximizes information per action potential and per neuron for the real visual world. They were successful in this regard. At an empirical level, they set about a massive exercise analyzing real world photographs to see just exactly how adjacent pixels of images are related. What this second exercise yielded, and this is a critical insight, was a matrix that shows the degree of relatedness between adjacent pixels in the real world. The numbers in each box in Figure 2B reveal how much knowing the state of that pixel tells you about the central pixel. This is, in the language of linear algebra, a *correlation matrix*. Now with that insight in hand, we can easily make sense of the model Simoncelli and his colleagues produced:

$$\mathrm{FR}_1 = \frac{\mathrm{Stim}_1}{\sigma^2 + \sum_i \omega_i \mathrm{Stim}_i}. \qquad (2)$$

The numerator remains the direct input to the neuron, but now the denominator takes all of the activity in nearby neighbors and multiplies each by the degree to which it allows one to predict the value of Stim_1. The sum $\omega_i \mathrm{Stim}_i$ aggregates the total knowledge one can derive from the activity of the neighbors about our pixel, and then the equation pulls that information out of the firing rate by dividing it away. The result is a maximally efficient signal, "cleaned" of redundant information.

What followed this demonstration were a host of papers showing that divisive normalization models accurately explained the firing rates of neurons in cortical sensory areas ranging from V1 to V2 and even in auditory and olfactory areas. (For a review of this literature, see Reynolds and Heeger 2009; Carandini and Heeger 2012.)

NORMALIZATION MODELS AND DECISION-MAKING

The observation that many cortical and subcortical systems appear to use a normalization-based encoding scheme led Kenway Louie and I to test the hypothesis in the late 2000s that cortical circuits important for decision-making might also use this encoding strategy (Louie et al. 2011). Over the preceding decade my colleagues and I had conducted detailed studies of the activity in the lateral intraparietal area (LIP) of the monkey parietal cortex. We had shown that the activity of these neurons often appears to encode decision variable–like signals (Platt and Glimcher 1999; Glimcher 2013). Each neuron in LIP, we found, encodes the desirability of looking at a particular point in extrapersonal space. We had hypothesized that LIP firing rates could be thought of as encoding a "priority map." The higher the firing rate, the more the animal wants to look at the location encoded by that neuron.

Recall that the *why* theories of economists like von Neumann tell us what an idealized decision-maker *should* do when she makes a choice: As the magnitude of a reward increases, a chooser should be more likely to pick that reward (it should have higher priority), and as the probability of being rewarded declines, a chooser should be less likely to pick that reward. These theories hypothesize that choosers behave exactly *as if* somewhere in their brains a numerical value called a decision variable is computed and represented for each option and the process of choosing is selecting the option having the highest decision variable. Working from those theories in the late 1990s, Michael Platt and I (Platt and Glimcher 1999) had shown that neurons in LIP do follow the general predictions of these *why* models, and we hypothesized that neurons like those in area LIP could be viewed as physically instantiating these decision variables, which my colleagues and I came to call *subjective values* (Fig. 3).

Of course we knew that the *why* theories of economists are imperfect predictors of behavior. We had always argued that the *why* models we used to explain cortical firing rates should only be considered approximations. It was with that caveat in mind that Louie and I formally examined the firing rates of LIP neurons with a cortical normalization model. To do that, we used a simplified form of the Heeger–Simoncelli model that John

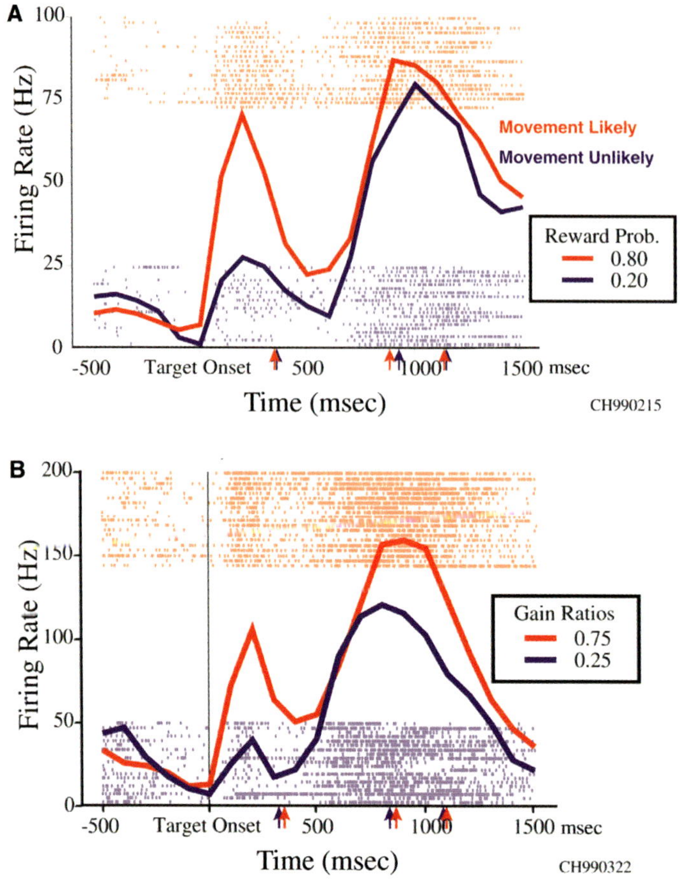

Figure 3. Firing rates for two neurons in the lateral intraparietal area (LIP). (*A*) Solid lines indicate average firing rates for a neuron as a function of time under two conditions. The red line indicates average response on trials in which there was a high probability of reward. The blue line indicates firing rate on trials with a low probability of reward. (*B*) A similar figure for varying reward magnitude. Red identifies trials on which the monkey expects a large reward, and blue identifies trials on which the monkey expects a small reward. (Adapted from Platt and Glimcher 1999.)

Reynolds and Heeger (Reynolds and Heeger 2009) had recently proposed:

$$\mathrm{FR}_1 = \frac{\mathrm{Baseline} + V_1}{\sigma^2 + \sum_i \omega_i V_i}, \tag{3}$$

which included a baseline firing rate and took as inputs the objective values of the rewards that the monkey (and his neurons) were evaluating. Recall that a core prediction of the normalization models is that stimuli in the surround influence firing rate through the denominator. One can see this influence as a process of decorrelation, achieved by multiplying the activity in each of the other neurons by a covariance matrix (ω). Entirely equivalently, one can see this as a process of "contextual suppression," which says the same thing in a less mathematical way. But how should one think about contextual suppression in a decision-making system? One might imagine that the other options currently under consideration by a chooser (like the other stimuli presented near a visual receptive field) form a kind of "context" that might modulate the firing rate of the neuron under study.

Consider a situation in which the option represented by neuron FR_1 yields a much larger reward (say 1 mL of water) than two other options being represented by two other neurons. Under those conditions, our neuron might be expected to have a very high firing rate, correctly signaling that the reward under consideration is large compared with the other two options under consideration. But under conditions in which the two other options are of much higher value than the 1 mL reward encoded by FR_1, then we might expect FR_1 to show a much reduced firing rate—even though the absolute value of the reward it encodes is the same. The key idea here is that the FR_1 neuron, if it follows the normalization model, should report the value of an option modulated by current context just as V1 neurons report the orientation of a stimulus modulated by local orientation context. (And, of course, both of these contextual modulations are a form of decorrelation.)

To test the hypothesis that LIP neurons do use a normalization-like encoding scheme, Louie and I examined the delay-period activity of LIP neurons while monkeys were presented with three options of variable magnitude (Louie et al. 2011). What we found was that the neurons were very much influenced by local context. And that this local context effect could be best described using a normalization model. Figure 4 shows this for a database of neurons and compares the ability of three models to ac-

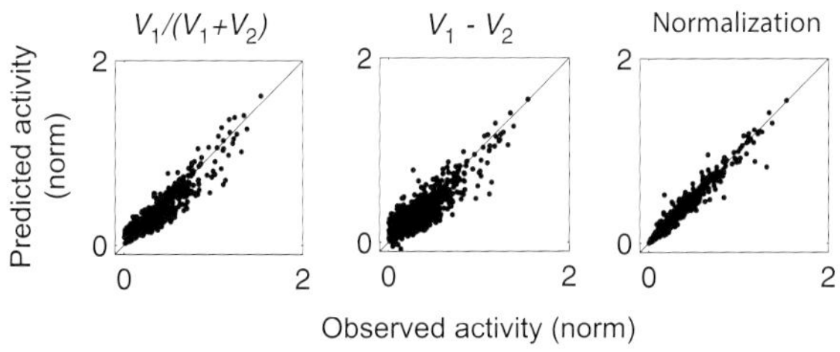

Figure 4. Comparison of predicted and observed firing rates for three models: the original relative value model of Platt and Glimcher (1999); the "difference model" underlying firing rates in the drift diffusion model; and the normalization model.

count for those delay-period firing rates. Firing rates are much better accounted for by the normalization model than by two competing models that have been widely used in the past.

THE "NORMALIZATION MODEL" CONSIDERED AS A NETWORK

A key idea of the normalization model is that neurons do not just represent their direct inputs, but that these inputs are modulated by the activity of their neighbors, a *contextual sensitivity* that maximizes their efficiency. Barlow, Heeger, and Simoncelli had considered this in a static sense: How should steady state firing rates incorporate context to maximize the information carried by each action potential and by each neuron? Simoncelli largely solved that problem, proving that a variant of the original model (one incorporating a covariance matrix) maximizes information. As Tom LoFaro, Louie, and I worked on this problem, however, we began to wonder how real neural circuits accomplish this decorrelation (LoFaro et al. 2014). Figure 5 shows the simplest possible network for computing this kind of normalization model. The values of two options, V_1 and V_2, serve as excitatory inputs to encoding neurons R_1 and R_2. As in Figure 1, the R neurons receive an inhibitory input (assumed to be divisive although that is not critical) that pools together the outputs of all of the R neurons. (In this network there are only two

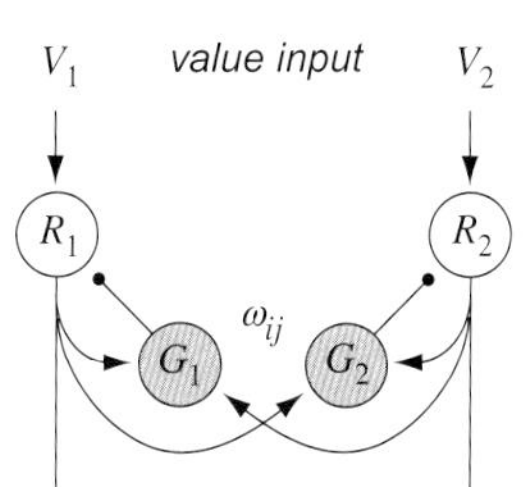

Figure 5. A simple network for divisive normalization. Value inputs (V) enter excitatory neurons R, the activity of which is reduced by a divisive inhibitory interneuron, G. For simplicity, the model shown here includes only two pairs of neurons and thus supports pairwise choice. Extension of the model to n pairs yields results identical to the ones reported here for two-pair models.

R neurons, though we have shown that our findings generalize to large numbers of R–G neuron pairs.)

The excitatory neurons of the network are its backbone. They get excitatory inputs that we think of as "value" signals, but which in the visual cortex one might think of as the contrasts of pixels. A network of inhibitory neurons receives inputs from the excitatory neurons. One can think of the strengths of the synapses connecting the excitatory neurons to the inhibitory neurons as implementing the covariance matrix of Equation 2, although for simplicity here we will just treat those synapses as having a fixed unit strength. That is all there is to this network. To examine how such a simple network would behave dynamically we need to represent how firing rates change when inputs change with simple differential equations:

$$R\text{ neuron}: \quad \tau\frac{dR}{dt} = -R + \frac{V_1}{1+G_1}, \tag{4}$$

$$G\text{ neuron}: \quad \tau\frac{dG}{dt} = -G + \sum_i \omega_i R_i. \tag{5}$$

Notice that these equations describe how neuronal firing rates change from one moment to the next, they report the first derivative of firing rate with respect to time. The first equation represents the first derivative of the firing rate for an R neuron, and the second equation represents it for a G neuron. τ is a time constant for these neurons.[1] Notice that the R neuron responds with an increase in firing rate when the strength of the input V increases. It responds with a (divisive) decrease in firing rate when the inhibitory neuron G increases its firing rate.

To better understand the behavior of such a network, we can select a value for V_1 and then solve the equations to see how the firing rate of each neuron would evolve over time for any possible preexisting firing rate in the R and the G neuron. Figure 6 plots a graph of that mathematical "experiment." The arrows at each point on the graph show how firing rates evolve from that point when $V_1 = 10$. On such a plot, called a phase plane, we can look for all of the places where (for a given input V_1) the first derivative of the R neuron or the G neuron goes to zero. These two lines, called *null clines*, show where each neuron is stable. Wherever these lines intersect, where neither neuron changes, the network is in a stable *equilibrium* state. Interestingly, we were able to prove in a quite general way that these dynamic networks always have one (and only one) stable equilibrium. Once the input is set, the network evolves toward a unique equilibrium (regardless of how many R–G neuron pairs the network contains).

To understand how important this fact is, we need to rewrite the equations above in terms of actual firing rates for R neurons. Equation 6 shows such a restating of these

[1]Here we use a single time constant for both kinds of neurons, but our findings are robust to the selection of time constant. We often set the time constant for the R neurons to 10 msec and for the G neurons to 5 msec to mirror the membrane time constants of excitatory and inhibitory cortical neurons.

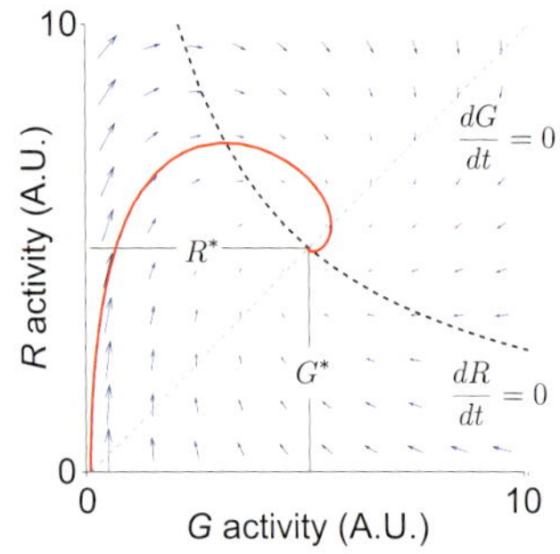

Figure 6. A phase plane trajectory for a single pair network. The red line indicates the firing rate path when the value input is abruptly stepped to a positive value. Dashed lines identify the null clines for the two neurons. The intersection of the dashed lines marks the unique equilibrium state for the network: *divisive normalization*.

equations, but rather than being plotted in continuous time we plot it here in discrete time (this discrete form representation is equivalent at equilibrium to the continuous time form in Equations 4 and 5):

$$R_t = \frac{V_1}{1 + \sum_{t=0}^{-\infty} \sum_i \delta^t \omega_i R_{i_t}}. \tag{6}$$

What you can see is that the firing rate of an R neuron at time t reflects the current excitatory input (V_1) divided by a denominator that may look complicated, but is not. In the denominator is a double sum. The right-hand part of the sum is familiar. It is simply the activity of the inhibitory neuron that computes a (covariance matrix) weighted sum of the R neurons (rather than the V inputs) in the network: $\sum_i \omega_i R_i$. To capture the dynamics, though, we have added one additional feature, a sum over time. The outer sum and the δ (which is a number <1) cause the equation to weight current value-related activity heavily (when t is small) and allows older value-related activity to decay away (as t grows into the past). The function of this "temporal summation" is simply to exponentially decrease the effects of old inputs on the network—to make old inputs decay away. We can think of the size of δ as representing the time constant of the network.

To take our understanding one step further, let us algebraically separate the denominator into two parts: The current activity of the excitatory network of R cells and the past activity of those same cells. The next equation shows that

$$R_t = \frac{V_1}{\left[1 + \sum_{t=-1}^{-\infty} \sum_i \delta^t \omega_i R_{i_t}\right] + \sum_i \omega_i R_{i_o}}, \tag{7}$$

$$R = \frac{V_1}{[\sigma^2] + \sum_i \omega_i V_i}. \tag{8}$$

Now compare the dynamic equation (Equation 7) with the original static Heeger–Simoncelli shown in Equation 8. The two equations are almost identical—the only big difference is that the constant σ^2 has turned into something much more complicated and dynamic, as shown in the bracketed parts of Equation 7. That in itself is an important point; when you build a divisive network of almost any kind you have something that generates as an output much like the original normalization models. To put that another way, normalization models are what divisive (and many subtractive) networks naturally do.[2]

Now consider the term that has replaced σ^2. This is the term that captures all of the temporal dynamics of the system. When the firing rate is changing but the inputs are constant, it is because this part of the equation is changing as time passes. What is critical about that is: *Whenever the stuff in the bracket is not changing—when the stuff in the bracket is a* ***constant*** *and the inputs are fixed—the system is at equilibrium and becomes a Heeger–Simoncelli-type model.* The Heeger–Simoncelli model is, it turns out, a special case of a divisive network. It is what a divisive network does when it comes to rest.

Recall from Figure 1A that the value of σ^2 controls how steeply the firing rate function rises. A small constant here causes the function to rise quickly, and a large constant causes it to rise slowly. So what is this constant in the general form? It is simply an exponentially weighted sum of past activity derived from past input values—what we would call past "options" in decision-making. In more formal terms, it is an "adaptive expectation." *It is the expected average value of the upcoming inputs, based on an analysis of recent history.* What the "constant" does in this dynamic network is to adjust the firing rate function so it has the greatest possible resolution near the current expectation. What that means is that these classes of networks are much more powerful and general than one might have expected. At equilibrium they computes a form of divisive normalization. Their denominators contain a "tuning constant" that maximizes the firing rate function's sensitivity around the network's expected input.

First, that tells us that the static equations of the Heeger and Simoncelli forms in fact describe the unique equilibrium state of *nearly any network of this class*. If you have an excitatory input to your excitatory neuron and a pooling inhibitory network, you wind up with a network that at equilibrium computes divisive normalization.

NORMALIZATION NETWORKS AND NEURONAL DYNAMICS

Interestingly, these dynamic networks make an additional prediction. Recall that R neurons receive excitatory value input (Vs) and inhibitory input from the G neurons, G, but the G input is delayed, reflecting the time it takes for the inhibitory network to become active. One can see this in Figure 6, which shows the R neuron overshooting before being pulled down by the G neuron.

[2] Note, however, that the denominator in the Heeger form (Equation 8) has Vs in it. The denominator here (Equation 7) has Rs in it. We have shown that this distinction does not make much of a difference at equilibrium. It does make a significant difference with regard to network dynamics. Louie, LoFaro, Webb, and I have shown that the actual dynamics of LIP neurons are much more closely approximated by this form, a point discussed next and in Louie et al. (2014).

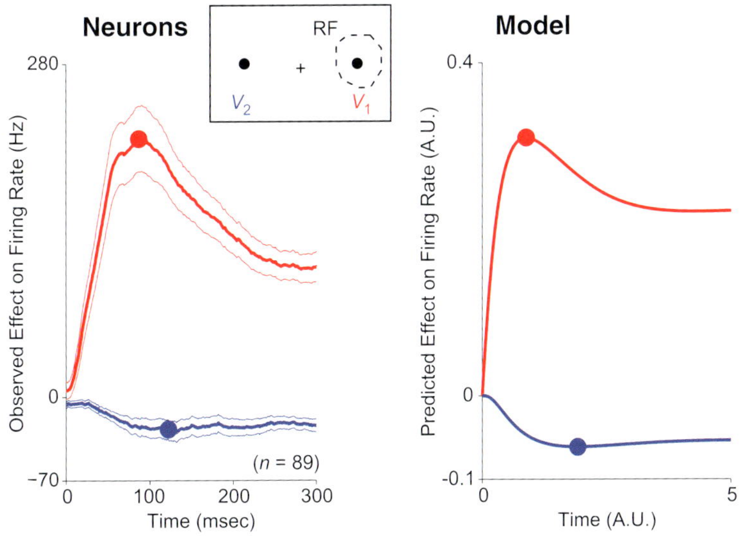

Figure 7. Excitatory and inhibitory contributions to population firing rates in area LIP. (*Left*) Observed data from an LIP population. (*Right*) The prediction of the network model that achieves divisive normalization at equilibrium.

Thus, *R* neurons respond to a step change in *V* with a transient burst-like increase before they settle down to stable "delay-period" activity. In real life, area LIP neurons have long been known to show what is often called a "visual burst" before sustained delay-period activity. What the model reveals is that for divisive (or subtractive) networks of this kind there is nothing "visual" about the "visual burst." The early rise in firing rate is the natural network response when its input goes up. The "burst" is the off-equilibrium response of the network before it settles to a sustained equilibrium.

Louie and I (Fig. 7; Louie et al. 2014) examined the dynamics of LIP neurons to see how closely their "visual burst" could be predicted by these kinds of model. What we found was that the magnitude and timing of the burst and the structure of its later suppression were exactly what would be expected of such a network.

NORMALIZATION NETWORKS AND CHOICE BEHAVIOR

Normalization networks capture the idea that neurons have limited capacity and that they distribute this capacity to maximize their efficiency. But their limited capacity necessarily means that they perform imperfectly as the amount of information they encode grows. Although we have not focused on this issue yet, the source of that capacity limitation is neuronal noise. If firing rates could encode input values to 20 decimal places, then they would have no real capacity limit. The fact that firing rate variance (roughly) scales with mean rate, however, means that these neurons are quite limited in their precision.

Consider a network representing two high-valued options, each of almost equal value and a third low-value (and undesirable) option. Figure 8A shows a firing rate *distribution* for three *R* neurons encoding the values of these three options at equilibrium. Each of the colored curves represents the distribution of observed firing rates over time. What we can see is that the network does a good job of separating (decorrelating) the two highest-valued options, even given that the firing rates fluctuate randomly within the colored distributions. Consider, however, what happens as we increase the value of the undesirable option (Fig. 8B). The firing rate functions begin to overlap—the decorrelation begins to fail. And thus it gets less and less likely that the network will uniquely identify the highest-valued option as option number grows.[3]

To see whether choice behavior degrades in exactly this way, Ryan Webb, LoFaro, Louie, and I (Louie et al. 2013; Webb et al. 2014) developed a series of human and animal choice experiments in which we searched for anomalies like the one described above. Perhaps surprisingly, we found that each of the predictions our models made accurately described an anomalous choice behavior found in humans and in monkeys that defied traditional economic analysis. This is a key point: Our model predicts that the predictions of traditional economic theory are wrong because traditional theory neglects the cost of representation. In any real system, where neurons are costly and capacity is limited, choice accuracy will have to trade off representational costs against benefits. What this suggests is that the gap between the *whys* of the economists and the *whats* of the psychologists really reflects a misspecification of the costs of neural computation.

CONCLUSION

Neuroeconomic models of human and animal decision-making emerged as an effort to reconcile economic theory and psychological observation. Economists had

[3]Louie and I showed that this is true for essentially any noise regime. That is a somewhat counterintuitive fact covered in Louie et al. (2011).

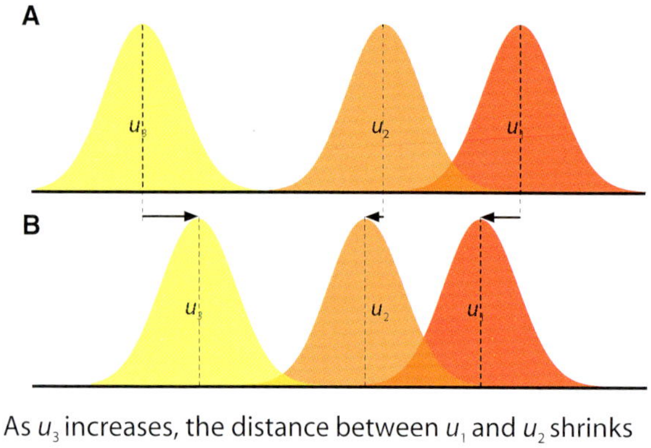

Figure 8. Firing rate distributions for normalization models show a sensitivity to the value of irrelevant options. See text.

long argued that choice was a process of optimization and had developed lovely theories of how that optimization should work. But what the economists had failed to take into account was the fact that neural computation is costly and neural representational capacity is limited. Psychologists had correctly noted the failures of traditional economic theory but offered little deep explanation for why these anomalies arise. What has become clear in the last decade is that models that incorporate real constraints faced by the nervous system can close much of this gap. These models recognize that perfect choice behavior requires an infinitely costly nervous system. Real animals operating under evolutionary pressure accept errors whenever correcting those errors would be more costly (in terms of a larger or more precise nervous system) than beneficial. The study of human decision-making stands poised to use neuroscience to develop a next-level understanding of the *whys*, *whats*, and *hows* of human behavior.

REFERENCES

Allais M. 1953. Behavior of the rational man before risk—Criticism of American school postulates and axioms. *Econometrica* **21:** 503–546.

Barlow HB. 1961. Possible principles underlying the transformation of sensory messages. In *Sensory communication* (ed. Rosenblith WA), MIT Press, Cambridge, MA.

Carandini M, Heeger DJ. 2012. Normalization as a canonical neural computation. *Nat Rev Neurosci* **13:** 51–62.

Glimcher PW. 2013. Value-based decision-making. In *Neuroeconomics: Decision-making and the brain* (ed. Glimcher PW, Fehr E), Academic, New York.

Heeger DJ. 1992. Normalization of cell responses in cat striate cortex. *Vis Neurosci* **9:** 181–197.

Heeger DJ, Simoncelli EP, Movshon JA. 1996. Computational models of cortical visual processing. *Proc Natl Acad Sci* **93:** 623–627.

Kahneman D, Tversky A. 1979. Prospect theory—Analysis of decision under risk. *Econometrica* **47:** 263–291.

LoFaro T, Louie KL, Webb R, Glimcher PW. 2014. The temporal dynamics of cortical normalization models in decision-making tasks. *Lett Biomath* **1:** 209–220.

Louie K, Grattan LE, Glimcher PW. 2011. Reward value-based gain control: Divisive normalization in parietal cortex. *J Neurosci* **31:** 10627–10639.

Louie K, Khaw MW, Glimcher PW. 2013. Normalization is a general neural mechanism for context-dependent decision making. *Proc Natl Acad Sci* **110:** 6139–6144.

Louie K, LoFaro T, Webb R, Glimcher PW. 2014. Dynamic divisive normalization predicts time-varying value coding in decision-related circuits. *J Neurosci* **34:** 16046–16047.

Neumann J, Morgenstern O. 1944. *Theory of games and economic behavior*. Princeton University Press, Princeton, NJ.

Olshausen BA, Field DJ. 1997. Sparse coding with an overcomplete basis set: A strategy employed by V1? *Vis Res* **37:** 3311–3325.

Platt ML, Glimcher PW. 1999. Neural correlates of decision variables in parietal cortex. *Nature* **400:** 233–238.

Ratliff F, Hartline HK. 1959. The responses of *Limulus* optic nerve fibers to patterns of illumination on the receptor mosaic. *J Gen Physiol* **42:** 1241–1255.

Reynolds JH, Heeger DJ. 2009. The normalization model of attention. *Neuron* **61:** 168–185.

Samuelson PA. 1947. *Foundations of economic analysis*. Harvard University Press, Cambridge, MA.

Schwartz O, Simoncelli EP. 2001. Natural signal statistics and sensory gain control. *Nat Neurosci* **4:** 819–825.

Shannon CE, Weaver W. 1949. *The mathematical theory of communication*. University of Illinois Press, Urbana, IL.

Simon HA. 1955. A behavioral model of rational choice. *Q J Econ* **69:** 99–118.

Simoncelli EP. 2003. Vision and the statistics of the visual environment. *Curr Opin Neurobiol* **13:** 144–149.

Wainwright MJ, Schwartz O, Simoncelli EP. 2002. Natural image statistics and divisive normalization: Modeling nonlinearity and adaptation in cortical neurons. In *Probabilistic models of the brain: Perception and neural function* (ed. Rao RPN, Olshausen BA, Lewicki MS), pp. 203–222. MIT Press, Cambridge, MA.

Webb R, Glimcher PW, Louie KL. 2014. *Rationalizing context-dependent preferences: Divisive normalization and neurobiological constraints on choice*. SSRN Electronic Library 2462895.

Spontaneous Decisions and Free Will: Empirical Results and Philosophical Considerations

JOANA RIGATO,[1] MASAYOSHI MURAKAMI,[2] AND ZACHARY MAINEN[2]
[1]*Center for Philosophy of Sciences of the University of Lisbon, Faculty of Sciences, 1749-016, Lisbon, Portugal*
[2]*Champalimaud Neuroscience Programme, Champalimaud Centre for the Unknown, 1400-038, Lisbon, Portugal*
Correspondence: *zmainen@neuro.fchampalimaud.org; mjrigato@fc.ul.pt*

Spontaneous actions are preceded by brain signals that may sometimes be detected hundreds of milliseconds in advance of a subject's conscious intention to act. These signals have been claimed to reflect prior unconscious decisions, raising doubts about the causal role of conscious will. Murakami et al. (2014. *Nat Neurosci* **17:** 1574–1582) have recently argued for a different interpretation. During a task in which rats spontaneously decided when to abort waiting, the authors recorded neurons in the secondary motor cortex. The neural activity and relationship to action timing was parsimoniously explained using an integration-to-bound model, similar to those widely used to account for evidence-based decisions. In this model, the brain accumulates spontaneously occurring inputs voting for or against an action, but only commits to act once a certain threshold is crossed. The model explains how spontaneous decisions can be forecast (partially predicted) by neurons that reflect either the input or output of the integrator. It therefore presents an explicit hypothesis capable of rejecting the claim that such predictive signals imply unconscious decisions. We suggest that these results can inform the current debate on free will but must be considered with caution.

In classical experiments on self-initiated actions in humans, Kornhuber and Deecke (1965) discovered a "readiness potential" that could be recorded from the scalp electroencephalogram (EEG). In brain areas such as the supplementary motor area (SMA), these signals appear as slowly ramping activity starting as much as 1 sec in advance of a voluntary movement. Benjamin Libet et al. (1983), in a now famous follow-up of this work, showed that the agent's conscious awareness of an intention to act occurs much later—only 200 msec before the act itself (Fig. 1). Although these experiments have been criticized on empirical grounds, including the ability to accurately measure the timing of conscious awareness (van de Grind 2002), similar results by other researchers have revealed both single neuron and fMRI signals substantially preceding spontaneous movements (Haggard and Eimer 1999; Lau et al. 2004; Soon et al. 2008; Fried et al. 2011), in some cases by several seconds (Soon et al. 2008).

This line of experiments (which we will hereafter refer to as "Libet-type" experiments) has led to significant attention and debate. Libet originally interpreted the experiments as implying that decisions are made at an unconscious level and only later do they become available to the conscious self. Given the usual demand that a free decision be made consciously, this interpretation seems to pose a grave threat to free will. But it has been greeted with diverse opinions and arguments. Libet himself argued that free will existed but only in the possibility of vetoing the action in that short gap between awareness and movement (Libet 1985). In the field of social psychology it was used to defend the claim that conscious will is an illusion (Wegner 2002). In philosophy, the tendency has been to diminish the relevance of Libet's work, even by philosophers who are skeptical about free will (Dennett 1991; Levy 2014). Among those who defend the existence of free will, Mele (2009, 2014a,b) has questioned the claims attributed to Libet's experiments, arguing that there is no proof that the readiness potentials are identical to an intention to act and that the type of decisions involved in these experiments is not representative of the kinds of important decisions about which one should care about free will.

Neuroscience may have something to contribute to this debate, as it has been studying the neural basis of decision-making for more than a decade (Gold and Shadlen 2007). The psychology and neurobiology of simple perceptual decisions such as discrimination tasks (Roitman and Shadlen 2002; Wang 2002; Mazurek et al. 2003), reaction time tasks (Hanes and Schall 1996), or simple value-based choices such as choosing one of two food items (Krajbich et al. 2010) have converged to a common if not consensual mechanistic hypothesis. According to this hypothesis, decisions are the result of the accumulation of evidence. Evidence may correspond to sensory signals or to more abstract variables such as subjective value. When a certain evidence threshold or bound is reached, the decision for the corresponding choice is made. Choices may correspond to observable actions or commitments to act in the future. Formal mathematical accounts of this sort are referred to as "integration-to-bound" models.

Spontaneous decisions, like starting to move out of one's own initiative, have been less studied than evi-

 doi: 10.1101/sqb.2014.79.024810

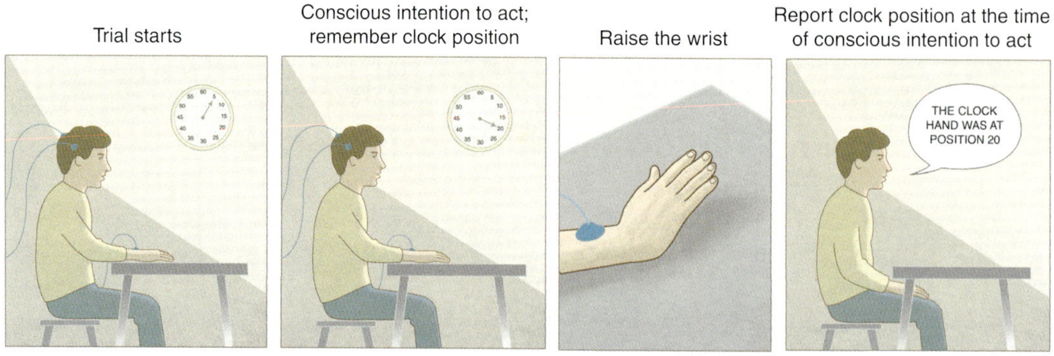

Figure 1. The spontaneous action generation task in Libet's experiment (1983). A human subject is instructed to spontaneously flex the wrist of his/her right hand at any time while looking at a clock-like visual display. After the trial, the subject reports the time he/she became consciously aware of his/her intention to act. (Adapted from Haggard 2008, with permission from Macmillan Publishers Ltd.)

dence-based decision-making and may seem initially to correspond to a different type of decision. Self-initiated acts do have characteristics that might lie outside the normal issues addressed by decision models. A first issue is how to account for the extreme unpredictability in the timing of spontaneous actions. A second, and related, issue is when there are not cues from the environment or well-defined values at stake, what plays the role of "evidence" driving the decision? Then most perplexingly, is there room for a role for the "self" or agent in this model?

The primary aim of this review is to consider what empirical results in neuroscience, by attempting to address issues such as these, may have to say about the nature of free will. The central reference will be the recent work of Murakami et al. (2014), who presented new experimental evidence about the neural origins of spontaneous decisions in rats. We have taken as our mission the nontrivial task of providing an account accessible to both scientists and philosophers, without doing injustice to either by oversimplification. Surprisingly, perhaps, we will suggest that even studies in rats might be illuminating to the case of free will in humans. Nevertheless, great care must be taken in making the leap from empirical data to metaphysical conjectures.

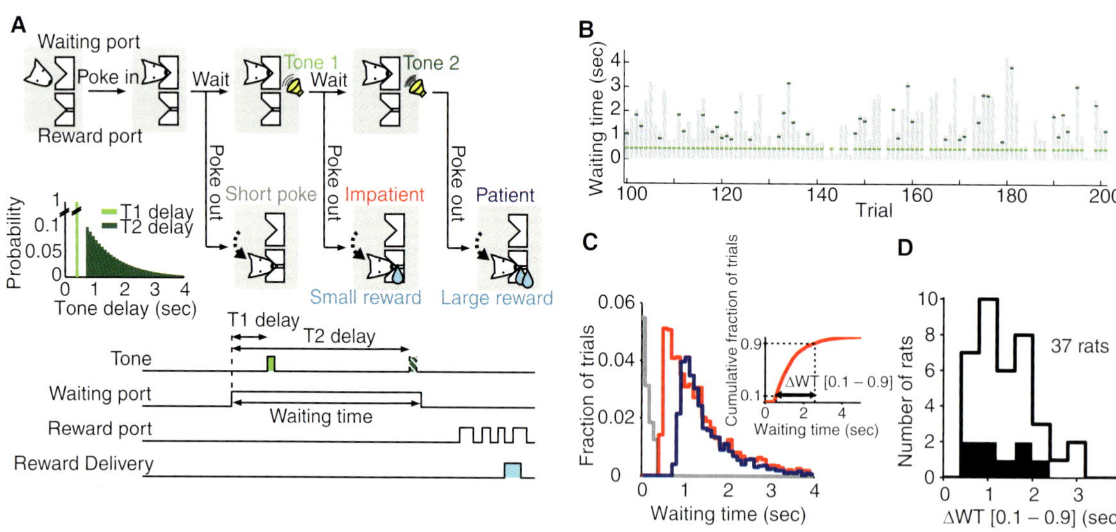

Figure 2. The waiting task and waiting behavior. (*A*) Schematic diagram of trial events in the rat waiting task (*top*). In each trial, the rat is required to wait for tone(s) and move to the reward port to obtain water reward. If the rat fails to wait for Tone 1 (T1), the reward is not available. If the rat waits for T1 but leaves the port before Tone 2 (T2), a small reward is available. If the rat waits until T2, a large reward is available. Probability distributions of the delays of T1 (light green) and T2 (dark green) are shown in the *inset*. Time line of the task events (*bottom*). The light green rectangle indicates presentation of T1; dark green rectangle, T2; light blue rectangle, reward. Tone 2 is represented by a hatched rectangle to indicate it was not played in the impatient trials. (*B*) Snapshot of the waiting behavior. The waiting period in each trial is indicated by a gray bar. Light green ticks represent the presentation of T1; dark green ticks, T2. (*C*) Waiting time histograms of short poke trials (gray), impatient trials (red), and patient trials (blue) for an example rat. The histograms show data pooled across sessions. *Inset*: Cumulative histogram of waiting times in impatient trials from this rat. The arrow indicates the range from 10th to 90th percentile waiting times (ΔWT [0.1–0.9]), which represents the variability of waiting time for this rat. (*D*) Distribution of ΔWT [0.1–0.9], which shows the variability of waiting time across rats. Filled bars indicate rats used in electrophysiology experiments. (Adapted from Murakami et al. 2014.)

SPONTANEOUS ACTIONS FROM RODENTS TO HUMANS

Murakami et al.'s experiment went as follows (see Fig. 2). A thirsty rat initiated waiting by inserting its snout into the waiting port. Shortly thereafter, a first tone was played, after which the rat could garner a small amount of water reward at the reward port. Alternatively, if the rat successfully waited for a second tone, played at a random delay, it would get a larger reward. Of particular interest were those trials in which the rat responded between the first and second tones. In those "impatient" trials rats sometimes responded quickly, but often waited for a second or more before leaving the port. These leaving decisions were said to be "self-initiated" or "spontaneous" because of the high variability in their timing despite otherwise identical conditions.

The authors then recorded neuronal activity in the rat secondary motor cortex. Of the 385 neurons recorded in eight rats, 27 were identified as having "ramp-to-threshold" activity, which means that their firing rate ramped up during the waiting period, reaching a constant threshold just before movement initiation (Fig. 3). Moreover, the timing of threshold crossing was correlated with waiting time in such a way that when the rat waited longer the neurons' firing rate ramped up more slowly. The characteristic activity of these ramp-to-threshold neurons suggested that they might serve as an internal trigger for the behavioral response in the rat. An additional 64 neurons were identified as "transient neurons," because their activity also correlated with the waiting time, but they fired in a brief burst rather than a ramp (Fig. 4A). Remarkably, these transient neurons fired in correlation with waiting time even during a period extending to well over 1 sec before the actual movement. Some of the transient neurons were positively correlated with waiting time, whereas others were negatively correlated (Fig. 4B,C).

To attempt to explain the relationships between the observed electrophysiological recordings and the behavior, a computational model was created (Fig. 5). In this "integration-to-bound" model, a set of transiently active "units" constitute the input to the integrator, and the ramping neurons are taken to reflect the output of the integrator. When the integrator crosses a certain positive threshold,

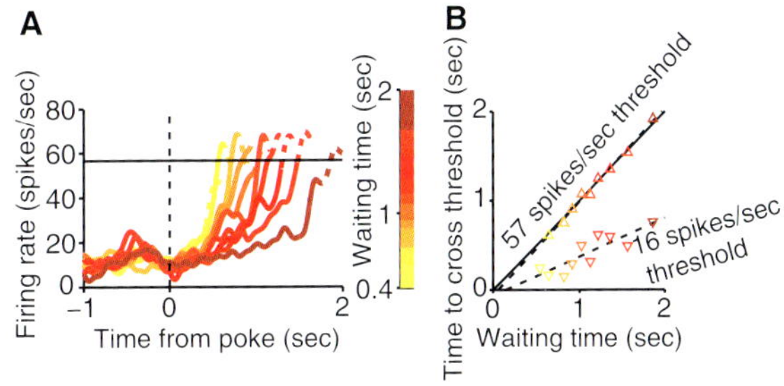

Figure 3. Waiting time predictive activity of ramp-to-threshold neurons. (*A*) Spike density functions (SDFs) of an M2 neuron in different waiting time trials, aligned to poke-in and smoothed with a Gaussian filter (SD = 50 msec). Impatient trials are grouped according to the waiting time, indicated by the color scale. Dashed lines in SDFs indicate times when the rat already left the port in some of the trials in that group. A threshold of 57 spikes/sec is indicated by the horizontal solid line. (*B*) Time to cross a given threshold firing level as a function of mean waiting time. The analyses with the highest and lowest thresholds with significant correlation (57 spikes/sec, triangle, and 16 spikes/sec, inverted triangle, respectively) are shown. Dashed lines indicate the linear regression line for the time to cross a threshold as a function of waiting time. The color represents the waiting time group as indicated by the color scale in *A*. (Adapted from Murakami et al. 2014.)

the decision to abort waiting is taken. Each input unit is activated at a certain time point at a level that varies randomly from trial to trial, matching the transient neuron data. Each unit is assumed to connect to the integrator with a different positive or negative weight and therefore promotes shorter or longer waiting accordingly.

This integration-to-bound model could parsimoniously account for the data collected at both the neuronal and the behavioral level. Most importantly, the properties of the ramp-to-threshold neurons strongly resembled the output of the integrator in the model, and the model could explain the correlation between both ramp-to-threshold and transient neurons' activity and waiting time. Moreover, the authors noticed that if transient neurons were assumed to fire independently from each other, the correlation of each one's activity with waiting time systematically decreased to zero as the number of transient neurons increased. Instead, by introducing a common noise source

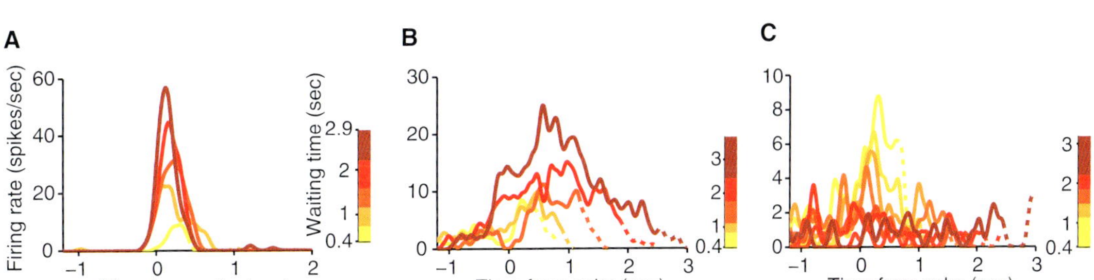

Figure 4. Waiting time predictive activity of M2 transient neurons. (*A–C*) Spike density functions (SDFs) of three M2 neurons. (*A*) An M2 neuron that shows phasic activation at the beginning of waiting and whose firing rate is positively correlated with waiting time. (*B*) An M2 neuron that shows sustained activation during waiting and whose firing rate is positively correlated with waiting time. (*C*) An M2 neuron that shows negative correlation between the firing rate and the waiting time. The format is the same as is in Figure 3A. (Adapted from Murakami et al. 2014.)

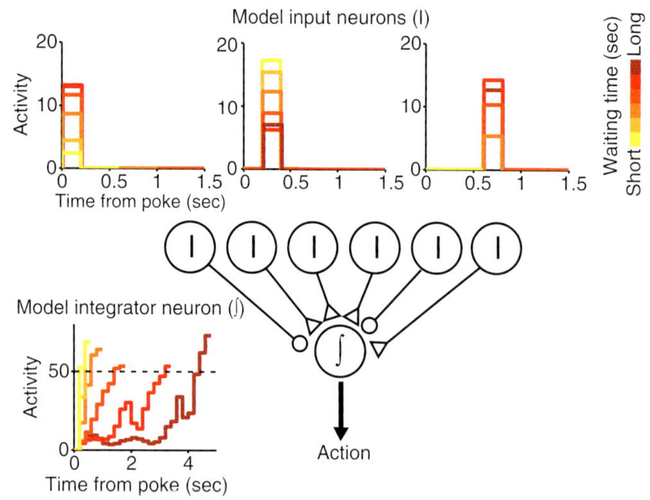

Figure 5. A schematic diagram of an integration-to-bound model for spontaneous actions. In the model, multiple input neurons (I) connect to one temporal integrator neuron (∫) with variable synaptic weights (circles indicate inhibitory weights and triangles indicate excitatory weights). *Inset* panels show activity of example model neurons (*top* three panels are example input neurons and the *bottom left* panel is an integrator neuron, the same format as in Fig. 3A). (Adapted from Murakami et al. 2014.)

by which transient units became correlated with each other, the contribution of each single unit became much more significant, even at large population sizes. Interneuronal correlations with the predicted magnitude were in fact observed in M2, an independent measurement that supports the merits of the model.

The integration-to-bound model suggests a critical perspective on the causal relationship between the two kinds of neural data and the behavioral response. The causal connection is mediated by the integrator circuit, which can be thought of as analogous to a ballot system during an election. Transient neurons are considered as input to the integrator. They play the role played by sensory evidence in perceptual decisions, "voting" for or against giving up waiting. Each one has some causal bearing in the timing of the action, but none can be said to determine it. The ramping neurons' activity, considered as output of the integrator, reflect a cumulative tally of votes. But if read out at a level less than the decision threshold, the tally is only suggestive and not determinant of an outcome. Even a tally reaching very close to threshold might happen to meander back away from it. The appearance of ramping is a result of averaging of trajectories that hit the threshold. The dynamics of individual trials are tortuous paths reflecting the random bombardment of votes both for and against. The origin of trial-by-trial variability in behavior is due to the variability in transient neurons' firing together with their strong interneuronal correlations.

With these data, Murakami et al. show the integration-to-bound model to be a useful explanatory tool in relating the timing of spontaneous decisions to the neuronal level, providing a hypothesis about what is taking place in the rats' brains. We suggest that they can potentially inform us about choices in Libet-type experiments as well. There are different levels of comparison in which we can see the parallels between the Murakami and Libet experiments. First of all, the similarity of behavior: Spontaneous wrist movements in humans and spontaneous waiting aborts in rats are *prima facie* similar (although not identical) kinds of "self-initiated" behaviors, and both occur with highly unpredictable timing. The second parallel is the similarity of shape and time course between ramping neurons' activity recorded in rat secondary motor cortex and readiness potentials recorded over the homologous human supplemental motor area (Fig. 6). Fried's (2011) study, showing something like ramping neurons in humans per-

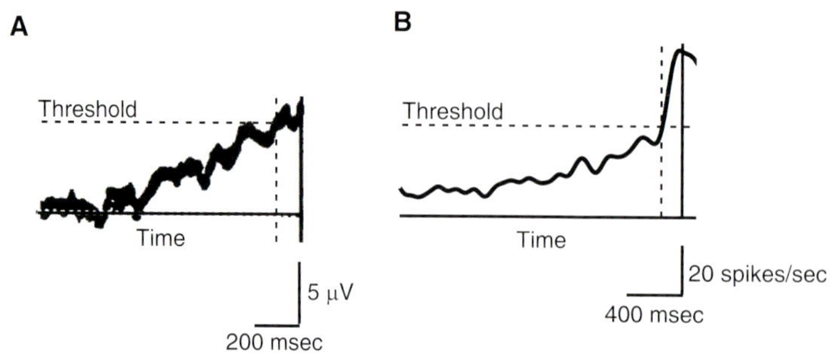

Figure 6. Ramping activity in human and rat motor cortex preceding spontaneous action. (*A*) An example EEG trace recorded from the vertex cortex corresponding to the supplementary motor area. The EEG trace is an average of 40 trials, aligned to the movement onset (solid vertical line). In this session, the subject felt a conscious intention to act, on average, 118 msec before the movement. A dashed horizontal line indicates a hypothetical threshold line for the conscious intention to act. (Adapted from Libet et al. 1983, by permission of Oxford University Press.) (*B*) Example single neuron activity recorded in rat M2 during spontaneous giving up trials. The trace is an average of 10 trials with the longest waiting times (1.7–2.0 sec), aligned to the give-up time (solid vertical line). A hypothetical threshold line (dashed horizontal line) and the time the activity crosses the threshold (dashed vertical line) are indicated. (Adapted from Murakami et al. 2014.)

forming the Libet task, is also strongly suggestive of this parallel. The third parallel is the applicability of the integration-to-bound model to describing the data. The case for the rodent data has been made above. In the case of Libet-type experiments, Schurger et al. (2012) have recently made this case strongly at the behavioral level.

A final very intriguing parallel between the two sets of experiments is the timing of threshold crossing in the rat experiments and the moment of subjective intention to act reported by Libet's subjects (Fig. 6). Both occur at ~200 msec before the movement. Accepting a similar set of mechanisms between the two experiments, it could be hypothesized that threshold-crossing corresponds to the moment of decision both at the neural and the mental level. Clearly, however, concrete evidence that threshold-crossing of neurons with ramping activity corresponds to the moment of awareness of decision, in one species or the other, would be needed to justify this idea.

PHILOSOPHY AND NEUROSCIENCE OF FREE WILL

The relevance of these results for the understanding of free will, defined as the agent's ability to control her decisions and actions, may not be entirely straightforward but we will argue for their importance nonetheless. Similarly to the now common distinction between the easy and the hard problems of consciousness (Chalmers 1995), we believe it can be useful to distinguish "easy" and "hard" problems of free will (Shariff et al. 2008). The "hard" questions of free will are the ones that have typically concerned philosophers for 25 centuries and seem to remain squarely out of reach by scientific experiments. Can a person act freely in a deterministic world? (A world is deterministic if, given a full description of all its elements and laws at t_1, only one possible state can follow at t_2. This means that, given a certain cause, only one effect can follow. A world is said to be indeterministic if this is sometimes not the case.) Can anyone make sense of an indeterministically caused action that is not random? In other words, is it coherent to think that a person might make a decision based on reasons and still have the possibility to act otherwise given the exact same circumstances and laws of nature?

In contrast, the "easy" problems are in principle scientific in nature, and today's neuroscience can contribute at least to illuminating the empirical aspects that are related to these issues. Is the variability that we perceive in the behavior of biological systems originated within the organisms themselves or is it just the deterministic output of undetectable changes in the environment? Do we have evidence that brain processes at the neuronal level are sensitive to genuine indeterminacy originated at a more fundamental level (the quantum domain, say)? What do we know about how the brain weighs alternatives and decides what to do when facing different possibilities of action? What does Libet's readiness potential represent in the psychological process of decision-making? What is the neural basis of the self who endorses the decision?

The relationship between the mental level, where deliberation is described in terms of beliefs and desires, and the neuronal level, where scientists measure spikes per second, has been and remains an object of great debate in philosophy. However, no one denies that what can be provisionally proven or disproven about the brain must be somehow taken into account when we speak about the mind. It is generally accepted that the mind "supervenes" on the brain: That is, that there cannot be changes at the mental level without there being corresponding changes at the neural level. This is not considered a strong thesis, as it regards only the interdependence of mind and brain, not their nature nor how one produces the other, but it is enough to justify the importance of understanding how the physiological substrate works in order to know the limits of what philosophers are entitled to conjecture. It would be pointless to argue for the philosophical plausibility of a certain account of free will if its empirical commitments were totally implausible from a scientific point of view.

However, one must be aware of several aspects of the philosophical concept of free will before purporting to infer any conclusions from what is possible to observe in experiments such as this one.

First of all, the philosophical literature about free will has intricately associated it with moral responsibility (Strawson 1962; Kane 1996; Pereboom 2001, 2014). To consider that to be free is to be fit for responsibility attributions is the standard way of framing the problem of free will, and one that only occasionally has been called into question (Steward 2012). In contrast, experiments such as Libet's or Murakami's have obviously nothing to do with morality. Yet, we believe morally neutral decisions are just as relevant for the debate as any other, for what matters for free will is the agent's control over her choice or her lack thereof.

Second of all, the idea that free will is an exclusively human ability is often taken for granted from the start (Clarke 2003), and that of course would prevent us from considering that animal models can provide us with anything other than evidence for the evolutionary antecedents of this privilege of ours. We do not favor this view, however, for there is no obvious reason why this should be taken for granted from a scientific standpoint. Even if there are relevant differences, we are likely to learn something about our situation from the evolutionarily common aspects we share with other animals.

Moreover, when trying to use these studies as arguments in the philosophical debate about free will, one must be very careful with terminology. What philosophers call "reasons for action" neuroscientists call "evidence"; what philosophers call "intentional," neuroscientists call "goal-directed"; and, most importantly, while philosophers (in the context of this debate) take "decision" to mean a conscious and active formation of an intention to act, usually following a deliberative process, neuroscientists apply the term both to the commitment made after some sort of reasoning in the case of evidence-based decisions *and* to more general mechanisms of settling an indeterminate matter, such as where to turn the head. It is also important to note that, in simple "decisions," the indeterminacy that precedes choice may be only epistemic (i.e., it may depend only on the contingent limitations of an external observer's knowledge of the stimulation). In fact, the problem of determinism, from an empirical point of view, is mainly an epistemic issue, whereas the sort of indeterminism that philosophers consider important is also ontological (i.e., it regards the underlying nature of reality). Also for that reason, philosophers and neuroscientists tend to be talking about different things when discussing these matters.

PHILOSOPHICAL IMPLICATIONS OF MURAKAMI'S RESULTS

Reinterpreting Libet's Readiness Potential

The arguments based on Libet-type cases typically use as a premise the fact that an unconscious ramping activity at the neuronal level is correlated with the decision made consciously by the agent in such a way that we are able to predict that decision to some degree. Since Libet et al. interpret the readiness potential as part of the decision, they suppose this observation to *mean* that the decision was unconsciously made by the agent's brain, which becomes aware of it only some time later. The gap between the start of the brain's predictive activity and the subjective awareness of the urge to move is said to be as long as 10 sec in some of the experiments (Soon et al. 2008). This is considered to undermine free will because it is hard to see how the agent might control what she is not aware of.

In the above experiment, Murakami et al. defend a model according to which the ramping activity of neurons reflects the accumulation of input in favor or against a certain spontaneous action, to which the agent is decisively committed only when a certain threshold is crossed. Until that moment, the action might be predicted with a certain degree of accuracy, like election polls, but the outcome is open until the very last moment when the decision is made—that is, when threshold is crossed, and overt action takes place immediately after. The authors hypothesized that the ramping activity of the integrator, which in the case of the waiting task in rats is reflected in the ramping neurons in the secondary motor cortex, is, in Libet's experiment, reflected in the slowly building readiness potential.

This model and its use in Murakami's experiment fleshes out what Alfred Mele (2014b) has argued when discussing putative neuroscientific evidence that free will is just an illusion: "when the rise starts about half a second before the muscle burst in the main experiment, the beginning of the EEG reading—or the first half of it—is correlated with something that precedes an intention rather than with an intention itself" (Mele 2014b, p.19).

Free Will in Rodents?

The relevance of experiments with rodents for the debate around human free will might be called into question by philosophers who would maintain that free will is a specifically human feature. But so far evidence has confirmed the success of integration-to-bound models in explaining simple evidence-based decisions not only in nonhuman primates and in humans, but also in rats and even in invertebrates (e.g., DasGupta et al. 2014), consistent with a wealth of biological data arguing for evolutionary continuity in biological systems. With the present data, Murakami et al. added a further argument for extending the case from perceptual or value-based judgments to self-initiated actions. So, from rats to monkeys to humans, the same type of process seems to take place when simple decisions are made, both when they are based on evidence provided by the environment and when they are spontaneously made.

From Waiting Tasks to Rational Decisions

Another question has to do with the relevance of this type of task, given that, in the philosophical literature, free will is taken to be the agent's ability to control a choice that is made for reasons. What philosophers mean by "reasons" in this context are beliefs and desires, and it may not be entirely clear how such reasons translate into factors in spontaneous decision tasks. In humans, these tasks typically involve apparently arbitrary urges. Agents are usually asked to flex their wrists or press a button whenever they feel like it, and they are explicitly instructed not to plan their action in advance. There is no extrinsic reason why it should be preferable to flex the wrist *now* rather than *later*, apart from maybe boredom or the intention to keep a random pattern. It may thus be objected that Libet-type experiments do not probe the right kinds of decisions because they are without reasons.

With respect to this issue, Murakami's task and interpretation would seem to help the case for relevance. Murakami's rats had conflicting motivations related to the content of the decisions themselves: the desire to drink plus the certainty of the small reward versus the possibility of a larger reward, despite the inconvenience of having the snout poked in while waiting for the delayed tone. One could thus argue that their experience is more closely related to the human decisions that philosophers discuss as free than to the tasks that have been tested in Libet-type experiments. And the fact that the rats hesitated before leaving the waiting port in impatient trials, while moving promptly in patient trials, was evidence for how much they had learned the task and understood what could be expected. In other words, one can argue that subjects in Murakami's waiting task did have beliefs—that water would be delivered in the reward port, that a larger amount would be given if they waited for the second tone—as well as desires—above all, the desire to drink. Therefore, their decisions were not random but made for reasons. On the other hand, although these reasons are clearly relevant to the rats' behavior in general, they do not necessarily dictate the *precise timing* of giving up in a particular trial, which varies substantially and randomly even given apparently similar level of thirst and so on. Thus, the precise moment of giving up in a particular time would seem to be a less-reasoned decision, much as is argued for the Libet experiments.

Dual Theories of Decision-Making

The main trend in neurobiology is to assume that "the path from simple decisions to complex ones may be more straightforward than it appears" (Gold and Shadlen 2007, p. 562). However, further evidence that the integration-to-bound decision model applies to truly "deliberative" decision-making may be needed in order to more reliably move on to the sort of cases philosophers enjoy discussing. Are rapid perceptual decisions truly representative of the process by which, for instance, a human agent decides to change jobs and move with her family to a different city? Is the rational, abstract, and linguistically mediated consideration of reasons in this case just a more complex version of what animals such as Murakami's rats experience? It has indeed been argued that complex human choices, like deciding to marry someone, are at a different qualitative level with respect to more simple ones. In particular, Kahneman's defense (2011) of a two-systems approach to judgment and choice in which a fast, automatic, and often unconscious System 1 is distinguished from the slow, effortful, and controlled System 2, has a long and influential history in psychology (Tversky and Kahneman 1974; Kahneman et al. 1982; Gigerenzer et al. 1999). In the context of moral psychology, a dual-process theory has also been proposed (Greene et al. 2004), suggesting that competing subsystems (emotional versus cognitive) in the brain are responsible for moral judgments under different situations.

It is tempting to think that decisions based on deliberation are more likely to be free than rapid choices based on "gut," in the sense that the longer process of weighing different reasons, such as very abstract ones like moral questions, seems to allow for more control by the agent's conscious self than an automatic mechanism. However, processes underlying System 1, despite being, by definition, not conscious or effortful, are associated with experiences of agency and choice, just like the ones underlying System 2. Also, there is not yet any evidence for such a dichotomy at the neurophysiological level (Sugrue et al. 2005), which would be fundamental given our assumption of the supervenience of the mind on the brain. Moreover, we do not have reasons to believe a putative deliberative system would be less deterministic than its counterpart based on fast heuristics.

All or Nothing at All

The questions raised by the differences between Murakami's rats and human agents undergoing complex deliberative processes remind us how the use of neuroscientific results in philosophical discussions about free will needs to be conducted with caution. However, one thing that becomes clear when we review the results from this experiment and others is that the same integration-to-bound model that has successfully been applied to simple decisions in humans and monkeys allows for a parsimonious explanation in the case of rats and, most probably, in the case of Libet-type subjects. So the similarity between the neural patterns in all these cases makes it likely that either there is a continuity between all of them and the more complex cases considered by philosophers when discussing free will, or none of them is relevant for the discussion—Libet cases included.

CONCLUSION

The integration-to-bound model and the evidence provided by Murakami et al.'s results of its general applicability to situations of self-initiated decisions in biological systems is a good challenge to the common idea that neuroscience is inimical to free will. In cases of human actions, it provides us a hypothetical explanation for the timing of the phenomenological experience of conscious decision, identified as the moment of threshold crossing within the integrator mechanism. At the same time, and importantly, it accounts for the existence of causal antecedents that make one decision or another more likely. The outcome is neither inevitable beforehand nor random.

Some free will defenders argue that Libet-type experiments are not representative of the sort of situations that are typically under discussion in the philosophical context. We believe this argument is open for discussion, but if Murakami and his co-workers are right, it is actually irrelevant. Libet's threat to free will has been overrated simply because it wrongly assumed that because a choice is preceded by a signal it follows that a signal is always followed by a choice. Instead, even though the choice can be forecast, it remains open until decision is made. Together with growing consensus in favor of an inner source of variability in biological systems (Brembs 2011), these results leave the door open even for the most demanding indeterministic accounts of free will.

REFERENCES

Brembs B. 2011. Towards a scientific concept of free will as a biological trait: Spontaneous actions and decision-making in invertebrates. *Proc R Soc B* **278:** 930–939.

Chalmers D. 1995. Facing up to the problem of consciousness. *J Conscious Studies* **2:** 200–219.

Clarke R. 2003. *Libertarian accounts of free will.* Oxford University Press, New York.

DasGupta S, Ferreira CH, Miesenböck G. 2014. FoxP influences the speed and accuracy of a perceptual decision in *Drosophila. Science* **344:** 901–904.

Dennet D. 1991. *Consciousness explained.* Little, Brown, Boston.

Fried I, Mukamel R, Kreiman G. 2011. Internally generated preactivation of single neurons in human medial frontal cortex predicts volition. *Neuron* **69:** 548–562.

Gigerenzer G, Todd PM, the ABC Research Group. 1999. *Simple heuristics that make us smart.* Oxford University Press, New York.

Gold JI, Shadlen MN. 2007. The neural basis of decision making. *Annu Rev Neurosci* **30:** 535–574.

Greene JD, Nystrom LE, Engell AD, Darley JM, Cohen JD. 2004. The neural bases of cognitive conflict and control in moral judgment. *Neuron* **44:** 389–400.

Haggard P. 2008. Human volition: Towards a neuroscience of will. *Nat Rev Neurosci* **9:** 934–946.

Haggard P, Eimer M. 1999. On the relation between brain potentials and the awareness of voluntary movements. *Exp Brain Res* **126:** 128–133.

Hanes DP, Schall JD. 1996. Neural control of voluntary movement initiation. *Science* **274:** 427–430.

Kahneman D. 2011. *Thinking fast and slow*. Farrar, Straus & Giroux, New York.

Kahneman D, Slovic P, Tversky A, ed. 1982. *Judgment under uncertainty: Heuristics and biases*. Cambridge University Press, New York.

Kane R. 1996. *The significance of free will*. Oxford University Press, New York.

Kornhuber HH, Deecke L. 1965. Hirnpotentialanderungen bei wilkürbewegungen und passiven bewegungen des menschen: Bereitschaftspotential und reafferente potentiale. *Pflugers Arch Gesamte Physiol Menschen Tiere* **284:** 1–17.

Krajbich I, Armel C, Rangel A. 2010. Visual fixations and the computation and comparison of value in simple choice. *Nat Neuroscience* **13:** 1292–1298.

Lau HC, Rogers RD, Haggard P, Passingham RE. 2004. Attention to intention. *Science* **303:** 1208–1210.

Levy N. 2014. *Consciousness and moral responsibility*. Oxford University Press, Oxford.

Libet B. 1985. Unconscious cerebral initiative and the role of conscious will in voluntary action. *Behav Brain Sci* **8:** 529–566.

Libet B, Gleason CA, Wright EW, Pearl DK. 1983. Time of unconscious intention to act in relation to onset of cerebral activity (Readiness-Potential). *Brain* **106:** 623–642.

Mazurek ME, Roitman JD, Ditterich J, Shadlen MN. 2003. A role for neural integrators in perceptual decision making. *Cereb Cortex* **13:** 1257–1269.

Mele AR. 2009. *Effective intentions: The power of conscious will*. Oxford University Press, New York.

Mele AR. 2014a. *A dialogue on free will and science*. Oxford University Press, New York.

Mele AR. 2014b. *Free. Why science hasn't disproved free will*. Oxford University Press, New York.

Murakami M, Vicente MI, Costa GM, Mainen ZF. 2014. Neural antecedents of self-initiated actions in secondary motor cortex. *Nat Neurosci* **17:** 1574–1582.

Pereboom D. 2001. *Living without free will*. Cambridge University Press, New York.

Pereboom D. 2014. *Free will, agency and the meaning of life*. Oxford University Press, Oxford.

Roitman JD, Shadlen MN. 2002. Response of neurons in the lateral intraparietal area during a combined visual discrimination reaction time task. *J. Neurosci* **22:** 9475–9489.

Schurger A, Sitt JD, Dehaene S. 2012. An accumulator model for spontaneous neural activity prior to self-initiated movement. *Proc Natl Acad Sci* **109:** e2904–e2913.

Shariff AF, Schooler J, Vohs KD. 2008. The hazards of claiming to have solved the hard problem of free will. In *Are we free? Psychology and free will* (ed. Baer J, Kaufman JC, Baumeister RF), pp. 181–204. Oxford University Press, New York.

Soon CS, Brass M, Heinze HJ, Haynes JD. 2008. Unconscious determinants of free decision in the human brain. *Nat Neurosci* **11:** 543–545.

Steward H. 2012. *A metaphysics for freedom*. Oxford University Press, Oxford.

Strawson PF. 1962. Freedom and resentment. *Proc Br Acad* **48:** 187–211.

Sugrue LP, Corrado GS, Newsome WT. 2005. Choosing the greater of two goods: Neural currencies for valuation and decision making. *Nat Rev Neurosci* **6:** 363–375.

Tversky A, Kahneman D. 1974. Judgment under uncertainty: Heuristics and biases. *Science* **185:** 1124–1131.

van de Grind W. 2002. Physical, neural, and mental timing. *Conscious Cogn* **11:** 241–264.

Wang XJ. 2002. Probabilistic decision making by slow reverberation in cortical circuits. *Neuron* **36:** 955–968.

Wegner D. 2002. *The illusion of conscious will*. MIT Press, Cambridge, MA.

Predicting the Accuracy of a Decision: A Neural Mechanism of Confidence

CHRISTOPHER R. FETSCH,[1] ROOZBEH KIANI,[2] AND MICHAEL N. SHADLEN[1]
[1]*Howard Hughes Medical Institute, Department of Neuroscience and Kavli Institute for Brain Science, Columbia University, New York, New York 10032*
[2]*Center for Neural Science, New York University, New York, New York 10003*
Correspondence: crf2129@columbia.edu

The quantitative study of decision-making has traditionally rested on three key behavioral measures: accuracy, response time, and confidence. Of these, confidence—defined as the degree of belief, prior to feedback, that a decision is correct—is least well understood at the level of neural mechanism, although recent years have seen a surge in interest in the topic among theoretical and systems neuroscientists. Here we review some of these developments and highlight a particular candidate mechanism for assigning confidence in a perceptual decision. The mechanism is appealing because it is rooted in the same decision-making framework—bounded accumulation of evidence—that successfully explains accuracy and reaction time in many tasks, and it is validated by neurophysiology and microstimulation experiments.

The merits of self-knowledge were well known to the ancients, as reflected in the Confucian saying: "When you know a thing, to hold that you know it; and when you do not know a thing, to allow that you do not know it—that is knowledge" (*Analects*, 2:17). A rudimentary but important form of self-knowledge is the degree of certainty, or confidence, one has in a decision (e.g., whether it is correct or will lead to a desired outcome). In many instances,[1] a judgment of confidence boils down to the question: How reliable was the evidence on which my decision was based? Here, evidence is meant to include external events transduced by the senses, as well as internal signals derived from memory. It seems likely that evidence from multiple sources, along with other factors such as biases and prior expectations, converges onto an internal representation that governs the choice. As we will show, the same process that converts evidence to a choice can additionally explain not only the time required to make the decision but also the confidence associated with it.

Confidence has obvious practical and social value in our daily lives. Any time you attempt to convince someone that something is true, or justify a course of action, your degree of certainty guides the interaction. Confidence serves to modulate learning rates and the allocation of attentional resources (Dayan et al. 2000), and it enables online regulation of the decision process itself (Vickers 1979). More generally, confidence is crucial for behavior in complex environments where sequences of decisions must be made without immediate feedback. This type of problem is common to most animals, as it includes decisions about foraging (explore or exploit?) and predator avoidance (hide or flee?). Yet in common parlance it also carries an air of subjectivity that would seem to preclude its study in animal models. This is a challenge if we wish to study confidence at a mechanistic level using the full toolkit of neuroscience. Fortunately, recent advances have begun to break through this barrier. Researchers interested in the neurobiology of decision-making have taken on the problem in earnest, buoyed by the propagation of new behavioral paradigms that allow an assessment of confidence in nonverbal animals. One might say that, in neuroscience at least, confidence no longer plays a "Cinderella role..., overlooked as an interesting variable in its own right" (Vickers 2001).

Psychologists have long recognized two main sources of information for generating confidence judgments: evidence strength and decision time. The relative emphasis on the latter has waxed and waned over the decades, and recent studies in both humans (e.g., Rahnev et al. 2011; Ko and Lau 2012) and animals (Kepecs et al. 2008; Komura et al. 2013) have employed signal-detection models that do not involve the temporal dimension at all. However, most decisions are based on streams of evidence rather than single observations and generally incorporate a termination rule that implements the trade-off between speed and accuracy. Thus, the decision process is naturally conceived as the accumulation of evidence over time, formalized with a variety of models based on sequential analysis (Stone 1960; Laming 1968; Link and Heath 1975; Ratcliff 1978; Vickers 1979). These models have been highly successful in explaining the speed and accuracy of decisions in multiple settings, and they find support in neurophysiological observations (Gold and Shadlen 2007; Shadlen and Kiani 2013). We will describe recent work (Kiani and Shadlen 2009; Fetsch et al. 2014; Kiani et al. 2014a) that

[1]Naturally, we are unable to cover every sense of the word confidence (but see Conclusion).

 doi: 10.1101/sqb.2014.79.024893

extends the bounded accumulation framework in a way that—building on insights by Vickers and others—parsimoniously explains confidence and is consistent with neurophysiology. We will also touch on a few lesser known implications of these advances—the nature of noise, biases, and other forms of confidence—and propose some future directions.

EVIDENCE AND TIME: ORIGINS OF THE PSYCHOPHYSICAL STUDY OF CONFIDENCE

Confidence was first approached in a quantitative manner by a second generation of experimental psychologists (Peirce and Jastrow 1884; Fullerton and Cattell 1892) who began to question the assumption of their forebears—Fechner, Weber, and Wundt—of a "just-noticeable" difference or increment. The strong version of this idea, known as absolute- or high-threshold theory, asserts that variations in the internal representation of a stimulus (or difference between stimuli) cannot influence a percept unless they result in the crossing of an absolute threshold. Below the threshold, or within a just-noticeable increment, changes in stimulus intensity or difference magnitude affect only the probability of threshold crossing, not the quality of the sensation. By the end of the 19th century, however, it became evident that the nervous system permits smooth gradations of "clearness" in discrimination (Fullerton and Cattell 1892), an intuitive concept that opened the door to the rigorous measurement and modeling of subjective confidence judgments.[2]

For confidence to be useful, it ought to predict accuracy on average, at least weakly.[3] Indeed, the empirical relationship between confidence and accuracy inspired the first attempt to describe confidence using an equation, namely as the scaled logit of probability correct (Peirce and Jastrow 1884). Later experiments (Garrett 1922; Johnson 1939; Festinger 1943; Pierrel and Murray 1963) found that in fact confidence is more tightly coupled to evidence strength or discriminability than to accuracy per se. In parallel, another school of thought was emerging based on an observed inverse relationship between confidence and the time taken to reach a decision (reaction time, RT; Henmon 1911; Seward 1928; Kellogg 1931; Volkmann 1934; Johnson 1939; Audley 1960). The consistency of the trend led some authors to conclude that the level of confidence might be dictated entirely by RT (Volkmann 1934; Reed 1951) or, relatedly, the number of contradicting observations leading up to the choice (Audley 1960). According to this view, confidence bears a positive relationship with evidence strength and accuracy solely as a by-product of being inversely related to time, because those other quantities are also inversely related to time.

To reconcile these findings, Vickers emphasized the distinction between the two ways a decision process may be terminated (Vickers 1979; Vickers and Packer 1982). Intuitively, if the experimenter controls the stimulus duration, then longer trials will tend to deliver more information and result in greater accuracy. In this case ("time-limited," Fig. 1A,B), confidence should increase as a function of time. If instead the subject decides when to terminate the decision ("information-limited," Fig. 1C), longer decisions will be associated with lower discriminability and hence lower confidence. This insight, and its agreement with data (e.g., Irwin et al. 1956; Audley 1960), led Vickers to posit the balance-of-evidence (BE) hypothesis (Vickers 1972), in which confidence is a function of the difference between accumulated evidence totals favoring the chosen and unchosen option. With BE, a single mechanism can be responsible for confidence judgments in both types of task yet produce the opposite relationship with trial duration.

Importantly, the BE hypothesis implies that the primary relationship is between evidence strength and confidence; the correlation between RT and confidence is incidental (Vickers 2001)—a property shared by more recent work as well (Rolls et al. 2010; Zylberberg et al. 2012; Ratcliff and Starns 2013). In contrast, our recent findings (Kiani and Shadlen 2009; Fetsch et al. 2014; Kiani et al. 2014a) support a role for time in and of itself, as explained below.

A TALE OF TWO THEORIES

From the last century of psychophysical inquiry emerged two general frameworks that form the backbone of the modern study of confidence: signal-detection theory (SDT) and sequential analysis (SA). According to SDT (Peterson and Birdsall 1953; Tanner and Swets 1954; van Meter and Middleton 1954), an observation is drawn from one of two overlapping distributions, and the goal of the observer is to determine which distribution. In detection tasks, the two distributions represent "noise" and "signal plus noise," whereas in discrimination tasks they might represent the two alternatives being discriminated. In either case, the decision is made by converting the noisy observation to a decision variable (DV) and comparing it with a criterion. Because the DV is monotonically related to the probability of making the observation given the state of the world (Swets et al. 1961; Gold and Shadlen 2001), its distance from the criterion is a reasonable proxy for confidence. The seeds of this idea actually predated formal SDT with the theory of Cartwright and Festinger (1943), but it matured in the ensuing decades among researchers across several disciplines (Norman and Wickelgren 1969; Kinchla and Collyer 1974; Ferrell and Mcgoey 1980; Treisman and Faulkner 1984).

SDT-based models of confidence remain highly influential in psychology and neuroscience (Higham 2007;

[2]It also presaged the more flexible criterion that underlies signal-detection theory (SDT; see below).

[3]If we define a decision as a commitment to a proposition or hypothesis, then confidence is the subjective (Bayesian) probability—for a single decision—that the hypothesis is true, given the data. This differs from accuracy, which is the objective (frequentist) probability—across multiple decisions—that the chosen hypothesis was true. The reason these quantities are not equivalent, except under certain conditions, has recently been clarified by Drugowitsch et al. (2014).

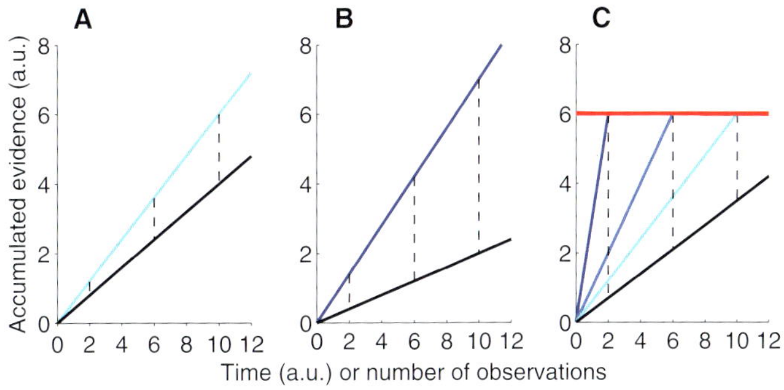

Figure 1. Confidence as the balance of evidence (BE) predicts different relationships with decision time depending on how the decision terminates. (*A*,*B*) Accumulated evidence totals favoring two options, as a function of time or number of observations, are depicted schematically as lines of different slope (blue = chosen option, black = unchosen). When the environment or experimenter dictates the time of commitment (*A*,*B*), (e.g., by soliciting a choice after 2, 6, or 10 observations), the BE (length of vertical dashed lines) increases with elapsed time. The pattern holds for both low (*A*) and high (*B*) discriminability. (*C*) If on the other hand the decision is terminated when accumulated evidence reaches a threshold (red line), BE decreases with elapsed decision time, because the latter is inversely related to discriminability (three different levels are shown as shades of blue). Note that this relationship breaks down when negative increments are allowed (i.e., if accumulators are anticorrelated; see description of Kiani et al. [2014a] below for details). (Adapted from Vickers and Packer 1982.)

Kepecs et al. 2008; Rounis et al. 2010; Rahnev et al. 2011, 2012; Ko and Lau 2012; Komura et al. 2013; Lak et al. 2014; Scott et al. 2014). However, these models are fundamentally limited: they cannot explain the time it takes to reach a decision nor the effects of decision time on confidence mentioned above. Of course, all models simplify to some degree, but in this case time is not just a detail that can safely be ignored for convenience. As we shall see, deliberation time is a key factor in the determination of confidence when there is a speed-accuracy trade-off at play—that is, when the decision-maker terminates a decision based on accumulated evidence.[4] In such settings, not only does time play a role in computing confidence, but also the key prediction of "distance from criterion" is reversed (see below). This is not to minimize the relevance of SDT, nor to rule out a distance-from-criterion mechanism in all cases (e.g., when the task involves little or no deliberation; Zariwala et al. 2013). But in line with the psychophysical tradition, it seems clear that a general neural theory of confidence must include the dimension of time.

A natural starting point for such a theory can be found within the family of models involving sequential analysis, a statistical decision process that can be viewed as an extension of SDT into the time domain (Fig. 2A). Enthusiasm for SA models—which include bounded evidence accumulation, drift-diffusion, and race models—derives from their effectiveness in explaining choice and RT distributions across many domains[5] (Busemeyer 1985; Luce 1986; Reddi et al. 2003; Ratcliff and Smith 2004; Krajbich and Rangel 2011; DasGupta et al. 2014), as well as their robust neural correlates (Smith and Ratcliff 2004; Gold and Shadlen 2007; Heekeren et al. 2008; Ploran et al. 2011). When monkeys are trained to report the outcome of a decision with an eye movement, neurons related to oculomotor planning show persistent activity resembling the accumulation of noisy evidence toward a threshold or bound (Hanes and Schall 1996; Shadlen and Newsome 1996; Kim and Shadlen 1999; Horwitz and Newsome 2001; Roitman and Shadlen 2002; Ratcliff et al. 2003). In the lateral intraparietal area (LIP), this activity reflects not only the mean (Roitman and Shadlen 2002; Huk and Shadlen 2005; Kiani et al. 2008) but also the variance and autocorrelation (Churchland et al. 2011) of an evolving DV that explains choice and RT. As shown next, our recent work extends the bounded accumulation framework to explain how the brain might arrive at a prediction about the accuracy of a decision.

BOUNDED EVIDENCE ACCUMULATION UNIFIES CHOICE, RT, AND CONFIDENCE IN A COMMON FRAMEWORK

A feature of SA models is their explicit representation of both time and evidence, the principal ingredients for establishing a level of confidence. A key insight is that the amount of accumulated evidence (i.e., the state of the DV) at the time of the choice is associated with a particular probability that the decision will be correct (Fig. 2B). Kiani and Shadlen (2009) hypothesized that the brain uses implicit knowledge of this association, or mapping, to establish a degree of confidence in the choice, harnessing the same decision-making machinery that gives rise to the choice itself and the time it takes to decide.

Coming up with a way to test the hypothesis at the level of neurophysiology was not trivial: How does one ask a laboratory animal how confident it is? Fortunately, some promising answers to this question had recently been

[4]Note that the brain may apply such a termination rule even when the environment or experimenter controls stimulus duration (Kiani et al. 2008).

[5]Why do evidence accumulation/drift-diffusion models provide such good fits to choice and RT data even when the task does not involve a continuous stream of evidence to integrate? This is an important question for future work, but we speculate the common feature is the sequential nature of updating (e.g., repeated queries of a mnemonic representation).

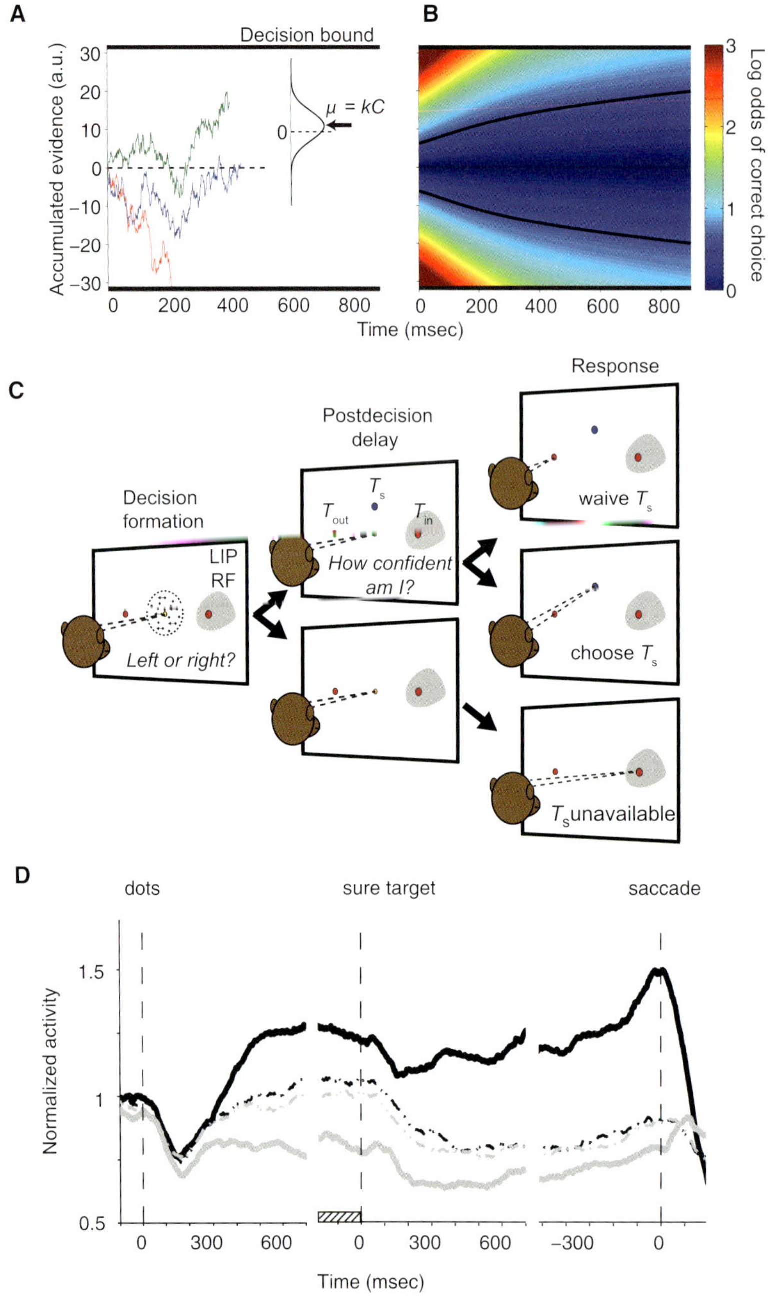

Figure 2. Model, behavioral task, and neural representation of a DV that explains confidence. (*A*) Depiction of bounded accumulation model (i.e., one-dimensional drift-diffusion) in which momentary evidence is drawn from a normal distribution with mean (μ) proportional to motion strength (coherence, *C*). Three example trials are shown (colored traces). Decision formation terminates when the stimulus is turned off (green and blue trials) or when the accumulated evidence reaches an upper or lower decision bound (red trial). (*B*) The model assumes the brain has implicit knowledge of the probability of being correct, plotted as the logarithm of the posterior odds (color), as a function of accumulated evidence and decision time. Black contours indicate a criterion on log-odds that partitions the space into low- and high-confidence regions. (*C*) Postdecision wagering (PDW) task (see text). Circle in *left* panel represents the random-dot motion stimulus. Red spots indicate direction targets, one of which was aligned with the response field (RF, gray region) of a neuron in area LIP. Blue spot is the "sure-bet" target (T_s). (*A–C*, Reproduced from Fetsch et al. 2014, with permission from Elsevier.) (*D*) LIP responses predict direction choices and PDW. Solid black and gray traces show averaged activity for trials ending in a saccadic choice into (T_{in}) or out of (T_{out}) the RF, respectively. Dashed traces represent trials in which T_s was chosen (black and gray, motion toward T_{in} or T_{out}, respectively). Figure shows only the random half of trials in which T_s was presented during the delay period. The monkey did not know whether the T_s option would be available until this time. (*D*, Reproduced from Kiani and Shadlen 2009, with permission from AAAS.)

furnished by a number of innovative studies in comparative psychology (for reviews, see Hampton 2009; Smith 2009; Terrace and Son 2009; Kepecs and Mainen 2012). Inspired by these advances in behavioral methodology, Kiani and Shadlen (2009) trained monkeys to report their confidence in a decision about the direction of visual motion (left or right) using a form of a postdecision wagering (PDW). Following the perceptual decision phase, on a random half of trials, the monkeys were allowed to refrain from reporting the direction choice and instead select a "sure bet" target (T_S; Fig. 2C). Choosing the sure bet yielded a guaranteed reward but one that was smaller than for a correct direction choice. The monkey could not predict whether he would have the opportunity to choose T_S until well after the visual stimulus was turned off. Indeed, the task was explicitly designed so that the monkey had to form a binary decision about the motion on each trial, even if on a particular trial they might have hoped that T_S would be shown. Crucially, the monkey's accuracy was greater on trials when the sure bet was offered but waived compared with when it was not offered. The improvement in accuracy held within each level of difficulty and for a subset of trials with identical motion stimuli (Kiani and Shadlen 2009). This strongly suggests that the decision to opt out was based on an assessment of the reliability of internal sensory evidence, not simply a classification of trial difficulty or the presence of lapses of attention.[6]

If the DV that underlies the choice also governs confidence, then a putative neural representation of the DV should predict postdecision wagering behavior. This was found to be the case. To appreciate the result, recall that LIP neurons represent the accumulated evidence favoring the saccadic choice target in their response field (RF). This activity ramps upward preceding choices into the RF and decreases or remains flat preceding choices out of the RF. Prior to sure-bet choices (Kiani and Shadlen 2009), the firing rates were intermediate (Fig. 2D), analogous to the intermediate levels of evidence associated with lower confidence in the model (Fig. 2B). This trend remained after accounting for stimulus difficulty (motion strength and viewing duration) and was apparent in both the rate of rise of activity during motion viewing and the sustained level of activity during the delay period before the sure-bet target was presented. In control experiments with the eventual sure-bet target location aligned with the RF, firing rates did not deviate from baseline during the stimulus or delay periods, regardless of trial outcome. In other words, the neurons also did not behave as though it were a three-choice task, suggesting that the intermediate activity was truly a signature of lower confidence in the binary perceptual decision.

[6]Despite these reassurances, the task has a drawback that choice and confidence are not overtly reported on the same trials. This was rectified in subsequent experiments in humans (Kiani et al. 2014a; see below) and monkeys (Kiani et al. 2011). For discussions of the pros and cons of different methods for assessing confidence and metacognition in animals, see Smith et al. (2012), Kepecs and Mainen (2012), Hampton (2009), Middlebrooks and Sommer (2011), and Terrace and Son (2009).

A CAUSAL TEST OF THE THEORY

These findings were consistent with the hypothesis that a common mechanism of bounded evidence accumulation underlies all three pillars of choice behavior in decision theory: choice (i.e., accuracy), decision time, and confidence. Conveniently, our understanding of the neural representation of sensory evidence in the motion discrimination task allowed a powerful causal test of the hypothesis. The momentary evidence for this task is known to reside in motion-sensitive regions of extrastriate visual cortex, especially the middle temporal (MT) and medial superior temporal (MST) areas. Electrical microstimulation (μS) of MT/MST biases monkeys' choices (Salzman et al. 1990; Celebrini and Newsome 1995) and RTs (Ditterich et al. 2003) in a manner consistent with a change in the strength of momentary evidence (Fig. 3A). Based on these studies, we reasoned that μS combined with PDW would provide a sensitive test of the idea that confidence arises from the same neural mechanism as choice and RT.

Monkeys performed the direction discrimination task with PDW as in the previous study (Kiani and Shadlen 2009), but during motion viewing, on half the trials, we applied μS to a cluster of neurons selective for one of the two alternatives. As previously shown (Salzman et al. 1990, 1992; Celebrini and Newsome 1995), μS caused the monkey to choose the preferred direction of the stimulated neurons more frequently, manifesting as a horizontal shift of the choice function (Fig. 3B). The key finding was that μS altered confidence in a similar way, mimicking a change in the visual motion itself (Fetsch et al. 2014). Specifically, the bell-shaped function relating the proportion of sure-bet choices to motion strength (Fig. 3D) shifted along the motion axis in a similar fashion as the choice function, and the effects on choice and confidence were well-correlated across sessions (Fig. 3C). Moreover, μS did not prevent the improvement in sensitivity when the sure bet was offered but waived (Fig. 3B). This implies that the brain's ability to assess the reliability of evidence and opt out accordingly was not disrupted, despite the artificial pattern of activity induced by microstimulation. In terms of the model, this means μS did not alter the internal mapping between accumulated evidence and confidence (Fig. 2B), nor the criteria applied to that mapping for opting out.

Taken together with previous work (Salzman et al. 1990; Ditterich et al. 2003; Hanks et al. 2006), this study provides a quantitative, causal link between the activity of neurons in visual cortex and the three core behavioral indices in decision-making. The results support the idea that integration of noisy sensory signals gives rise to a common quantity that underlies choice, decision time, and confidence. That one can perturb all three is not surprising, but it is remarkable that the perturbation can be quantitatively reconciled such that the amount of change in choice is commensurate with specific changes in PDW (Fetsch et al. 2014) and RT (Ditterich et al. 2003; Hanks et al. 2006). Because μS alters neural activity in a manner that is unlike natural vision, one might have expected it to generate an anomalous experience

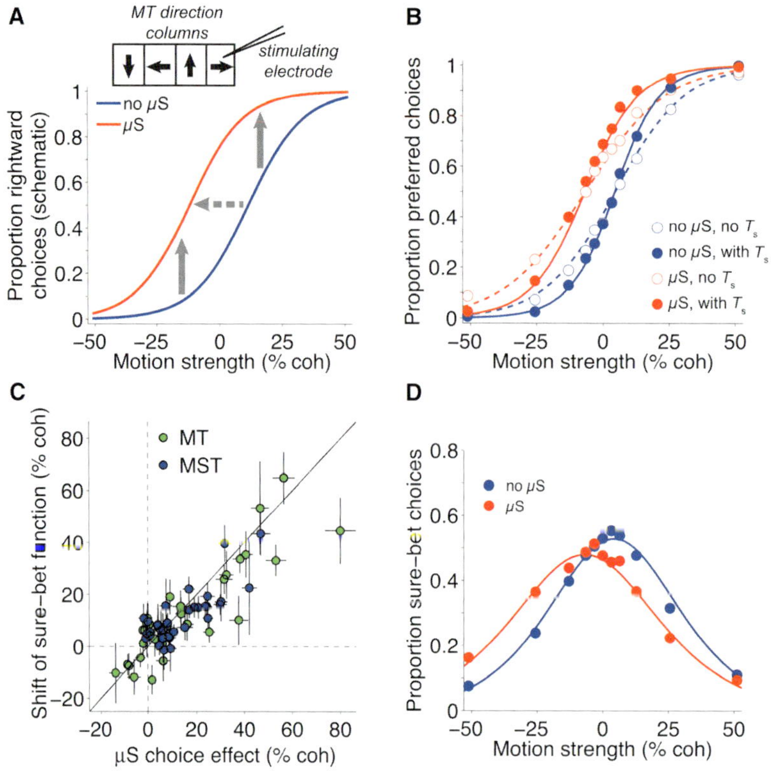

Figure 3. MT/MST microstimulation (μS) affects confidence in a manner similar to a change in visual motion. (*A*) (*Top*) Illustration of experimental approach (Salzman et al. 1990) that exploits the columnar architecture of MT to inject a relatively pure directional signal (in this case, rightward). Red and blue curves show schematically the proportion of rightward choices by the monkey as a function of motion strength (percent coherence; positive = rightward), for μS and no-μS trials, respectively. Electrical μS in this example causes the monkey to choose rightward more often (upward arrows). The effect on choice and RT (Ditterich et al. 2003, not shown) mimics a horizontal shift along the motion strength axis (dashed arrow). (*B*) Proportion of preferred-direction choices as a function of motion strength (positive coh = preferred direction of neurons at the stimulation site) for 63 sessions in two monkeys (Fetsch et al. 2014), plotted separately for the four conditions of the 2 × 2 design: μS present (red) or absent (blue), and with (solid) or without (dashed) the sure bet available. (*C*) Magnitude of μS effects on choice and confidence (horizontal "shifts")—plotted separately for stimulation sites in area MT (green) and MST (blue)—were correlated across sessions ($r = 0.87$). Error bars represent SE of the fitted parameters quantifying the shift. (*D*) Proportion of sure bet (T_s) choices as a function of motion strength for all sessions. Smooth curves in *B* and *D* represent fits/predictions of the bounded evidence-accumulation model (see text). (*B–D*, Reproduced from Fetsch et al. 2014, with permission from Elsevier.)

leading to a net decrease in confidence. It is also conceivable that, having been compelled to choose the preferred direction of the stimulated neurons more often, the monkey would have experienced an unusually high degree of confidence in those choices. Instead, the model—which implements μS essentially as an offset to motion strength—was able to *predict* the shift of the choice function and the degree of improvement on waived-sure-bet trials (Fig. 3B, red curves) based on a fit to the remaining observations.

One may notice in Figure 3D that the sure-bet curve on μS trials (red) is not a translated replica of the no-μS curve (blue). There is a slight decrease in the probability of a sure-bet choice, most prominently at the weaker motion strengths, indicating an increase in confidence with μS. The most plausible explanation for this feature of the data is a counterintuitive one: It suggests μS adds a small amount of noise to the sensory representation. Of course an increase in noise should reduce the animal's accuracy (sensitivity), and indeed this is consistent with the subtle flattening[7] of the choice function in Figure 3B (red vs. blue solid curves). But why should this cause an increase in confidence? In the context of evidence accumulation, increasing the variability of momentary evidence leads the DV to "diffuse" further away from the starting point, which in our model (Fig. 2B) implies greater confidence and fewer sure-bet choices. Notably, this explanation requires that the brain does not "know" that the variance of the evidence has changed: It applies the same mapping to the accumulated evidence on all trials, hence greater confidence with μS despite lower sensitivity. This paradoxical result is also predicted

[7]This effect is not obvious by eye but is statistically reliable. A similar flattening was observed by Salzman et al. (1992), an effect they attributed to the spread of stimulation current to neighboring columns with different preferred directions of motion. Although they referred to this as adding noise, it is better construed as a reduction of the signal strength. This phenomenon may have accounted for part of the decrease in sensitivity we observed, but not the effect on confidence, which suggests an increase in noise itself.

by SDT-based models and at least one predecessor (Cartwright and Festinger 1943; Vickers 1979), and similar observations have been made in human subjects undergoing transcranial magnetic stimulation (Rahnev et al. 2012). We are currently exploring this phenomenon further by measuring MT/MST activity and PDW behavior in response to a visual manipulation designed to mimic the effect of adding noise to the DV (Zylberberg et al. 2014).

ELAPSED DECISION TIME INFORMS THE SENSE OF CONFIDENCE

The success of the modeling exercise in the microstimulation study (Fetsch et al. 2014) and earlier LIP recording study (Kiani and Shadlen 2009) was due to a critical feature of the model: The mapping between evidence and degree of belief depends on elapsed deliberation time. The time dependence comes from the mixture of motion strengths in the experiment: The subject does not know the reliability of evidence a priori, and so the passage of time itself becomes indicative of weaker evidence. However, owing to a lack of RT measurements in these studies, a more direct test of the role of time was warranted.

Kiani et al. (2014a), therefore, conducted a psychophysical experiment in which all three components of the decision were measured simultaneously. In this study, human subjects performed a direction discrimination task and could report their decision (up or down) as soon as they were ready. They did so with an upward or downward eye movement, but were instructed to vary the horizontal position of their saccadic end point based on how confident they felt in the decision. The pattern of saccadic end points indicated that, for all six subjects, motion strength and RT had independent leverage on confidence ratings (Fig. 4A). Because these three measures are correlated with accuracy and with each other, establishing an independent role for time required a stimulus manipulation that could prolong the decision without affecting accuracy. This was achieved by presenting a reversal of the animation sequence, which effectively undid evidence that had been supplied in the preceding ~160 msec. The manipulation had the effect of prolonging the decision without a significant change in accuracy, and on those trials the subjects reported lower confidence (Fig. 4B). The result suggests that the brain uses elapsed time as a source of information about signal reliability, even when controlling for the amount of accumulated evidence (Kiani et al. 2014a).

Explaining these findings required a more complicated model than the one-dimensional bounded accumulation process depicted in Figure 2A. This is because for a single accumulator with symmetric bounds, decisions are always terminated at a particular level of evidence, leaving nothing to covary with accuracy (and confidence) except RT. For example, if the bounds were flat (i.e., stationary as a function of time) the only possible arbiter of confidence would be decision time. Confidence would not depend on motion strength,[8] but this is not supported by the data. A solution (which happens to be more biologically plausible) is to posit separate accumulators for different options—a race between two diffusion processes (Fig. 4C; Mazurek et al. 2003; Churchland et al. 2008; Niwa and Ditterich 2008)—and a calculation of certainty based on the status of the losing accumulator (Fig. 4D). Intuitively, if the race between accumulators was a close call, it means the net evidence supporting the chosen option was weak and we should be less confident. In some ways the model resembles Vickers' BE hypothesis (Fig. 1C), except that negative increments are allowed, and the races are anticorrelated, because stimulus evidence for left is negative evidence for right. The degree of this anticorrelation effectively determines the degree to which time and motion strength account for confidence.

A critical feature of this study (Kiani et al. 2014a) was the simultaneous report of choice and confidence by a single ballistic (saccadic) eye movement, ensuring that both were guided by the same information. In this case the model predicts—and the data show—that confidence is an increasing function of evidence strength on both correct and error trials (Fig. 4E; Kiani et al. 2014a). This contradicts a key prediction of SDT-based models, that confidence on error trials should decrease with evidence strength or discriminability (Fig. 5A,B). The logic of the SDT prediction is that for strong stimuli the distributions representing the two alternatives are only slightly overlapping, such that any error must have arisen from an observation very close to the criterion, hence with low confidence (see also Vickers et al. 1985). We think this pattern has been oversold as a defining feature of confidence and its neural correlates (Insabato et al. 2010; Rolls et al. 2010; Komura et al. 2013). Consider that most experiments solicit a confidence judgment after the decision has been reported (Baranski and Petrusic 1998), allowing information "in the pipeline" to contribute to confidence but not the choice (Pleskac and Busemeyer 2010). The use of such information has been demonstrated in the study of changes of mind (Fig. 6A; Rabbitt 1966; Resulaj et al. 2009; Song and Nakayama 2009; Kiani et al. 2014b). This would allow the rare error associated with an easy stimulus to undergo correction, leading to low confidence on errors with stronger stimuli. In fact, we can reproduce the pattern predicted by SDT in a simple simulation of bounded accumulation in which the DV is allowed to continue evolving and affect confidence after the decision bound is reached (Fig. 6B). Thus, the superficial similarity between SDT and some publised experimental results does not indicate validity of SDT-based models over SA-based alternatives. When confidence and choice are reported at different times, such results must be interpreted with caution to avoid incorrect conclusions about the computation of certainty and its neural mechanisms.

[8]This prediction is surprisingly robust even if the bounds collapse over time, which they do, typically.

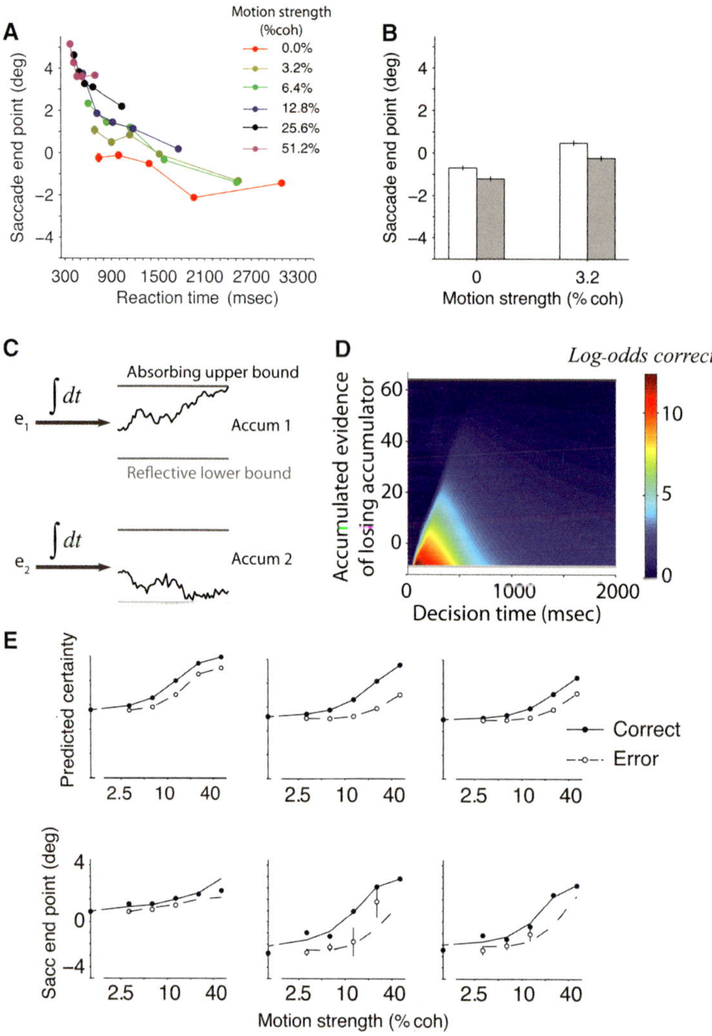

Figure 4. Confidence is informed by evidence and decision time. (*A*) Across-subject average ($N = 6$) of saccade end points, a psychophysical report of confidence, plotted as a function of RT and color-coded by motion strength. Positive end points indicate greater confidence. (*B*) Average saccade end point on trials with (gray) and without (white) a "reverse pulse" that increased RT without a significant change in accuracy (not shown). Error bars represent SEM. (*C*) In an extension of the model in Figure 2A, two accumulators compete by integrating noisy momentary evidence in favor of the two choices. The accumulator that first reaches the absorbing bound dictates the choice and decision time. (*D*) Confidence in the model is governed by the status of the losing accumulator, which at any given time is associated with a particular log-odds of being correct (color). Each level of evidence is associated with less confidence as time elapses. (*E*) Theoretical predictions (*top* row) and model fits (*bottom* row) of the confidence, or certainty, for three representative subjects, plotted as a function of motion strength and separated by trial outcome (correct vs. error). Model parameters were first estimated by fitting each subject's RT distributions irrespective of choice and confidence, and then used to predict the general form of the certainty functions and to fit their observed saccade end points. (Reproduced with permission from Kiani et al. 2014a.)

CONCLUSION AND FUTURE DIRECTIONS

It seems neuroscience may soon be in a position to help resolve decades of debate in mathematical psychology about the mechanism of confidence judgments. This optimism should be tempered somewhat by the reality that the brain is flexible enough to use different mechanisms depending on context or past experience. Nevertheless, a few core principles are emerging that should help frame the discussion and guide future investigations. From our perspective, these principles include the use of time as a proxy for evidence reliability (see also Hanks et al. 2011) and the idea that a common neural mechanism may be responsible for choice, RT, and confidence.

Some key unanswered questions remain. We think confidence involves the conversion of a DV to degree of belief based on both the magnitude of the DV and deliberation time. But how does this happen? Along with accumulated evidence, neurons in LIP represent information about elapsed time (Leon and Shadlen 2003; Janssen and Shadlen 2005; Maimon and Assad 2006) and thus could participate in computations involving both quantities. On the other hand, temporal signals related to urgency (Churchland et al. 2008; Cisek et al. 2009) or dynamic

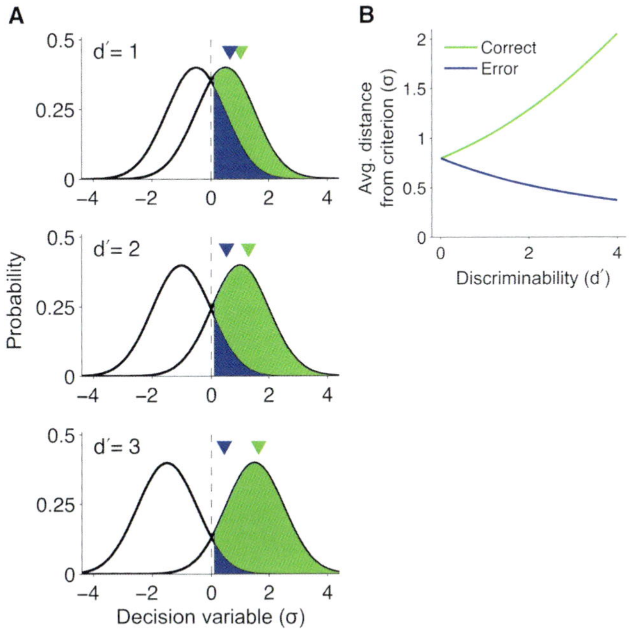

Figure 5. Confidence as the distance from criterion in signal-detection theory (SDT). (*A*) The overlapping Gaussians represent probability distributions of a noisy observation under two competing hypotheses (i.e., a leftward or rightward stimulus was present). The panels depict different levels of discriminability defined by the difference in means of the distributions in units of their standard deviation (d′). An observation to the right of the decision criterion (dashed line) will be correct if arising from the rightward distribution (green) or will be an error if from the leftward distribution (blue). Confidence in these choices could be computed as the distance from the criterion, the means of which are shown by the inverted triangles. Depiction is consistent with a hypothetical two-interval task, but the same logic applies in the single-interval case. (*B*) The average distance from criterion increases with discriminability for correct choices (green), but decreases for errors (blue). (See also Fig. 8g in Kim and Shadlem [1999].)

bias (Hanks et al. 2011) may need to be discounted or subtracted at later stages in order to recover the true level of evidence that informs confidence. Moreover, LIP represents a DV for decisions about where to move the eyes, but for other effectors or types of decisions, the readout mechanism that establishes a sense of confidence must be targeted elsewhere.

Interestingly, our microstimulation results (Fetsch et al. 2014) suggest there is at least one signal that is not subtracted prior to establishing a level of certainty. When we cause monkeys to choose the preferred direction of the stimulated neurons more frequently, they adopt a compensatory "null choice" bias to respond in the opposite direction (Fig. 5B, lower; Salzman et al. 1992). This is believed to reflect a stubborn prior: The brain does not adjust its expectation that the two choice targets are equally likely to be rewarded. Thus, when faced with a preponderance of, say, rightward choices induced by μS, it adds a leftward bias so that the proportion of each choice remains ~50%. The compensatory bias may therefore be akin to a bias induced by asymmetry in reward expectation (Feng et al. 2009). For example, on trials without μS in which the monkey viewed 0% coherent motion, he tended to choose the null direction (Fig.

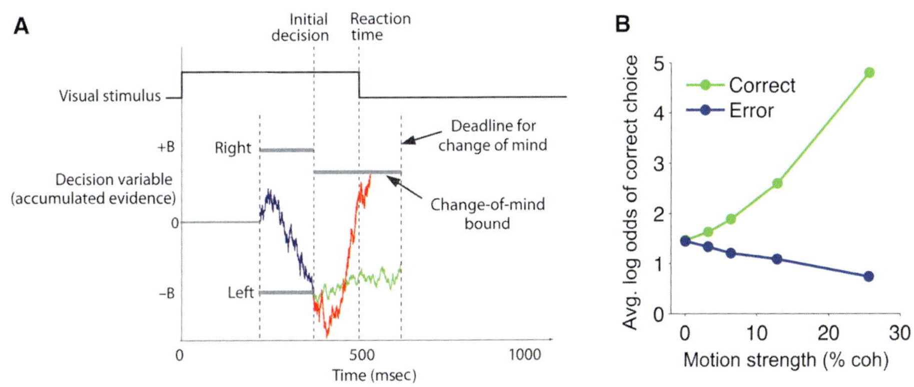

Figure 6. Confidence on errors could in some cases be explained by a process akin to change-of-mind. (*A*) Depiction of a hypothetical "rightward" trial. Because of noise, the accumulated evidence first reaches the bound signaling "left," and movement is initiated toward the wrong choice target. However, evidence in the processing pipeline continues to accrue favoring rightward, and if the decision is reported with a nonballistic movement (i.e., a reach of the arm), this additional evidence can cause a revision of the initial movement trajectory toward the correct choice. (*B*) If the initial choice in *A* is not revised, but the experiment solicits a confidence report following the choice, the extra evidence in the pipeline would lead to lower confidence on errors when evidence is strong, as observed in some experiments and asserted as a signature of confidence within an SDT framework. The graph (compare with Fig. 5B) shows results of a simulated one-dimensional bounded accumulation process where evidence was allowed to accumulate for 250 msec after bound crossing. Symbols show average log-odds associated with the final level of evidence as a function of motion strength on correct trials (green) and errors (blue). (*A*, Modified, with permission, from Resulaj et al. 2009.)

3B, blue curves shifted to the right), yet we do not believe he perceives motion in this direction. He is simply biased. It turns out this bias is also reflected in PDW: the monkey opts out most frequently for stimuli supporting an equal probability of choosing either direction (Fig. 3D, blue curve shifted to the right). In terms of the model, this says that the mapping (Fig. 2B) does not shift to account for the bias, just as it does not shift in response to μS. Thus, a distinction to keep in mind is whether a behavioral assay measures confidence about the state of the world (is the motion leftward?) or about whether the decision is appropriate (should I choose leftward?). PDW can indicate confidence in the state of the world only to the extent that subjects' expectations reflect it. For most of our experiments the connection is tight, but there are many ways it can be loosened.

Beyond this distinction, other forms of confidence are involved in foraging (Hayden et al. 2011), monitoring performance and conflict (Stuphorn et al. 2000; Botvinick et al. 2001; Ito et al. 2003), and ranking unchosen options (Hayden et al. 2009; Rushworth et al. 2012). These processes may involve neural representations of confidence-like quantities that are disentangled from the decision process (Lak et al. 2014) and therefore independent of modality and effector. We emphasize that the mechanism we propose for computing confidence in a perceptual decision is not incompatible with such a task-independent representation (de Gardelle and Mamassian 2014; Heereman and Heekeren 2014). Rather our experiments point toward a stage of processing in which certainty emerges directly from a decision process—consisting of operations on evidence and (in many cases) elapsed time—which may then support elaboration to a broader array of cognitive functions such as those listed above. Exactly where and how these and other forms of (un)certainty (e.g., Bromberg-Martin and Hikosaka 2011; Monosov and Hikosaka 2013) are transformed and combined in the brain will be an area of great future interest for experimenters.

José Saramago once wrote that "strictly speaking, we do not make decisions, decisions make us." Indeed, one might say that it is through our decisions that we differentiate ourselves as moral agents (Kane 2002), and it is through this window that neuroscience illuminates the philosopher's study of ethics (Shadlen and Roskies 2012). More generally, we feel that decision-making offers a window on the neurobiological underpinnings of a variety of aspects of cognition that have little to do with decision-making. In particular, in the study of confidence we encounter the topic of metacognition. When the monkey chooses or waives the sure bet, it is now indicating something about the decision process—something beyond the choice itself. One might be tempted to dismiss the metacognitive designation in light of the nuts-and-bolts account of the process emerging from our experiments. We would regard that as mistaken, however, for explaining a phenomenon, especially in cognitive neuroscience, is not explaining that phenomenon away. Just as vision arises via neural mechanisms, so too do many of our treasured cognitive functions.

Turning to vision itself, it is tempting to ask if the monkeys experience electrical microstimulation as vision. Pioneers of this experimental approach (Salzman et al. 1992; Romo et al. 1998) speculated that the subjective experience during μS resembles that of natural sensory input, an intriguing proposal given that the manipulation directly affects only a small portion of the circuit and does so in an artificial manner. In our experiment (Fetsch et al. 2014), this question is really whether the experience of μS plus random-dot motion is like seeing another random-dot stimulus (e.g., one that appears more "rightward"-ish). We would like to ask a monkey what it feels like on the trials with μS. When we displayed a stimulus with no net motion accompanied by μS of rightward-preferring neurons, did this feel like seeing motion to the right? We cannot pose these questions to a monkey, but the PDW paradigm might come as close as possible. Consider that the monkeys had an opportunity to use the sure bet to tell us something is peculiar on stimulation trials. They did not do that, yet we know that they will indicate a decrease in confidence if we increase the current (Fetsch et al. 2014). More importantly, μS does not appear to affect the way that the brain consults the mapping of DV (and elapsed time) to render a metacognitive decision to choose or forgo the sure bet. This is a strong test of what it means to experience visual motion plus artificial μS as equivalent to visual motion plus a variant of visual motion. It does not rise to the subjective, personal aspect of "feels like," but it comes close.

ACKNOWLEDGMENTS

The authors' research highlighted in this essay was supported by the Howard Hughes Medical Institute (HHMI), the Human Frontiers Research Program (HFSP), National Eye Institute Grant EY11378 to M.N.S., National Eye Institute Core Grant P30EY01730, National Center for Research Resources Grant RR00166, a Sloan Research Fellowship to R.K., and a Whitehall Foundation Research Grant to R.K. We thank Ariel Zylberberg, Bill Newsome, and Daniel Wolpert for helpful discussions.

REFERENCES

Audley RJ. 1960. A stochastic model for individual choice behavior. *Psychol Rev* **67:** 1–15.

Baranski JV, Petrusic WM. 1998. Probing the locus of confidence judgments: Experiments on the time to determine confidence. *J Exp Psychol Hum Percept Perform* **24:** 929–945.

Botvinick MM, Braver TS, Barch DM, Carter CS, Cohen JD. 2001. Conflict monitoring and cognitive control. *Psychol Rev* **108:** 624–652.

Bromberg-Martin ES, Hikosaka O. 2011. Lateral habenula neurons signal errors in the prediction of reward information. *Nat Neurosci* **14:** 1209–1216.

Busemeyer JR. 1985. Decision making under uncertainty: A comparison of simple scalability, fixed-sample, and sequential-sampling models. *J Exp Psychol Learn Mem Cogn* **11:** 538–564.

Cartwright D, Festinger L. 1943. A quantitative theory of decision. *Psychol Rev* **50:** 595–621.

Celebrini S, Newsome WT. 1995. Microstimulation of extrastriate area MST influences performance on a direction discrimination task. *J Neurophysiol* **73:** 437–448.
Churchland AK, Kiani R, Shadlen MN. 2008. Decision-making with multiple alternatives. *Nat Neurosci* **11:** 693–702.
Churchland AK, Kiani R, Chaudhuri R, Wang XJ, Pouget A, Shadlen MN. 2011. Variance as a signature of neural computations during decision making. *Neuron* **69:** 818–831.
Cisek P, Puskas GA, El-Murr S. 2009. Decisions in changing conditions: The urgency-gating model. *J Neurosci* **29:** 11560–11571.
DasGupta S, Ferreira CH, Miesenböck G. 2014. FoxP influences the speed and accuracy of a perceptual decision in *Drosophila*. *Science* **344:** 901–904.
Dayan P, Kakade S, Montague PR. 2000. Learning and selective attention. *Nat Neurosci* **3:** 1218–1223.
de Gardelle V, Mamassian P. 2014. Does confidence use a common currency across two visual tasks? *Psychol Sci* **25:** 1286–1288.
Ditterich J, Mazurek ME, Shadlen MN. 2003. Microstimulation of visual cortex affects the speed of perceptual decisions. *Nat Neurosci* **6:** 891–898.
Drugowitsch J, Moreno-Bote R, Pouget A. 2014. Relation between belief and performance in perceptual decision making. *PLoS ONE* **9:** e96511.
Feng S, Holmes P, Rorie A, Newsome WT. 2009. Can monkeys choose optimally when faced with noisy stimuli and unequal rewards? *PLoS Comput Biol* **5:** e1000284.
Ferrell WR, Mcgoey PJ. 1980. A model of calibration for subjective probabilities. *Organ Behav Hum Perf* **26:** 32–53.
Festinger L. 1943. Studies in decision: I. Decision-time, relative frequency of judgment and subjective confidence as related to physical stimulus difference. *J Exp Psychol* **32:** 291–306.
Fetsch CR, Kiani R, Newsome WT, Shadlen MN. 2014. Effects of cortical microstimulation on confidence in a perceptual decision. *Neuron* **83:** 797–804.
Fullerton GS, Cattell JM. 1892. *On the perception of small differences: With special reference to the extent, force and time of movement*. Philosophy Series No 2. University of Pennsylvania, Philadelphia.
Garrett HE. 1922. A study of the relation of accuracy to speed. *Arch Psychol* **8:** 1–105.
Gold JI, Shadlen MN. 2001. Neural computations that underlie decisions about sensory stimuli. *Trends Cogn Sci* **5:** 10–16.
Gold JI, Shadlen MN. 2007. The neural basis of decision making. *Annu Rev Neurosci* **30:** 535–574.
Hampton RR. 2009. Multiple demonstrations of metacognition in nonhumans: Converging evidence or multiple mechanisms? *Comp Cogn Behav Rev* **4:** 17–28.
Hanes DP, Schall JD. 1996. Neural control of voluntary movement initiation. *Science* **274:** 427–430.
Hanks TD, Ditterich J, Shadlen MN. 2006. Microstimulation of macaque area LIP affects decision-making in a motion discrimination task. *Nat Neurosci* **9:** 682–689.
Hanks TD, Mazurek ME, Kiani R, Hopp E, Shadlen MN. 2011. Elapsed decision time affects the weighting of prior probability in a perceptual decision task. *J Neurosci* **31:** 6339–6352.
Hayden BY, Pearson JM, Platt ML. 2009. Fictive reward signals in the anterior cingulate cortex. *Science* **324:** 948–950.
Hayden BY, Pearson JM, Platt ML. 2011. Neuronal basis of sequential foraging decisions in a patchy environment. *Nat Neurosci* **14:** 933–939.
Heekeren HR, Marrett S, Ungerleider LG. 2008. The neural systems that mediate human perceptual decision making. *Nat Rev Neurosci* **9:** 467–479.
Heereman J, Heekeren H. 2014. A task-independent neural representation of subjective certainty. *Program No. 587.10. 2014 Neuroscience Meeting Planner*. Society for Neuroscience, Washington, DC.
Henmon VAC. 1911. The relation of the time of a judgment to its accuracy. *Psychol Rev* **18:** 186.
Higham PA. 2007. No special K! A signal detection framework for the strategic regulation of memory accuracy. *J Exp Psychol Gen* **136:** 1–22.
Horwitz GD, Newsome WT. 2001. Target selection for saccadic eye movements: Prelude activity in the superior colliculus during a direction-discrimination task. *J Neurophysiol* **86:** 2543–2558.
Huk AC, Shadlen MN. 2005. Neural activity in macaque parietal cortex reflects temporal integration of visual motion signals during perceptual decision making. *J Neurosci* **25:** 10420–10436.
Insabato A, Pannunzi M, Rolls ET, Deco G. 2010. Confidence-related decision making. *J Neurophysiol* **104:** 539–547.
Irwin FW, Smith WAS, Mayfield JF. 1956. Tests of two theories of decision in an "expanded judgment" situation. *J Exp Psychol* **51:** 261–268.
Ito S, Stuphorn V, Brown JW, Schall JD. 2003. Performance monitoring by the anterior cingulate cortex during saccade countermanding. *Science* **302:** 120–122.
Janssen P, Shadlen MN. 2005. A representation of the hazard rate of elapsed time in macaque area LIP. *Nat Neurosci* **8:** 234–241.
Johnson DM. 1939. Confidence and speed in the two-category judgement. *Arch Psychol* **34:** 1–53.
Kane R. 2002. Free will: New directions for an ancient problem. In *Free will* (ed. Kane R), pp. 222–246. Blackwell, Oxford.
Kellogg WN. 1931. The time of judgment in psychometric measures. *Am J Psychol* **43:** 65–86.
Kepecs A, Mainen ZF. 2012. A computational framework for the study of confidence in humans and animals. *Philos Trans R Soc Lond B Biol Sci* **367:** 1322–1337.
Kepecs A, Uchida N, Zariwala HA, Mainen ZF. 2008. Neural correlates, computation and behavioural impact of decision confidence. *Nature* **455:** 227–231.
Kiani R, Shadlen MN. 2009. Representation of confidence associated with a decision by neurons in the parietal cortex. *Science* **324:** 759–764.
Kiani R, Hanks T, Shadlen M. 2008. Bounded integration in parietal cortex underlies decisions even when viewing duration is dictated by the environment. *J Neurosci* **28:** 3017–3029.
Kiani R, Shadlen MN, Newsome WT. 2011. Belief and time: Response dynamics of lateral intraparietal neurons represent choice certainty. *Program No. 17.06. 2011 Neuroscience Meeting Planner*. Society for Neuroscience, Washington, DC.
Kiani R, Corthell L, Shadlen MN. 2014a. Choice certainty is informed by both evidence and decision time. *Neuron* **84:** 1329–1342.
Kiani R, Cueva CJ, Reppas JB, Newsome WT. 2014b. Dynamics of neural population responses in prefrontal cortex indicate changes of mind on single trials. *Curr Biol* **24:** 1542–1547.
Kim J-N, Shadlen MN. 1999. Neural correlates of a decision in the dorsolateral prefrontal cortex of the macaque. *Nat Neurosci* **2:** 176–185.
Kinchla RA, Collyer CE. 1974. Detecting a target letter in briefly presented arrays: A confidence rating analysis in terms of a weighted additive effects model. *Percept Psychophys* **16:** 117–122.
Ko Y, Lau H. 2012. A detection theoretic explanation of blindsight suggests a link between conscious perception and metacognition. *Philos Trans R Soc Lond B Biol Sci* **367:** 1401–1411.
Komura Y, Nikkuni A, Hirashima N, Uetake T, Miyamoto A. 2013. Responses of pulvinar neurons reflect a subject's confidence in visual categorization. *Nat Neurosci* **16:** 749–755.
Krajbich I, Rangel A. 2011. Multialternative drift-diffusion model predicts the relationship between visual fixations and choice in value-based decisions. *Proc Natl Acad Sci* **108:** 13852–13857.
Lak A, Costa GM, Romberg E, Koulakov AA, Mainen ZF, Kepecs A. 2014. Orbitofrontal cortex is required for optimal waiting based on decision confidence. *Neuron* **84:** 190–201.
Laming DRJ. 1968. *Information theory of choice-reaction times*. Academic, Oxford.

Leon MI, Shadlen MN. 2003. Representation of time by neurons in the posterior parietal cortex of the macaque. *Neuron* **38:** 317–327.
Link SW, Heath RA. 1975. A sequential theory of psychological discrimination. *Psychometrika* **40:** 77–105.
Luce RD. 1986. *Response times: Their role in inferring elementary mental organization.* Oxford University Press, Oxford, UK.
Maimon G, Assad JA. 2006. A cognitive signal for the proactive timing of action in macaque LIP. *Nat Neurosci* **9:** 948–955.
Mazurek ME, Roitman JD, Ditterich J, Shadlen MN. 2003. A role for neural integrators in perceptual decision making. *Cereb Cortex* **13:** 1257–1269.
Middlebrooks PG, Sommer MA. 2011. Metacognition in monkeys during an oculomotor task. *J Exp Psychol Learn Mem Cogn* **37:** 325–337.
Monosov IE, Hikosaka O. 2013. Selective and graded coding of reward uncertainty by neurons in the primate anterodorsal septal region. *Nat Neurosci* **16:** 756–762.
Niwa M, Ditterich J. 2008. Perceptual decisions between multiple directions of visual motion. *J Neurosci* **28:** 4435–4445.
Norman DA, Wickelgren WA. 1969. Strength theory of decision rules and latency in retrieval from short-term memory. *J Math Psychol* **6:** 192–208.
Peirce CS, Jastrow J. 1884. On small differences in sensation. *Mem Natl Acad Sci* **3:** 73–83.
Peterson WW, Birdsall TG. 1953. *The theory of signal detectability.* Technical Report No. 13, Electronic Defense Group. University of Michigan, Ann Arbor, MI.
Pierrel R, Murray CS. 1963. Some relationships between comparative judgment, confidence, and decision-time in weightlifting. *Am J Psychol* **76:** 28–38.
Pleskac TJ, Busemeyer JR. 2010. Two-stage dynamic signal detection: A theory of choice, decision time, and confidence. *Psychol Rev* **117:** 864–901.
Ploran EJ, Tremel JJ, Nelson SM, Wheeler ME. 2011. High quality but limited quantity perceptual evidence produces neural accumulation in frontal and parietal cortex. *Cereb Cortex* **21:** 2650–2662.
Rabbitt PM. 1966. Error correction time without external error signals. *Nature* **212:** 438.
Rahnev D, Maniscalco B, Graves T, Huang E, de Lange FP, Lau H. 2011. Attention induces conservative subjective biases in visual perception. *Nat Neurosci* **14:** 1513–1515.
Rahnev DA, Maniscalco B, Luber B, Lau H, Lisanby SH. 2012. Direct injection of noise to the visual cortex decreases accuracy but increases decision confidence. *J Neurophysiol* **107:** 1556–1563.
Ratcliff R. 1978. A theory of memory retrieval. *Psychol Rev* **85:** 59–108.
Ratcliff R, Smith PL. 2004. A comparison of sequential sampling models for two-choice reaction time. *Psychol Rev* **111:** 333–367.
Ratcliff R, Starns JJ. 2013. Modeling confidence judgments, response times, and multiple choices in decision making: Recognition memory and motion discrimination. *Psychol Rev* **120:** 697–719.
Ratcliff R, Cherian A, Segraves M. 2003. A comparison of macaque behavior and superior colliculus neuronal activity to predictions from models of two-choice decisions. *J Neurophysiol* **90:** 1392–1407.
Reddi BA, Asrress KN, Carpenter RH. 2003. Accuracy, information, and response time in a saccadic decision task. *J Neurophysiol* **90:** 3538–3546.
Reed JB. 1951. *The speed and accuracy of discriminating differences in hue, brilliance, area, and shape.* Technical Report, Special Devices Center. US Navy.
Resulaj A, Kiani R, Wolpert DM, Shadlen MN. 2009. Changes of mind in decision-making. *Nature* **461:** 263–266.
Roitman JD, Shadlen MN. 2002. Response of neurons in the lateral intraparietal area during a combined visual discrimination reaction time task. *J Neurosci* **22:** 9475–9489.
Rolls ET, Grabenhorst F, Deco G. 2010. Decision-making, errors, and confidence in the brain. *J Neurophysiol* **104:** 2359–2374.
Romo R, Hernández A, Zainos A, Salinas E. 1998. Somatosensory discrimination based on cortical microstimulation. *Nature* **392:** 387–390.
Rounis E, Maniscalco B, Rothwell JC, Passingham RE, Lau H. 2010. θ-burst transcranial magnetic stimulation to the prefrontal cortex impairs metacognitive visual awareness. *Cogn Neurosci* **1:** 165–175.
Rushworth MF, Kolling N, Sallet J, Mars RB. 2012. Valuation and decision-making in frontal cortex: One or many serial or parallel systems? *Curr Opin Neurobiol* **22:** 946–955.
Salzman CD, Britten KH, Newsome WT. 1990. Cortical microstimulation influences perceptual judgements of motion direction. *Nature* **346:** 174–177.
Salzman CD, Murasugi CM, Britten KH, Newsome WT. 1992. Microstimulation in visual area MT: Effects on direction discrimination performance. *J Neurosci* **12:** 2331–2355.
Scott RB, Dienes Z, Barrett AB, Bor D, Seth AK. 2014. Blind insight: Metacognitive discrimination despite chance task performance. *Psychol Sci* **25:** 2199–2208.
Seward GH. 1928. Recognition time as a measure of confidence, an experimental study of redintegration. *Arch Psychol* **16:** 1–50.
Shadlen MN, Kiani R. 2013. Decision making as a window on cognition. *Neuron* **80:** 791–806.
Shadlen MN, Newsome WT. 1996. Motion perception: Seeing and deciding. *Proc Natl Acad Sci* **93:** 628–633.
Shadlen MN, Roskies AL. 2012. The neurobiology of decision-making and responsibility: Reconciling mechanism and mindedness. *Front Neurosci* **6:** 56.
Smith JD. 2009. The study of animal metacognition. *Trends Cogn Sci* **13:** 389–396.
Smith PL, Ratcliff R. 2004. Psychology and neurobiology of simple decisions. *Trends Neurosci* **27:** 161–168.
Smith JD, Couchman JJ, Beran MJ. 2012. The highs and lows of theoretical interpretation in animal-metacognition research. *Philos Trans R Soc Lond B Biol Sci* **367:** 1297–1309.
Song JH, Nakayama K. 2009. Hidden cognitive states revealed in choice reaching tasks. *Trends Cogn Sci* **13:** 360–366.
Stone M. 1960. Models for choice-reaction time. *Psychometrika* **25:** 251–260.
Stuphorn V, Taylor TL, Schall JD. 2000. Performance monitoring by the supplementary eye field. *Nature* **408:** 857.
Swets JA, Tanner J, Wilson P, Birdsall TG. 1961. Decision processes in perception. *Psychol Rev* **68:** 301.
Tanner WP, Swets JA. 1954. A decision-making theory of visual detection. *Psychol Rev* **61:** 401–409.
Terrace HS, Son LK. 2009. Comparative metacognition. *Curr Opin Neurobiol* **19:** 67–74.
Treisman M, Faulkner A. 1984. The setting and maintenance of criteria representing levels of confidence. *J Exp Psychol Hum Percept Perform* **10:** 119.
van Meter D, Middleton D. 1954. Modern statistical approaches to reception in communication theory. *Trans IRE Prof Group Inform Theory* **4:** 119–145.
Vickers D. 1972. Some general features of perceptual discrimination. In *Psychological aspects of driver behaviour* (ed. Asmussen EG). Institute for Road Safety Research, S.W.O.V., Voorlung, Netherlands.
Vickers D. 1979. *Decision processes in visual perception.* Academic Press, New York.
Vickers D. 2001. Where does the balance of evidence lie with respect to confidence? In *Proceedings of the Seventeenth Annual Meeting of the International Society for Psychophysics*, pp. 148–153**. Pabst Science Publishers, Lengerich, Germany.
Vickers D, Packer J. 1982. Effects of alternating set for speed or accuracy on response time, accuracy and confidence in a unidimensional discrimination task. *Acta Psychol* **50:** 179–197.

Vickers D, Smith P, Burt J, Brown M. 1985. Experimental paradigms emphasising state or process limitations: II. Effects on confidence. *Acta Psychol* **59:** 163–193.

Volkmann J. 1934. The relation of time of judgment to certainty of judgment. *Psychol Bull* **31:** 672–673.

Zariwala HA, Kepecs A, Uchida N, Hirokawa J, Mainen ZF. 2013. The limits of deliberation in a perceptual decision task. *Neuron* **78:** 339–351.

Zylberberg A, Barttfeld P, Sigman M. 2012. The construction of confidence in a perceptual decision. *Front Integr Neurosci* **6:** 79.

Zylberberg A, Fetsch CR, Sigman M, Shadlen MN. 2014. The influence of the strength and variability of sensory evidence on confidence in a perceptual decision. *Program No. 435.16. 2014 Neuroscience Meeting Planner*. Society for Neuroscience, Washington, DC.

Internal States and Behavioral Decision-Making: Toward an Integration of Emotion and Cognition

Ann Kennedy,[1] Kenta Asahina,[3,5] Eric Hoopfer,[1,2,5] Hidehiko Inagaki,[4,5] Yonil Jung,[1,5] Hyosang Lee,[1,5] Ryan Remedios,[1,5] and David J. Anderson[1,2]

[1]*Division of Biology, California Institute of Technology, Pasadena, California 91125*

[2]*Howard Hughes Medical Institute, Pasadena, California 91125*

Correspondence: wuwei@caltech.edu

Social interactions, such as an aggressive encounter between two conspecific males or a mating encounter between a male and a female, typically progress from an initial appetitive or motivational phase, to a final consummatory phase. This progression involves both changes in the intensity of the animals' internal state of arousal or motivation and sequential changes in their behavior. How are these internal states, and their escalating intensity, encoded in the brain? Does this escalation drive the progression from the appetitive/motivational to the consummatory phase of a social interaction and, if so, how are appropriate behaviors chosen during this progression? Recent work on social behaviors in flies and mice suggests possible ways in which changes in internal state intensity during a social encounter may be encoded and coupled to appropriate behavioral decisions at appropriate phases of the interaction. These studies may have relevance to understanding how emotion states influence cognitive behavioral decisions at higher levels of brain function.

Survival and reproduction are mediated by innate, goal-directed activities (LeDoux 2012; Sternson 2013), such as feeding, drinking, mating, fighting, and escape from a predator. Each such activity is characterized by its own complex collection of actions,[6] modulated by sensory information from multiple modalities. Furthermore, these actions are often organized in a characteristic sequence or progression, from an initial appetitive or motivational phase, to a more directed investigative phase, to a final consummatory phase (Fig. 1A; Berridge 2004). For example, in the case of an agonistic encounter between conspecific males, the appetitive phase includes investigation to determine the sex and assess the relative size or strength of the opponent; this is often followed by threat displays and then finally overt attack or retreat (Miczek et al. 2007). This progression is also accompanied by an increasing risk of injury, illness, or death; therefore, such behavioral transitions involve cost–benefit decisions (Gillette et al. 2000; Hirayama et al. 2012).

Progression through these different phases of an aggressive or sexual encounter is typically associated with an escalating internal state of motivation, arousal, or drive (for concepts and definitions, see Berridge 2004; LeDoux 2012). The concept of drive has proven useful to behavioral neuroscience and fits with our intuitive experience of behavior—not as a sequence of independent motor acts, but as a continuous escalation of a single state (e.g., from annoyance to anger to rage). But it is difficult to determine through observation alone whether drives are a component of neural computation, playing a causal role in the control of behavioral decisions, or simply epiphenomena. Konrad Lorenz formulated a "hydraulic" metaphor to explain how different behavioral actions might be successively released under the control of an escalating internal drive (or motivational) state, with different drives for different behaviors (Fig. 1B; Lorenz and Leyhausen 1973; Berridge 2004). Although this model invoked a causal role for the level of drive states in controlling transitions or "decisions" between different behaviors, it made no predictions as to how such a role might be implemented in neural hardware.

In a view complementary to Lorenz's Tinbergen proposed that behavioral decisions are made in a hierarchical manner: More generic choices between competing or "opponent" activities (e.g., to engage in fighting or mating) are made before more specific selections of a particular action (e.g., between a threat display and a bite) (Fig. 1C; Tinbergen 1950). He further suggested that each decision stage could be mediated by a group of mutually inhibitory "centers" in the brain, with successive decision stages connected in a feed-forward manner to form a behavioral decision tree (Tinbergen 1950, 1951).

Although Tinbergen's model is useful as a heuristic, it leaves open a number of important theoretical issues. First, it does not explain why different actions controlled by "third-level" command centers progress through a characteristic sequence during an escalating encounter nor suggest how such progressions might be controlled. Second, it does not distinguish whether the higher-order ("second-level") centers that control different activities (mating, fighting) exert some control over decisions be-

[6]The word "activity" is used here operationally to describe an ongoing behavior, such as "aggression"; "action" is used to describe a specific motor program executed during that activity, such as biting (Anderson and Perona 2014).

Present Addresses: [3] The Salk Institute, La Jolla, California 92037.[4]HHMI Janelia Research Campus, Ashburn, Virginia 20147.
[5]These authors are listed alphabetically.

 doi: 10.1101/sqb.2014.79.024984

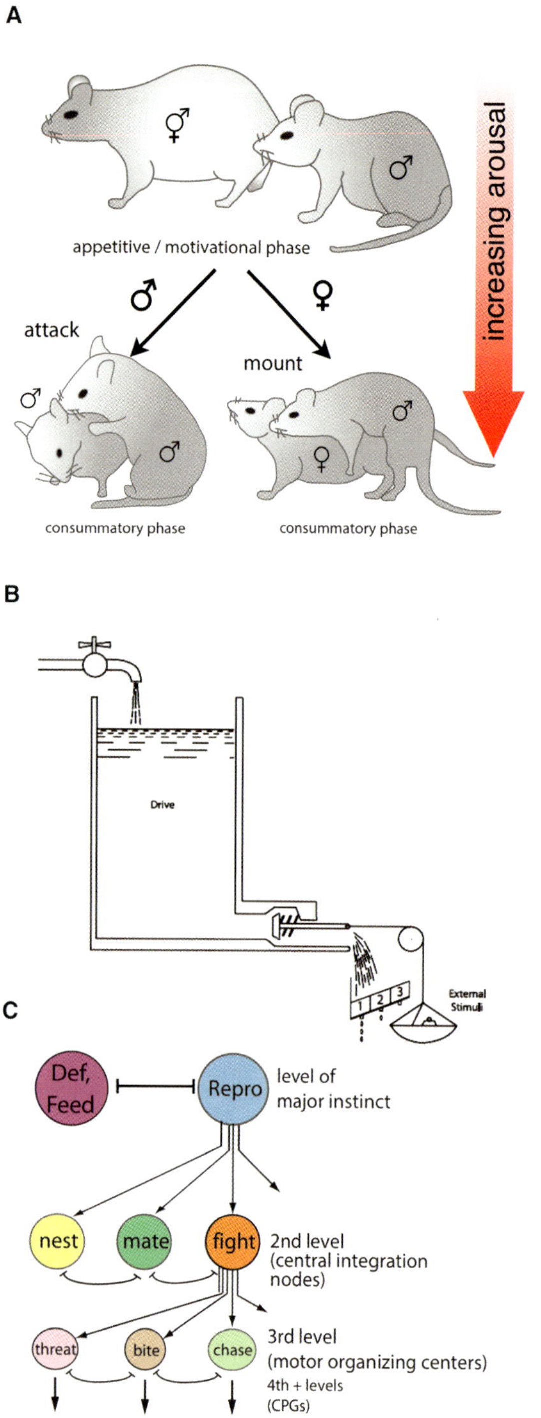

Figure 1. (*A*) Progression from investigative to consummatory phases of social behavior. (*B*) Lorenz's "hydraulic" metaphor for the control of behavior by internal drive states. (*C*) Tinbergen's hierarchical model for behavioral decisions (Tinbergen 1950). (*B*, Reprinted from Berridge 2004, with permission from Elsevier; *C*, modified from Anderson 2012.).

tween actions dictated by third-level command centers, or rather simply act as a switch or gate, that permits such activities to occur. Finally, it does not integrate this hierarchical view of behavioral decision-making with any sort of internal drive or arousal states during a social interaction or consider whether and how such states might contribute to the control of such progressive decisions.

The question of how escalating internal states influence decisions between competing behaviors or actions, while fascinating, has not been widely investigated at the level of neural mechanism. To approach this subject, a number of basic questions need to be addressed. How are behavioral states, such as arousal or motivation, encoded or instantiated in the brain (Berridge 2004; Pfaff et al. 2005)? How is the escalation of these states implemented? Are these states behavior-specific, or generic (Devidze et al. 2006)? At what node(s) in a decision hierarchy or network do such states exert their influence on behavioral decisions? How are these influences exerted mechanistically? And is the level or intensity of such escalating states important in the actual control of behavioral decisions?

Here, we summarize our efforts to begin to approach these questions, in the context of innate behaviors, with a focus on aggression. We have pursued parallel studies in the mouse and in the vinegar fly *Drosophila melanogaster* in an effort to identify underlying evolutionarily conserved principles of neural circuit function across species. Our hope is that these studies may also prove relevant to understanding the link between internal states and decision-making in higher organisms, such as the integration of emotion and cognition in humans (Salzman and Fusi 2010).

STATE-DEPENDENT CONTROL OF SOCIAL BEHAVIORS IN *DROSOPHILA*

A Neuron and a Neuropeptide That Control Aggressive Arousal

Evidence from many systems suggests that neuromodulators, such as biogenic amines or neuropeptides, are important regulators of internal states and behavior (Insel and Young 2000; Nassel and Winther 2010; Bargmann 2012; Marder 2012; Taghert and Nitabach 2012). As a first step toward understanding the neural coding of a state of aggressiveness, we carried out a screen for neuropeptide-secreting neurons that control agonistic behavior in *Drosophila* (Asahina et al. 2014). This screen identified a small cluster (three to five cells) of male-specific, fruitless (FruM)-expressing neurons (Fig. 2A–C) that contain the neuropeptide *Drosophila* tachykinin (DTK, Fig. 2A, inset) and that profoundly influence levels of aggressiveness. Thermogenetic activation of these neurons increased aggression (Fig. 2D), whereas inhibition of these neurons strongly reduced fighting (Fig. 2E; Asahina et al. 2014). Overexpression of the *DTK* peptide in DTK-expressing neurons potentiated the aggression-promoting effect of activating these cells, whereas deletions in the *DTK* gene reduced it (Fig. 2F). Importantly, activation of these neurons had no effect on courtship or mating behavior. Together, these data identified a cell type and a gene that play a major role in controlling inter-male aggression in flies. Interestingly, homologs of DTK such as Substance P have been implicated in aggression in mammals

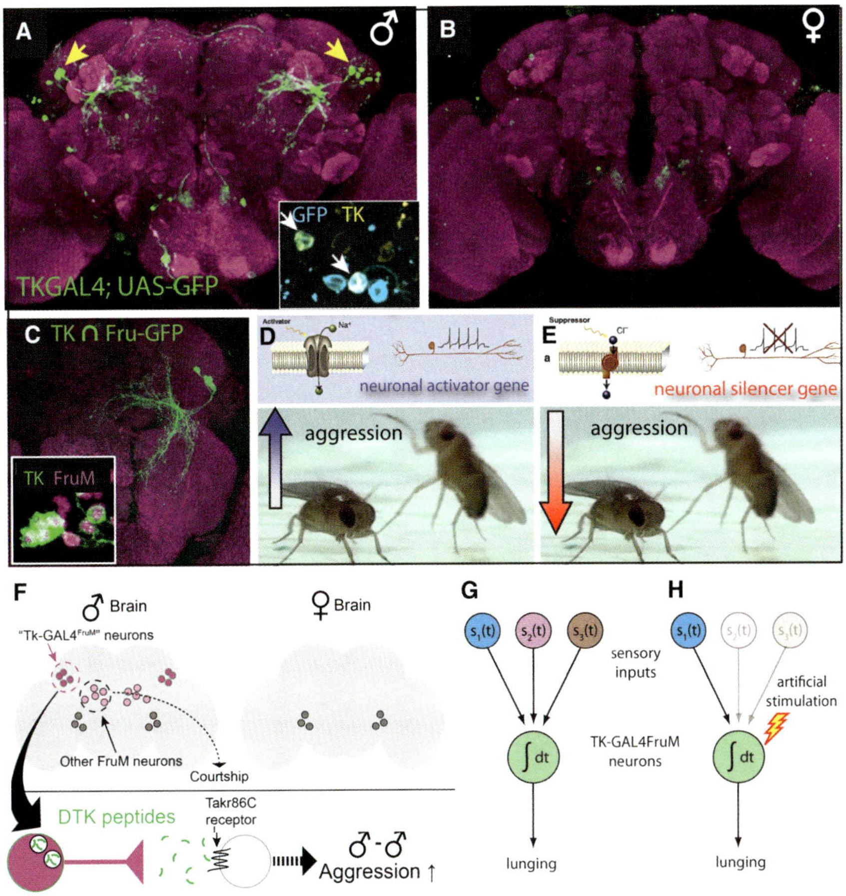

Figure 2. (*A*,*B*) Identification of a population of aggression-promoting, sexually dimorphic neurons in male flies that express the neuropeptide *Drosophila* tachykinin (DTK) (*A*, *inset*). Yellow arrowheads (*A*) indicate cell bodies. (*C*) TK-GAL4 neurons express the male-specific sex-determination factor Fru^{M} (*inset*). (*D*,*E*) Activation of TK-$GAL4^{FruM}$ neurons increases aggression (*D*), whereas inhibition of these neurons decreases aggression (*E*). (*F*) The aggression-promoting function of TK-$GAL4^{FruM}$ neurons is enhanced by release of DTK peptides. (*G*) TK-$GAL4^{FruM}$ neurons are proposed to integrate the effects of multiple classes of sensory stimuli over time ($s_1(t)$, $s_2(t)$, etc.). (*H*) Artificial enhancement of TK-$GAL4^{FruM}$ neuronal activity can bypass the requirement for some of these sensory stimuli. (*D*,*E*, Cartoons modified from Borst 2009, with permission from Elsevier; *F*, reproduced from Asahina et al. 2014.)

(including humans), by genetic and pharmacological studies (Shaikh et al. 1993; Katsouni et al. 2009; Coccaro et al. 2012).

Do DTK^{FruM} neurons simply direct a motor program of aggression or do they regulate an internal aggression-promoting state, such as motivation or arousal? Arousal often involves an increased sensitivity to sensory cues, such that the threshold for release of a stimulus-driven behavior is reduced (Van Swinderen and Andretic 2003, 2011; Devidze et al. 2006). Interestingly, artificial activation of DTK^{FruM} neurons could evoke aggression even in the absence of sensory cues that are normally required for aggression, such as male-specific pheromones (Fig. 2G,H; de la Paz Fernandez et al. 2010; Wang and Anderson 2010; Wang et al. 2011). In some cases, activation of DTK^{FruM} neurons supplemented with excess DTK peptide could even promote attack toward a moving fly-sized inanimate object (Asahina et al. 2014). These data suggest that DTK^{FruM} neurons promote a state of arousal or motivation that is apparently aggression-specific. Our previous work has provided evidence of other behavior-specific forms of arousal in *Drosophila* (Lebestky et al. 2009).

We have recently suggested that emotions are internal states that are characterized by certain general properties, or emotion primitives, common to different emotions in a single species and that apply to emotions across species (Anderson and Adolphs 2014). These properties include persistence, scalability, valence, stimulus degeneracy, and generalization (trans-situationality). Neuropeptides have properties that could contribute to encoding some of these emotion state features (Flavell et al. 2013; for review, see Insel and Young 2000; Berridge 2004; Nassel and Winther 2010; Bargmann 2012). For example, the concentration of DTK could encode the scalability (intensity) of an aggressive state, whereas slowly decaying levels of the peptide, determined by its rate of degradation, could underlie a persistent state of aggressiveness. The relationship between *DTK* gene dosage and the aggression-promoting activity of TK^{FruM} neuron activation suggests that levels of DTK might control the intensity of aggression, but whether these neurons can induce a persistent state of aggressiveness is not yet clear. Work from Kravitz and others indicates that biogenic amines, such as octopamine, dopamine, and serotonin, also influence levels of aggressiveness in *Drosophila* (Dierick and Greenspan 2007; Zhou et al. 2008; Mundiyanapurath et al. 2009; Alekseyenko et al. 2010, 2013; Certel et al. 2010; Andrews et al. 2014), as well as in other arthropod species (Stevenson et al. 2005; for review, see Kravitz and Huber 2003), and likely contribute to aggressive state control as well.

Neurons That Control Persistent Social Behavior States in *Drosophila*

The ability to identify neurons that control persistent behavioral states requires a means of activating such neurons in freely moving animals, with millisecond time resolution. This requirement is, of course, afforded by optogenetics; however, the use of optogenetics in *Drosophila* has been limited by the inability of blue light to penetrate the flies' cuticle. Recently, this obstacle has been overcome by the development and application of red-shifted opsins such as ReaChR (Fig. 3A) and Chrimson (Lin et al. 2013; Inagaki et al. 2014; Klapoetke et al. 2014).

In a proof-of-principle application of this technology, we performed optogenetic stimulation of different populations of interneurons that control wing extension, which mediates male courtship song (Kohatsu et al. 2011; von Philipsborn et al. 2011); (Fig. 3B,C; for review, see Yamamoto et al. 2014). Strikingly, activation of P1 interneurons in single male flies evoked probabilistic and persistent wing extension that lasted for many minutes following stimulus termination (Fig. 3D; Inagaki et al. 2014). In contrast, activation of a descending interneuron, pIP10 (Fig. 3B; von Philipsborn et al. 2011), evoked wing extension in a deterministic manner time-locked to the onset and offset of photostimulation (Fig. 3E). Similar observations were made independently using IR laser-based transient thermogenetic activation of P1 and pIP10 (Bath et al. 2014). Interestingly, transient activation of P1 neurons in pairs of male flies evoked persistent aggression (Fig. 3F; E Hoopfer and DJ Anderson, unpubl.). These results indicate that P1 neurons can regulate aggression as well as courtship (Fig. 3G), although whether their influence on aggression is direct or indirect remains to be determined.

The persistent effect of P1 neuron activation on courtship raised the question of whether these neurons might be involved in internal states of arousal or motivation associated with such reproductive behaviors. Social isolation is known to increase both courtship and aggression in flies (Wang et al. 2008; Liu et al. 2011), as well as in many other animal species (Koike et al. 2009; Toth et al. 2011). Strikingly, social isolation potentiated the effect of optogenetic activation of P1, but not of pIP10 neurons, to promote wing extension (Inagaki et al. 2014). Moreover, this effect was associated with an increased excitability of P1 neurons, as determined by combined optogenetic stimulation and calcium imaging experiments. These data suggest that P1 neurons could form part of the neural substrate for an internal state change that reflects social experience and that affects multiple social behaviors (Wang et al. 2008; Liu et al. 2011). The persistent behavioral effects of P1 activation might then serve to perpetuate the influence of past environmental conditions on behavior.

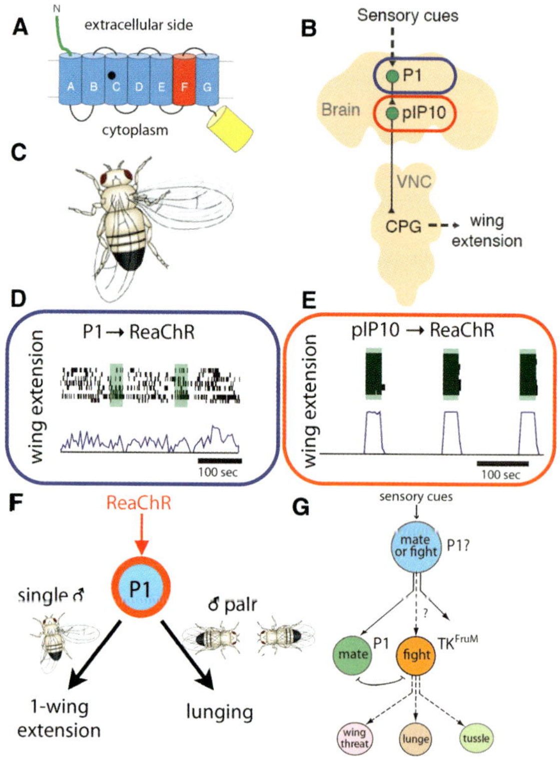

Figure 3. (*A*) Schematic illustrating the red-shifted opsin ReaChR (Lin et al. 2013; Inagaki et al. 2014). (*B,D*) Cartoon illustrating two classes of interneurons that control wing extension (*C*) in male *Drosophila*. (*D,E*) Optogenetic activation of P1 and pIP10 neurons evokes persistent and time-locked wing extension in single *Drosophila* males, respectively. (*F*) Schematic illustrating dual effect of P1 activation by ReaChR to evoke courtship behavior (1-wing extension) in single males (*left*), and aggressive behavior in pairs of males (*right*) (E Hoopfer et al. in prep.) (*G*) Schematic illustrating possible circuit relationship between P1 neurons and TK-GAL4^{FruM} neurons. (*B–C*, Modified from von Philipsborn et al. 2011; *D,E*, modified from Inagaki et al. 2014.)

NEURAL CONTROL OF AGGRESSION IN MICE

Classic experiments by Hess in the late 1920s demonstrated that electrical stimulation of the hypothalamus could elicit aggressive displays in cats (Hess 1928), such as the "affecktiven Abwehr-reaktion" ("affective defensive reaction"), in which the animal exhibited hissing, baring of teeth, and laid-back ears characteristic of a state of "rage" (Hess and Brügger 1943). This transformative experiment suggested that artificial stimulation of a specific brain region could evoke an emotion state and its associated behavioral expression.

Hess's findings raised three major questions: (1) What neurons are sufficient to evoke aggression when artificially stimulated? (2) Are those same neurons required for normal aggressive behavior? (3) Are those neurons normally active during aggressive behavior? (Kruk 2014). Despite almost a century of follow-up research (for review, see Kruk 1991; Siegel et al. 1999; Siegel 2004; Adams 2006; Miczek et al. 2007; Nelson and Trainor 2007; Anderson 2012; Falkner and Lin 2014; Yang and Shah 2014), until recently, these questions have remained unanswered.[7] They could potentially be addressed by bringing to bear powerful genetic tools for neural circuit manipulations available in mice. Surprisingly, however, in the 70 yr since Hess's initial paper, there was no report of brain-stimulated aggression in mice,[8] despite the fact that such manipulations are routine in other rodent species such as rats and hamsters (Kruk 2014).

Identification of an Aggression Locus in the Mouse Hypothalamus

To identify neurons activated during aggression, we carried out multi-electrode, single-unit recordings from the ventrolateral subdivision of the ventromedial hypothalamus (VMHvl), in freely behaving male mice (Lin et al. 2011). We chose VMHvl because it overlaps the so-called hypothalamic attack area (HAA) identified by microstimulation experiments in rats (Kruk et al. 1988; Hrabovszky et al. 2005) and because it is labeled strongly by c-fos following aggression (Newman 1999; Veening et al. 2005; Lin et al. 2011). Extracellular recording in VMHvl revealed that it contains neurons with dynamic and heterogeneous responses during social encounters, with some units activated during male–male (Fig. 4A) or male–female (Fig. 4B) interactions, or both (Fig. 4C; Lin et al. 2011); a small number of units were active exclusively during attack or mounting (Fig. 4D). Thus, VMHvl contains a mixture of neurons active during different phases and different types of social encounters (Fig. 4E).

To investigate a causal role for VMHvl neurons in aggression, we optogenetically activated these cells, using channelrhodopsin-2 (ChR2), in an anatomically restricted (but not cell-type-specific) manner (Lin et al. 2011). Photostimulation elicited time-locked attack toward males, females, and inanimate objects, with a latency of a few seconds. Chemogenetic inhibition using an ivermectin-gated chloride channel (Slimko et al. 2002; Lerchner et al. 2007) reversibly reduced agonistic behavior (Lin et al. 2011). However, the temporal resolution of this method was insufficient to determine whether these neurons were required during attack per se. Surprisingly,

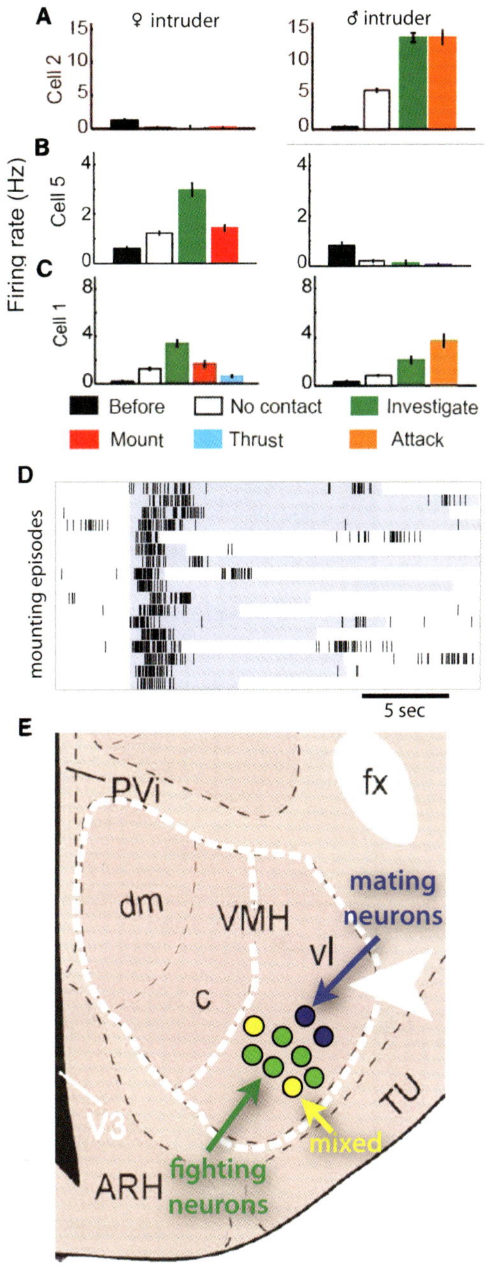

Figure 4. (*A–C*) Average firing rates (in 5-sec bins) for three individual units recorded in the VMHvl of a male mouse during the indicated social behavior epochs (established by manual annotation of videos at 30 Hz) (Lin et al. 2011). (*D*) Raster plot of an individual unit in VMHvl during mounting. Each row illustrates a separate trial. Shading indicates periods where mounting occurred. (*E*) Schematic illustrating mixed population of neurons in VMHvl that are responsive during male social behavior. (Modified from Anderson 2012.)

[7]The main reasons for this are that (1) the electrical stimulation techniques used did not have sufficient resolution to distinguish relevant neurons from fibers-of-passage; (2) lesion experiments in relevant hypothalamic regions yielded conflicting results; and (3) methods for recording the activity of single units in freely moving animals were not available.

[8]We now know that this is because the experiment does not work. Because the mouse brain is much smaller than the rat brain, electrical current delivered to sites containing attack neurons spreads to adjacent sites that promote defensive behaviors, which are dominant to social behaviors such as fighting and mating (at least when artificially stimulated). The advantage of optogenetic activation is that it allows restriction of activation to cell bodies, by combining focal injection of viruses with localized photostimulation.

neither manipulation caused any change in male–female mating behavior, despite the presence of female-responsive units in VMHvl (Lin et al. 2011).

Genetic Identification of Hypothalamic Aggression Neurons

The foregoing experiments indicated that VMHvl contains neurons involved in attack, but they did not identify these cells. They also did not account for the function of VMHvl neurons activated by females; in principle, such neurons could promote mating, inhibit aggression, or both (Anderson 2012). To address these issues, we sought to identify molecular markers for these neurons. We found that VMHvl contains a subset (~40%) of neurons that express the type 1 estrogen receptor (Esr1; Fig. 5A), ~30% of which are activated during attack as determined by double-labeling for c-fos (Fig. 5H; Lee et al. 2014). To manipulate the function of these neurons, we generated gene-targeted mice expressing Cre recombinase in *Esr1*$^+$

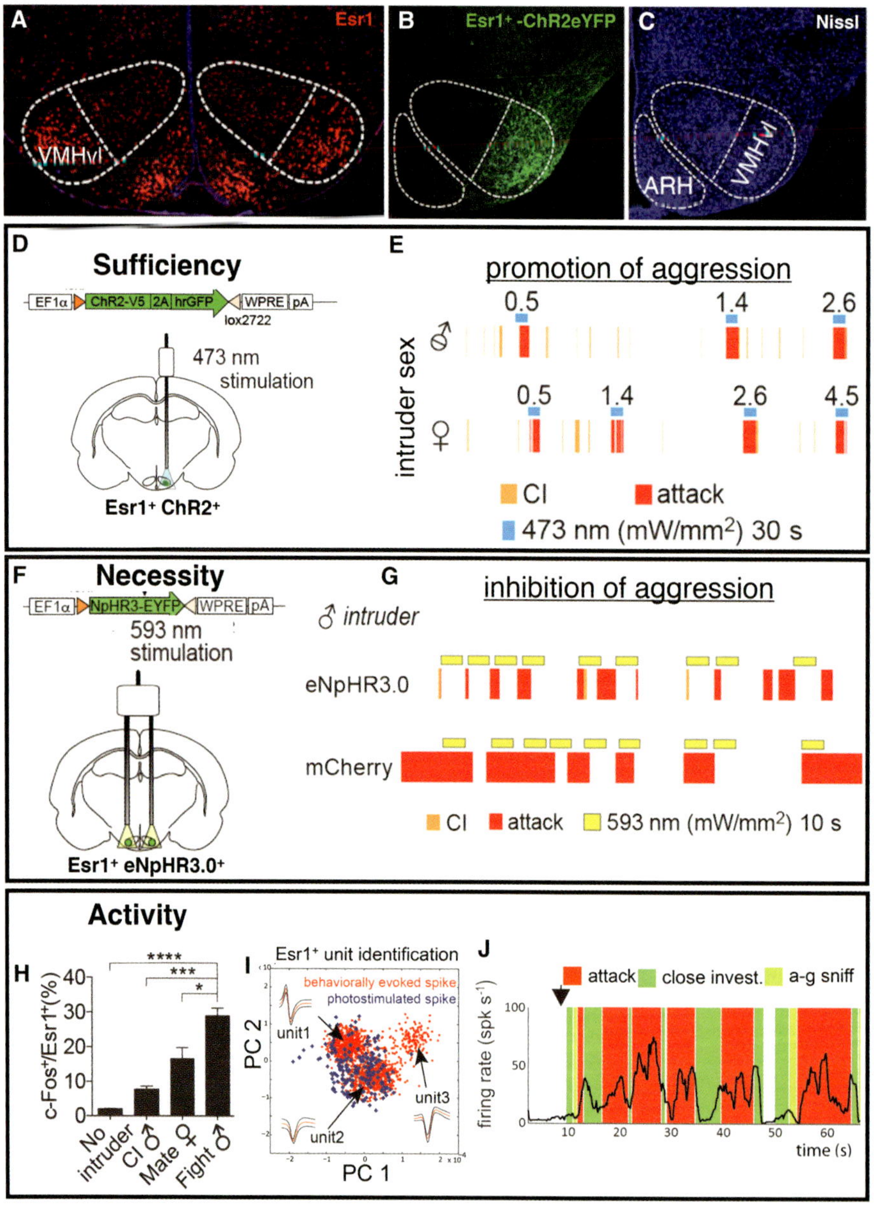

Figure 5. (*A*) Expression of Esr1 in VMHvl. (*B*) Expression of eYFP in Esr1$^+$ neurons in an *Esr1-Cre* mouse injected in VMHvl with a Cre-dependent AAV encoding ChR2-eYFP. (*C*) Nissl staining of the same section as in *B*. ARH, arcuate nucleus. (*D,E*) Optogenetic activation of VMHvl Esr1$^+$ neurons promotes attack. (*F,G*) Optogenetic inhibition of VMHvl Esr1$^+$ neurons inhibits attack compared to control (mCherry). (*H*) Induction of c-FOS in Esr1$^+$ neurons following different social interactions. CI, close investigation. (From Lee et al. 2014.) (*I,J*) An Esr1$^+$ neuron expressing ChR2 in VMHvl, identified by optogenetic phototagging using a chronically implanted optrode array, is active during attacks (*J*). (*I*) Plot illustrating overlap in principal component space between waveforms of optogenetically (blue) and behaviorally (red) evoked spikes from Esr1$^+$ neurons in VMHvl. PC, principal component.

neurons, allowing cell-type-specific expression of optogenetic effectors in VMHvl (Fig. 5B,C).

Optogenetic activation of $Esr1^+$ neurons in VMHvl indicated that they were sufficient to promote attack (Fig. 5D,E), whereas activation of $Esr1^-$ neurons was not (Lee et al. 2014). Time-resolved optogenetic inhibition of $Esr1^+$ neurons interrupted ongoing attack (Fig. 5F,G), indicating that their activity is required during this consummatory phase of an aggressive encounter, confirming and extending earlier loss-of-function studies (Lin et al. 2011; Sano et al. 2013; Yang et al. 2013).[9] Finally, preliminary experiments using optogenetic phototagging (Lima et al. 2009) to identify $Esr1^+$ neurons in multi-electrode recordings indicate that these cells exhibit increased spiking rates during attack (Fig. 5I–J; R Remedios and DJ Anderson, unpubl.). Together, these data indicate that $Esr1^+$ neurons in VMHvl fulfill the three major criteria necessary to identify them as attack neurons (Kruk 2014): they are sufficient to elicit attack when artificially stimulated; they are necessary for naturally occurring attack; and they are active during normal aggressive behavior.

Scalable Control of Social Behavior by VMHvl Esr1$^+$ Neurons

Systematic variation of optogenetic stimulation parameters revealed, unexpectedly, that although high levels of stimulation evoked attack, lower levels[10] evoked nonagonistic social behaviors, including sniffing or close investigation (CI) and attempted mounting (Figs. 6A, 7A; Lee et al. 2014). Mounting was evoked toward normal and castrated males as well as females, with equal efficacy. Remarkably, in some cases, the social behaviors evoked within a single stimulation session could be shifted, simply by increasing the light intensity used for photostimulation, from mounting to attack (Fig. 6A,B). The threshold for mounting and CI was similar, and approximately six- to sevenfold lower than that required to evoke attack (Fig. 6C).

Together, these data suggest that optogenetic activation of VMHvl $Esr1^+$ neurons evokes different social behaviors in a scalable manner (Lee et al. 2014): Low-intensity stimulation evokes appetitive and sexual behavior—social investigation and mounting—whereas higher intensities evoke attack (Figs. 6B, 7A). Interestingly, we have recently observed similar results following optogenetic stimulation of GABAergic neurons in the medial amygdala (Hong et al. 2014), a structure that projects indirectly to VMHvl (Canteras et al. 1992, 1995; Swanson 2000). We have also observed scalable control of different defensive behaviors—avoidance, freezing, and flight—following stimulation of a different subpopulation of VMH neurons located in its dorsomedial subdivision (Kunwar et al. 2015). These observations suggest that scalable control may be a general feature of hypothalamic cell populations controlling innate motivated behaviors. Although such stimulation intensity-dependent changes in evoked behavior have been previously observed using electrical methods (von Holst and von Saint Paul 1960, 1962), the use of optogenetic tools to restrict activation to genetically identified neuronal cell bodies rules out the possibility that such behavioral switches simply reflect current spread to neighboring brain regions, a possibility that could not be excluded in early studies.

Role of Esr1$^+$ Neurons in Male–Male Social Interactions

Several lines of evidence suggest that the relationship between increasing optogenetic stimulation of VMHvl $Esr1^+$ neurons, and the change in evoked behavior from investigation to attack, reflects a control mechanism that operates during naturally occurring male–male social encounters. First, the fraction of $Esr1^+$ neurons that are activated ($c\text{-}fos^+$) following attack is approximately five- to sixfold higher than after close investigation without attack (Fig. 5H; Lee et al. 2014). Second, electrophysiological recordings in VMHvl indicate that the average spiking rate ramps up quickly during the progression from CI to attack and that investigation is more likely to be followed by attack when the spiking rate is high (Fig. 6E; Lin et al. 2011; Falkner et al. 2014). Finally, and most importantly, time-resolved optogenetic inhibition delivered during the initial phase of a social encounter interrupts close investigation (Fig. 6D; Lee et al. 2014). Thus, inhibition of $Esr1^+$ neurons can interrupt either close investigation or ongoing attack (Fig. 7B), depending on when photostimulation is delivered. Together, these data are consistent with the idea that the behavioral progression during a male–male aggressive encounter is controlled in a scalable, threshold-dependent manner, according to the level of activity among VMHvl $Esr1^+$ neurons (Fig. 6B).

Role of Esr1$^+$ Neurons in Male–Female Social Interactions

A role for VMHvl in male sexual behavior is unexpected, as this function has traditionally been assigned to the MPO (Simerly 2002; Yang and Shah 2014). Optogenetic stimulation experiments should be interpreted with caution, because artificial patterns of activation could produce abnormal behaviors. Male mounting has been suggested to be a "default" social behavior (Stowers

[9]Previous studies indicated that killing progesterone receptor (PR)-expressing neurons in VMHvl (which overlap $Esr1^+$ neurons by 90%) partially reduced aggression (Yang et al. 2013), as did RNAi-mediated knockdown of Esr1 in VMHvl (Sano et al. 2013). However, because of the long delay between the manipulation and behavioral testing (2–3 wk), these studies could not distinguish a permissive requirement for VMHvl, e.g., in the detection of chemosensory cues essential for male–male recognition, from a requirement during attack per se.

[10]"Lower levels" were achieved by (1) injecting fewer ChR2-expressing viral particles; (2) incubating the animals for a shorter period of time following viral injection (which results in lower levels of ChR2 expression); (3) reducing the light intensity used in photostimulation; and (4) reducing the frequency of photostimulation (although reduced frequency was less effective than reduced light intensity). These data suggest that it is the number of activated cells, and the level of activity per cell, which is responsible for the observed behavioral differences.

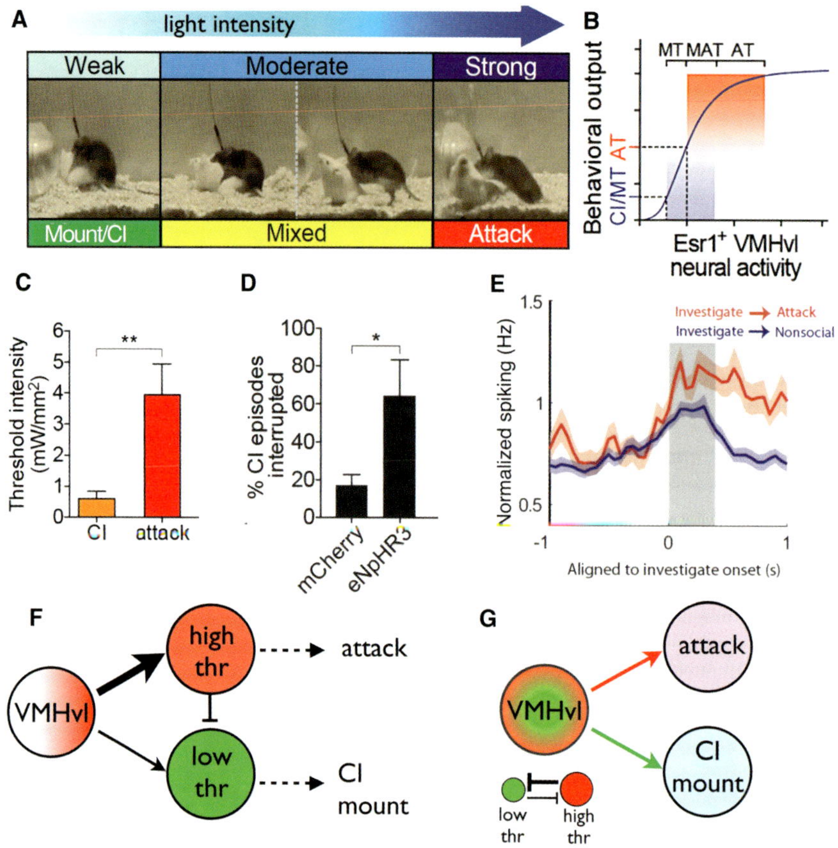

Figure 6. (*A,B*) Social behavior evoked by optogenetic stimulation of VMHvl Esr1$^+$ neurons switches from close investigation (CI) and mounting at low light intensity, to attack at high intensity. (*C*) Threshold intensities required to elicit CI versus attack. (*D*) Interruption of naturally occurring close investigative behavior (CI) by optogenetic inhibition of Esr1$^+$ neurons delivered during sniffing (vs. mCherry control). (*E*) Average spiking activity during the last 400 msec of sniffing preceding attack (red) or nonsocial behavior (blue). A higher level of spiking during the sniffing period predicts subsequent attack. (*F,G*) Alternative models to explain intensity dependence of optogenetically evoked behaviors from Esr1$^+$ neurons. (*F*) The level of activity among a common population of VMHvl Esr1$^+$ neurons activates behavior-specific downstream centers with low versus thresholds for activation. Thickness of arrows indicates relative level of activity in VMHvl output required to produce behavior. (*G*) Distinct and mutually inhibitory subsets of VMHvl Esr1$^+$ neurons controlling mounting (green) or attack (red) have low versus high thresholds for activation, respectively; output from the dominant population (red vs. green arrows) determines behavioral outcome. (*A–D*, Modified from Lee et al. 2014; *E*, modified from Falkner et al. 2014, with permission from the Society for Neuroscience.)

et al. 2002); thus, low-intensity stimulation of VMHvl Esr1$^+$ neurons could lead to mounting by default. Nevertheless, genetic ablation of progesterone receptor (PR)-expressing VMHvl neurons (Yang et al. 2013) (which highly overlap Esr1$^+$ neurons) or knockdown of Esr1 mRNA in VMHvl (Sano et al. 2013) both significantly reduced male mounting toward females. These data are consistent with the idea that VMHvl does play a role in normal male sexual behavior. However, acute optogenetic inactivation of Esr1$^+$ neurons did not interrupt normal male–female mounting *in media res* (Fig. 7B; Lee et al. 2014).[11] Interestingly, the average spiking rate in VMHvl decreased dramatically as male–female sexual encounters progressed from investigation and mounting to intromission and ejaculation (Lin et al. 2011). It is possible, therefore, that the role of Esr1$^+$ neurons in male sexual behavior is limited to the detection of chemosensory cues required for recognition of females, rather than for male copulatory behavior per se.

Our finding that mounting was evoked by weaker optogenetic stimulation of VMHvl Esr1$^+$ neurons than was required to trigger attack was also unexpected. However, it is consistent with the observation that fewer of these neurons are c-fos$^+$ following an encounter with a female, than with a male (Lin et al. 2011; Lee et al. 2014; Fig. 5H). Does this mean that the decision of whether to fight or mate is controlled simply by the level of population activity among VMHvl Esr1$^+$ neurons? It is possible that a low level of output from VMHvl activates low-threshold neurons in a downstream target structure that mediates CI or mounting, whereas a higher level of output

[11]Negative results must also be interpreted with caution, however, especially in the case of loss-of-function manipulations. The inability to arrest mounting by optogenetic inhibition of Esr1$^+$ neurons could reflect a redundant role for other structures (e.g., MPO), or incomplete inhibition of activity. Genetic ablation (Yang et al. 2013), although it lacks temporal resolution, completely eliminates the activity of a given population of neurons.

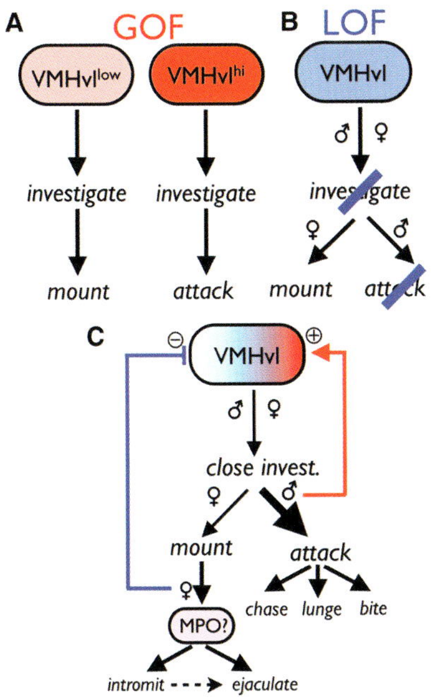

Figure 7. (*A*,*B*) Summary of gain-of-function (*A*, GOF) and loss-of-function (*B*, LOF) phenotypes obtained by optogenetic manipulation of VMHvl Esr1$^+$ neurons. Blue bars in *B* indicate behaviors interrupted by optogenetic inhibition. (*C*) Schematic illustrating a hypothetical view of the function of Esr1$^+$ neurons in VMHvl during a normal male–male or male–female social encounter. Thickness of arrows denotes relative level of output (spiking rate and/or number of active output neurons required to produce behavior). Blue versus red shading in VMHvl indicates gradual decrease (blue) or increase (red) in average spiking rate during an evolving male–female (blue) or male–male (red) social encounter. Colored arrows indicate behavioral feedback. See text for details.

activates high-threshold neurons in a different target that mediates attack; the latter could then inhibit the former (Fig. 6F). Alternatively, the intensity-dependent effects of optogenetic stimulation might reflect the existence of different, sex-specific subpopulations of Esr1$^+$ neurons, with female-specific neurons activated at a lower threshold than male-specific neurons (Lee et al. 2014).[12] If so, then asymmetric and reciprocal inhibition between these Esr1$^+$ neuronal subpopulations could determine whether output from these neurons promotes mating or fighting (Fig. 6G).

Whether the optogenetic results (Fig. 7A) reflect intensity coding or cellular heterogeneity, our electrophysiological recordings suggest a dynamic and time-evolving role for VMHvl neurons during a natural social encounter (Fig. 7C). Most of the neurons activated during both male–female and male–male encounters are most active at the initial phases of a social encounter (Lin et al. 2011). Perhaps these neurons promote approach and close (anogenital) investigation of a conspecific to identify intruder gender, which requires detection of short-range, sex-specific pheromonal cues (Fig. 7C, "close invest"; Brennan and Zufall 2006; Dulac and Wagner 2006). The recognition of a male would promote further social investigation, ramping up the activity of VMHvl neurons in a positive-feedback manner, until the threshold for attack was reached (Fig. 7C, red arrow). The recognition of a female would promote initial attempts at mounting, with further interactions driving a reduction of activity in VMHvl and initiating the engagement of other structures, such as the MPO, that may promote the consummatory phase of a sexual encounter (Fig. 7C, blue arrow). This speculative scenario incorporates both sex-specific subpopulations and dynamic changes in population activity among VMHvl Esr1$^+$ neurons during social encounters.

[12]Note, however, that optogenetic stimulation of Esr1$^+$ neurons in VMHvl slices, together with calcium imaging using GCaMP6s, did not induce a different pattern of neuronal activity at low versus high stimulation intensities, but rather simply increased the number of active neurons and the average level of activity per cell (Lee et al. 2014).

SOCIAL BEHAVIOR CONTROL CENTERS: RHEOSTATS OR SWITCHES?

The foregoing data indicate that artificial activation of VMHvl (and MeApd) neurons at different intensities can evoke different types of social behavior at different thresholds (Hong et al. 2014; Lee et al. 2014) and that the average spiking rate among VMHvl neurons increases as animals progress from investigative to attack behavior (Lin et al. 2011; Falkner et al. 2014). Together, these findings raise the intriguing possibility that VMHvl encodes the intensity of an escalating state of arousal, motivation, or drive, in a graded manner according to its level of activity (Fig. 8A, "social behavior control center"). This increasing level of activity may reflect changes in sensory input as a social encounter evolves over time (e.g., via an increase in the concentration of olfactory cues during close investigation) (Fig. 8A, "sensory inputs"). Such escalation could, in turn, be used to control transitions between different behaviors by activating lower-level command centers controlling these behaviors (Fig. 8A; B_1, B_2, etc.) at different thresholds, perhaps according to the level of tonic inhibition imposed on such centers (Fig. 8A, "action selection"). Such a model is analogous to the size principle proposed by Henneman to explain the progressive recruitment of larger motor units during muscle engagement (Henneman 1985). In this way, VMHvl could both encode the intensity of an internal state and couple the intensity of that state to behavioral decisions. Such a model integrates the Lorenzian view of behavioral control by drive state intensity (Fig. 1B) with the Tinbergian view of hierarchical control of behavioral decision-making (Fig. 1C).

Alternatively, VMHvl could act simply as a permissive switch, or gate, for social behavior, whereas action selection would be controlled by sensory inputs that act directly on the lower-level command centers (Fig. 8B). In this scenario, the level of motivation or arousal would have to be encoded elsewhere in the brain and might be a consequence, rather than a cause, of escalating behavior. The essential difference between these models is that in one case, the control center simply determines whether or not aggression or mating will occur (Fig. 8B), whereas in

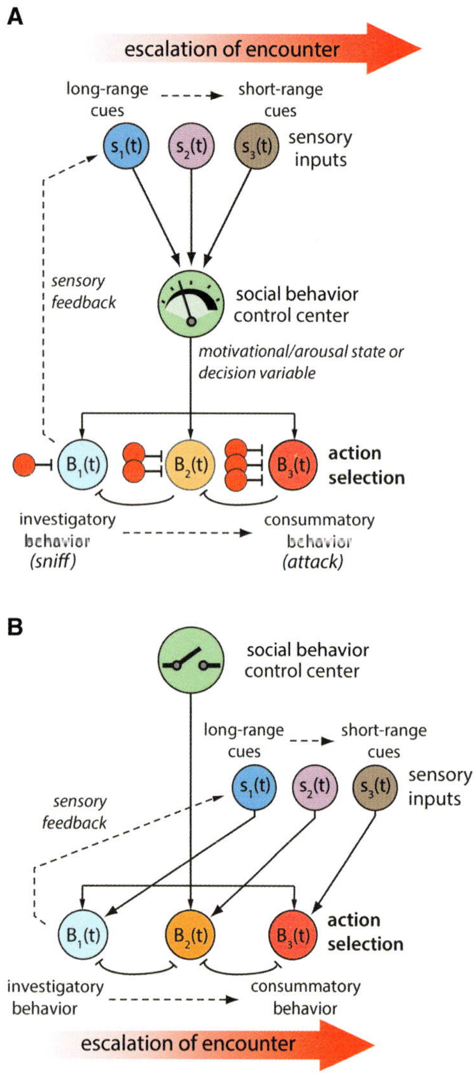

Figure 8. Alternative models for the control of action selection during an escalating social encounter. (*A*) Neurons in a social behavior control center (green) function like a rheostat, which encodes the intensity of an escalating internal state (arousal, motivation, or drive) according to the integrated sensory input converging on this center. Common outputs from this center project to lower-level command centers controlling distinct behavioral actions occurring at different times ($B_1(t)$, $B_2(t)$, etc.); each command center receives the same amount of input from the higher-level center. Command centers controlling investigatory (appetitive) to consummatory actions are held under progressively higher levels of tonic inhibition (red circles). Therefore, release from this inhibition occurs with progressively stronger output from the higher-level center, as the intensity of the internal state associated with the social encounter escalates. (*B*) Alternative model in which the social behavioral control center functions simply as a switch or gate to allow agonistic behavior to occur, and different behavioral actions are controlled by the direct influences of changing sensory cues on lower-level command centers during an evolving social encounter. Models that combine features of *A*,*B* are not illustrated but also possible.

the other case, it also controls the type of lower-level behavioral action that is selected (Fig. 8A). Distinguishing between these alternatives will require new methods for imaging neural population activity in freely behaving animals (Ghosh et al. 2011; Ziv et al. 2013; Jennings et al. 2015), more specific tools for identifying and functionally manipulating neuronal subtypes and their connectivity (Zeng and Madisen 2012; Oh et al. 2014), as well as the application of modeling and theory. The combination of such approaches should yield new mechanistic insights into the neural control of animal behavior and may shed light on the interaction between emotion states and decision-making in humans.

ACKNOWLEDGMENTS

We thank members of the Anderson laboratory, past and present, for their contributions to the data and ideas discussed in this article, including Todd Anthony, Vivian Chiu, Brian Duistermars, Tyler Gibson, Weizhe Hong, Dong-Wook Kim, Prabhat Kunwar, Kiichi Watanabe, Allan Wong, and Moriel Zelikowsky. A long-term collaboration with Prof. Pietro Perona, and his students Piotr Dollar, Carlos Gonzalez, and Eyrún Eyolfsdottir, has been essential in developing automated methods for behavioral analysis used in this work. We also acknowledge Dr. Hongkui Zeng and her colleagues at the Allen Institute for Brain Sciences for their ongoing collaborative contributions. We thank Gina Mancuso for administrative assistance and Celine Chiu for laboratory management. Portions of this work were supported by grants from the Ellison Medical Foundation, the Gordon and Betty Moore Foundation, the Simons Foundation, the Paul G. Allen Family Foundation, and National Institutes of Health grants MH070053, MH085082, and DA031389. D.J.A. is an Investigator of the Howard Hughes Medical Institute.

REFERENCES

Adams DB. 2006. Brain mechanisms of aggressive behavior: An updated review. *Neurosci Biobehav Rev* **30:** 304–318.

Alekseyenko OV, Lee C, Kravitz EA. 2010. Targeted manipulation of serotonergic neurotransmission affects the escalation of aggression in adult male *Drosophila melanogaster*. *PLoS One* **5:** e10806.

Alekseyenko OV, Chan Y-B, Li R, Kravitz EA. 2013. Single dopaminergic neurons that modulate aggression in *Drosophila*. *Proc Natl Acad Sci* **110:** 6151–6156.

Anderson DJ. 2012. Optogenetics, sex, and violence in the brain: Implications for psychiatry. *Biol Psychiatry* **71:** 1081–1089.

Anderson DJ, Adolphs R. 2014. A framework for studying emotions across species. *Cell* **157:** 187–200.

Andrews JC, Fernández MP, Yu Q, Leary GP, Leung AK, Kavanaugh MP, Kravitz EA, Certel SJ. 2014. Octopamine neuromodulation regulates Gr32a-linked aggression and courtship pathways in *Drosophila* males. *PLoS Genet* **10:** e1004356.

Asahina K, Watanabe K, Duistermars BJ, Hoopfer E, González CR, Eyjólfsdóttir EA, Perona P, Anderson DJ. 2014. Tachykinin-expressing neurons control male-specific aggressive arousal in *Drosophila*. *Cell* **156:** 221–235.

Bargmann CI. 2012. Beyond the connectome: How neuromodulators shape neural circuits. *Bioessays* **34:** 458–465.

Bath DE, Stowers JR, Hormann D, Poehlmann A, Dickson BJ, Straw AD. 2014. FlyMAD: Rapid thermogenetic control of neuronal activity in freely walking *Drosophila*. *Nat Methods* **11:** 756–762.

Berridge KC. 2004. Motivation concepts in behavioral neuroscience. *Physiol Behav* **81:** 179–209.

Borst A. 2009. *Drosophila*'s view on insect vision. *Curr Biol* **19:** R36–R47.

Brennan PA, Zufall F. 2006. Pheromonal communication in vertebrates. *Nature* **444:** 308–315.

Canteras NS, Simerly RB, Swanson LW. 1992. Connections of the posterior nucleus of the amygdala. *J Comp Neurol* **324:** 143–179.

Canteras NS, Simerly RB, Swanson LW. 1995. Organization of projections from the medial nucleus of the amygdala: A PHAL study in the rat. *J Comp Neurol* **360:** 213–245.

Certel SJ, Leung A, Lin C-Y, Perez P, Chiang A-S, Kravitz EA. 2010. Octopamine neuromodulatory effects on a social behavior decision-making network in *Drosophila* males. *PLoS One* **5:** e13248.

Coccaro EF, Lee R, Owens MJ, Kinkead B, Nemeroff CB. 2012. Cerebrospinal fluid substance P-like immunoreactivity correlates with aggression in personality disordered subjects. *Biol Psychiatry* **72:** 238–243.

de la Paz Fernandez M, Chan Y-B, Yew J, Billeter J-C, Dreisewerd K, Levine J, Kravitz E. 2010. Pheromonal and behavioral cues trigger male-to-female aggression in *Drosophila*. *PLoS Biol* **8:** e1000541.

Devidze N, Lee A, Zhou J, Pfaff D. 2006. CNS arousal mechanisms bearing on sex and other biologically regulated behaviors. *Physiol Behav* **88:** 283–293.

Dierick HA, Greenspan RJ. 2007. Serotonin and neuropeptide F have opposite modulatory effects on fly aggression. *Nat Genet* **39:** 678–682.

Dulac C, Wagner S. 2006. Genetic analysis of brain circuits underlying pheromone signaling. *Annu Rev Genet* **40:** 449–467.

Falkner AL, Lin D. 2014. Recent advances in understanding the role of the hypothalamic circuit during aggression. *Front Syst Neurosci* **8:** 168.

Falkner AL, Dollar P, Perona P, Anderson DJ, Lin D. 2014. Decoding ventromedial hypothalamic neural activity during male mouse aggression. *J Neurosci* **34:** 5971–5984.

Flavell SW, Pokala N, Macosko EZ, Albrecht DR, Larsch J, Bargmann CI. 2013. Serotonin and the neuropeptide PDF initiate and extend opposing behavioral states in *C. elegans*. *Cell* **154:** 1023–1035.

Ghosh KK, Burns LD, Cocker ED, Nimmerjahn A, Ziv Y, Gamal AE, Schnitzer MJ. 2011. Miniaturized integration of a fluorescence microscope. *Nat Methods* **8:** 871–878.

Gillette R, Huang RC, Hatcher N, Moroz LL. 2000. Cost-benefit analysis potential in feeding behavior of a predatory snail by integration of hunger, taste, and pain. *Proc Natl Acad Sci* **97:** 3585–3590.

Henneman E. 1985. The size-principle: A deterministic output emerges from a set of probabilistic connections. *J Exp Biol* **115:** 105–112.

Hess WR. 1928. Stammganglien-Reizversuche. Berichte der gesamten. *Physiologie* **42:** 554–555.

Hess WR, Brügger M. 1943. Das subkortikale Zentrum der affecktiven Abwehr-reaktion. *Helv Physiol Acta* **1:** 33–52.

Hirayama K, Catanho M, Brown JW, Gillette R. 2012. A core circuit module for cost/benefit decision. *Front Neurosci* **6:** 123.

Hong W, Kim DW, Anderson DJ. 2014. Antagonistic control of social versus repetitive self-grooming behaviors by separable amygdala neuronal subsets. *Cell* **158:** 1348–1361.

Hrabovszky E, Halász J, Meelis W, Kruk MR, Liposits Z, Haller J. 2005. Neurochemical characterization of hypothalamic neurons involved in attack behavior: Glutamatergic dominance and co-expression of thyrotropin-releasing hormone in a subset of glutamatergic neurons. *Neuroscience* **133:** 657–666.

Inagaki HK, Jung Y, Hoopfer ED, Wong AM, Mishra N, Lin JY, Tsien RY, Anderson DJ. 2014. Optogenetic control of *Drosophila* using a red-shifted channelrhodopsin reveals experience-dependent influences on courtship. *Nat Methods* **11:** 325–332.

Insel TR, Young LJ. 2000. Neuropeptides and the evolution of social behavior. *Curr Opin Neurobiol* **10:** 784–789.

Jennings JH, Ung RL, Resendez SL, Stamatakis AM, Taylor JG, Huang J, Veleta K, Kantak PA, Aita M, Shilling-Scrivo K, et al. 2015. Visualizing hypothalamic network dynamics for appetitive and consummatory behaviors. *Cell* **160:** 516–527.

Katsouni E, Sakkas P, Zarros A, Skandali N, Liapi C. 2009. The involvement of substance P in the induction of aggressive behavior. *Peptides* **30:** 1586–1591.

Klapoetke NC, Murata Y, Kim SS, Pulver SR, Birdsey-Benson A, Cho YK, Morimoto TK, Chuong AS, Carpenter EJ, Tian Z, et al. 2014. Independent optical excitation of distinct neural populations. *Nat Methods* **11:** 338–346.

Kohatsu S, Koganezawa M, Yamamoto D. 2011. Female contact activates male-specific interneurons that trigger stereotypic courtship behavior in *Drosophila*. *Neuron* **69:** 498–508.

Koike H, Ibi D, Mizoguchi H, Nagai T, Nitta A, Takuma K, Nabeshima T, Yoneda Y, Yamada K. 2009. Behavioral abnormality and pharmacologic response in social isolation-reared mice. *Behav Brain Res* **202:** 114–121.

Kravitz EA, Huber R. 2003. Aggression in invertebrates. *Curr Opin Neurobiol* **13:** 736–743.

Kruk MR. 1991. Ethology and pharmacology of hypothalamic aggression in the rat. *Neurosci Biobehav Rev* **15:** 527–538.

Kruk MR. 2014. Hypothalamic attack: A wonderful artifact or a useful perspective on escalation and pathology in aggression? A viewpoint. *Curr Top Behav Neurosci* **17:** 143–188.

Kruk M, Vanderpoel A, Lammers J, Haccou P, Olivier B. 1988. Ethopharmacology of aggression induced by stimulation in the rats hypothalamus. *Aggress Behav* **14:** 138–138.

Kunwar P, Zelikowsky M, Remedios R, Cai H, Yilmaz M, Meister M, Anderson DJ. 2015. Ventromedial hypothalamic neurons control a defensive emotion state. *Elife* **4**. doi: 10.77554/eLife.06633.

Lebestky T, Chang JS, Dankert H, Zelnik L, Kim YC, Han KA, Wolf FW, Perona P, Anderson DJ. 2009. Two different forms of arousal in *Drosophila* are oppositely regulated by the dopamine D1 receptor ortholog DopR via distinct neural circuits. *Neuron* **64:** 522–536.

LeDoux J. 2012. Rethinking the emotional brain. *Neuron* **73:** 653–676.

Lee H, Kim D-W, Remedios R, Anthony TE, Chang A, Madisen L, Zeng H, Anderson DJ. 2014. Scalable control of mounting and attack by $Esr1^+$ neurons in the ventromedial hypothalamus. *Nature* **509:** 627–632.

Lerchner W, Xiao C, Nashmi R, Slimko EM, van Trigt L, Lester HA, Anderson DJ. 2007. Reversible silencing of neuronal excitability in behaving mice by a genetically targeted, ivermectin-gated Cl-channel. *Neuron* **54:** 35–49.

Lima SQ, Hromádka T, Znamenskiy P, Zador AM. 2009. PINP: A new method of tagging neuronal populations for identification during in vivo electrophysiological recording. *PLoS One* **4:** e6099.

Lin D, Boyle MP, Dollar P, Lee H, Lein ES, Perona P, Anderson DJ. 2011. Functional identification of an aggression locus in the mouse hypothalamus. *Nature* **470:** 221–226.

Lin JY, Knutsen PM, Muller A, Kleinfeld D, Tsien RY. 2013. ReaChR: A red-shifted variant of channelrhodopsin enables deep transcranial optogenetic excitation. *Nat Neurosci* **16:** 1499–1508.

Liu W, Liang X, Gong J, Yang Z, Zhang YH, Zhang JX, Rao Y. 2011. Social regulation of aggression by pheromonal activation of Or65a olfactory neurons in *Drosophila*. *Nat Neurosci* **14:** 896–902.

Lorenz K, Leyhausen P. 1973. *Motivation of human and animal behavior; an ethological view*. Van Nostrand-Reinhold, New York.

Marder E. 2012. Neuromodulation of neuronal circuits: Back to the future. *Neuron* **76:** 1–11.

Miczek KA, de Almeida RM, Kravitz EA, Rissman EF, de Boer SF, Raine A. 2007. Neurobiology of escalated aggression and violence. *J Neurosci* **27:** 11803–11806.

Mundiyanapurath S, Chan Y-B, Leung AKW, Kravitz EA. 2009. Feminizing cholinergic neurons in a male *Drosophila* nervous system enhances aggression. *Fly* **3:** 179–184.

Nassel DR, Winther AM. 2010. *Drosophila* neuropeptides in regulation of physiology and behavior. *Prog Neurobiol* **92:** 42–104.

Nelson RJ, Trainor BC. 2007. Neural mechanisms of aggression. *Nat Rev Neurosci* **8:** 536–546.

Newman SW. 1999. The medial extended amygdala in male reproductive behavior. A node in the mammalian social behavior network. *Ann N Y Acad Sci* **877:** 242–257.

Oh SW, Harris JA, Ng L, Winslow B, Cain N, Mihalas S, Wang Q, Lau C, Kuan L, Henry AM, et al. 2014. A mesoscale connectome of the mouse brain. *Nature* **508:** 207–214.

Pfaff D, Westberg L, Kow L. 2005. Generalized arousal of mammalian central nervous system. *J Comp Neurol* **493:** 86–91.

Salzman CD, Fusi S. 2010. Emotion, cognition, and mental state representation in amygdala and prefrontal cortex. *Ann Rev Neurosci* **33:** 173–202.

Sano K, Tsuda MC, Musatov S, Sakamoto T, Ogawa S. 2013. Differential effects of site-specific knockdown of estrogen receptor α in the medial amygdala, medial pre-optic area, and ventromedial nucleus of the hypothalamus on sexual and aggressive behavior of male mice. *Eur J Neurosci* **37:** 1308–1319.

Shaikh MB, Steinberg A, Siegel A. 1993. Evidence that substance P is utilized in medial amygdaloid facilitation of defensive rage behavior in the cat. *Brain Res* **625:** 283–294.

Siegel A. 2004. *Neurobiology of aggression and rage*. CRC Press, Boca Raton, FL.

Siegel A, Roeling TA, Gregg TR, Kruk MR. 1999. Neuropharmacology of brain-stimulation-evoked aggression. *Neurosci Biobehav Rev* **23:** 359–389.

Simerly RB. 2002. Wired for reproduction: Organization and development of sexually dimorphic circuits in the mammalian forebrain. *Ann Rev Neurosci* **25:** 507–536.

Slimko EM, McKinney S, Anderson DJ, Davidson N, Lester HA. 2002. Selective electrical silencing of mammalian neurons in vitro by the use of invertebrate ligand-gated chloride channels. *J Neurosci* **22:** 7373–7379.

Sternson SM. 2013. Hypothalamic survival circuits: Blueprints for purposive behaviors. *Neuron* **77:** 810–824.

Stevenson PA, Dyakonova V, Rillich J, Schildberger K. 2005. Octopamine and experience-dependent modulation of aggression in crickets. *J Neurosci* **25:** 1431–1441.

Stowers L, Holy TE, Meister M, Dulac C, Koentges G. 2002. Loss of sex discrimination and male-male aggression in mice deficient for TRP2. *Science* **295:** 1493–1500.

Swanson LW. 2000. Cerebral hemisphere regulation of motivated behavior. *Brain Res* **886:** 113–164.

Taghert PH, Nitabach MN. 2012. Peptide neuromodulation in invertebrate model systems. *Neuron* **76:** 82–97.

Tinbergen N. 1950. The hierarchical organization of nervous mechanisms underlying instinctive behaviour. *Sym Soc Exp Biol* **4:** 305–312.

Tinbergen N. 1951. *The study of instinct*. Clarendon Press/Oxford University Press, New York.

Toth M, Mikics E, Tulogdi A, Aliczki M, Haller J. 2011. Post-weaning social isolation induces abnormal forms of aggression in conjunction with increased glucocorticoid and autonomic stress responses. *Horm Behav* **60:** 28–36.

van Swinderen B, Andretic R. 2003. Arousal in *Drosophila*. *Behav Processes* **64:** 133–144.

van Swinderen B, Andretic R. 2011. Dopamine in *Drosophila*: Setting arousal thresholds in a miniature brain. *Proc Biol Sci* **278:** 906–913.

Veening JG, Coolen LM, de Jong TR, Joosten HW, de Boer SF, Koolhaas JM, Olivier B. 2005. Do similar neural systems subserve aggressive and sexual behaviour in male rats? Insights from c-Fos and pharmacological studies. *Eur J Pharmacol* **526:** 226–239.

von Holst E, von Saint Paul U. 1960. On the functional organization of drives. *J Anim Behav* **11:** 1–20.

von Holst E, von Saint Paul U. 1962. Electrically controlled behavior. *Sci Am* **206:** 50–59.

von Philipsborn AC, Liu T, Yu JY, Masser C, Bidaye SS, Dickson BJ. 2011. Neuronal control of *Drosophila* courtship song. *Neuron* **69:** 509–522.

Wang L, Anderson DJ. 2010. Identification of an aggression-promoting pheromone and its receptor neurons in *Drosophila*. *Nature* **463:** 227–231.

Wang L, Dankert H, Perona P, Anderson DJ. 2008. A common genetic target for environmental and heritable influences on aggressiveness in *Drosophila*. *Proc Natl Acad Sci* **105:** 5657–5663.

Wang L, Han X, Mehren J, Hiroi M, Billeter J-C, Miyamoto T, Amrein H, Levine JD, Anderson DJ. 2011. Hierarchical chemosensory regulation of male-male social interactions in *Drosophila*. *Nat Neurosci* **14:** 757–762.

Yamamoto D, Sato K, Koganezawa M. 2014. Neuroethology of male courtship in *Drosophila*: From the gene to behavior. *J Comp Physiol A Neuroethol Sens Neural Behav Physiol* **200:** 251–264.

Yang CF, Shah NM. 2014. Representing sex in the brain, one module at a time. *Neuron* **82:** 261–278.

Yang CF, Chiang MC, Gray DC, Prabhakaran M, Alvarado M, Juntti SA, Unger EK, Wells JA, Shah NM. 2013. Sexually dimorphic neurons in the ventromedial hypothalamus govern mating in both sexes and aggression in males. *Cell* **153:** 896–909.

Zeng H, Madisen L. 2012. Mouse transgenic approaches in optogenetics. *Prog Brain Res* **196:** 193–213.

Zhou C, Rao Y, Rao Y. 2008. A subset of octopaminergic neurons are important for *Drosophila* aggression. *Nat Neurosci* **11:** 1059–1067.

Ziv Y, Burns LD, Cocker ED, Hamel EO, Ghosh KK, Kitch LJ, El Gamal A, Schnitzer MJ. 2013. Long-term dynamics of CA1 hippocampal place codes. *Nat Neurosci* **16:** 264–266.

Early Language Learning and the Social Brain

Patricia K. Kuhl
Institute for Learning & Brain Sciences, University of Washington, Seattle, Washington 98195
Correspondence: pkkuhl@u.washington.edu

Explaining how every typically developing child acquires language is one of the grand challenges of cognitive neuroscience. Historically, language learning provoked classic debates about the contributions of innately specialized as opposed to general learning mechanisms. Now, new data are being brought to bear from studies that employ magnetoencephalograph (MEG), electroencephalograph (EEG), magnetic resonance imaging (MRI), and diffusion tensor imaging (DTI) studies on young children. These studies examine the patterns of association between brain and behavioral measures. The resulting data offer both expected results and surprises that are altering theory. As we uncover what it means to be human through the lens of young children, and their ability to speak, what we learn will not only inform theories of human development, but also lead to the discovery of neural biomarkers, early in life, that indicate risk for language impairment and allow early intervention for children with developmental disabilities involving language.

The earliest phases of language acquisition invoke a special fascination because they allow us to examine humans' extraordinary abilities to acquire what is arguably our most unique skill—the ability to speak. Humans' capacity for language provoked classic debates on nature versus nurture by equally strong proponents of nativism (Chomsky 1959) and learning (Skinner 1957). Although we are far beyond those debates and now informed by a great deal of data about infants, their innate predispositions, and their incredible abilities to learn once exposed to natural language (Saffran et al. 2006; Kuhl 2010, 2011), we are only beginning to posit the mechanisms in the brain and mind of a child that enable language acquisition (e.g., Kuhl et al. 2014).

In this paper, I advance a new hypothesis—that in the earliest phases of language acquisition, infants combine a powerful set of domain-general computational skills with their equally extraordinary social skills to "crack" the speech code. Furthermore, I hypothesize that the social brain—in ways we have yet to understand—"gates" the computational mechanisms underlying learning in the domain of language. My goal here is to discuss recent results that advance our understanding of the mechanisms underlying the social influences on language learning in infants.

The assertion that social factors gate language learning, I argue, explains not only how typically developing children acquire language, but also why children with autism exhibit twin deficits in social cognition and language, and why nonhuman animals with impressive computational abilities do not acquire language. Moreover, the gating hypothesis may explain why social factors play a far more significant role than previously realized in human learning beyond language and throughout our lifetimes (Meltzoff et al. 2009).

Using methods of systems neuroscience tailored for use with very young children, we can now explore how the integration of information across specialized brain systems such as language and social cognition might take place, and thus provide valuable data about the potent role social interaction plays in language learning. These approaches, as well as others described here, could lead us toward a view of language acquisition in the human child that would be transformational.

THE LEARNING PROBLEM

Speech learning is a deep puzzle that our theories and machines struggle to solve. How do infants discover the sounds and words used in their particular language(s) when the most sophisticated computers cannot? What is it about the human mind that allows the young child, merely a year old, to understand the words that induce meaning in our collective minds, and to begin to use those words to convey their innermost thoughts and desires? Children's ability to express a thought through words is a breathtaking feat of the human mind.

In explaining a new approach to this problem, I will focus on perception of the elementary units of language, the consonants and vowels that make up words, and children's early word recognition, to show how our computation and social skills combine to form a very powerful learning mechanism. Interestingly, the new solution resembles neither Skinner's operant conditioning and reinforcement model of learning, nor Chomsky's nativist model. Instead, it is a view of the process that takes into account infants' innate skills to perceive all phonetic distinctions at birth, an ability that is not unique to humans (Kuhl and Miller 1975), and new data on children's extraordinary abilities to learn from computational and social mechanisms once exposed to speech (Kuhl et al. 2003).

 doi: 10.1101/sqb.2014.79.024802

SECOND LANGUAGE LEARNING EXHIBITS A "CRITICAL PERIOD" FOR LEARNING

A stage-setting concept for human language learning is the graph shown in Figure 1, redrawn from a study by Johnson and Newport on English grammar in speakers of Korean learning English (1989). The graph as rendered shows a simplified schematic of second language learning as a function of age.

The graph is surprising from the standpoint of human learning more generally. In the domain of language, infants and young children are superior learners when compared with adults, despite adults' cognitively superiority. Language is one of the classic examples of a "critical" or "sensitive" period in neurobiology (Johnson and Newport 1989; Newport et al. 2001; Knudsen 2004; Kuhl 2004; Bruer 2008).

Scientists are in agreement that this curve represents data across a wide variety of language-learning studies (Johnson and Newport 1989; Bialystok and Hakuta 1994; Neville et al. 1997; Flege et al. 1999; Weber-Fox and Neville 1999; Yeni-Komshian et al. 2000; Birdsong and Molis 2001; Mayberry and Lock 2003; Kuhl et al. 2005, 2008; EL Newport and T Supalla, unpubl.). The learning function describes our current understanding of second language learning, and reflects many individual sensitive periods for phonological, semantic, and grammatical learning, over time. Given widespread agreement on the fact that we do not learn equally well over the lifespan, theory is currently focused on attempts to explain this phenomenon. What accounts for adults' inability to learn a new language with the facility of an infant?

One of the candidate explanations is Lenneberg's hypothesis that development of the corpus callosum changes the brain in a way that affects learning (Lenneberg 1967; Newport et al. 2001). More recent hypotheses take a different perspective. Newport raised the "less is more" hypothesis, which suggests that infants' limited cognitive capacities actually allow superior learning of the simplified language spoken to infants (Newport 1990). Work in my laboratory has led me to advance the concept of *neural commitment*, the idea that neural networks are established early in infancy to detect the phonetic and prosodic patterns of speech (Kuhl 2004; Kuhl et al. 2014). The neural architecture is designed to maximize processing of the language experienced by the infant. Once established, the neural architecture for French or Tagalog, for example, impedes learning of new patterns that do not conform.

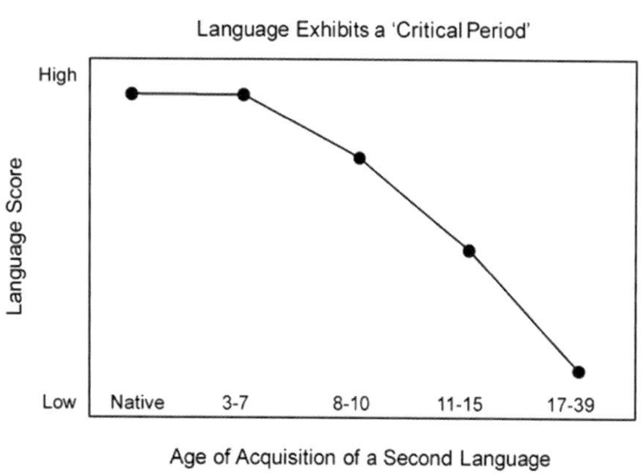

Figure 1. The relationship between age of acquisition of a second language and language skill. (Reprinted from Kuhl 2010; originally adapted from Johnson and Newport 1989, with permission from Elsevier.)

FOCAL EXAMPLE: PHONEME LEARNING

The world's languages contain approximately 600 consonants and 200 vowels (Ladefoged 2001). Each language uses a unique set of approximately 40 distinct elements, *phonemes*, which change the meaning of a word (e.g., from *bat* to *pat*). But phonemes are actually groups of nonidentical sounds, *phonetic units*, that are functionally equivalent in the language. The infant's task is to make some progress in figuring out the composition of the 40-odd phonemic categories before trying to acquire words on which these elementary units depend. For example, Japanese infants must learn to group the phonetic units *r* and *l* into a single phonemic category (Japanese *r*), whereas American infants must learn to uphold the distinction to separate *rake* from *lake*. Similarly, Spanish-learning infants must distinguish phonetic units critical to Spanish words (*bano* and *pano*), whereas English learning infants must combine them into a single category (English *b*). If infants were exposed only to the subset of phonetic units that will eventually be used phonemically to differentiate words in their language, the problem would be trivial. But infants are exposed to many more phonetic variants than will be used phonemically. Simple exposure to a category of sounds does not explain phonetic learning.

Learning to produce the sounds that will characterize infants as speakers of their "mother tongue" is equally challenging, and is not completely mastered until the age of 8 yr (Ferguson et al. 1992). Yet, by 10 mo of age, differences can be discerned in the babbling of infants raised in different countries (de Boysson-Bardies 1993), and in the laboratory, vocal imitation can be elicited by 20 wk (Kuhl and Meltzoff 1982). The speaking patterns we adopt early in life last a lifetime (Flege 1991).

My colleagues and I have suggested that this kind of indelible learning stems from a linkage between sensory and motor experience; sensory experience with a specific language establishes auditory patterns stored in memory that are unique to that language and these auditory representations guide infants' successive motor approximations until a match is achieved (Kuhl and Meltzoff 1996). The ability to imitate vocally may depend on the brain's social understanding mechanisms, those that form a human mirroring system for social understanding (Hari and Kujala 2009), and recent brain data obtained on infants support this view (Kuhl et al. 2014). We revisit the brain's sensorimotor coupling for speech later in this paper.

What enables the kind of learning we see in infants for speech? No machine in the world can derive the phonemic inventory of a language from natural language input (Rabiner and Huang 1993), although models improve when exposed to "motherese," the linguistically simplified and acoustically exaggerated speech that adults universally use when speaking to infants (de Boer and Kuhl 2003; see also Kuhl et al. 1997). The variability in speech input is simply too enormous; Japanese adults produce both English *r*- and *l*-like sounds, exposing Japanese infants to both sounds (Lotto et al. 2004; Werker et al. 2007). How do Japanese infants learn that these two sounds do not distinguish words, and that their differences should be ignored? Similarly, adult English speakers produce the Spanish *b* and *p*, exposing American infants to both categories of sound (Abramson and Lisker 1970). How do infants learn which sounds are important in distinguishing words in English? An important discovery in the 1970s showed that infants initially hear all these phonetic differences, but that this universal perceptual ability declines before the age of 1 yr (Eimas et al. 1971; Eimas 1975; Lasky et al. 1975; Werker and Lalonde 1988).

THE TIMING OF INFANT LEARNING

The transition from an early universal phonetic ability to language specific phonetic perception occurs between 6 and 12 mo of age, when nonnative phonetic perception declines (Werker and Tees 1984; Best and McRoberts 2003; Rivera-Gaxiola et al. 2005b; Tsao et al. 2006) and there is increased sophistication in the processing of native speech (Kuhl et al. 1992). Work in my laboratory showed that at the same time that nonnative perception for consonants declines, native-language speech perception shows a significant increase. We demonstrated that Japanese infants' discrimination of English *r–l* declined between 8 and 10 mo of age, whereas at the same time in development, American infants' discrimination of the same sounds showed an increase (Fig. 2; Kuhl et al. 2006).

Our work also demonstrated that native-language discrimination between 6 and 7 mo predicts the rate of language growth between 11 and 30 mo (Tsao et al. 2004; Conboy et al. 2005; Kuhl et al. 2005, 2008; Rivera-Gaxiola et al. 2005a). Intriguingly, although better performance on *native* contrast discrimination predicts more rapid growth in later language abilities, better performance on *non*native contrasts predicts slower language growth (Fig. 3; Kuhl et al. 2005, 2008). In other words, phonetic learning does not simply depend on auditory acuity, but also on the ability to learn from exposure to language and to attend to some acoustic features and not others.

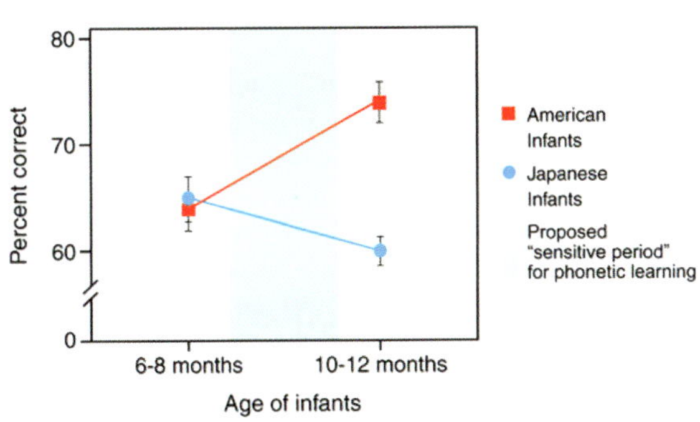

Figure 2. Effects of age on discrimination of the American English /ra-la/ phonetic contrast by American and Japanese infants at 6–8 and 10–12 mo of age. Mean percent correct scores are shown with standard errors indicated. (Reprinted from Kuhl 2010, with permission from Elsevier; originally adapted from Kuhl et al. 2006.)

Bilingual learners acquire two languages, using the same principles, although our data suggest that the sensitive period for learning is extended (García-Sierra et al. 2011). We are currently investigating the extended sensitive period in bilinguals: The neural system may remain "open" longer due to variability in overall language input or due to reduction in input per unit time for each of the native languages, thus extending the learning process (see Kuhl et al. 2008, for discussion).

Both monolingual and bilingual infants benefit from hearing "motherese" speech. Produced at a higher pitch, with exaggerated intonation contours and slower articulation, motherese is an ideal signal for infant learning. Produced by parents and caregivers all over the world when speaking to infants, motherese exaggerates the acoustic differences between phonetic units, making phonemes easier to differentiate (Kuhl et al. 1997). Our studies link exposure to motherese to advanced language development. We demonstrated that 11- and 14-mo-old infants exposed to a greater proportion of motherese utterances at home, and particularly during face-to-face social interactions, have advanced language development at the age of 2 yr (Ramírez-Esparza et al. 2014, 2015).

A COMPUTATIONAL SOLUTION TO PHONETIC LEARNING

What explains infant learning? Studies in the decade of the 1990s demonstrated that infants are capable of a new form of learning, referred to as "statistical learning" (Saffran et al. 1996). Statistical learning reflects implicit rather than explicit learning. It relies on the ability to automatically pick up and learn from the statistical regularities that exist in the stream of sensory information we process, and studies show that both phonetic learning (Maye et al. 2002) and early word learning (Saffran et al. 1996) are strongly influenced by this form of learning.

Recent studies show that infants are sensitive to the statistical distributions of sounds that they hear in ambient language, and that this affects perception. Although adult listeners hear /r/ and /l/ as either distinct (English speakers) or identical (Japanese), speakers of both languages produce highly variable sounds. Japanese adults produce both English r- and l-like sounds, so Japanese infants are exposed to both. Similarly, in Swedish there are 16 vowels, whereas English uses 10 and Japanese uses only five—but speakers of these languages produce a wide range of sounds. The mere presence of a particular

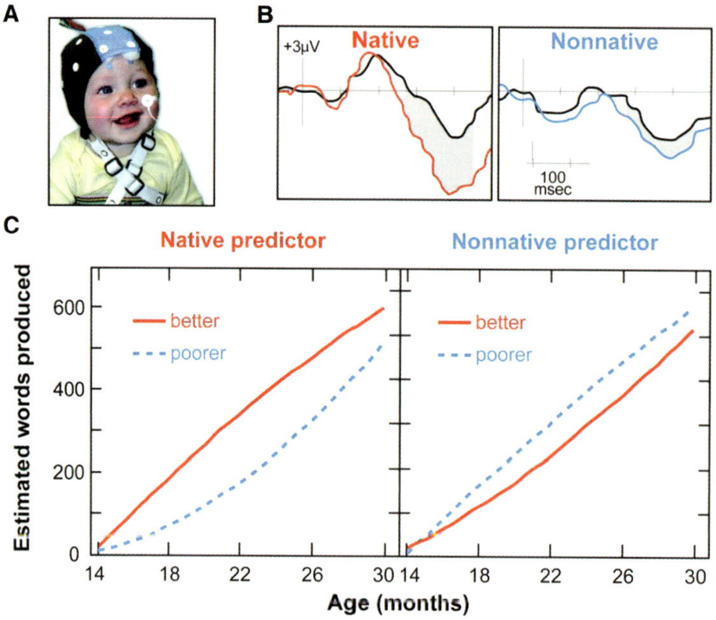

Figure 3. (*A*) A 7.5-mo-old infant wearing an ERP electrocap. (*B*) Infant ERP waveforms at one sensor location (CZ) for one infant are shown in response to a native (English) and nonnative (Mandarin) phonetic contrast at 7.5 mo. The mismatch negativity (MMN) is obtained by subtracting the standard waveform (black) from the deviant waveform (English = red; Mandarin = blue). This infant's response suggests that native-language learning has begun because the MMN negativity in response to the native English contrast is considerably stronger than that to the nonnative contrast. (*C*) Hierarchical linear growth modeling of vocabulary growth between 14 and 30 mo for MMN values of +1 S.D. and −1 S.D. on the native contrast at 7.5 mo (*C*, *left*) and vocabulary growth for MMN values of +1 S.D. and −1 S.D. on the nonnative contrast at 7.5 mo (*C*, *right*). Analyses show that both contrasts predict vocabulary growth and that the effects of better discrimination are reversed for the native and nonnative contrasts. (*A*, Reprinted from Kuhl 2010, with permission from Elsevier; *C*, adapted from Kuhl et al. 2008.)

sound, therefore, does not account for infant learning. However the distributional patterns of such sounds differ across languages. In each language, distributional frequency is high for phonetic units at the center of phonemic categories, and low at the borders between categories. Distributional patterns of sounds thus provide clues about the phonemic structure of a language. If infants are sensitive to the relative distributional frequencies of phonetic segments in the language that they hear, this could explain native-language phonetic learning.

Cross-cultural studies indicate that infants are sensitive to these statistical properties. Infants tested in Sweden and the United States at 6 mo of age show a unique response to the "prototypical" vowels in their language—those that are not only the distributional mean in productions of adults but are also perceptually preferred by adults; this response is not shown to foreign-language vowel prototypes (Kuhl et al. 1992). Moreover, when tested with very simple stimuli in the laboratory, infants have been shown to learn from distributional patterns in language input after short-term exposure to phonetic stimuli (Maye et al. 2002). Word learning is also supported by the detection of statistical regularities in speech, before infants understand word meanings (Saffran et al. 1996). Statistical learning is not limited to humans (Hauser et al. 2001), nor to speech; it operates for musical and visual patterns in the same way (Saffran et al. 1999; Fiser and Aslin 2002; Kirkham et al. 2002).

EFFECTS OF A SOCIAL CONTEXT ON SPEECH LEARNING

Although statistical learning provides a potential explanation for phonetic learning, studies in my laboratory suggest that learning from complex natural language requires more. Our studies indicate that statistical learning is not sufficient for phonetic learning from complex natural language input: Social interaction is necessary. In other words, infants appear to gain something critical from social context in natural language-learning situations that is not available in the absence of another human being (Kuhl et al. 2003).

We posed a simple question—can infants learn phonetically from first-time natural exposure to a foreign language at 9 mo? We wondered whether statistical learning at 9 mo required a long-term history of listening to that language—we reasoned that infant learning at this age might depend on the cumulative statistical distributions over the initial 9 mo of life. Alternatively, the transition might occur at 9 mo because a learning process initially became available to the child at the age of 9 mo.

We designed a foreign-language intervention to test whether learning the statistics of a new language would occur. Nine-month-old American infants listened to four different native speakers of Mandarin during 12 sessions scheduled over a 4- to 5-wk period. The foreign language "tutors" read books and played with toys in sessions that were unscripted. A control group was also exposed to language for 12 sessions but heard only English from native speakers. After infants in the experimental Mandarin exposure group and the English control group completed their sessions, all were tested with a Mandarin phonetic contrast that does not occur in English. Both behavioral and ERP methods were used. The results indicated that infants had a remarkable ability to learn from the live sessions—they performed significantly better on the Mandarin contrast when compared with the control group that heard only English. In fact, they performed equivalently to infants of the same age tested in Taiwan who had listened to the language for 10 mo (Kuhl et al. 2003).

Learning was robust and durable. Infants returned to the laboratory for their behavioral discrimination tests between 2 and 12 d after the final exposure session, with a median of 6 d, and for their ERP measures between 8 and 33 d following the last exposure session, with a median of 15 d. These delays allowed us to examine whether longer periods between exposure and test resulted in poorer Mandarin speech discrimination. A median-split approach was used to subdivide infants based on the median delay in days between exposure and test. The results indicated no significant differences between discrimination performance for infants above and below the median delay for either the behavioral or ERP brain measures. No "forgetting" of the Mandarin contrast occurred during the 2- to 33-d delay.

Infants' remarkable learning led us to test two additional conditions. In one, infants were exposed to the same information on the same schedule in the absence of a human being, via television. In another, infants heard an audiotaped presentation of the same information. The results showed that infants exposed to the same foreign-language material, at the same time in development, and at the same rate, but via standard television or via audiotape only, showed no learning—their performance equaled that of infants in the control group who had not been exposed to Mandarin at all (Fig. 4).

Thus, the presence of a human being interacting with the infant during language exposure, although not required for simpler statistical-learning tasks (Saffran et al. 1996; Maye et al. 2002), is critical for learning in

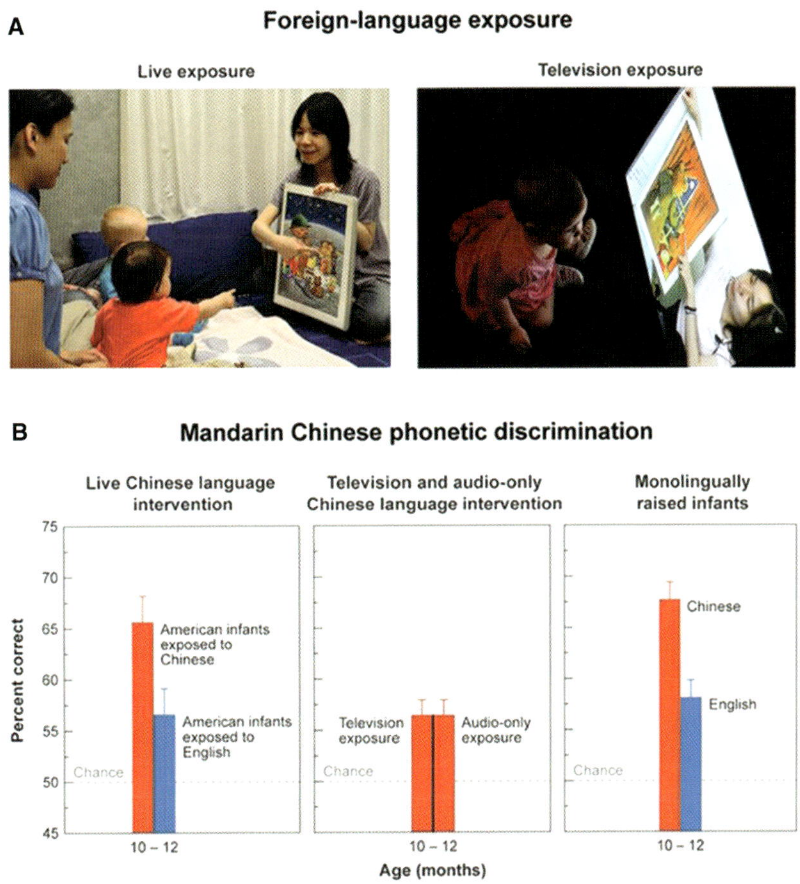

Figure 4. The need for social interaction in language acquisition is shown by foreign-language learning experiments. Nine-month-old infants experienced 12 sessions of Mandarin Chinese through (*A*) natural interaction with a Chinese speaker (*left*) or the identical linguistic information delivered via television (*right*) or audiotape (not shown). (*B*) Natural interaction resulted in significant learning of Mandarin phonemes when compared with a control group who experience interaction but heard only English (*left*). No learning occurred from television or audiotaped presentations (*middle*). Data for age-matched Chinese and American infants learning their native languages are shown for comparison (*right*). (*A*,*B*, Reprinted from Kuhl 2010, with permission from Elsevier; *B*, originally adapted from Kuhl et al. 2003.)

complex natural language-learning situations in which infants heard, on average, 33,000 Mandarin syllables, from a total of four different talkers, over a 4- to 5-wk period (Kuhl et al. 2003).

EXPLAINING THE EFFECTS OF SOCIAL INTERACTION ON LANGUAGE LEARNING

I proposed a *Social Gating Hypothesis* (Kuhl 2007) to explain how social interaction might influence language learning. Two broad mechanisms—motivational and informational—were cited as possible vehicles for social learning. They are not mutually exclusive. Recent evidence suggests that both play a role.

Regarding motivation as a mechanism, attention, and arousal affect learning in a wide variety of domains (Posner 2004). Could they impact infant learning during exposure to a new language? Infant attention, measured in our studies, was significantly higher in response to the live person than to either inanimate source (Kuhl et al. 2003). In recent tests, we examined whether increasing infants' attention to the screen would result in increased learning. The new experiment used stimulus recordings from the same television presentations we used in the original study, and a touch-screen TV to give infants control over the presentation of the Mandarin speaker—infants turned on a 20-sec video presentation simply by touching the screen (SR Lytle, A García-Sierra, and PK Kuhl, in prep.). Other details of the experiment were identical to the original—12 sessions were conducted between the ages of 9.5 and 10.5 mo of age. We expected that the contingent presentation of a 20-sec clip of speech would increase learning, although not equal to that produced with social exposure.

The experiment was also designed to test a second hypothesis, that the presence of another baby while learning—in other words, two-baby learning—would enhance performance over one-baby learning. The critical manipulation was that half the infants in the touch-screen experiment were tested as solo learners, whereas the other half were tested in two-baby pairs in which one of the infants was designated the "target" learner. We expected that the target babies in the two-baby situation would show higher learning than infants tested in the solo learning condition. As in our original experiments, infant learning was tested both neurally, using event-related potential (ERP) methods (see Kuhl 2011), and using our standard behavioral method, the conditioned head-turn (HT) technique (see Kuhl et al. 2003 for details).

The results show that contingent presentation of the recorded information does not produce behavioral evidence of learning. Neither the one- or two-baby conditions produced significant HT performance that exceeded chance.

However, the neural data show very clearly that two are better than one—infants tested in pairs rather than individually showed more negative ERP responses to the Mandarin contrasts, consistent with those of an adult native speaker. Moreover, detailed examination of ERP data from the one- versus two-baby learning sessions revealed an additional important result. ERP data from infants tested in the one- versus two-baby learning sessions showed two different patterns of ERP responses to phonetic stimuli that have been previously reported in the literature. One-baby learners showed a positive mismatch response (MMR) that our laboratory and others interpret as increased auditory attention to an acoustic stimulus, but not one associated with phonetic learning (Morr et al. 2002). Our data show that positive MMR responses become a more negative with increased exposure to a phonetic contrast (García-Sierra et al. 2011). Infants in the two-baby learning condition showed a negative MMR in response to the phonetic contrast, a response that resembles the negative MMR shown in phonetic tests on monolingual adults and infants (Näätänen et al. 1997). We interpret this to mean that infants tested in the two-baby condition are learning at a faster rate as a function of social interaction.

But what kind of social interaction produces this accelerated learning? Do infants in the two-baby learning condition collaborate, showing each other how to press the screen to produce the video? Or is it related to social arousal, produced by the mere presence of another infant? Detailed coding of the video and audio records by trained assistants who were blind to the learning outcomes support the social arousal explanation. We tracked the number of novel partners each target baby experienced during the paired sessions. We found a statistically significant relationship between the number of novel partners a baby experienced in paired sessions and the strength of that infant's negative MMR response to the Mandarin contrasts—the greater the number of novel partners, the stronger the negative MMR, indicating greater evidence of learning. Enhanced learning was not a function of the number of videos viewed by the infants, the duration of exposure to the videos, the number of screen touches, or infants' attention to the screen.

Infants paired with other babies while learning not only show enhanced brain measures when reacting to the foreign-language phonemes, paired babies show significantly higher numbers of speech vocalizations, supporting the idea that the social presence of another infant induces arousal, social communication, and increased learning. Our data suggest that infants are socially aroused in the presence of a peer baby, that their arousal increases when paired with a novel peer, and that social arousal increases learning in infants.

Social interaction also provides additional information, which helps explain learning enhancement in the social condition. During live exposure, tutors focus their visual gaze on pictures in the books or on the toys they talk about, and beginning at this point in development, infants' gaze tends to follow the speaker's gaze (Baldwin 1995; Brooks and Meltzoff 2002, 2015). Referential information is present in both the live and televised conditions, but it is more difficult to pick up via television, and is totally absent during audio-only presentations. We hypothesized that the degree to which infants in our social language-learning situation interact and engage socially with the tutor should correlate with learning. In studies testing this hypothesis, we exposed 9-mo-old infants to Spanish, and measured gaze following in the infants during the learning sessions. Spanish exposure confirmed learning for the phonetic units of Spanish (Conboy and Kuhl 2011), and also revealed that infants' eye-gaze behavior during Spanish exposure predicted the degree of language learning as shown by the ERP brain measures (Conboy et al. 2015). Infants' eye-gaze data show that when the tutor introduced new toys, infants who shifted attention between the tutor and the toy learned more than infants who simply gazed at the tutor, or at the toy.

BRAIN MECHANISMS UNDERLYING PHONETIC LEARNING

I hypothesized that social signals may activate brain mechanisms of social understanding that link perception and action (Hari and Kujala 2009). Neuroscience research focused on shared neural systems for perception and action have a long tradition in speech, and interest in "mirror systems" for social cognition (Rizzolatti and Craighero 2004; Pulvermuller 2005; Rizzolatti 2005) have reinvigorated this tradition.

Our recent MEG brain imaging work on infants indicates that the brain systems involved in speech production are activated when infants listen to speech (Kuhl et al. 2014). We tested infants who straddled the change from universal speech perception to language specific speech perception, at 7 and 11 mo (Fig. 5). We found that listening to speech at both ages activated not only auditory regions in the brain, but also regions involved in motor planning, Broca's area and the cerebellum. Developmental change was also observed—7-mo-old infants activated both auditory and motor areas equally for native and nonnative speech, whereas 11-mo-old infants show greater activation in auditory areas for native speech, and greater activation for nonnative speech in motor areas.

The study advanced our understanding of infant speech processing, and raised additional questions. First, at the earliest age tested (7 mo of age), both native and nonnative speech activated motor brain areas, and equivalently, so the activation of motor brain areas in response to speech at 7 mo of age is not limited to sounds that infants hear in ambient language. At this early age, infants' motor brain areas appear to be reacting to all sounds with speech-like qualities. We argue that infants are engaged in something like "analysis by synthesis," a concept with origins in the field of artificial intelligence (see Stevens and Halle 1967). Analysis by synthesis proposed that speech perception involves a dual "hypothesize and test" process. Bottom-up *analysis* and top-down *synthesis* jointly and actively constrain perceptual interpretation. On this account, listeners generate an *internal model* of the motor commands needed to produce a given auditory signal, in essence, a "guess" or prediction about the auditory input. On this view, an internally generated motor hypothesis is based on a listener's experience producing speech, and then tested against incoming auditory data. We argue that infants' brain responses are also indicative of the creation of internal motor models for speech.

How are 7-mo-old infants capable of creating motor models for speech? We argue that speech production experience—the early "cooing" that occurs by 12 wk of age and allows infants to imitate vowels by 20 wk (Kuhl and Meltzoff 1996)—yields a nascent auditory–articulatory map, an emergent "schema" that specifies generative rules relating articulatory movements to sound. This emerging auditory–articulatory map is likely highly abstract, but it allows infants to generate internal motor models when they listen to sounds with speech-like qualities, regardless of whether they have experienced those specific speech sounds or not. Thus, we posit that infants' nascent speech motor experience is the catalyst for the effects observed in infants as young as 7 mo of age.

Our MEG data also demonstrate developmental change. By the end of the first year, infants' auditory and motor brain areas pattern differently for native and nonnative sounds: Auditory areas show greater activation for native sounds, whereas motor areas show greater activation for nonnative sounds, suggesting that these two brain systems are coding different kinds of information in response to speech. We offer the following tentative explanation: Hearing native speech throughout the first year increases infants' auditory sensitivity to native speech-sound differences. At the same time, hearing speech strengthens sensorimotor pairings, because incoming native speech signals prompt internally generated speech motor models based on infants' prior speech production experience. Language experience thus serves to strengthen knowledge of native-language speech—both perceptual and motor. By the end of the first year, the emergent schema relating speech motor movements to sound would describe native speech quite well. However, nonnative speech would not be well described by the developing schema, making it more difficult to generate internal models for nonnative speech. Hence our finding that nonnative speech elicits greater cortical activation (indicating greater effort) in Broca's area and the cerebellum than native speech.

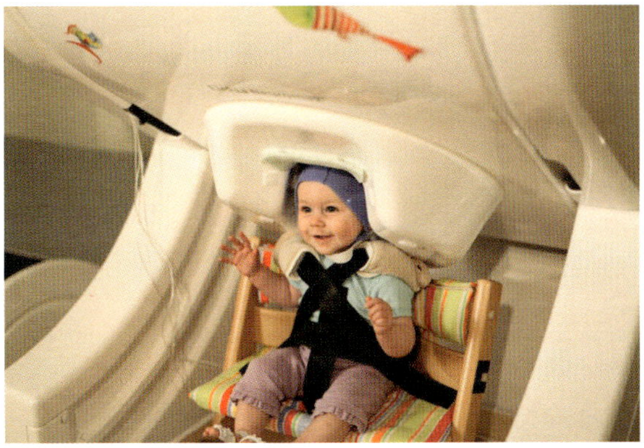

Figure 5. A 7-mo-old infant in the MEG machine listening to language. (Courtesy of P.K. Kuhl, UW Institute for Learning & Brain Sciences ©2011.)

Early models of developmental speech perception, such as my native language neural commitment (NLNC) concept (Kuhl 2004), described a process of "neural commitment" to the auditory patterns of native speech. Revisions in the model, named native language magnet-expanded (NLM-e) described emergent links between speech perception and production (Kuhl et al. 2008). The new data will allow further refinement of the model by suggesting *how* speech perception and speech production become linked early in development: Infants' brains respond to hearing speech by activating motor brain areas, co-registering perceptual experience and motor brain patterns. When infants generate vocalizations, they build a schema that predicts how moving their tongues and lips during vocalizations will modify the sounds they create. By 20 wk, infants can imitate pitch patterns (Kuhl and Meltzoff 1982) and vowels (Kuhl and Meltzoff 1996) that acoustically and perceptually correspond to the ones they hear, indicating knowledge of speech production. This articulatory knowledge, we argue, allows infants to generate internal motor models of speech as they listen to us talk. On this view, both auditory and motor components contribute to the developmental transition in speech perception that occurs at the end of the first year of life.

Theories of social understanding in adults (Hari and Kujala 2009) and infants (Marshall and Meltzoff 2014), suggest that humans evolved brain mechanisms to detect and interpret human actions, behaviors, movements, and sounds. The current data contribute to these views by demonstrating that auditory speech activates motor areas in the infant brain. Motor brain activation in response to others' communicative signals could assist broader development of social understanding in humans.

Our findings offer an opportunity to test children with developmental disabilities such as autism spectrum disorder (Kuhl et al. 2013; Stavropoulos and Carver 2013) whose social and language deficits are potentially associated with a decreased ability to activate motor brain systems in response to human signals. The acquisition of language requires social learning, and therefore speech signals may provide excellent "biomarkers" of early risk for children with autism spectrum disorder (ASD), and our data show promise in this regard (Kuhl et al. 2013).

NEUROBIOLOGICAL FOUNDATIONS OF SOCIAL LEARNING

Humans are not the only species in which communicative learning is affected by social interaction, and the parallels between song learning in birds and speech learning in infants has been a rich area for theory (Doupe and Kuhl 1999). The work of Allison Doupe and her colleagues has beautifully illustrated the role of social contexts on vocal learning in the zebra finch (Woolley and Doupe 2008; Kojima and Doupe 2011). Moreover, studies on the zebra finch have also shown that motor learning occurs earlier and is more sophisticated than previously thought (Raghav and Doupe 2013). Theorizing in these two closely related fields is likely to advance based on these kinds of cross-species comparisons.

CONCLUSION

Studies on the development of human speech and language offer a unique view of the human mind and brain, one made tractable through the design of behavior and brain experiments on learning. Innate predisposition, rapid and robust learning during "sensitive" periods, and brain systems that link action and perception all play a role in the developing story that connects the dots regarding infant acquisition of one or more languages. Studies at the phonetic level of speech—designed to evaluate infant learning of the sounds used to convey meaning in their language environment—are tractable and allow cross-species and cross-cultural developmental experiments. These show strong auditory predispositions that exist at birth, allowing infants to discriminate all possible phonetic variations used in the world's languages. Equally impressive is the demonstration that infants learn rapidly and robustly from linguistic exposure, and that the timing and the context are both critical. The power of social interaction to induce language learning in infants is new, and raises intriguing questions. Recent MEG brain imaging results on infants listening to speech show that hearing speech activates motor planning areas (Brocas and the cerebellum) in the infant brain. The emerging picture is complex. Infants begin life prepared for any language, and have an extraordinary capacity to learn from experience, especially optimized experience—motherese in a social context during sensitive periods, perhaps because action–perception systems in the human brain enable learning from other humans in a special way. Experiments on speech may advance theoretical explanations for the strong influence of the "social brain" on human learning more generally.

ACKNOWLEDGMENTS

The author and research were supported by a grant from the National Science Foundation's Science of Learning Program to the LIFE Center (SBE-0354453; P.K.K.) and by the Ready Mind Project. This paper updates data and theoretical positions expressed by Kuhl (2010, 2011).

REFERENCES

Abramson AS, Lisker L. 1970. Discriminability along the voicing continuum: Cross-language tests. In *Proceedings of the Sixth International Conference of Phonetic Sciences in Prague* (pp. 569–573). Academia, Prague.

Baldwin DA. 1995. Understanding the link between joint attention and language. In *Joint attention: Its origins and role in development* (ed. Moore C, Dunham PJ), pp. 131–158. Lawrence Erlbaum Associates, Hillsdale, NJ.

Best C, McRoberts GW. 2003. Infant perception of nonnative consonant contrasts that adults assimilate in different ways. *Lang Speech* **46:** 183–216.

Bialystok E, Hakuta K. 1994. *In other words: The science and psychology of second-language acquisition*. Basic Books, New York.

Birdsong D, Molis M. 2001. On the evidence for maturational constraints in second-language acquisitions. *J Mem Lang* **44:** 235–249.

Brooks R, Meltzoff AN. 2002. The importance of eyes: How infants interpret adult looking behavior. *Dev Psychol* **38:** 958–966.

Brooks R, Meltzoff AN. 2015. Connecting the dots from infancy to childhood: A longitudinal study connecting gaze following, language, and explicit theory of mind. *J Exp Child Psychol* **130:** 67–78.

Bruer JT. 2008. Critical periods in second language learning: Distinguishing phenomena from explanation. In *Brain, behavior and learning in language and reading disorders* (ed. Mody M, Silliman E), pp. 72–96. The Guilford Press, New York.

Chomsky N. 1959. Review of "Verbal behavior" by B.F. Skinner. *Language* **35:** 26–58.

Conboy BT, Kuhl PK. 2011. Impact of second-language experience in infancy: Brain measures of first- and second-language speech perception. *Dev Sci* **14:** 242–248.

Conboy BT, Rivera-Gaxiola M, Klarman L, Aksoylu E, Kuhl PK. 2005. Associations between native and nonnative speech sound discrimination and language development at the end of the first year. In *Supplement to the Proceedings of the 29th Boston University Conference on Language Development* (ed. Brugos A, Clark-Cotton MR, Ha S). http://www.bu.edu/linguistics/APPLIED/BUCLD/supp29.html

Conboy BT, Brooks R, Meltzoff AN, Kuhl PK. 2015. Social interaction in infants' learning of second-language phonetics: An exploration of brain-behavior relations. *Dev Neuropsychol* **40:** in press.

de Boer B, Kuhl PK. 2003. Investigating the role of infant-directed speech with a computer model. *Acoust Res Lett Online (ARLO)* **4:** 129–134.

de Boysson-Bardies B. 1993. Ontogeny of language-specific syllabic productions. In *Developmental neurocognition: Speech and face processing in the first year of life* (ed. de Boysson-Bardies B, de Schonen S, Jusczyk P, McNeilage P, Morton J), pp. 353–363. Kluwer, Dordrecht.

Doupe AJ, Kuhl PK. 1999. Birdsong and human speech: Common themes and mechanisms. *Annu Rev Neurosci* **22:** 567–631.

Eimas PD. 1975. Auditory and phonetic coding of the cues for speech: Discrimination of the /r–l/ distinction by young infants. *Percept Psychophys* **18:** 341–347.

Eimas PD, Siqueland ER, Jusczyk P, Vigorito J. 1971. Speech perception in infants. *Science* **171:** 303–306.

Ferguson CA, Menn L, Stoel-Gammon C, ed. 1992. *Phonological development: Models, research, implications*. York Press, Timonium, MD.

Fiser J, Aslin RN. 2002. Statistical learning of new visual feature combinations by infants. *Proc Natl Acad Sci* **99:** 15822–15826.

Flege JE. 1991. Age of learning affects the authenticity of voice-onset time (VOT) in stop consonants produced in a second language. *J Acoust Soc Am* **89:** 395–411.

Flege JE, Yeni-Komshian GH, Liu S. 1999. Age constraints on second-language acquisition. *J Mem Lang* **41:** 78–104.

García-Sierra A, Rivera-Gaxiola M, Percaccio CR, Conboy BT, Romo H, Klarman L, Ortiz S, Kuhl PK. 2011. Bilingual language learning: An ERP study relating early brain responses to speech, language input, and later word production. *J Phon* **39:** 546–557.

Hari R, Kujala M. 2009. Brain basis of human social interaction: From concepts to brain imaging. *Physiol Rev* **89:** 453–479.

Hauser MD, Newport EL, Aslin RN. 2001. Segmentation of the speech stream in a nonhuman primate: Statistical learning in cotton-top tamarins. *Cognition* **78:** B53–B64.

Johnson J, Newport E. 1989. Critical period effects in second language learning: The influence of maturation state on the acquisition of English as a second language. *Cogn Psychol* **21:** 60–99.

Kirkham NZ, Slemmer JA, Johnson SP. 2002. Visual statistical learning in infancy: Evidence for a domain general learning mechanism. *Cognition* **83:** B35–B42.

Knudsen EI. 2004. Sensitive periods in the development of the brain and behavior. *J Cogn Neurosci* **16:** 1412–1425.

Kojima S, Doupe AJ. 2011. Social performance reveals unexpected vocal competency in young songbirds. *Proc Natl Acad Sci* **108:** 1687–1692.

Kuhl PK. 2004. Early language acquisition: Cracking the speech code. *Nat Rev Neurosci* **5:** 831–843.

Kuhl PK. 2007. Is speech learning 'gated' by the social brain? *Dev Sci* **10:** 110–120.

Kuhl PK. 2010. Brain mechanisms in early language acquisition. *Neuron* **67:** 713–722.

Kuhl PK. 2011. Social mechanisms in early language acquisition: Understanding integrated brain systems supporting language. In *The Oxford handbook of social neuroscience* (ed. Decety J, Cacioppo J), pp. 649–667. Oxford University Press, Oxford.

Kuhl PK, Meltzoff AN. 1982. The bimodal perception of speech in infancy. *Science* **218:** 1138–1141.

Kuhl PK, Meltzoff AN. 1996. Infant vocalizations in response to speech: Vocal imitation and developmental change. *J Acoust Soc Am* **100:** 2425–2438.

Kuhl PK, Miller JD. 1975. Speech perception by the chinchilla: Voiced-voiceless distinction in alveolar plosive consonants. *Science* **190:** 69–72.

Kuhl PK, Williams KA, Lacerda F, Stevens KN, Lindblom B. 1992. Linguistic experience alters phonetic perception in infants by 6 months of age. *Science* **255:** 606–608.

Kuhl PK, Andruski JE, Chistovich IA, Chistovich LA, Kozhevnikova EV, Ryskina VL, Stolyarova EI, Sundberg U, Lacerda F. 1997. Cross-language analysis of phonetic units in language addressed to infants. *Science* **277:** 684–686.

Kuhl PK, Tsao F-M, Liu H-M. 2003. Foreign-language experience in infancy: Effects of short-term exposure and social interaction on phonetic learning. *Proc Natl Acad Sci* **100:** 9096–9101.

Kuhl PK, Conboy BT, Padden D, Nelson T, Pruitt J. 2005. Early speech perception and later language development: Implications for the 'critical period'. *Lang Learn Dev* **1:** 237–264.

Kuhl PK, Stevens E, Hayashi A, Deguchi T, Kiritani S, Iverson P. 2006. Infants show facilitation for native language phonetic perception between 6 and 12 months. *Dev Sci* **9:** 13–21.

Kuhl PK, Conboy BT, Coffey-Corina S, Padden D, Rivera-Gaxiola M, Nelson T. 2008. Phonetic learning as a pathway to language: New data and native language magnet theory expanded (NLM-e). *Philos Trans R Soc B* **363:** 979–1000.

Kuhl PK, Coffey-Corina S, Padden D, Munson J, Estes A, Dawson G. 2013. Brain responses to words in 2-year-olds with autism predict developmental outcomes at age 6. *PLoS ONE* **8:** e64967.

Kuhl PK, Ramirez RR, Bosseler A, Lin J-F, Imada T. 2014. Infants' brain responses to speech suggest analysis by synthesis: Developmental MEG studies. *Proc Natl Acad Sci* **111:** 11238–11245.

Ladefoged P. 2001. *Vowels and consonants: An introduction to the sounds of language*. Blackwell Publishers, Oxford.

Lasky RE, Syrdal-Lasky A, Klein RE. 1975. VOT discrimination by four to six and a half month old infants from Spanish environments. *J Exp Child Psychol* **20:** 215–225.

Lenneberg E. 1967. *Biological foundations of language*. Wiley, New York.

Lotto AJ, Sato M, Diehl R. 2004. Mapping the task for the second language learner: The case of Japanese acquisition of /r/ and /l/. In *From sound to sense* (ed. Slitka J, Manuel S, Matthies M), pp. C181–C186. MIT Press, Cambridge, MA.

Marshall PJ, Meltzoff AN. 2014. Neural mirroring mechanisms and imitation in human infants. *Philos Trans R Soc B* **369:** 20130620.

Mayberry RI, Lock E. 2003. Age constraints on first versus second language acquisition: Evidence for linguistic plasticity and epigenesis. *Brain Lang* **87:** 369–384.

Maye J, Werker JF, Gerken L. 2002. Infant sensitivity to distributional information can affect phonetic discrimination. *Cognition* **82:** B101–B111.

Meltzoff AN, Kuhl PK, Movellan J, Sejnowski T. 2009. Foundations for a new science of learning. *Science* **17:** 284–288.
Morr ML, Shafer VL, Kreuzer JA, Kurtzberg D. 2002. Maturation of mismatch negativity in typically developing infants and preschool children. *Ear Hear* **23:** 118–136.
Näätänen R, Lehtokoski A, Lennes M, Cheour M, Huotilainen M, Iivonen A, Vainio M, Alku P, Ilmoniemi RJ, Luuk A, et al. 1997. Language-specific phoneme representations revealed by electric and magnetic brain responses. *Nature* **385:** 432–434.
Neville HJ, Coffey SA, Lawson DS, Fischer A, Emmorey K, Bellugi U. 1997. Neural systems mediating American Sign Language: Effects of sensory experience and age of acquisition. *Brain Lang* **57:** 285–308.
Newport EL. 1990. Maturational constraints on language learning. *Cogn Sci* **14:** 11–28.
Newport EL, Bavelier D, Neville HJ. 2001. Critical thinking about critical periods: Perspectives on a critical period for language acquisition. In *Language, brain, and cognitive development: Essays in honor of Jacques Mehlter* (ed. Dupoux E), pp. 481–502. MIT Press, Cambridge, MA.
Posner MI, ed. 2004. *Cognitive neuroscience of attention*. Guilford Press, New York.
Pulvermuller F. 2005. Brain mechanisms linking language to action. *Nat Rev Neurosci* **6:** 574–582.
Rabiner LR, Huang BH. 1993. *Fundamentals of speech recognition*. Prentice Hall, Englewood Cliffs, NJ.
Raghav RR, Doupe AJ. 2013. Behavioral and neural signatures of readiness to initiate a learned motor sequence. *Curr Biol* **23:** 87–93.
Ramírez-Esparza N, García-Sierra A, Kuhl PK. 2014. Look who's talking: Speech style and social context in language input to infants are linked to concurrent and future speech development. *Dev Sci* **17:** 880–891.
Ramírez-Esparza N, García-Sierra A, Kuhl PK. 2015. The language you hear, the language you speak: A study with Spanish–English bilingual infants. *Child Dev* **8:** in press.
Rivera-Gaxiola M, Klarman L, García-Sierra A, Kuhl PK. 2005a. Neural patterns to speech and vocabulary growth in American infants. *Neuroreport* **16:** 495–498.
Rivera-Gaxiola M, Silva-Pereyra J, Kuhl PK. 2005b. Brain potentials to native and non-native speech contrasts in 7- and 11-month-old American infants. *Dev Sci* **8:** 162–172.
Rizzolatti G. 2005. The mirror neuron system and imitation. In *Perspectives on imitation: From neuroscience to social science—Volume 1: Mechanisms of imitation and imitation in animals* (ed. Hurley S, Chater N), pp. 55–76. MIT Press, Cambridge, MA.
Rizzolatti G, Craighero L. 2004. The mirror-neuron system. *Annu Rev Neurosci* **27:** 169–192.
Saffran J, Aslin R, Newport E. 1996. Statistical learning by 8-month old infants. *Science* **274:** 1926–1928.
Saffran JR, Johnson EK, Aslin RN, Newport EL. 1999. Statistical learning of tone sequences by human infants and adults. *Cognition* **70:** 27–52.
Saffran JR, Werker JF, Werner LA. 2006. The infant's auditory world: Hearing, speech, and the beginnings of language. In *Handbook of child psychology: Vol. 2, cognition, perception and language*, 6th ed. (series ed. Damon W, Lerner RM; vol. ed. Siegler R, Kuhn D), pp. 58–108. Wiley, New York.
Skinner BF. 1957. *Verbal behavior*. Appleton-Century-Crofts, New York.
Stavropoulos K, Carver L. 2013. Research review: Social motivation and oxytocin in autism—Implications for joint attention development and intervention. *J Child Psychol Psychiatry* **54:** 603–618.
Stevens KN, Halle M. 1967. Remarks on analysis by synthesis and distinctive features. In *Models for the perception of speech and visual form* (ed. Walthem-Dunn W), pp. 88–102. MIT Press, Cambridge, MA.
Tsao F-M, Lui H-M, Kuhl PK. 2004. Speech perception in infancy predicts language development in the second year of life: A longitudinal study. *Child Dev* **75:** 1067–1084.
Tsao F-M, Liu H-M, Kuhl PK. 2006. Perception of native and non-native affricate-fricative contrasts: Cross-language tests on adults and infants. *J Acoust Soc Am* **120:** 2285–2294.
Weber-Fox CM, Neville HJ. 1999. Functional neural subsystems are differentially affected by delays in second language immersion: ERP and behavioral evidence in bilinguals. In *Second language acquisition and the critical period hypothesis* (ed. Birdsong D). Lawerence Erlbaum, Mahwah, NJ.
Werker JF, Lalonde C. 1988. Cross-language speech perception: Initial capabilities and developmental change. *Dev Psychol* **24:** 672–683.
Werker JF, Tees RC. 1984. Cross-language speech perception: Evidence for perceptual reorganization during the first year of life. *Infant Behav Dev* **7:** 49–63.
Werker JF, Pons F, Dietrich C, Kajikawa S, Fais L, Amano S. 2007. Infant-directed speech supports phonetic category learning in English and Japanese. *Cognition* **103:** 147–162.
Woolley SC, Doupe AJ. 2008. Social context-induced song variation affects female behavior and gene expression. *PLoS Biol* **6:** e62.
Yeni-Komshian GH, Flege JE, Liu S. 2000. Pronunciation proficiency in the first and second languages of Korean–English bilinguals. *Bilingualism: Lang Cogn* **3:** 131–149.

Dopaminergic Dynamics Contributing to Social Behavior

LISA A. GUNAYDIN[1] AND KARL DEISSEROTH[2]

[1]*The Gladstone Institutes, University of California, San Francisco, California 94158*

[2]*Departments of Bioengineering and Psychiatry, Howard Hughes Medical Institute, Stanford University, Stanford, California 94305*

Correspondence: deissero@stanford.edu; lisa.gunaydin@gladstone.ucsf.edu

Social interaction is a complex behavior that is essential for the survival of many species, and it is impaired in a broad range of neuropsychiatric disorders. Several cortical and subcortical brain regions have been implicated in a variety of sociosexual behaviors, with pharmacological studies pointing to a key role of the neurotransmitter dopamine. However, little is understood about the real-time circuit dynamics causally underlying social interaction. Here, we consider current knowledge on the role of brain reward circuitry in same-sex social behavior and describe findings from new methods for probing how this circuitry governs social motivation in health and disease.

Social interaction is a challenging and highly integrative cognitive behavior that is essential for many mammalian species, and it is impaired in major psychiatric disorders such as autism, schizophrenia, social anxiety, and depression, with a broad range of potential etiologies. Several cortical and subcortical brain regions have been implicated in controlling social behavior, such as the prefrontal cortex, amygdala, striatum, dorsal raphe, and hypothalamus (Gingrich et al. 2000; Young et al. 2001; Leypold et al. 2002; Robinson et al. 2002; Liu and Wang 2003; Young and Wang 2004; Curtis and Wang 2005; Aragona et al. 2006; Lin et al. 2011; Robinson et al. 2011; Dölen et al. 2013; Yang et al. 2013; Felix-Ortiz and Tye 2014; Hong et al. 2014; Unger et al. 2015). In rodents, the majority of these studies have focused on sociosexual behaviors, such as pair bonding, aggression, and other behaviors related to sexual competition. However, comparatively little is known about the neural circuitry regulating adult same-sex, nonaggressive social interaction, which is of relevance for understanding circuits that may go awry in social-function disorders. Here, we discuss the role of dopaminergic circuitry in same-sex social interaction, highlighting recent findings from new optogenetic methods for probing endogenous and causal circuit dynamics underlying social motivation.

MESOLIMBIC CIRCUITRY IN NORMAL SOCIAL BEHAVIOR

The neurotransmitter dopamine (DA), produced in the ventral tegmental area (VTA), has long been known to play a role in the processing of both natural and conditioned rewards. The terminal region with the densest VTA DA projections is the ventral striatum, or nucleus accumbens (NAc), which is thought to encode reward-related signals from the VTA. The NAc comprises primarily the inhibitory projection neurons called medium spiny neurons (MSNs) that can be differentiated by the type of DA receptor they express: D1 or D2. These two subpopulations of NAc MSNs are thought to bidirectionally control reward (Lobo et al. 2010) and have been pharmacologically implicated in affiliative behaviors (Puglisi-Allegra and Cabib 1997; Young and Wang 2004). The NAc also receives inputs from other regions implicated in social behavior, such as the dorsal raphe, hypothalamus, and prefrontal cortex, as well as sensory inputs, and is thus poised to orchestrate the integration of diverse streams of socially relevant information into behavioral output.

Human genetic studies have shown a role for genes involved in the dopamine pathway in modulating social behavior. The nine-repeat allele of the DA transporter DAT1, thought to result in increased striatal DA, was associated with stronger social approach tendency in an implicit social approach-avoidance task (Enter et al. 2012). Interestingly, the authors observed a significantly stronger approach to images of happy faces, whereas avoidance of angry faces was not affected, consistent with a role for striatal DA in approach of appetitive socially relevant stimuli. Another study showed that administration of L-DOPA, a DA precursor, improved the ability of 10-repeat genotype subjects, assumed to have lower endogenous striatal DA, to learn about a partner's prosocial preferences (Eisenegger et al. 2013). In rats, studies using fast-scan cyclic voltammetry to record temporally precise DA release in postsynaptic targets found a sixfold increase in the frequency of DA transients throughout the dorsal and ventral striatum of rats investigating a novel conspecific (Robinson et al. 2002). Recording specifically from the NAc, they observed DA release upon orientation toward and initial contact with the conspecific, an effect that habituated upon subsequent presentations of the same conspecific (Robinson et al. 2011).

 doi: 10.1101/sqb.2014.79.024711

These studies, although largely correlative, suggest that DA may be involved in encoding or modulating social behavior.

SOCIAL COGNITION VERSUS SOCIAL MOTIVATION THEORIES OF AUTISM

Examining the changes that occur in the brains of patients with social-function disorders can inform us about the circuitry underlying normal behavior. One of the classical hypotheses put forward to explain the social deficits in autism spectrum disorders (ASDs) is the notion of a dysfunction in social cognition or executive control, with specific deficits in theory of mind, the ability to understand and ascribe mental states to others to explain and predict their behavior. These deficits in social cognition are thought to arise from abnormal information processing in higher-level executive control regions, such as the frontal cortex. For example, recent studies in human genetics and development have implicated cortical abnormalities in ASD (e.g., Willsey et al. 2013), and direct optogenetic modulation of neocortical excitation–inhibition balance was found to disrupt social behavior in mice (Yizhar et al. 2011). However, the idea of autism as a deficit in reward-related social motivation may be distinct and complementary to processes involved in cortical information processing, and thus is likely in its own right to be relevant to etiology, genetics, diagnosis, and investigation of potential avenues for treatment.

Social motivation encompasses the drive to seek social interaction and take pleasure in it. Many species, from humans to rodents, find social interaction rewarding, independent of familial or reproductive motivations. Subjects playing economic games take pleasure in cooperation and collaboration and will work to obtain social rewards even at the cost of higher individual economic outcome (Fehr and Camerer 2007). Neuroimaging during tasks such as the prisoner's dilemma show that the reward value of mutual cooperation with a human partner is associated with increased activation of the ventral striatum compared with cooperation with a computer partner, despite identical monetary gain in both situations (Fehr and Camerer 2007). Even toddlers show a strong preference for collaborative over individual access to reward (Chevallier et al. 2012). There is also evidence from animal studies suggesting that mice find social interaction rewarding outside the context of reproduction or parental care. Juvenile mice are equally motivated to investigate conspecifics regardless of their sex, and social conditioned place preference tests have shown that adolescent mice will strongly prefer an environment that has previously been paired with a conspecific to one without a conspecific, suggesting that a brief episode of social contact can have positive conditioning value even before animals reach sexual maturity (Panksepp et al. 2007).

The social motivation hypothesis of autism postulates that children with ASD do not find social stimuli rewarding, a primary deficit leading to later abnormal development of social skills and social cognition as a consequence of lacking social interest. This theory points to the dysregulation of subcortical circuits such as the mesolimbic pathway in the etiology of the disorder. Several neuroimaging studies have supported the notion of lacking social reward in ASD, accompanied by changes in striatal circuitry implicated in motivation. Several studies showed pronounced impairment in learning to choose social rewards compared with monetary rewards in ASD, which was associated with decreased frontostriatal response during social but not monetary reward learning (Scott-Van Zeeland et al. 2010; Lin et al. 2012). In typically developing children, the authors also found a positive correlation between ventral striatal activity and social reciprocity (Scott-Van Zeeland et al. 2010). Yet another study showed that oxytocin, a prosocial neuropeptide being explored as a potential therapy for ASD, enhances VTA activation to social reward and social punishment cues (Groppe et al. 2013). Genetic studies have also suggested that mutations in genes along the mesolimbic DA pathway are associated with ASD. A dopamine D1 receptor (DRD1) haplotype as well as DRD3 single-nucleotide polymorphisms have been associated with increased risk for ASD (Hettinger et al. 2008; Staal 2014). Mutations in the dopamine transporter (DAT) have also been associated with ASD (Bowton et al. 2014); DAT mutations that result in increased expression of the transporter, and thus likely lower synaptic levels of DA, are associated with increased social anxiety in ASD children (Gadow et al. 2008). Together, these data suggest a deficit in social reward processing in ASD accompanied by genetic and functional abnormalities in the mesolimbic DA pathway. An intriguing hypothesis arising from this research is that increasing social reward responsiveness in ASD, perhaps via manipulation of the dopaminergic mesolimbic pathway, may improve social learning and prevent the emergence of social cognitive deficits.

A CIRCUIT-BASED APPROACH TO SOCIAL MOTIVATION

However, DA pathway genetic polymorphisms are present in only a small fraction of ASD cases, and human neuroimaging studies are purely correlative in nature. To gain a better mechanistic understanding of the role of mesolimbic circuitry in normal and pathological social behavior, we can take advantage of the genetic tools available in mice. Previous behavioral pharmacology studies have shown that dopamine receptor agonists and antagonists can bidirectionally modulate social interaction (Puglisi-Allegra and Cabib 1997), suggesting a causal role of the neurotransmitter in social behavior. However, a circuit-level understanding of which DA cells, projections, and postsynaptic targets mediate these effects and how they interact in real time during behavior has remained largely unknown. This problem was recently addressed using all-optical readout and control of DA neurons in socializing mice to understand how genetically specified cells and projections are causally involved in driving social behavior (Gunaydin et al. 2014).

To observe socially relevant native patterns of neural activity in real time, the genetically encoded Ca^{2+} indicator GCaMP5, which changes its fluorescence properties upon elevation of Ca^{2+} levels inside the cell (a well-established proxy for neural activity), was used to record temporally precise Ca^{2+} transients (Akerboom et al. 2012; Chen et al. 2013) in VTA DA neurons (Gunaydin et al. 2014). To selectively target dopaminergic neurons in the VTA, a virus carrying Cre-dependent GCaMP5 was injected into the VTA of mice expressing Cre recombinase under the tyrosine hydroxylase (TH) promoter (TH::Cre mice). Developing and using a novel technique called fiber photometry, the investigators implanted a single optical fiber just above the VTA, which was used both to deliver excitation light and to collect bulk activity-dependent GCaMP fluorescent transients in the freely behaving animal. Excitation and emission fluorescence were spectrally separated using a dichroic, passed through a single band filter, and focused onto a photodetector to record real-time VTA activity as animals socialized with a novel conspecific in their home cage (Fig. 1A). There was a marked increase in VTA GCaMP signal as the test subject investigated the conspecific, specifically time-locked to appetitive approach and contact investigation of the other mouse, which habituated over time (Fig. 1B). These data show for the first time the real-time dynamics of VTA DA neurons during a complex social behavior, and although correlative, they suggested that experimentally prolonging these habituating VTA signals might be a way to increase the animal's social behavior.

Optogenetics enables us to causally link VTA DA activity to social interaction by expressing a Cre-dependent light-gated cation channel, channelrhodopsin-2 (ChR2), in the VTA of TH::Cre mice to enable temporally precise control over their activity using pulses of 473-nm blue light. Precise patterns of phasic photostimulation of these cells, which had previously been shown to evoke maximal levels of downstream DA release (Adamantidis et al. 2011), significantly increased the amount of time mice spent investigating a novel conspecific. This effect was blocked by DA receptor antagonism (although no genetic targeting strategy resolves cell types with complete specificity, these results confirmed DA dependence of the behavioral result, supporting prior immunostaining-based validation of this targeting methodology; Tsai et al. 2009). Conversely, inhibition of VTA DA neurons using the inhibitory chloride pump halorhodopsin (eNpHR3.0) significantly decreased interaction, showing bidirectional control of social behavior by VTA DA neurons (Fig. 2A,B). Importantly, modulation of VTA DA activity did not affect time spent investigating a novel inanimate object (Fig. 2C). These data showed for the first time an endogenous and causal role of VTA DA activity in driving social approach behavior.

Previous human and animal studies have suggested that the ventral striatum (or NAc), a primary target of reward-related VTA DA neurons, may be a key downstream region relevant to the processing of social reward and one dysregulated in autism. Selective optogenetic stimulation of the VTA-to-NAc projection was sufficient to reproduce the prosocial effect of VTA DA cell body stimulation, whereas stimulation of other VTA DA projections, such as to the PFC, did not affect social behavior, pointing to a critical role specifically for the VTA-NAc circuit (Fig. 2D,E). Electrophysiologically, the stimulation parameters that increased social behavior increased firing rate in the NAc (Fig. 3A,B). Animals were then exposed to the three-chamber test, an apparatus consisting of one "social chamber" with a caged novel conspecific on one side and a "neutral chamber" containing a caged inanimate object on the other, separated by an empty middle chamber. Recording in vivo during this behavioral test, higher NAc activity was also found when animals chose to explore the social chamber compared with the neutral one (Fig. 3C,D), suggesting that increased NAc activity is a correlate of native prosocial behavior independent of any exogenous neural manipulation, corroborating previous reports from human neuroimaging (Scott-Van Zeeland et al. 2010).

These electrophysiological recordings suggested that the increased NAc activity observed during prosocial

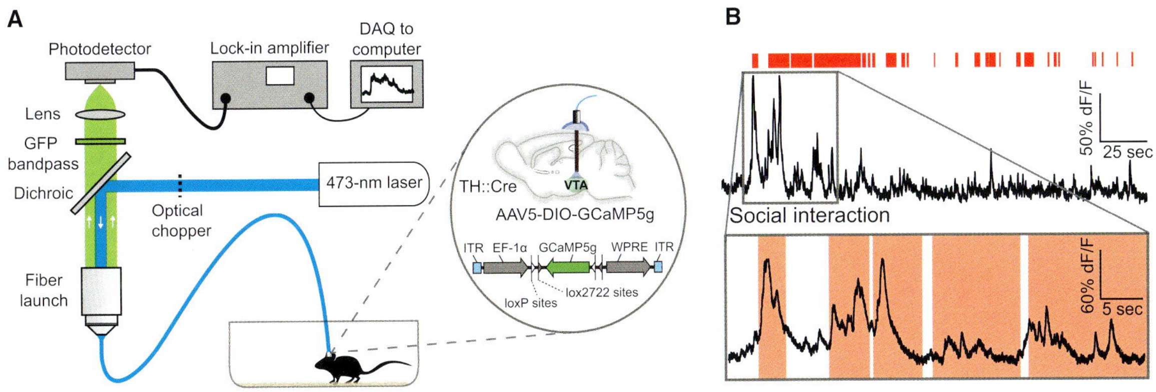

Figure 1. Optical recording of dopaminergic dynamics during social interaction. (*A*, *Left*) Fiber photometry setup. Light path for GCaMP fluorescence excitation and emission is through a single optical fiber implanted in the VTA. (*Right*) viral targeting of GCaMP5 to VTA DA neurons. (*B*, *Top*) Example trace of VTA DA activity in social behavior. Red dashes indicate interaction bouts. (*Bottom*) zoom-in of dashed interval relating VTA DA GCaMP signal and social interaction (red boxes). (Adapted from Gunaydin et al. 2014.)

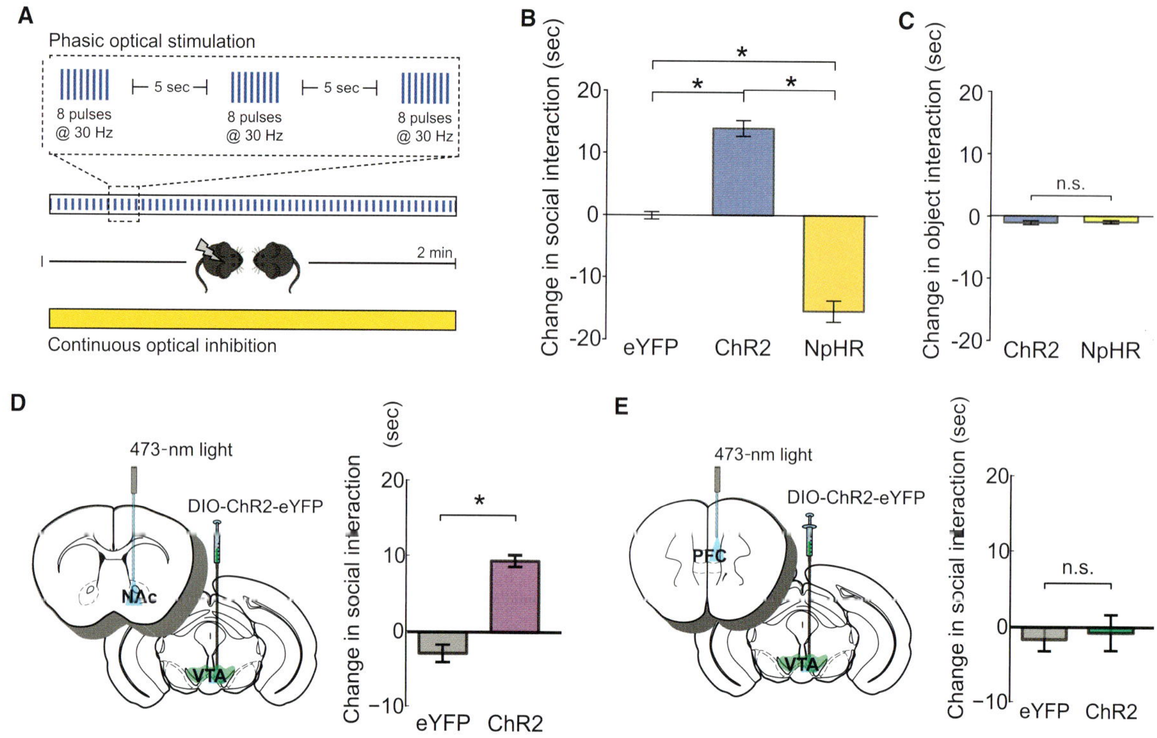

Figure 2. VTA modulation of social behavior. (*A*) Optical stimulation parameters for home cage interaction. For excitation, phasic bursts of blue light were delivered every 5 sec. For inhibition, continuous yellow light was delivered. (*B*) Summary of light-evoked changes in social interaction after bidirectional control of VTA DA neurons. Phasic stimulation of VTA cell bodies increased social interaction, whereas inhibition of VTA cell bodies decreased interaction. (*C*) Neither stimulation nor inhibition of VTA cell bodies significantly affected investigation of a control novel inanimate object. (*D*) Phasic stimulation of VTA-NAc projections increased social interaction in ChR2 animals (purple) compared with controls (gray). (*E*) Phasic stimulation of VTA-PFC projections had no effect on social interaction in ChR2 animals (blue) or controls (gray). (Adapted from Gunaydin et al. 2014.)

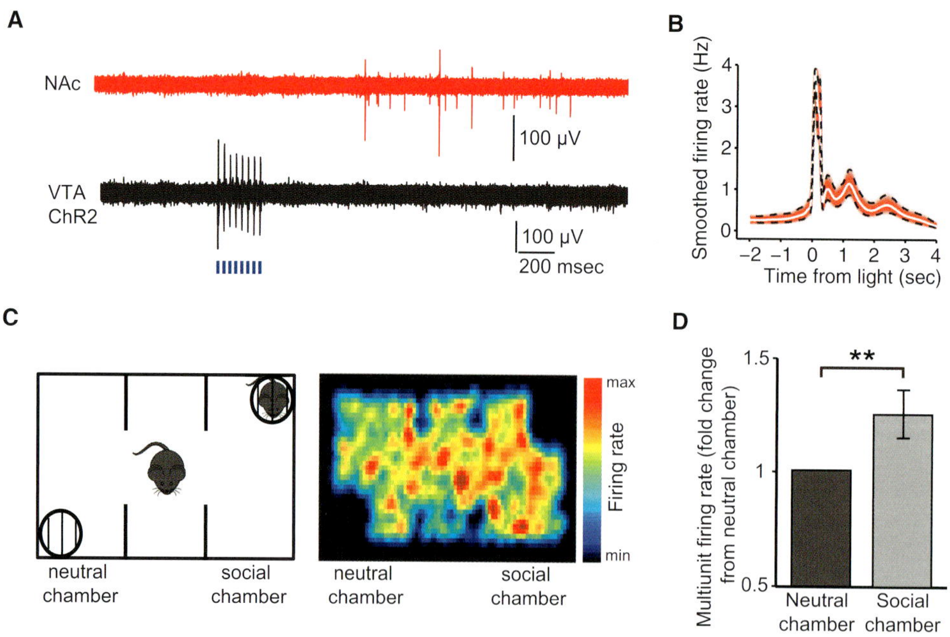

Figure 3. NAc electrophysiological correlates of increased social behavior. (*A*) Increase in NAc activity (red) evoked by VTA stimulation (black). (*B*) PSTH showing light-evoked increase in NAc firing with one burst of VTA stimulation. (*C*) Heat map showing firing rate of NAc neurons in freely moving animals exploring neutral and social environments. Warmer colors indicate higher firing rate. (*D*) NAc spiking is higher in the social environment. (Adapted from Gunaydin et al. 2014.)

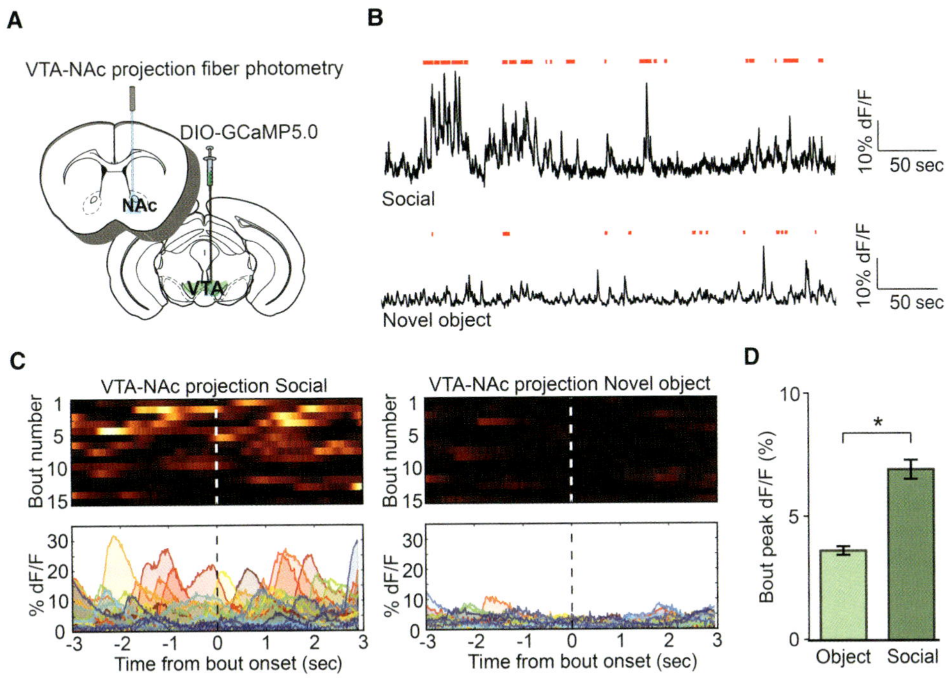

Figure 4. Fiber photometry of DA projection activity in NAc during social interaction. (*A*) Fiber photometry of VTA projections in NAc. (*B*) VTA projection activity during social (*top*) and novel object investigation (*bottom*; interaction bouts in red). (*C*) Heat maps (*top*) and peri-event plots (*bottom*) of NAc projection fluorescence aligned to start of interaction bout for social or novel object investigation. For heat maps, warmer colors indicate higher fluorescence signal; for peri-event plots, warmer colors indicate earlier interaction bouts. (*D*) NAc projections largely recapitulate social signals in VTA, with lower response to a novel object. (Adapted from Gunaydin et al. 2014.)

behavior was likely driven by increased VTA input. However, until the advent of fiber photometry, no technique existed to directly measure activity in a set of genetically defined afferent projections to a region. By expressing GCaMP5 in the VTA and implanting the recording fiber-optic in the NAc (Fig. 4A), fluorescent transients were detected in VTA-NAc projections during epochs of social interaction, recapitulating the increased activity seen in VTA cell bodies, and demonstrating for the first time the activity of genetically defined, projection-specific inputs to a region during social behavior (Fig. 4B–D). Further pharmacological and optogenetic investigation showed that downstream D1 neurons in the NAc mediated this prosocial effect of increased VTA input. This work showed a causal role for reward circuitry in driving social behavior and opened the door to further investigation of specific mechanisms within the VTA-NAc circuit that may go awry in social-function disorders such as autism and could potentially one day be harnessed therapeutically to augment the rewarding nature of social stimuli in these disorders.

INTEGRATING MOTIVATIONAL AND COGNITIVE CONTROL OF SOCIAL BEHAVIOR

In addition to motivational factors, social behavior requires rapid integration and updating of complex stimuli that are used to guide appropriate actions for initiation and maintenance of interaction, likely mediated by higher-level cognitive areas such as the prefrontal cortex (PFC). Another optogenetic study using direct manipulation of prefrontal microcircuit elements showed a crucial role of this region in regulating social behavior. Yizhar et al. (2011) developed a novel channelrhodopsin variant called the stabilized step-function opsin (SSFO) for long-timescale modulation of cortical activity using a brief pulse of blue light before assessing social interaction. One advantage of using the SSFO to study the causal relationships between PFC circuit elements and behavior was that its long-lasting depolarization facilitated social behavioral assessment without requirement for the fiber-optic during behavior, because a single pulse of blue light before behavioral testing is sufficient to cause activation of cells for the duration of the assay. They found that activating excitatory neurons in the mPFC with the SSFO caused a dramatic impairment in social behavior, as stimulated animals spent significantly less time investigating a novel conspecific. This social impairment was accompanied by an increase in power of high-frequency γ oscillations in the mPFC, a pathological signature observed in patients with autism (Orekhova et al. 2007). Concurrent elevation of activity in inhibitory parvalbumin (PV)-expressing local interneurons partially rescued the social deficit, demonstrating that excitatory/inhibitory balance in mPFC plays a causal role in modulating social interaction.

Although activation of the mesolimbic DA pathway drove an increase in social behavior, activation of the mesocortical pathway interestingly had no effect on social behavior, but instead drove aversion and anxiety-related behaviors (Gunaydin et al. 2014). Accordingly, there is no current evidence that VTA projections play a causal role within the PFC in modulating the PFC's important role in social behavior regulation, although certainly this mesocortical circuit could provide subthreshold modulation of the relevant circuitry or other glutamatergic cortical inputs, whereas under these conditions, activation alone is not sufficient to alter social behavior. It is also possible that in situations of high stress and anxiety, this circuit may serve to negatively regulate social behavior. Future studies are certainly needed to better understand how subcortical and cortical circuits work together to control social motivation and cognition, and which specific aspects of social interaction (e.g., initiation, maintenance, and reward) are controlled by each circuit and cell type. In this regard, fiber photometry and optogenetics together will be useful for combinatorial readout and control of multiple independent cell populations in behaving animals and hold great promise for beginning to unravel how other subcortical regions work in a coordinated fashion to regulate normal and pathological social behavior.

ACKNOWLEDGMENTS

We are grateful to our coauthors on Gunaydin et al. (2014), from which the figures and related text were adapted, as well as to the sources of support described therein (including the National Institutes of Health, the Defense Advanced Research Projects Agency, the Simons Foundation, and the Gatsby Foundation), and to all of the members of the Deisseroth lab.

REFERENCES

Adamantidis AR, Tsai HC, Boutrel B, Zhang F, Stuber GD, Budygin EA, Touriño C, Bonci A, Deisseroth K, de Lecea L. 2011. Optogenetic interrogation of dopaminergic modulation of the multiple phases of reward-seeking behavior. *J Neurosci* **31:** 10829–10835.

Akerboom J, Chen TW, Wardill TJ, Tian L, Marvin JS, Mutlu S, Calderón NC, Esposti F, Borghuis BG, Sun XR, et al. 2012. Optimization of a GCaMP calcium indicator for neural activity imaging. *J Neurosci* **32:** 13819–13840.

Aragona BJ, Liu Y, Yu YJ, Curtis JT, Detwiler JM, Insel TR, Wang Z. 2006. Nucleus accumbens dopamine differentially mediates the formation and maintenance of monogamous pair bonds. *Nat Neurosci* **9:** 133–139.

Bowton E, Saunders C, Reddy IA, Campbell NG, Hamilton PJ, Henry LK, Coon H, Sakrikar D, Veenstra-VanderWeele JM, Blakely RD, et al. 2014. SLC6A3 coding variant Ala559Val found in two autism probands alters dopamine transporter function and trafficking. *Transl Psychiatry* **4:** e464.

Chen TW, Wardill TJ, Sun Y, Pulver SR, Renninger SL, Baohan A, Schreiter ER, Kerr RA, Orger MB, Jayaraman V, et al. 2013. Ultrasensitive fluorescent proteins for imaging neuronal activity. *Nature* **499:** 295–300.

Chevallier C, Kohls G, Troiani V, Brodkin ES, Schultz RT. 2012. The social motivation theory of autism. *Trends Cogn Sci* **16:** 231–239.

Curtis JT, Wang Z. 2005. Ventral tegmental area involvement in pair bonding in male prairie voles. *Physiol Behav* **86:** 338–346.

Dölen G, Darvishzadeh A, Huang KW, Malenka RC. 2013. Social reward requires coordinated activity of nucleus accumbens oxytocin and serotonin. *Nature* **501:** 179–184.

Eisenegger C, Pedroni A, Rieskamp J, Zehnder C, Ebstein R, Fehr E, Knoch D. 2013. DAT1 polymorphism determines L-DOPA effects on learning about others' prosociality. *PLoS One* **8:** e67820.

Enter D, Colzato LS, Roelofs K. 2012. Dopamine transporter polymorphisms affect social approach-avoidance tendencies. *Genes Brain Behav* **11:** 671–676.

Fehr E, Camerer CF. 2007. Social neuroeconomics: The neural circuitry of social preferences. *Trends Cogn Sci* **11:** 419–427.

Felix-Ortiz AC, Tye KM. 2014. Amygdala inputs to the ventral hippocampus bidirectionally modulate social behavior. *J Neurosci* **34:** 586–595.

Gadow KD, Roohi J, DeVincent CJ, Hatchwell E. 2008. Association of ADHD, tics, and anxiety with dopamine transporter (DAT1) genotype in autism spectrum disorder. *J Child Psychol Psychiatry* **49:** 1331–1338.

Gingrich B, Liu Y, Cascio C, Wang Z, Insel TR. 2000. Dopamine D2 receptors in the nucleus accumbens are important for social attachment in female prairie voles (*Microtus ochrogaster*). *Behav Neurosci* **114:** 173–183.

Groppe SE, Gossen A, Rademacher L, Hahn A, Westphal L, Gründer G, Spreckelmeyer KN. 2013. Oxytocin influences processing of socially relevant cues in the ventral tegmental area of the human brain. *Biol Psychiatry* **74:** 172–179.

Gunaydin LA, Grosenick L, Finkelstein JC, Kauvar IV, Fenno LE, Adhikari A, Lammel S, Mirzabekov JJ, Airan RD, Zalocusky KA, et al. 2014. Natural neural projection dynamics underlying social behavior. *Cell* **157:** 1535–1551.

Hettinger JA, Liu X, Schwartz CE, Michaelis RC, Holden JJ. 2008. A DRD1 haplotype is associated with risk for autism spectrum disorders in male-only affected sib-pair families. *Am J Med Genet B Neuropsychiatr Genet* **147B:** 628–636.

Hong W, Kim DW, Anderson DJ. 2014. Antagonistic control of social versus repetitive self-grooming behaviors by separable amygdala neuronal subsets. *Cell* **158:** 1348–1361.

Leypold BG, Yu CR, Leinders-Zufall T, Kim MM, Zufall F, Axel R. 2002. Altered sexual and social behaviors in trp2 mutant mice. *Proc Natl Acad Sci* **99:** 6376–6381.

Lin D, Boyle MP, Dollar P, Lee H, Lein ES, Perona P, Anderson DJ. 2011. Functional identification of an aggression locus in the mouse hypothalamus. *Nature* **470:** 221–226.

Lin A, Rangel A, Adolphs R. 2012. Impaired learning of social compared to monetary rewards in autism. *Front Neurosci* **6:** 143.

Liu Y, Wang ZX. 2003. Nucleus accumbens oxytocin and dopamine interact to regulate pair bond formation in female prairie voles. *Neuroscience* **121:** 537–544.

Lobo MK, Covington HE III, Chaudhury D, Friedman AK, Sun H, Damez-Werno D, Dietz DM, Zaman S, Koo JW, Kennedy PJ, et al. 2010. Cell type-specific loss of BDNF signaling mimics optogenetic control of cocaine reward. *Science* **330:** 385–390.

Orekhova EV, Stroganova TA, Nygren G, Tsetlin MM, Posikera IN, Gillberg C, Elam M. 2007. Excess of high frequency electroencephalogram oscillations in boys with autism. *Biol Psychiatry* **62:** 1022–1029.

Panksepp JB, Jochman KA, Kim JU, Koy JJ, Wilson ED, Chen Q, Wilson CR, Lahvis GP. 2007. Affiliative behavior, ultrasonic communication and social reward are influenced by genetic variation in adolescent mice. *PLoS One* **2:** e351.

Puglisi-Allegra S, Cabib S. 1997. Psychopharmacology of dopamine: The contribution of comparative studies in inbred strains of mice. *Prog Neurobiol* **51:** 637–661.

Robinson DL, Heien ML, Wightman RM. 2002. Frequency of dopamine concentration transients increases in dorsal and ventral striatum of male rats during introduction of conspecifics. *J Neurosci* **22:** 10477–10486.

Robinson DL, Zitzman DL, Smith KJ, Spear LP. 2011. Fast dopamine release events in the nucleus accumbens of early adolescent rats. *Neuroscience* **176:** 296–307.

Scott-Van Zeeland AA, Dapretto M, Ghahremani DG, Poldrack RA, Bookheimer SY. 2010. Reward processing in autism. *Autism Res* **3:** 53–67.

Staal WG. 2014. Autism, DRD3 and repetitive and stereotyped behavior, an overview of the current knowledge. *Eur Neuropsychopharmacol* doi: 10.1016/j.euroneuro.2014.08.011.

Tsai HC, Zhang F, Adamantidis A, Stuber GD, Bonci A, de Lecea L, Deisseroth K. 2009. Phasic firing in dopaminergic neurons is sufficient for behavioral conditioning. *Science* **324:** 1080–1084.

Unger EK, Burke KJ CF Jr, Bender KJ, Fuller PM, Shah NM. 2015. Medial amygdalar aromatase neurons regulate aggression in both sexes. *Cell Rep* **10:** 453–462.

Willsey AJ, Sanders SJ, Li M, Dong S, Tebbenkamp AT, Muhle RA, Reilly SK, Lin L, Fertuzinhos S, Miller JA, et al. 2013. Coexpression networks implicate human midfetal deep cortical projection neurons in the pathogenesis of autism. *Cell* **155:** 997–1007.

Yang CF, Chiang MC, Gray DC, Prabhakaran M, Alvarado M, Juntti SA, Unger EK, Wells JA, Shah NM. 2013. Sexually dimorphic neurons in the ventromedial hypothalamus govern mating in both sexes and aggression in males. *Cell* **153:** 896–909.

Yizhar O, Fenno LE, Prigge M, Schneider F, Davidson TJ, O'Shea DJ, Sohal VS, Goshen I, Finkelstein J, Paz JT, et al. 2011. Neocortical excitation/inhibition balance in information processing and social dysfunction. *Nature* **477:** 171–178.

Young LJ, Wang Z. 2004. The neurobiology of pair bonding. *Nat Neurosci* **7:** 1048–1054.

Young LJ, Lim MM, Gingrich B, Insel TR. 2001. Cellular mechanisms of social attachment. *Horm Behav* **40:** 133–138.

Cognitive Skills Needed for Social Hierarchies

Russell D. Fernald
Biology Department, Stanford University, Stanford, California 94305
Correspondence: rfernald@stanford.edu

Dominance hierarchies are ubiquitous in social species that require social cognition to maintain. Status may be established initially through physical conflict but is maintained by social signals between individuals that depend critically on the relative social status of those interacting. How do individuals collect information they need to modulate their behavior? Using a particularly suitable fish model system living in a complex social environment, we describe how the social context of behavior shapes the brain and, in turn, alters the behavior of animals as they interact. These fish observe social interactions carefully to gather information vicariously that guides future behavior. Social opportunities produce rapid changes in gene expression in key brain nuclei, and both social success and failure produce changes in neuronal cell size and connectivity in reproductive centers of the brain. It remains unknown how social information is transduced into cellular and molecular changes. Understanding the cellular and molecular changes underlying animal cognition will yield unique insights into how the brain works.

It seems likely that cognition evolved in the service of social behavior because living in groups requires social awareness among individuals. A ubiquitous feature of such social systems are dominance hierarchies through which both social and sexual status are communicated directly among animals during social intercourse via chemical, visual, auditory, and postural signs in various combinations. Although many animal signals convey information about external events such as food, danger, nest sites, etc., more commonly animals signal information about themselves, their species and individual identity, and their social or sexual status. Dominance hierarchies, typically organized around male status, are most common where males will compete vigorously for high rank. This is because the perks of high status include increased access to food, reproductive opportunity, and improved health outcomes. This is not so for low-ranking animals who have limited access to food, a suppressed reproductive system as well as limited reproductive opportunities, and adverse health effects (e.g., Sapolsky 2005). The ubiquity of social status can be seen in its representation in brain structures including in humans (Zink et al. 2008), and social status in macaques has been suggested to covary with particular neural circuits (Noonan et al. 2014). Because psychopathology and social disorders like autism typically reflect a failure to act in socially acceptable ways, understanding the mechanisms responsible for normal social cognition may ultimately suggest methods for remediation.

To study cognition, we need a model system that can be assessed under controlled laboratory conditions mimicking the natural habitat but allowing careful experimentation without compromising behavioral interactions. Importantly, to study cognitive abilities in a social setting, we need to have animals living in natural social groups. Fishes, which comprise ~50% of all extant species, are a natural choice for research on social behavior for several reasons. Fish were the common ancestor of tetrapods so that all vertebrates share genetic features, including the brain. This means that similarities in brain structures are due to common ancestry, and, indeed, fish brain regions have well-described homologs with other vertebrate brains. Second, because fish species split from tetrapods into their own independent radiation now occupying a vast range of ecological niches, they have also evolved sensory systems exquisitely tuned to their particular environments. These include the usual senses—vision, olfaction, taste, and hearing—but also mechanosensory detection (e.g., lateral line organ), external taste buds, and electroreceptor systems that have driven evolution of specialized brain structures. Importantly, among fish species, virtually every known kind of social system has evolved from monogamy to harems to sex-changing animals.

Cognitive skills in various fish species have been shown in several domains including acquisition of foraging skills (Brännäs and Eriksson 1999), tool use (Timms and Keenleyside 1975; Paśko 2010), spatial memory, and manipulation of the environment (Hughes and Blight 1999). Examples of social intelligence in fish have been measured by how they interact in group-living environments (Balshine-Earn et al. 1998), enhance offspring survival with biparental care (Gross and Sargent 1985; Hourigan 1989; Alonzo et al. 2001; Van den Berghe and McKaye 2001), cooperate in hunting (Diamant and Shpigel 1985; Vail et al. 2013), and share information about predator inspection (Pitcher et al. 1986).

MODEL SOCIAL SYSTEM

To understand the mechanisms underlying social cognition, my laboratory analyzes the social behavior of a cichlid fish species from Lake Tanganyika, East Africa, *Astatotilapia burtoni* (formerly *Haplochromis burtoni*).

 doi: 10.1101/sqb.2014.79.024752

The *A. burtoni* male hierarchial social system requires particular social skills for success in maintaining a high status. Changes in social status cause changes in body color and behavior, as well as to neural connections related to reproduction. There are several reasons why *A. burtoni* is a valuable and unique species for understanding the role of social cognition: (a) The social system is based on resource guarding, and its natural state can be reliably and accurately replicated in the laboratory; (b) male status is signaled phenotypically via bright coloration, including a dark bar through the eye, making high-status animals easy to distinguish and their behavior readily quantifiable (see Fig. 1); (c) in this species, as in all vertebrates, gonadotropin-releasing hormone (GnRH1) neurons in the brain ultimately control reproduction (in *A. burtoni* these neurons are directly regulated by male social status and their size and connectivity changes with social status); (d) we can readily measure behavior in conjunction with levels of circulating hormones, tissue-specific peptides, and DNA expression; and (e) the *A. burtoni* genome has been sequenced (Brawand et al. 2014), so we can use transgenic and other genomic technologies.

SOCIAL SYSTEM OF *A. BURTONI*: CONSEQUENCES OF STATUS

Astatotilapia burtoni males live as one of two reversible, socially controlled phenotypes: reproductively competent dominant (D) males and reproductively incompetent nondominant (ND) males (see Fig. 1). D males are brightly colored, aggressively defend territories, and actively court females (Fernald and Hirata 1977). In striking contrast, ND males have a dull coloration, mimic female behavior, and school with females and other ND males, except when fleeing from an attacking D male.

These obvious external differences are reflected in major physiological and neural adaptations in response to differences in social status. As animals transition from one phenotype to the other, some changes, including expression of the black bar through the eye, brightening of the body color, and switching behavioral repertoires, are expressed in minutes (Fig. 1), whereas concomitant physiological and neural changes happen over days.

ND males attend closely to the unfolding social scene, assessing when they might be able to gain a territory by defeating a resident male. When this happens, there is typically a dramatic fight during which males engage in mouth-to-mouth biting, hitting each other with their bodies, and nipping at each other's fins. If the ND male successfully defeats the resident, he rapidly turns on his bright body colors (Fernald and Hirata 1977; Burmeister et al. 2005) and will quickly begin performing the 17 distinct behaviors characteristic of dominant males upon his social ascent. Over a few days, the reproductive system of the ascending male is remodeled rendering the male reproductively competent, changes that are evident at several levels along the hypothalamic-pituitary-gonadal (HPG) axis (Maruska and Fernald 2014). In *A. burtoni*, as in all vertebrates, reproduction is controlled by gonadotropin-releasing hormone (GnRH1) containing neurons in the hypothalamus that deliver the eponymously named GnRH decapeptide to the pituitary. When a male ascends (ND→D), delivery of this molecule sets in motion a cascade of actions ultimately resulting in reproductive competence and release of sex steroids from the gonads. The GnRH neurons increase in volume by eightfold (Davis and Fernald 1990), extend their dendrites (Fernald 2012), and rapidly increase production of GnRH mRNA (Burmeister et al. 2005) and GnRH peptide (White et al. 2002). However, when a D male is moved into a social system with larger D males (>5% longer), it abruptly loses its color (<1 min) and joins other ND males and

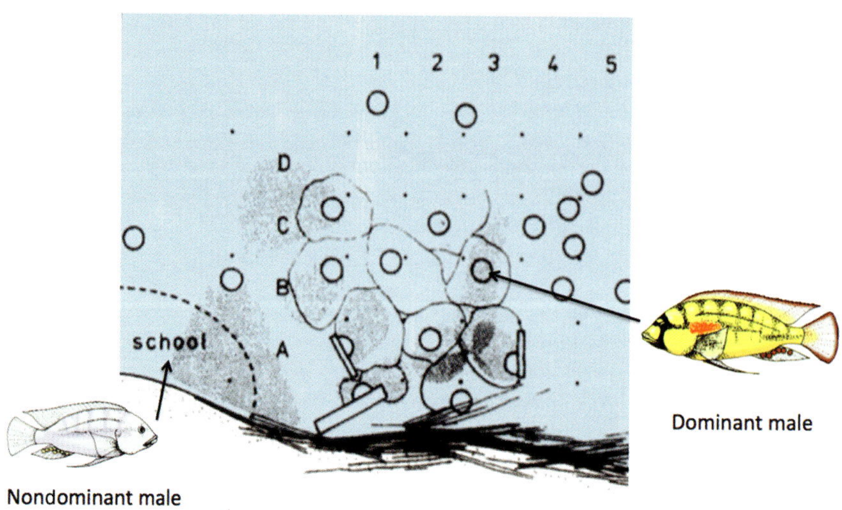

Figure 1. Sketch of an observation area in Lake Tanganyika, Burundi, Africa. Solid dots are grid stakes spaced ~50 cm apart that are labeled (1–5; A–D) for identification. Dark circles represent spawning pit locations of dominant males. Lighter colored outlines circumscribe the territories of individuals. Male territories are arrayed over the food source of detritus. Nondominant males and females school together near the territorial area. (Based on Fernald and Hirata 1977.)

females in a school. Its GnRH-containing neurons in the preoptic area (POA) shrink to one-eighth their volume and produce less GnRH mRNA and peptide, causing hypogonadism and loss of reproductive competence (~2 wk) (Davis and Fernald 1990; Francis et al. 1993). Similarly, androgen, estrogen, and GnRH receptor mRNA expression levels depend on social status (Au et al. 2006; Burmeister et al. 2007; Harbott et al. 2007), as do electrical properties of the GnRH neurons themselves (Greenwood and Fernald 2004).

ATTENTION HIERARCHIES IN MALE *A. BURTONI*

Animals in social groups monitor the behavior of conspecifics and use their observations to guide their ongoing behavior. For example, individuals may attend to male fighting, females choosing mates, or many other ongoing social interactions. In all social hierarchies, subordinate animals attend very closely to the behavior of dominant animals in what has been named an "attention hierarchy" (Chance and Larson 1976). Animals watch higher-ranking animals carefully presumably to calibrate their potential opportunities of social ascent. Attention hierarchies have been identified in humans, particularly in groups of children (Boulton and Smith 1990), where individuals modulate their behavior depending on their own status relative to that of others. Within a hierarchy, when a high-ranking individual threatens or attacks a lower-ranking individual, that individual often then subsequently attacks an individual of still lower rank (Vaughn and Waters 1981). In addition to humans, attention hierarchies have been described in baboons and mandrills (Emory 1976), as well as in reptiles (Summers et al. 2005).

To understand attention hierarchies in *A. burtoni*, we videotaped groups of individually marked animals ($n = 20$/group; four replicates) and quantified interactions between dominant and nondominant males (Desjardins et al. 2012). We found that D and ND males never behaved aggressively at the same time. Even more interesting was that when the D male was out of view in a shelter, the ND males that were larger and attempting to ascend behaved aggressively and even courted females, behaviors that never occurred when the D male could see the ND male (Fig. 2).

In this example, each time the dominant animal is out of view, the intermediate male attacks another ND male until the dominant male reappears. When the dominant male returns to the scene, he attacks within a few seconds but does not specifically target fish that have been aggressive to others in his absence (Desjardins et al. 2012).

Our data show that the ND males are attending to the D males and altering their behavior by acting aggressively, which is not possible when the D male is present. In addition, these males on occasion will approach and court females when the D male cannot see them, another behavior not possible when a D male is present. These behaviors reflect a sophisticated social calculus in which ND males are doing the most they can to increase their chances of becoming dominant. It is possible that they are also learning about being dominant through watching, a skill we have also reported elsewhere (Alcazar et al. 2014).

Based on the attention hierarchy described above, we reasoned that animals might alter their behavior when being observed by conspecifics, consistent with the notion that they could deceive other individuals that are watching them. To test this idea, we designed experiments in which two dominant male fish were in a tank, separated by a clear, watertight barrier, and both could be observed from a third compartment. When two size-matched dominant males are placed in these compartments, they will fight through the transparent barrier for the 20-min duration of the experiments while they are observed from a noninteracting audience. We asked how the behavior of the fighting animals would vary as a function of the composition of the audience and found that the animals fight harder when watched by a gravid female and less hard when watched by a larger male (Fig. 3; Desjardins et al. 2012).

In an analysis of eye movements in *A. burtoni*, we observed that when a D male was in the presence of ND males and he oriented his eyes toward ND males, they would typically move out of his way (Fernald 1985). This anticipation of the actions of the D male contributes to the attention hierarchy of males within the social hierarchy and suggests that the ND males attend to specific attributes of the male behavior.

CAN MALES BE DECEPTIVE?

Can *A. burtoni* males be deceptive? This is an outstanding question in animal cognition because it is often taken as a hallmark of cognitive skills. To ask this question, we used a novel paradigm in which two differently sized males share a tank, divided in half by a clear, watertight divider and a black removable divider (Chen and Fernald 2011). The general idea is to see how a small D male responds to a larger D male when he can only see the larger fish. To do this, we placed a half terra-cotta pot, cut in half lengthwise, so that it allowed the males on each side to occupy a "shared" shelter, although with the black divider in place, neither animal knew the other animal was present (Fig. 4). One male was ~4× larger than the other and each male had an appropriately sized female in their hemitank.

Both the small and large fish were habituated to the tank with the opaque barrier in place for 2 d during which time each behaved like a normal dominant male in its territory: excavating gravel from the hemi-pot, courting the female in their half of the tank, and leading her back to the shelter and performing typical courtship and territorial male behaviors, all of which were quantified and exactly like normal D males. On the third day, the opaque barrier was lifted, and although there was no physical or chemical contact possible, the larger male made several "attacks" toward the small male, which quickly lost his coloration, including his eyebar. This is typical behavior for an animal losing his territorial status and was con-

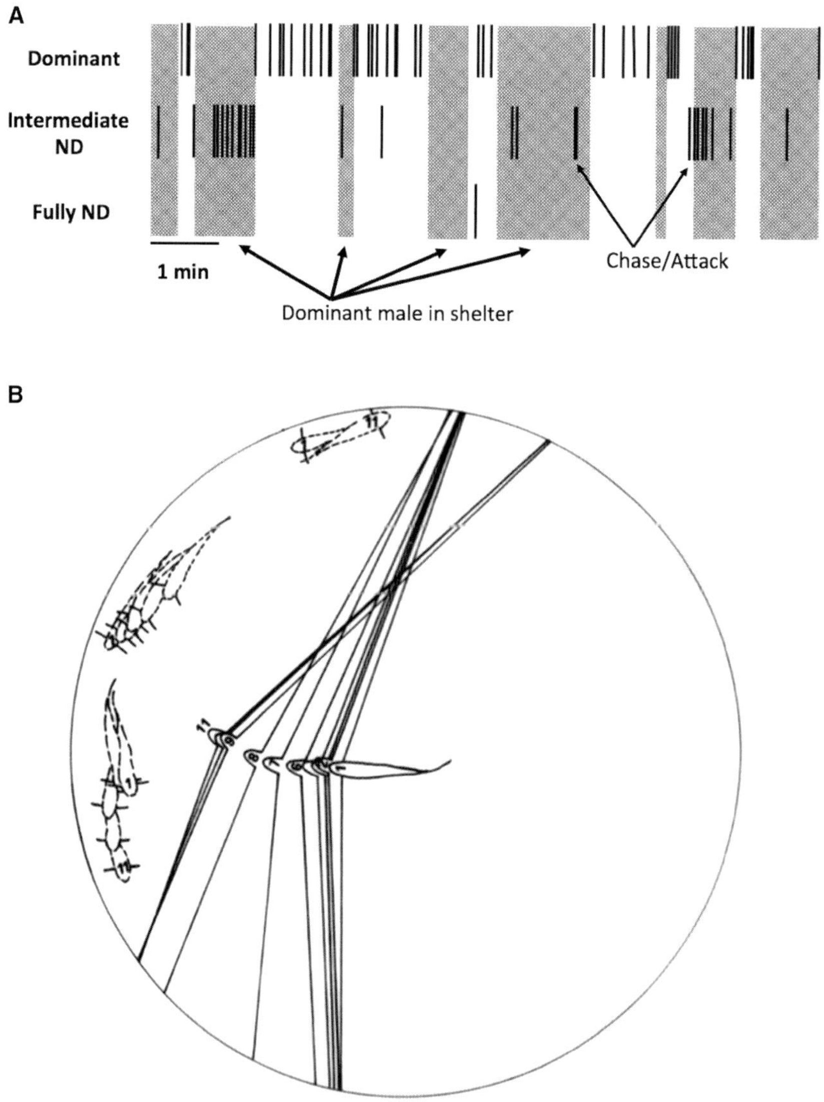

Figure 2. (*A*) Schematic illustration of typical dominant male behavior in the presence of an intermediate male attempting to ascend to dominance. Large rectangles represent the dominant male in his shelter, and small black bars show when an individual chases or attacks another fish. Note that intermediate males only attack other animals when they cannot be seen, because the dominant male is in his shelter (from data presented in Desjardins et al. 2012). (*B*) An example of eye movements during social interactions. Fish have been drawn from film images at 160-msec intervals with the first through 11th frames labeled. Relative eye locations of the dominant animal are shown by lines extending stalks attached to the eyes at an arbitrary angle behind the central axis of the eye. Note that the ND animals move out of the region being approached by the dominant male well before he arrives. (Redrawn from Fernald 1985.)

firmed by quantifying their behavior. Indeed, the smaller males essentially abandoned their part of the shelter, digging a new pit remote from the site of the flowerpot shelter. Interestingly, this suppression of dominance, based entirely on visual signals, as reflected by the behavioral quantification (Chen and Fernald 2011), also reduced the expression of androgen hormones, but only for the first 3 d. Seven days after the black barrier was lifted, the smaller animals recovered their hormone expression levels and other brain markers of dominance while maintaining the coloration of an ND male. Moreover, they could be seen courting their females when out of view of the dominant male. So the effects of the visual suppression resulted in changes in the expression of aggressive, territorial behavioral responses by the smaller male but not in sustained physiological changes. This suggests that the smaller males uncoupled physiological changes in circulating hormones from their effects on outward appearance, seemingly presenting a false outward appearance not consonant with internal changes. This appears to be an example of deceptive behavior on the part of the male, allowing him to continue his courtship but not be influenced by the larger male. One can assume that the smaller animal had learned that the clear barrier prevented the larger male from actual physical attacks, and this recognition led the smaller animals to

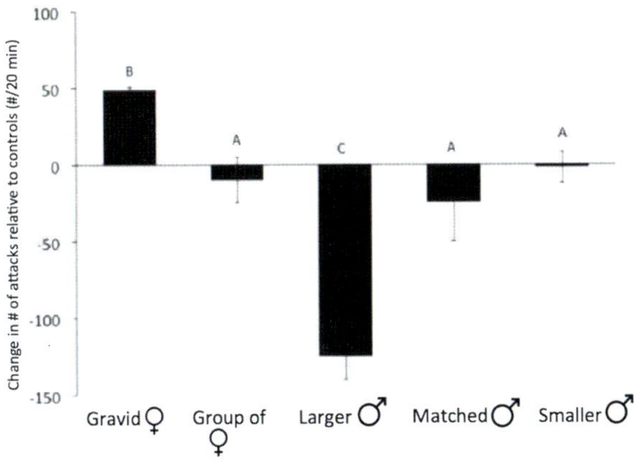

Figure 3. Aggressive acts (biting and ramming) displayed by a pair of males being watched by audiences composed of different kinds of conspecifics. Note that males were significantly ($\alpha = 0.05$) more aggressive when being watched by a gravid female, and significantly less aggressive when being watched by a larger male (marked with B and C, respectively). In contrast, differences in aggression when being watched by a group of females, matched size male, or smaller males did not differ from control conditions (marked by A). (Redrawn from Desjardins et al. 2012.)

this novel strategy. The smaller ND male attended carefully to the D male when carrying out his behavior consistent with the attention hierarchy described above.

TRANSITIVE INFERENCE BY MALES

An ongoing goal of ND *A. burtoni* males is to ascend to dominance, triggering the physiological changes required to become reproductively competent. But how do ND males assess their chances to make this change? In their natural habitat, Lake Tanganyika, colonies of *A. burtoni* may range in size from a few dozen animals to more than 100, depending on the size of the feeding substrate (Fernald and Hirata 1977). ND males could, in principle, fight each D male in a colony to identify which might be likely targets for a takeover of their territories. However, fighting with tens of animals to find a territory holder weak enough to beat would obviously be prohibitive. As described above, the attention hierarchy evident during social encounters suggested *A. burtoni* might have observational skills that would allow them to predict the outcome of male–male encounters. That is, could males infer their chances of winning a fight simply from watching other animals fight? The logic of this process, known as transitive inference, is that if you know that A is taller than B and B is taller than C, you can infer that A is taller than C by constructing a virtual cognitive hierarchy without needing to see A, B, and C lined up for comparison. This ability was one of the developmental milestones first described by Piaget (1928) and since identified in humans older than 3, as well as in nonhuman primates (McGonigle and Chalmers 1977; Gillian 1981; Rapp et al.1996), rats (Davis 1992; Roberts and Phelps 1994), and birds (von Fersen et al. 1991; Steirn et al.1995; Bond et al. 2003; Weiss et al. 2010).

To discover if *A. burtoni* had the ability to infer fighting abilities from watching them fight, we asked whether bystander males could synthesize information from observed pairwise fights into an implied hierarchy of male fighting abilities. We did this by having bystander fish observe pairwise staged fights between five size-matched males (A–E) in which A>B, B>C, C>D, D>E with the implied hierarchy of A>B>C>D>E (Grosenick et al. 2007). Fights were staged by moving one rival into another rival's territory, which resulted in the intruder animal losing. For fights among the control animals, there was no implied hierarchy (e.g., A = B = C = D = E). The bystander males were trained on pairwise fights, and we tested their preference between rivals they had never seen together before. That is, we tested whether the animal chose B or D as a winner. They consistently chose D as the weaker animal based on the prior data

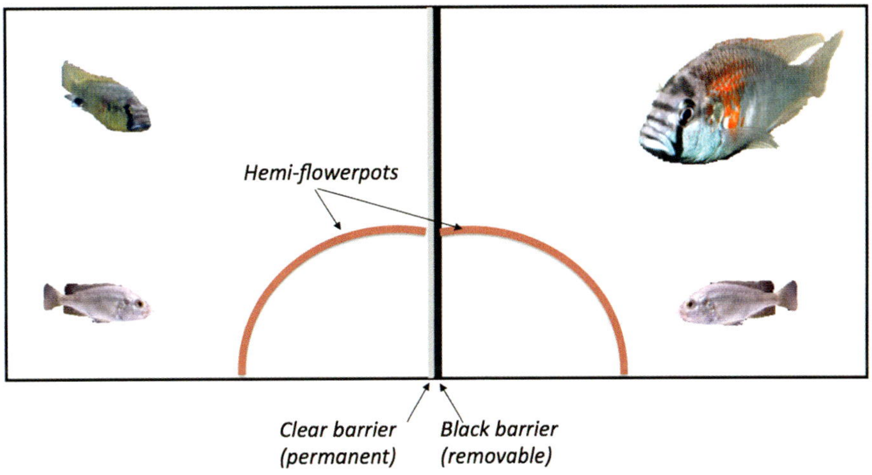

Figure 4. Front view of the aquarium (45 L) divided in half with a watertight, clear divider (gray midline), and a removable opaque barrier (black midline). There is a small male fish (*left* compartment) and the large male fish (~ 4× larger in the *right* compartment). The half terra cotta pot (red curved line) was cut so that both fish "shared" the same shelter, although they were not aware of each other's presence. This "shared" shelter was hemisected by both center dividers. A layer of gravel covered the bottom of the tank.

(Oliveira et al. 1998; Clement et al. 2005), showing that animals will indicate a choice by moving toward the rival they perceive to be weaker.

The fact that these fish can perform transitive inference in an important situation, choosing which male to attack, is consistent with the behavioral needs and ecological context in their natural habitat. In the temporary shore pools and estuaries of their native habitat, there is regular disruption of established territories by movement of hippopotamuses, wind, and predation (Fernald and Hirata 1977). So being able to judge their rivals based on featural representations, independent of context, would be invaluable for them to increase their chances of reproductive success. We have shown that social ascent upon gaining a territory is swift and activates many behavioral, physiological, and molecular processes, allowing the ascending male a chance at reproductive success (Burmeister et al. 2005; Maruska and Fernald 2014). It seems likely that transitive inference could be found in other colony-living animals that face similar constraints on reproduction. This would require designing experimental tests that exploit a natural context and behavioral elements related to behavioral acts in the animal's natural life.

The experiments described above relied entirely on visual information being provided to the individual, although chemosensory information is also essential for many animal species including fish. We have also tested whether *A. burtoni* uses other sensory channels when communicating with conspecifics. Specifically, we injected dye into D males and tested whether these individuals used urine pulses as a part of their sensory signaling. We found that D males increased their urination along with territorial behaviors when they were visually exposed to another male (Fig. 5; Maruska and Fernald 2012). This study of contextual chemosensory urine signaling showed that males could distinguish female reproductive states using visual cues alone, suggesting that urine signals very likely play a complementary role to visual signaling.

GENOMIC CONSEQUENCES OF FEMALE MATE CHOICE

For females of many species, information about potential mates changes their behavior dramatically, which makes sense because female mate choice is extremely important for the survival of their offspring. Given its importance, we wondered what cognitive activity might accompany behavioral and physiological changes in females in response to mate choice. Specifically, how does the female brain respond to social information about potential mates? This experimental question depends on deciding both how and where to look for a signal in the brain in response to female mate choice. We chose to measure brain sites that were marked by changes in gene expression using immediate-early genes (IEGs), *egr-1* and *c-fos*, as a female chooses a mate. IEGs are inducible transcription factors that comprise a part of the first wave of gene expression induced in neurons being activated. Extensive experimental work has shown that a range of natural experiences including sensory stimuli can induce IEG expression and, consequently, it has been used in mammals and birds (e.g., Rusak et al. 1990; Mello et al. 1992) to identify social responses. More recently, we showed that *egr-1* is highly conserved in *A. burtoni* and that functionally it responds robustly within 30 min of stimulation (Burmeister and Fernald 2005) and that *c-fos* is also a valuable genetic signal indicating brain responses in *A. burtoni*. As for the brain location, we hypothesized that the conserved vertebrate social behavior network (SBN) would be a logical place to look. The SBN was originally described by Newman (1999) as a collection of brain nuclei implicated in a variety of social behaviors including male mating behavior, female sexual behavior, parental behavior, and aggressive behavior. Anatomical homologs to this network have subsequently been identified in birds and fish (Goodson and Bass 2002; Goodson 2005) and are likely to exist in other vertebrates as well. It was unknown, however, whether they might also respond to social information as well as to behavioral actions. In previous experiments on female mate choice, we showed that reproductively ready (i.e., gravid) females prefer to associate with dominant, reproductively active males, whereas nongravid females prefer nondominant, nonreproductive males (Clement et al. 2005).

Using a similar paradigm, we placed females in a tank with one male at each end, behind a clear, watertight Plexiglas barrier so the female received only visual information from the males. We quantified the female's preference based on her interactions and proximity to that male over a 20-min period. Following this, we staged a fight between the two males in one of the male compartments chosen at random. Subsequently, the chosen male either won or lost that fight. Our control condition was for the female to choose between two males and not see a subsequent fight. We hypothesized that different IEG expression patterns would be generated by the condition of females seeing their chosen male win or lose a fight. We then mapped brain gene expression patterns, comparing the mRNA levels of *egr-1* and *c-fos* in six brain nuclei comprising the SBN (Desjardins et al. 2010).

Surprisingly, females seeing their selected male win or lose a fight produced dramatically different brain IEG expression patterns. Females who saw their preferred male win a fight activated brain nuclei associated with

Figure 5. A urine pulse (arrow) released from a D male. A D male, injected with dye was exposed visually to another D male (after Maruska and Fernald 2012).

Switched/total	Preferred male wins	Preferred male loses
Females saw a fight	2/13	11/12
Females didn't see a fight	0/10	1/9

Figure 6. Tabulation of female *A. burtoni* choices after they saw their preferred male either win or lose a fight (*top* row) as compared with their choices after not seeing a fight (*bottom* row). After seeing her preferred male win a fight, the female rarely switches her choice ($P = 0.0002$; Fisher's exact test), but after seeing her preferred male lose, she nearly always switches her choice ($P = 0.0004$; Fisher's exact test) (Klausner et al. in prep.).

reproduction and reproductive behavior. Specifically the anterior hypothalamus, ventromedial hypothalamus, preoptic area, and the periaqueductal gray all had significantly more activation. In striking contrast, females who saw their preferred male lose a fight had a much higher expression of the IEGs that were in a part of the brain associated with anxiety, the lateral septum (Desjardins et al. 2010). A remarkable aspect of these results is that the females were responding to visual information alone, because they did not mate with the male, suggesting that the social information had caused changes in activation of key brain areas. It is important to remember that the IEG expression we measured is surely only a very small part of the total brain response and hence is just a glimpse of the genomic response to social information. However, the differential responses in specific brain areas show that females are activating their brains based on visual information and may use this to guide decisions about what to do.

One additional question is how this information might inform the female's choice of a mate. In a separate experiment, we performed the same protocol, but after the female had chosen and seen the staged fight, she had to choose again. In this second choice, if she had seen her preferred male lose, she almost always switched her choice, and if he won, she rarely switched her choice (Fig. 6).

CONCLUSION

This review of research about the social behavior in *A. burtoni* has highlighted the important role of cognition in regulating the social life in this fish species. The *A. burtoni* social system is centered on a hierarchy of dominant males guarding a food resource, and we described some of the cognitive skills required for males and females to navigate this social landscape. Our observations in the field and in laboratory experiments have revealed that these animals collect information on conspecifics and use that information to decide what to do in particular circumstances. This is both surprising and yet, on reflection, obvious. Knowing what to do next in a social context is important. Both males and females use information gathered through observation to guide their behavior. We have followed this information into the brain to identify where it acts using immediate-early genes to mark active areas. As expected, males can change quickly what they do depending on what they perceive is happening in their surroundings, and these changes cause corresponding changes in the brain in specific cells, receptors, and circuits, thereby preparing the brain of the animal for a phase of life in a new status. How social information is transduced into cell and molecular changes in the brain, however, remains a mystery.

ACKNOWLEDGMENTS

I am grateful for the many people in my laboratory over the years for their inspired contributions and hard work. This work was supported by NS 034950, NIMH 087930, and NSF IOS-0923588 to R.D.F.

REFERENCES

Alcazar RM, Hilliard AT, Becker L, Bernaba M, Fernald RD. 2014. Brains over brawn: Experience overcomes a size disadvantage in fish social hierarchies. *J Exp Biol* **217:** 1462–1468.

Alonzo JJ, McKaye KR, van den Berghe EP. 2001. Parental defense of young by the convict cichlid, *Archocentrus nigrofasciatus*, in Lake Xiloá, Nicaragua. *J Aquaric Aquat Sci* **9:** 208–228.

Au TM, Greenwood AK, Fernald RD. 2006. Differential social regulation of two pituitary gonadotropin-releasing hormone receptors. *Behav Brain Res* **170:** 342–346.

Balshine-Earn S, Neat FC, Reid H, Taborsky M. 1998. Paying to stay or paying to breed? Field evidence for direct benefits of helping behavior in a cooperatively breeding fish. *Behav Ecol* **9:** 432–438.

Bond AB, Kamil AC, Balda RP. 2003. Social complexity and transitive inference in corvids. *Anim Behav* **65:** 479–487.

Boulton MJ, Smith PK. 1990. Affective bias in children's perceptions of dominance relationships. *Child Dev* **61:** 221–229.

Brännäs E, Eriksson T. 1999. Floating, switching, or nonswitching as different behaviours when Arctic charr (*Salvelinus alpinus*) are visiting two feeding tanks. *Can J Fish Aquat Sci* **56:** 1068–1077.

Brawand D, Wagner CE, Li YI, Malinsky M, Keller I, Fan S, Simakov O, Ng AY, Lim ZW, Bezault E, et al. 2014. The genomic substrate for adaptive radiation in African cichlid fish. *Nature* **513:** 375–381.

Burmeister SS, Fernald RD. 2005. Evolutionary conservation of the *egr-1* immediate-early gene response in a teleost. *J Comp Neurol* **481:** 220–232.

Burmeister SS, Jarvis ED, Fernald RD. 2005. Rapid behavioral and genomic responses to social opportunity. *PLoS Biol* **3:** e363.

Burmeister SS, Kailasanath V, Fernald RD. 2007. Social dominance regulates androgen and estrogen receptor gene expression. *Horm Behav* **51:** 164–170.

Chance MRA, Larsen RR, eds. 1976. *The social structure of attention*. Wiley, New York.

Chen CC, Fernald RD. 2011. Visual information alone changes behavior and physiology during social interactions in a cichlid fish (*Astatotilapia burtoni*). *PLoS ONE* **6:** e20313.

Clement TS, Parikh V, Schrumpf M, Fernald RD. 2005. Behavioral coping strategies in a cichlid fish: The role of social status and acute stress response in direct and displaced aggression. *Horm Behav* **47:** 336–342.

Davis H. 1992. Transitive inference in rats (*Rattus norvegicus*). *J Comp Psychol* **106:** 342–349.

Davis MR, Fernald RD. 1990. Social control of neuronal soma size. *J Neurobiol* **21:** 1180–1188.

Desjardins JK, Klausner JQ, Fernald RD. 2010. Female genomic response to mate information. *Proc Natl Acad Sci* **107:** 21176–21180.

Desjardins JK, Hofmann HA, Fernald RD. 2012. Social context influences aggressive and courtship behavior in a cichlid fish. *PLoS ONE* **7:** e32781.

Diamant A, Shpigel M. 1985. Interspecific feeding associations of groupers (Teleostei: Serranidae) with octopuses and moray eels in the Gulf of Eilat (Aqaba). *Environ Biol Fishes* **13:** 153–159.

Emory GR. 1976. Aspects of attention, orientation, and status hierarchy in mandrills (*Mandrillus sphinx*) and Gelada baboons (*Theropithecus gelada*). *Behaviour* **59:** 70–87.

Fernald RD. 1985. Eye movements in the African cichlid fish, *Haplochromis burtoni*. *J Comp Physiol A* **156:** 199–208.

Fernald RD. 2012. Social control of the brain. *Annu Rev Neurosci* **35:** 133–151.

Fernald RD, Hirata NR. 1977. Field study of *Haplochromis burtoni*: Habitats and co-habitants. *Environ Biol Fishes* **2:** 299–308.

Francis RC, Soma K, Fernald RD. 1993. Social regulation of the brain-pituitary-gonadal axis. *Proc Natl Acad Sci* **90:** 7794–7798.

Gillian DJ. 1981. Reasoning in the chimpanzee: II. Transitive inference. *J Exp Psychol Anim Behav Process* **7:** 87–108.

Goodson JL. 2005. The vertebrate social behavior network: Evolutionary themes and variations. *Horm Behav* **48:** 11–22.

Goodson JL, Bass AH. 2002. Vocal-acoustic circuitry and descending vocal pathways in teleost fish: Convergence with terrestrial vertebrates reveals conserved traits. *J Comp Neurol* **448:** 298–322.

Greenwood AK, Fernald RD. 2004. Social regulation of the electrical properties of gonadotropin-releasing hormone neurons in a cichlid fish (*Astatotilapia burtoni*). *Biol Reprod* **71:** 909–918.

Grosenick L, Clement TS, Fernald RD. 2007. Fish can infer social rank by observation alone. *Nature* **445:** 429–432.

Gross MR, Sargent RC. 1985. The evolution of male and female parental care in fishes. *Am Zool* **25:** 807–822.

Harbott LK, Burmeister SS, White RB, Vagell M, Fernald RD. 2007. Androgen receptors in a cichlid fish, *Astatotilapia burtoni*: Structure, localization, and expression levels. *J Comp Neurol* **504:** 57–73.

Hourigan TF. 1989. Environmental determinants of butterflyfish social systems. *Environ Biol Fishes* **25:** 61–78.

Hughes RN, Blight CM. 1999. Algorithmic behaviour and spatial memory are used by two intertidal fish species to solve the radial maze. *Anim Behav* **58:** 601–613.

Maruska KP, Fernald RD. 2012. Contextual chemosensory signaling in an African cichlid fish. *J Exp Biol* **215:** 68–74.

Maruska KP, Fernald RD. 2014. Social regulation of gene expression in the African cichlid fish *Astatotilapia burtoni*. In *The Oxford handbook of molecular psychology* (ed. Canli T), pp. 1–22. Oxford University Press, Oxford.

McGonigle BO, Chalmers M. 1977. Are monkeys logical? *Nature* **267:** 694–696.

Mello CV, Vicario DS, Clayton DF. 1992. Song presentation induces gene expression in the songbird forebrain. *Proc Natl Acad Sci* **89:** 6818–6822.

Newman SW. 1999. The medial extended amygdala in male reproductive behavior. A node in the mammalian social behavior network. *Ann NY Acad Sci* **877:** 242–257.

Noonan MP, Sallet J, Mars RB, Neubert FX, O'Reilly JX, Andersson JL, Mitchell AS, Bell AH, Miller KL, Rushworth MF. 2014. A neural circuit covarying with social hierarchy in macaques. *PLoS Biol* **12:** e1001940.

Oliveira RF, M PK, Burford FRL, Custódio MR, Latruffe C. 1998. Functions of mudballing behaviour in the European fiddler crab *Uca tangeri*. *Anim Behav* **55:** 1299–1309.

Paśko Ł. 2010. Tool-like behavior in the sixbar wrasse, *Thalassoma hardwicke* (Bennett, 1830). *Zoo Biol* **29:** 767–773.

Piaget J. 1928. *Judgment and reasoning in the child*. Kegan, Paul, Trench & Trubner, London.

Pitcher TJ, Green DA, Magurran AE. 1986. Dicing with death: Predator inspection behavior in minnow *Phoxinus phoxinus* shoals. *J Fish Biol* **28:** 439–448.

Rapp PR, Kansky MT, Eichenbaum H. 1996. Learning and memory for hierarchical relationships in the monkey: Effects of aging. *Behav Neurosci* **110:** 887–897.

Roberts WA, Phelps MT. 1994. Transitive inference in rats: A test of the spatial coding hypothesis. *Psychol Sci* **5:** 368–374.

Rusak B, Robertson HA, Wisden W, Hunt SP. 1990. Light pulses that shift rhythms induce gene expression in the suprachiasmatic nucleus. *Science* **248:** 1237–1240.

Sapolsky RM. 2005. The influence of social hierarchy on primate health. *Science* **308:** 648–652.

Steirn JN, Weaver JE, Zentall TR. 1995. Transitive inference in pigeons: Simplified procedures and a test of value transfer theory. *Anim Learn Behav* **23:** 76–82.

Summers CH, Forster GL, Korzan WJ, Watt MJ, Larson ET, Overli O, Höglund E, Ronan PJ, Summers TR, Renner KJ, et al. 2005. Dynamics and mechanics of social rank reversal. *J Comp Phys A Neuroethol Sens Neural Behav Physiol* **191:** 241–252.

Timms AM, Keenleyside MH. 1975. The reproductive behavior of *Aequidens paraguayensis* (Pisces, Cichlidae). *Z Tierpsychol* **39:** 8–23.

Vail AL, Manica A, Bshary R. 2013. Referential gestures in fish collaborative hunting. *Nat Commun* **4:** 1765.

Van den Berghe EP, McKaye KR. 2001. Reproductive success of maternaland biparental care in a Nicaraguan cichlid fish, *Parachromis dovii*. *J Aquaricult Aquat Sci* **9:** 49–65.

Vaughn BE, Waters E. 1981. Attention structure, sociometric status and dominance: Interrelations, behavioral correlates and relationships to social competence. *Dev Psychol* **17:** 275–288.

von Fersen L, Wynne CDL, Delius JD, Staddon JER. 1991. Transitive inference formation in pigeons. *J Exp Psychol Anim Behav Process* **17:** 334–341.

Weiss BM, Kehmeier S, Schloegl C. 2010. Transitive inference in free-living greylag geese, *Anser anser*. *Anim Behav* **79:** 1277–1283.

White SA, Nguyen T, Fernald RD. 2002. Social regulation of gonadotropin-releasing hormone. *J Exp Biol* **205:** 2567–2581.

Zink CF, Tong Y, Chen Q, Bassett DS, Stein JL, Meyer-Lindenberg A. 2008. Know your place: Neural processing of social hierarchy in humans. *Neuron* **58:** 273–283.

Summary: Cognition in 2014

TERRENCE J. SEJNOWSKI[1,2]
[1]Howard Hughes Medical Institute, the Salk Institute for Biological Studies, La Jolla, California 92037
[2]Division of Biological Sciences, University of California at San Diego, La Jolla, California 92093
Correspondence: terry@salk.edu

The goal of the 79th Cold Spring Harbor Symposia on Quantitative Biology on Cognition held on May 28–June 2, 2014 was to survey recent advances in cognitive neuroscience and assess future prospects. The symposium succeeded beyond the dreams of the organizers and the participants were treated to an extraordinarily rich feast of 58 long talks, six short talks, and 137 posters. Equally important to the success of the symposium was the perfect setting for informal scientific exchange between 260 participants generously provided by the Cold Spring Harbor Laboratory. The sense that emerged from the symposium was that a threshold had been crossed and a new era in the study of cognition was underway. My attempt here will be to capture that sense of awakening, to trace the strands that gave rise to it, and to access its implications for future discoveries.

WHAT HAS CHANGED?

An earlier Cold Spring Harbor Symposia on The Brain in 1990 covered a much broader range of topics that included molecular mechanisms for signaling, neural development, and sensory and motor systems, as well as cognitive neuroscience. The sweep of the meeting reflected the emergence of neuroscience as a young but vibrant discipline and in particular the impact of molecular biology on many of the difficult problems that previously had resisted explanations. As Mike Stryker, who gave the Summary at that meeting put it, "Is Biology Ready to Take on the Brain?" The meeting revealed how productive the interactions were between neuroscience and the rest of biology.

The papers on cognitive neuroscience from the 1990 symposium reflected the progress that had been made using single-unit recording techniques, which revealed the properties of single neurons, particularly in the visual system and hippocampus. The focus on single neurons dominated thinking in that era and studies using electroencephalogram and event-related potential had much poorer resolution. Brain imaging techniques were in their infancy, with pioneering work on positron emission tomography beginning at the University of Washington (Petersen et al. 1988) and the invention of functional magnetic resonance imaging (fMRI) at Bell Labs (Ogawa et al. 1990). Anatomical techniques for staining neurons and tract tracing were classical, some going back to Golgi and Ramón y Cajal in the 19th century. The molecular revolution that had already swept through other areas of neuroscience, the cloning of ion channels and gene expression during development, had not yet reached systems neuroscience.

Francis Crick (1979) wrote a prescient essay that foretold what the impact of techniques from molecular genetics would have on systems neuroscience:

> For example, a method that would make it possible to inject one neuron with a substance that would then clearly stain all the neurons connected to it, and not others, would be invaluable. So would a method by which all neurons of just one type could be inactivated, leaving the other more or less unaltered.

And, indeed, this has come to pass, as reflected in major advances that have been made in the last several years using viruses (Wall et al. 2010) and optogenetics (Zhang et al. 2006). Our ability to target specific types of neurons, to optically record, stimulate, and suppress their activity, has greatly accelerated research because of both the quantity of data we can collect and the new types of experiments that can be done by manipulating neural systems with exquisite selectivity.

Another major shift since 1990 is that more laboratories are using advanced data analysis techniques and computational models to guide their research. Only a few talks at the 1990 Brain Symposium used computational approaches but more than half of the talks at the 2014 Cognition Symposium incorporated them. Here again, Crick (1979) had anticipated how theory and experiment might be integrated:

> Hence an important role for theory in neurobiology is not merely trying to create correct and detailed theories of neural processes (which may be an extremely difficult task) but pointing to which features it would be most useful to study and in particular to measure, to see what kind of theory is needed.

We now have theoretical frameworks for how populations of neurons encode and store information, which are being used to analyze recordings from large-scale populations of neurons.

But on one prediction Crick (1979) missed the mark:

> It is no use asking for the impossible, such as, say, the exact wiring diagram for a cubic millimeter of brain tissue and the way all its neurons are firing.

 doi: 10.1101/sqb.2014.79.024885

What Crick failed to predict was the rapid advances in techniques for working out the wiring diagrams of neural circuits. John White and Sydney Brenner had laboriously traced out and manually reconstructed the wiring diagram for *Caenorhabditis elegans* from electron microscope cross sections, a bit of 21st century science in the 20th century. Large-scale production of serial electron micrographs and machine learning have now made it feasible to automate and scale-up the reconstruction of ever-larger neural circuits. Similarly, advances in recording techniques have made it possible to record simultaneously from hundreds of neurons.

WHERE ARE WE TODAY?

New techniques (Fig. 1) have made it possible to probe with much higher precision than was previously possible how neural circuits give rise to complex behaviors. It is not possible to summarize all of the results that were presented at the Symposium here, but instead I have selected a few representative examples that illustrate what has been accomplished, arranged according to the spatial scale of the measurements.

Genetically Encoded Indicators

An elegant example of how sensory information is transformed into actions is found in *C. elegans*, which has a relatively short chain of neurons between sensory receptors and motoneurons. Neurons in *C. elegans* are small and difficult to record from with microelectrodes because of body turgor. But now activity patterns can be followed with genetically encoded calcium-sensitive dyes. Cornelia Bargmann showed that sensory neurons reliably responded to chemical signals, but the next neuron down the chain displayed response variability that was highly correlated with reversal behavior. As she put it, "One of the most reliable aspects of behavior is its unreliability." We can now pinpoint where the variability is introduced. The ability to probe all the neurons in the circuit and to manipulate as well as observe them is providing a powerful way to divine the logic of neural circuits.

Genetic Fingerprints

The brain has more different cell types than any other organ by several orders of magnitude and we are far from having a complete catalog. This is changing rapidly as techniques from molecular genetics are being used to identify cell types by their pattern of transcription factors, which regulate gene expression. Thomas Jessell showed how this can be done to distinguish more than a dozen types of inhibitory interneurons in the spinal cord, using advanced statistical techniques for sorting through many different combinations of transcription factors. Other sources of cell identity may come from the pattern of DNA methylation, which also regulates gene expression. The DNA in neurons has a special pattern of methylation not found in other organs of adults (Lister et al. 2013).

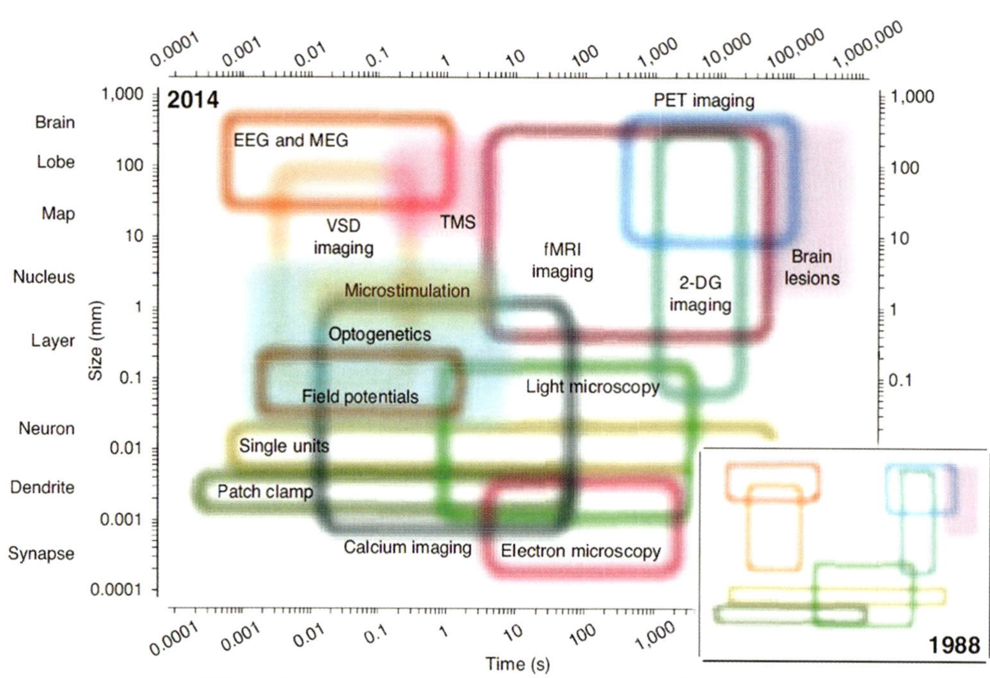

Figure 1. The space–time domain of methods in neuroscience in 2014, in contrast with a similar space–time map in 1988 (*inset*). Each colored region represents the useful domain of spatial and temporal resolution for one method available for the study of the brain. Open regions represent measurement techniques, filled regions, perturbation techniques. The gaping holes in the 1988 diagram were filled in with functional magnetic resonance imaging (fMRI), calcium imaging, optogenetics, and transcranial magnetic stimulation (TMS). (Reprinted, with permission, from Sejnowski et al. 2014.)

Neural Circuits

Hassenstein and Reichardt (1956) proposed a simple neural circuit that could compute the direction of motion of a visual stimulus moving across a fly eye. Despite heroic efforts over the next 50 years progress was slow because of the complexity of the neuropil in the fly eye. In just the last few years, Alexander Borst, following in Reichardt's footsteps, used cell-specific optogenetic techniques and connectomic reconstruction to identify the components of the circuit. Nature has implemented a version of the Hassenstein and Reichardt model, separating the ON and OFF pathways, in essence devoting twice the machinery necessary to compute the direction of motion because of biophysical limitations. This shows how an apparently complex circuit may implement a relatively simple algorithm. It also illustrates how much progress can be made very quickly if the right tools are available to an experimenter to answer the question.

Dynamic Stability

Invertebrate neural systems such as the stomatogastric ganglion of the lobster were once considered "simpler systems," but it is now apparent that they have complex mechanisms that allow small numbers of neurons to generate a wide range of complex dynamical behaviors that are difficult to predict from the spatial distribution of dozens of types of nonlinear ion channels. Eve Marder asked how the observed robustness in behavior is possible given the heterogeneity of the underlying mechanisms. She showed that neuromodulators are a source of circuit stability. This important principle could also be demonstrated in a "simple neural model," which provides a framework for exploring why neuromodulation is so effective at stabilizing network properties.

Large-Scale Recordings

Much can be gained by recording from all the neurons in a circuit that controls a behavior. William Kristan recorded with voltage-sensitive dyes from many of the neurons in leech ganglia that are involved in the decision to swim or crawl. Some neurons are involved in both behaviors, with different patterns of activity. The switch between the two patterns occurs rapidly, in less than one swim cycle, which is a signature of what theorists have called "attractor" states that were analyzed in simple models of neural networks in the 1980s. Evidence for attractors has also been found in the hippocampus, where recurrent networks settle into states that represent different places in the environment (Colgin et al. 2010).

Spike Timing Measurements

We know a great deal about brain activity in response to sensory stimuli, but very little about internally generated brain activity that has neither a sensory nor a motor correlate. The discovery of replay in the hippocampus has given us a rare glimpse into what the brain does offline, when it is no longer in a closed loop with the sensory environment. Loren Frank reported that previously experienced patterns of activity in the hippocampus of awake rats are replayed and that disruption of the replay interferes with learning and memory-guided decision-making. Using large-scale recordings of hippocampal neurons, György Buzsáki showed that these internally generated self-perpetuating sequences are supported by oscillations and precise spike timing. These discoveries open up a new window on internal brain states.

Activity Tagging

Psychologists have long postulated that human memory was encoded in the brain as an "engram" (Bruce 2001) and although many laboratories focusing on single neurons were able to make progress studying the mechanisms for synaptic plasticity, the engram remained an elusive concept; that is, until molecular tools became available to tag the neurons in an engram. Susumu Tonegawa told us about an experiment in which he was able to tag neurons involved in fear conditioning with channelrhodopsin-2, which allowed him to reinstate the engram by shining blue light on the tagged neurons (Redondo et al. 2014). Many other classical concepts based primarily on behavioral experiments can now be traced to specific neural circuits and brain systems.

Language Development

Language is a quintessential human behavior that develops rapidly during the first 2 years of life. Brain imaging has made it possible to follow the maturation of the cortical circuits responsible for language. Patricia Kuhl showed that structural MRI features can predict a baby's future language performance. Magnetoencephalography revealed that when infants listen to speech, brain activation occurs not only in auditory areas but also in brain regions involved in speech motor planning. This is evidence for a close collaboration between the sensory and motor systems in generating language, something that had been long suspected.

Virtual Reality

Computer games have become more popular than television in some age groups and it is of some concern to know how this affects the brain, especially of developing children. Daphne Bavelier reported that some games enhance sensory, attentional, and cognitive skills. These generalize beyond the specific skills learned in the games. Adam Gazzaley has taken this even further by showing how personalized close-loop virtual reality games can have therapeutic value and sustainable cognitive enhancement. This opens up a new window into human learning that could have major implications for education and rehabilitation.

Thought

One topic in cognition was notable by it absence at the symposium. At the 1990 Brain Symposium Francis Crick and Christof Koch presented "Some Reflections on Visual

Awareness." No one did more to draw our attention to the problem of consciousness in the 20th century than Crick (Sejnowski 2015). The topic of consciousness never came up in the formal talks at the 2014 Cognition Symposium. Zachary Mainen, however, addressed free will in his talk, which concluded that brain recordings that are correlated with decisions may reflect evidence accumulation rather than the decision itself. As techniques get better for testing causality by manipulating patterns of activity in neural populations it may become possible for us to probe the mechanisms underlying thoughts and actions.

WHERE WILL WE BE IN 2025?

On 3 April 2013, President Obama announced the BRAIN Initiative, whose goal is to develop innovative neurotechnologies. The working group that advised the Director of the National Institutes of Health (NIH) for the BRAIN (Brain Research through Advancing Innovative Neurotechnologies) Initiative prepared a detailed plan and goals. The final report, which was accepted by the NIH on June 5, 2015 (BRAIN 2025: http://www.braininitiative.nih.gov/2025/index.htm), a few days after the symposium, identified seven major goals.

1. *Discovering diversity:* Identify and provide experimental access to the different brain cell types to determine their roles in health and disease.
2. *Maps at multiple scales:* Generate circuit diagrams that vary in resolution from synapses to the whole brain.
3. *The brain in action:* Produce a dynamic picture of the functioning brain by developing and applying improved methods for large-scale monitoring of neural activity.
4. *Demonstrating causality:* Link brain activity to behavior with precise interventional tools that change neural circuit dynamics.
5. *Identifying fundamental principles:* Produce conceptual foundations for understanding the biological basis of mental processes through development of new theoretical and data analysis tools.
6. *Advancing human neuroscience:* Develop innovative technologies to understand the human brain and treat its disorders; create and support integrated human brain research networks.
7. *From BRAIN Initiative to the brain:* Integrate new technological and conceptual approaches produced in Goals (1)–(6) to discover how dynamic patterns of neural activity are transformed into cognition, emotion, perception, and action in health and disease.

The NIH BRAIN Initiative and complementary programs at the National Science Foundation (NSF) and the Defense Advanced Research Projects Agency (DARPA) are under way and are accelerating the pace of innovation in cognitive neuroscience. The overall goal is to scale up current techniques so that by 2025 it will be possible to have connectomic wiring diagrams for all areas of the mouse brain, to be able to record simultaneously from a million neurons, and to develop sophisticated behavioral paradigms that engage multiple brain systems under more natural conditions. These advances will generate enormous data sets that will require new experimental designs, the development of new algorithms for analyzing the data, new ways to model the data, and new theories for interpreting the results (Sejnowski et al. 2014). Some of these data sets will be made available to the community, just as the availability of large genomic data sets, crystallographic data, and astronomical data from the Hubble Space Telescope have enhanced their usefulness.

All of these goals will depend on scaling up the resolution, speed, and comprehensiveness of behavioral, anatomical, physiological, and genetic techniques by factors of thousands. We are already on an exponential rise in the number of neurons that can be simultaneously recorded, with data sets that include calcium measurements from 1000 neurons in the hippocampus with 10-msec resolution (Deisseroth and Schnitzer 2013) and 80,000 neurons in the zebrafish with 1-sec resolution (Keller et al. 2015). Machine learning, which provides a set of powerful algorithms for analyzing data in high-dimensional spaces, has made it possible to automatically track behavior with high spatial and temporal resolution (Anderson and Perona 2014). However, each algorithm makes statistical assumptions and care must be taken to confirm that they are satisfied by the brain data.

What cannot be predicted today is what impact the development of major new techniques could have on questions that we can ask about the brain. For example, at the symposium Tony Zador outlined a method for automating the collection of data from long-range tracts in the brain using short random sequences of RNA as barcodes, taking advantage of viruses to deliver the barcodes to neurons in one area and collect them in other brain areas. This would take advantage of advances in sequencing techniques to solve anatomical problems and could make it possible to rapidly, cheaply, and routinely obtain high-quality connectomic data. If this works it will revolutionize anatomical studies.

Another area where we can expect rapid progress to be made is in the study of human cognition. In many ways humans are the best model system to study humans. Imaging techniques allow us to observe human brain activity noninvasively and direct recording from the cortex and subcortical structures in patients with epilepsy are giving us new insights into language and memory consolidation. The current standards in human electrical recordings and deep brain stimulation is 30 years behind the state of the art in model systems such as monkeys and mice. The Food and Drug Administration, which approves drugs and devices, has a fast track for Investigational Device Exemptions (IDE), which should lead to a quantum leap in recording from and stimulation of human brains. DARPA has launched two major programs to develop better devices for closed loop recording and stimulation for treating neuropsychiatric disorders (SUBNET) and memory recovery (RAM).

Finally, animal models of human mental disorders and degenerative brain diseases have not translated into effective drugs for treating humans. For example, drugs that have been developed at great expense for Alzheimer's disease based on mouse models have failed in clinical trials. There is thus a great need for transgenic nonhuman primate models of neurological diseases and neuropsychiatric disorders, which is already under way and should be widely available by 2025 (Belmonte et al. 2015). This could lead to a better understanding of these brain diseases and disorders at the circuit level and more precise ways to target the underlying dysfunctions.

CONCLUSION

A question came up in discussion with Bruce Stillman at the symposium picnic: "Is Biology Ready to Take on Cognition?" The conclusion from the 1990 Brain symposium was that biology was ready to take on the brain, and the question was whether biology was ready to take on the cognitive functions of the brain. What surprised me and others at the conference was the extent to which cognition has already become a part of biology and this will accelerate with the BRAIN Initiative. We can anticipate that within the near future tools and techniques will become available that will in time illuminate the biological mechanisms underlying language, thought, free will, and consciousness. A new neurotechnology industry will be needed to translate these advances into medical devices that can alleviate suffering from brain dysfunctions.

REFERENCES

Anderson DJ, Perona P. 2014. Toward a science of computational ethology. *Neuron* **84:** 18–31.

Belmonte JC, Callaway EM, Churchland P, Caddick SJ, Feng G, Homanics GE, Lee KF, Leopold DA, Miller CT, Mitchell JF, et al. 2015. Brains, genes and primates. *Neuron* **86:** 617–631.

Bruce D. 2001. Fifty years since Lashley's In Search of the Engram: Refutations and conjectures. *J Hist Neurosci* **10:** 308–318.

Colgin LL, Leutgeb S, Jezek K, Leutgeb JK, Moser EI, McNaughton BL, Moser MB. 2010. Attractor-map versus autoassociation based attractor dynamics in the hippocampal network. *J Neurophysiol* **104:** 35–50.

Crick FH. 1979. Thinking about the brain. *Sci Am* **241:** 219–232.

Deisseroth K, Schnitzer MJ. 2013. Engineering approaches to illuminating brain structure and dynamics. *Neuron* **80:** 568–577.

Hassenstein B, Reichardt W. 1956. Systemtheoretische Analyse der Zeit-, Reihenfolgen- und Vorzeichenauswertung bei der Bewegungsperzeption des Ruesselkaufers Chlorophanus. *Z Naturforsch* **11:** 513–524.

Keller PJ, Ahrens MB, Freeman J. 2015. Light-sheet imaging for systems neuroscience. *Nat Methods* **12:** 27–29.

Lister R, Mukamel EA, Nery JR, Urich M, Puddifoot CA, Johnson ND, Lucero J, Huang Y, Dwork AJ, Schultz MD, et al. 2013. Global epigenomic reconfiguration during mammalian brain development. *Science* **341:** 1237905.

Ogawa S, Lee TM, Kay AR, Tank DW. 1990. Brain magnetic resonance imaging with contrast dependent on blood oxygenation. *Proc Natl Acad Sci* **87:** 9868–9872.

Petersen SE, Fox PT, Posner MI, Mintun M, Raichle ME. 1988. Positron emission tomographic studies of the cortical anatomy of single-word processing. *Nature* **331:** 585–589.

Redondo RL, Kim J, Arons AL, Ramirez S, Liu X, Tonegawa S. 2014. Bidirectional switch of the valence associated with a hippocampal contextual memory engram. *Nature* **513:** 426–430.

Sejnowski TJ. 2015. Consciousness. *Daedalus* **144:** 123–132.

Sejnowski TJ, Churchland PS, Movshon JA. 2014. Putting big data to good use in neuroscience. *Nat Neurosci* **17:** 1440–1441.

Wall NR, Wickersham IR, Cetin A, De La Parra M, Callaway EM. 2010. Monosynaptic circuit tracing in vivo through Cre-dependent targeting and complementation of modified rabies virus. *Proc Natl Acad Sci* **107:** 21848–21853.

Zhang F, Wang LP, Boyden ES, Deisseroth K. 2006. Channelrhodopsin-2 and optical control of excitable cells. *Nat Methods* **3:** 785–792.

DORCAS CUMMINGS LECTURE

Dorcas Cummings Lecture

Patricia Smith Churchland

Dr. Patricia Churchland presented the Dorcas Cummings lecture entitled "The Brains Behind Morality" to friends and neighbors of Cold Spring Harbor Laboratory and Symposium participants on Saturday, May 31, 2014. Dr. Churchland is Professor Emerita of Philosophy at the University of California, San Diego, and an adjunct Professor at the Salk Institute.

It's a great, great pleasure to be at Cold Spring Harbor Laboratory and a special honor to be speaking to the wider community of Cold Spring Harbor.

Today, I want to talk a little bit about morality. I'm going to begin with thinking in the normal way that we do about natural selection. What we know, of course, is that animals who are well-adapted to the particular environment they live in, if they can compete and prosper in that environment, are more likely to pass on their genes. Those who do not do so well will not pass on their genes.

We all understand in a very straightforward way the kind of competition that exists both within a species and across species in order to survive and reproduce. All animals have the neural wherewithal for organizing themselves to feed, to flee in the face of danger, to fight if they have to, and to reproduce. In neuroscience, these have been called the 4 Fs—feeding, fleeing, fighting, and reproduction.

If you look at it this way, then there's a certain sense in which value is right in the very basic structure of the organism, but it is a value that is very egocentric. It is very selfish. This has prompted some evolutionary biologists—for example, Richard Dawkins and Ed Wilson—to say that we are not just with respect to our genes selfish, but with respect to our social temperaments, we are fundamentally selfish. Consequently, morality or the capacity to incur a cost to benefit others has to somehow be imposed from without. It has to be shoved on us or scared into us.

Interestingly, I think, Darwin had a very different view. Darwin took the view that there are many species of social animal and that in the human case, our sense of morality or our conscious depends on essentially three things: our social instincts (our fundamental biology), habits and skills that we acquire in our lifetime, and reason, or as we might say, the capacity to problem solve.

Of course, Darwin didn't have the advantage that we now do of knowing something about the brain. Almost nothing was known at that time. What is new, I think, in the field of social neuroscience is that we're beginning to understand the biological basis for sociality, and that gives us really striking and interesting clues about the biological platform on which the superstructure of laws and norms and so forth can be built.

In thinking about this, it's useful to take advantage of what the bones and stones people, the anthropologists, have told us is the best hypothesis about early humans. As you know, there have been many species of humans. *Homo sapiens* has been on the planet only about 200,000 or 250,000 years. For most of that time, we lived in very small groups. We hunted, we gathered. There was probably some interaction for trade, mainly trade of females, but it wasn't until about 10,000 years ago with the advent of agriculture that we began to see large communities develop. It was really only very recently that we see anything like an established or a systematic religion.

Whatever norms and social rules and practices these people had, they grew up within the context of these very small groups where everybody knew everybody else. From the data on hunter–gatherer groups existing in the last 150 years, it would appear that they had social practices regarding the really essential things—how to resolve conflicts, what to do in times of dire straits when the resources were very low, what to do with orphans, and basically how to share food. That is not terribly complex, but those are the basic and fundamental moral issues that would have arisen in this context.

We think that social species evolved many times but under rather different organizing principles. There are, of course, social fish like the discus fish. There are many species of social insects, but sociality in mammals is very different. It has a kind of flexibility. It has, as we think of it in our own case, a kind of emotional component that is probably not there in insects and fish. This is likely also true of birds, perhaps by separate evolution, perhaps by common ancestor.

Now, you might be thinking, "What exactly do you mean by morality?" and, "What do we mean by honesty, or by friendliness?" or whatever. I'm going to draw on some very old—that is, 1960s—work by Eleanor Rosch, which I think is terribly important in orienting us in the right kind of way to these subconcepts. Her basic finding

 doi: 10.1101/sqb.2014.79.21

is quite simple, which is that our workaday concepts, our everyday concepts, by and large have a radial structure. What she meant by that was that there are prototypes at the center, and then there are fuzzy boundaries, and then in between, there's everything else.

The standard case to think of here is the concept of vegetable. It turns out that if you ask people, "What's the first vegetable that comes to your mind?" most people say "carrot", and almost everybody thinks carrot is a vegetable. On the other hand, part way out in the radial structure are radishes. Some people think, "They're just garnishes. Who would eat such a thing?" Way, way out on the fuzzy boundary, are things like, say, wild mushrooms. You might find them in the supermarket with the other vegetables. Does that really make them a vegetable?

The important understanding that Rosch had here is that for most purposes, it really doesn't matter that we have necessary and sufficient conditions for defining vegetable. I do not care whether parsley really is a vegetable. For most of our communicative purposes, we don't care. In science, we often do care, and we do want precise definitions.

I'm bringing this up because, of course, I think moral concepts are like this, too. They're like this in a way in which the concept of a dwelling or a house is like this. That is, that there are prototypes relative to a culture. What the Inuit regard as a prototypical house is different from what you would regard in this community as a prototypical house, or the house I grew up in. It's culturally relative, although there is, as Wittgenstein would have said, a kind of family resemblance running through them.

Now, this is interesting in the present context because I want to say that social categories are like that, including the category of moral and its correlative, not moral. What it is to be a friend, to be honest, to be kind, to be brave, to be trustworthy—these are all radial concepts. Sometimes, we disagree about the boundary cases; often, we perfectly well agree on the prototypical cases.

I think the same is basically true of what we count as moral, especially within a culture, but of course, we are going to see cultural differences. Across cultures, it tends to be the case that we find a lot of agreement on the prototypes. Most people think that it's deeply immoral to neglect or beat or abuse children, but then you get out into the boundaries, and there are other things that may be considered moral or may not. Sometimes, they matter, and sometimes less so. That's the background point I want to make about the category of moral and immoral. You might wonder whether there is a possibility of making it more precise, as we might make the notion of a protein more precise. I don't know. It kind of depends.

The hypothesis that I'm interested in here, given developments in social neuroscience, is that at least in the case of mammals, sociability turns out to be a basic value—and that's as a result of natural selection, and I'll explain why in a minute—and that the hub of the story depends on a peptide, oxytocin, a very ancient peptide, as it turns out. It has a sibling peptide, vasopressin, and they're both extremely important. When I say it's the hub, I don't mean, of course, that it's the only thing, but it plays a very crucial role and is probably regulatory with regard to some of the other neurochemicals.

Now, norms, which many people think typify morality and ethics, almost certainly emerge as a result of interactions of the young within the social context and involve the reward system in the brain. I'll say a little bit more about that. I think of norms, those social practices which may or may not be explicit, they may just be picked up. For example, here's one that's picked up when you first meet someone, and you're having a conversation, how close is it appropriate to stand? Most of us could not actually say. We might not even know that we know that, but we do *do* it, and we notice when there is a violation of that social norm, the norm that's relative to our own group.

The question then is if mammals are somewhat special in this regard, how did that come to be? What were the evolutionary pressures that brought about mammals at all, and what is it that brought about their sociality? Now, I want to talk a little bit about food. Before there were mammals, there were lots and lots of reptiles, and reptiles, of course, are cold-blooded. At some point around 200, 250 million years ago, there appeared warm-blooded creatures, not yet mammals but warm-blooded reptiles.

This was important because it meant they could hunt at night when all the other creatures had to wait until the sun warmed them up so they could be busy. Moreover, the warm-blooded creatures could hunt those sleeping reptiles. They could also avoid predation by those who were sleeping. In addition, they could go to very new environments, where it might be cold in general, and they could count on their thermoregulation to keep them functional.

The downside of being warm-blooded is that gram for gram you have to eat 10 times as much as your cold-blooded friends. It also means that a given patch of territory might support many lizards but few rodents or foxes. So a very different ecology emerges once you've got warm-blooded animals and Mother Nature basically said, "How are we going to make this all work?" One of the things evolution chanced upon is that it is hugely advantageous if these warm-blooded creatures have the capacity to learn—not just learn a little bit but really learn a lot about the causal structure of the world. This means they will have intellectual and cognitive advantages. Additionally, the brains can tune themselves up to a variety of different environments, so when food runs out, they can go elsewhere, even to cold climates. If the brains of the very young are so plastic (able to learn so much), when they are born their brains need to be very immature. Thus, there's a trade-off between a learning capacity and immaturity at birth. And, surprisingly perhaps, this immaturity at birth turns out to be the basis for sociality.

The cortex, of course, is unique to mammals. Within it is this really amazing cellular organization, very well-defined layers with specific cell types and specific wiring in each layer. Reptiles don't have anything like it.

We do not really understand how cortex evolved. Of course, one of the great things about cortex, as we know, is that it gives us flexibility because you're born very immature, and cortex tunes itself up to the environment, whether it's the particular physical environment— you're in the arctic or in the Brazilian jungle—or the social environment, so you come to acquire the appropriate habits.

We sing the praises of cortex but subcortical structures are absolutely crucial and a critical part of the reward system. You're not going to acquire any skills or habits—in fact, you're not going to make good use of cortex—unless those structures are in place and unless the right kinds of loops between cortex and those deeper structures are in place.

Having a big cortex means you can learn a lot of stuff, simple stuff or more complex, more abstract. And the more abstract you get, the more deeply you can penetrate the causal structure of the world, and the further into the future you can envisage to predict how things are likely to turn out.

Let's go back to newborn dependence. If the newborns are very immature, Mother Nature has a problem. With reptiles, she's got these fundamentally egocentric animals, and now she's got these babies that are born highly dependent. How is she going to arrange for them to mature? Although we do not entirely understand the evolutionary steps involved here, we do understand the outcome. Which is that over evolutionary time, there was a kind of restructuring of the hypothalamus in the brain, in tandem with the development of this amazing temporary organ, the placenta. The placenta and the hypothalamus coordinate their hormonal communication.

An important backfill here is oxytocin. Its presence in the nervous system long precedes the evolution of mammals. In pre-mammalian creatures such as frogs, it was involved in ejecting eggs from the female and sperm ejection in the male. But in mammals, it was put to new jobs. In particular, it seems to have been absolutely crucial for lactation—for milk ejection.

In all nervous systems, the hypothalamus is so critical for feeding, mating, attacking, and defending. In the mammalian brain, oxytocin is released, and the upshot is the mother strongly bonds to the offspring, who in turn strongly bond to the mother. What does bonding mean here? It means that just as the mother cares for her *own* food and warmth and safety, she cares for the food and warmth and safety of the other—the infant. It's as though the infant is an extension of her—its interests are *her* interests. This arrangement is a very interesting genetic trick that takes us from the realm of the truly egocentric to the realm of caring for another in almost the same way as you care for yourself.

While she was at it, Mother Nature rigged it so that you feel pleasure when you are with the child and you feel pain when you are separated. She gave the offspring vocalization so they would squeal when they were separated or when they were hungry. We all feel pain when the offspring squeal, and we take corrective action. During the evolution of mammahan care, there were many modifications, but the basic and very big thing involved caring for others, enjoying and wanting and liking being with them, feeling pain at separation, feeling pain at their pain, and taking corrective action.

This seems to me to be a huge change in the evolution of living things. All of a sudden, we care for these others, we are prepared to sacrifice for them. You might say, "We care for these others because they carry our genes." Fine, but parents incur a huge cost in order to care for the others, and the love and sacrifice are entirely real.

So we have this shift from being entirely egocentric to caring for others. How do we then get to morality, something that is recognizably within the domain of morality as we understand it, such as food sharing?

To me, at any rate, the clues come from the prairie voles. There are many species of voles, and I'm going to characterize two, the montane voles and the prairie voles. The montane voles are the stereotype of a rodent. The males and the females meet, they mate, then go on their separate ways.

Prairie voles are different. They meet, they mate, and they're bonded for life. What does that mean? It means that they stay together a lot. They do not like to be separated. It means the male guards the nest, which does not happen among montane voles. It means the male helps to rear the pups, huddling over them and protecting them. It means that they get depressed if they are separated. This bonding tends to happen immediately after the first mating and becomes even stronger after the first litter is born.

I should also just say that pair-bonding does not imply sexual exclusivity, and this is true in the 95% of bird species that are also bonded for life, such as Canada geese. Although it does not imply sexual exclusivity, much of the sexual activity does take place with the bonded mate.

In addition to that, prairie voles are generally very social. If you have a basket of them and you distribute them around a room, before very long, they come together. They just like to be together. Montane voles are not like that at all. Moreover, prairie voles live in these large communities, and the siblings frequently help in taking care of infants. We're seeing sociality that goes beyond just the bare mother–offspring relationship, and that probably involves a relatively small genetic change that is relative to the large genetic changes that result in care for offspring.

In order to understand the difference between montane vole and prairie vole brains, Larry Young and Sue Carter and a number of other people looked at their brains, staining them for receptors for oxytocin and receptors for vasopressin. The structure they were interested in is part of the reward system, the nucleus accumbens. In the prairie vole, there's a much greater density of receptors for oxytocin. The animals might all have the same amount of oxytocin in the brain, but the density of receptors on neurons to which the oxytocin binds is much greater. Same for vasopressin except in a different reward structure—the ventral pallidum.

The early discovery was only a correlation between receptor density and social behavior, but the neuroscientists took all of the additional steps needed to determine whether or not there was a causal relationship, and found there almost certainly was. I should, of course, also acknowledge the many other neurochemicals involved, including serotonin, dopamine, the endogenous opioids, and the endogenous cannabinoids, the latter probably contribute to making us feel good in those happy social conditions.

What actually does oxytocin do? A number of things in the mammalian brain, I should say. There seems to be a contrast or opposition between oxytocin and stress hormones. To a first approximation, when oxytocin levels go up, stress hormones go down. Part of why you feel good in a social context amongst friends is that stress hormones go down. You don't feel anxious. When stress hormones go down, other things can also be regulated, and we think that the endogenous cannabinoids and the endogenous opioids are probably playing a role.

It looks like this extra part of sociality might come for free as a result of getting mate bonding, which just then spills over into feeling good about being in a social context in general, or it might actually involve a different set of receptors and different genes. There might be a slightly different story there, but we're not entirely sure.

Let's turn to marmosets, the tiny South American monkeys. They bond for life and the fathers take care of infants to a greater extent than females, and even alloparent orphans in the community. Surprisingly, perhaps, chimpanzees will do this, too. There is some evidence that the oxytocin story is important in the social behavior of other primates.

The basic thesis that I think emerges is that attachment and trust are the platform for moral values. Once that is in place, then other behavior such as cooperation and mutual defense can emerge. I should just say, too, that although this story seems very nice at this point, mammals are not, of course, free of all selfish considerations, and individuals within a group do not cease to be competitive, as with siblings. In all social animals, from time to time, there is within-group competition, but in wolves, for example, rough competition between individuals within a group is generally discouraged.

We know this is also true in some hunter–gatherer groups. It was definitely true amongst the Inuit until they were westernized. They allowed very formal kinds of games where especially the boys within a family could knock each other about, but they were not allowed to go at it in an informal way. You can see why that practice would emerge. There they are, living on the knife edge of survival, and you can't have a whole lot of energy wasted on stupid things like fighting. Every hunter and fisher is needed, so injuring or laming is unproductive.

Part of what emerges from this story is that you might not need specific genes for every kind of social behavior, such as cooperation or food sharing. Some forms of cooperation are just going to emerge because the animals are smart enough to see that, "These are my friends, and I can trust them, and we'll do better if we work together."

Wolves, for example, in the Yukon, call to attract other individuals from the pack. They know when there is a move afoot to go after prey. They trot off in a line and find a grizzly that has brought down a caribou. Possible food here.

Only by working together, however, can they drive off the grizzly. They will harass and harass and harass until finally the bear gives up, and then they feed. It's a very effective strategy. Wolves also use cooperation to bring down a deer or a caribou themselves and it's a most astonishingly organized phenomenon.

In scavenging, wolves sometimes get help from ravens. A raven can fly over the woods and see quickly where action is happening. When a raven knows there's some grizzly having downed a caribou way off, the raven return to the wolf pack and alert the wolves by flapping around and attracting attention. The wolves have come to know what that means—it's not genetic, it's learned—and the raven leads the wolves to the kill. What's in it for the raven? The answer is that once the wolves are settled into the caribou carcass, he goes and gets his pals, and they come with their razor sharp, long, terrifying beaks, and they harass the wolves until they drive the wolves off. Everyone, save for the caribou, wins.

This story involves a nested set of cooperation ventures: The wolves cooperate with each other, the raven cooperates with the wolves, and then the raven cooperates with other ravens in the flock. But these are different forms of cooperation. The wolves like and trust each other—they are bonded. Ditto for the ravens in the flock. But the ravens and the wolves are not bonded to each other. They have just figured out a system whereby wolves and ravens mutually benefit. It's a way to get good caribou meat.

With regard to rules and norms, many are like the how-close-to-stand norm. We abide by them, our behavior is guided by them, but they are implicit. When groups become much larger than 150 or so members, there may be a need for explicit rules. Nevertheless, it's very interesting that a lot of conventions, methods, social practices can be picked up just by imitation, by watching, by modeling yourself after what the other individual is doing. You may not even be aware that you're doing it yourself.

In the first instance, in these small social groups of early humans, *Homo erectus*, *Homo neanderthalenis*, and *Homo sapiens,* such loosely defined norms probably sufficed for much of what we needed for social stability, but with the advent of agriculture and the emergence of large towns where not everybody knew everybody else, norms began to be laid down, discussed, and pronounced from on high.

Also, humans at about that time really began to pick up on the fact that it was tremendously advantageous to trade stuff, especially to trade tools. If I make hooks, and you make nets, and we trade, suddenly my tool box has doubled. Once they began to realize that, trade really got going. Then there needed to be rules and norms and practices for what constituted fair trade and how those kinds of practices were enforced.

I haven't yet talked about the dark side, to do with between-group hostility. Anthropologists say they really do not know very much about between-group hostility in ancient prehistoric times. Some anthropologists think it was unlikely to have been common because there's not much advantage in it. If no one has very much, why would I go on a raid to kill and steal—what would I get out of it? I don't have very much but I don't need very much. But of course, once wealth began to be accumulated, there is an incentive. Once there's wealth, once there's power, once there are inventions that allow you to get the upper hand, then we begin to see in full flower the emergence of aggression.

I find aggression really interesting and surprising and amazing. It's much older than other-caring, and in that sense, it's much more ancient than sociability. How exactly it plays into our lives, into our daily lives, why it emerges the way it does, is puzzling and interesting.

Not a great deal is understood about the neurobiology of aggression and in particular about this really rather disturbing but apparently factual relationship between aggression and pleasure. It's clearly there, but why? Why is hating especially bonding within a group of male adolescents or female adolescents? Why is hating pleasurable? There is speculation predation needs to be pleasurable in order to pull it off because it's pretty scary. If you're a wolf, and you're going after caribou, the likelihood of injury is high. Maybe you have to be in a state of excitement in order to be able to pull it off. This is just speculation, and we really don't know why the relationship is there, when in fact it is. David Anderson's lab at Caltech is studying aggression, and I hope to learn much in the coming years that will help us understand its neurobiological nature.

I want to end with a cheesy example, but I like it. We're always taught that orangutans are loners, and in the wild they are indeed solitary. That's largely because they are vegetable eaters, they need a very large territory, so in the wild they tend to be social only to the degree that the mums and the offspring stay together for 2 or 3 years. After that, they tend to be loners.

People have thought, "Isn't that interesting? They're probably not very social. They probably don't derive pleasure from company of others in the way that, say, baboons or humans do." But it seems that if conditions are right and resources are plentiful and available, the pleasure circuitry for enjoying the company of others in a trusting situation is all there, and it just spills over. In the Iowa Rescue Center for primates, Rusty the orang was just hanging out by himself until this old dog showed up and wandered into the facility, and they have been inseparable since.

YouTube has a plethora of examples of mother pigs suckling dogs and mother dogs suckling pigs, of an orphaned baby rhino becoming very attached to lambs, of owls and birds playing joyfully, of a bear, a lion, and a tiger cuddling together in close harmony. It's almost as though as mammals with our large brains we don't have to be so inflexible that we have to care precisely about every bit of energy going into making sure that our genes are passed on. It is fine to play and to enjoy the company of those who will not contribute to your inclusive fitness.

Does this happen in the wild? Again, if conditions are right, if times are prosperous, and the living is easy, animals can be very social.

There are wild dolphins off the coast of Florida which swim up and down the canals. A group of them found a dog stranded in a canal that had very steep sides, and the dog could not get out. It had been there for many hours, was becoming very weak. The dolphins went into the canal. The dog climbed on them and they take him to a place where there's a lot of people. The dolphins collectively make a huge amount of noise. The people come out, dog hops off, happy ending to the story.

This behavior is interesting because it's not conducive to passing on genes. It's got to be that circuits are there for pleasure and doing good and liking others and caring for others, and that the circuitry is sloppy. It's probably served sociality quite well that it's sloppy, even though, of course, there is great variability within a species of how social individuals are.

CONVERSATIONS AT THE SYMPOSIUM

A Conversation with David Anderson

INTERVIEWER: ANN GOLDSTEIN

Senior Editor, Cell Press

David Anderson is the Seymour Benzer Professor of Biology at the California Institute of Technology and an Investigator of the Howard Hughes Medical Institute.

Ann Goldstein: More than 10 years ago you switched your research focus from trying to understand neurodevelopment and how stem cells became neurons and glia to studying neuronal behavior. What motivated the switch?

Dr. Anderson: It's a combination of boredom and looking for new sources of excitement and stimulation. I think I do my best work when a field is new and there's more room to flail around and try things before the problem gets very reduced and focused.

Ann Goldstein: What was the most surprising thing you encountered as you made the switch?

Dr. Anderson: How difficult it was and how long it took. It was like trying to change an oil tanker in midcourse. It's not like you can walk into the lab one day and just ask everybody working on development to leave and hire a new group of equally talented people in a new field where nobody knows who you are.

In some ways it was like starting all over again as an assistant professor and having to build a new reputation and a new field from scratch.

The other thing I hadn't anticipated is how long it would take to get the tools we needed to do the kind of experiments I wanted to do—which was to use molecular genetic tools to mark and map and manipulate different populations of neurons.

We worked on developing some of that technology, but it just wasn't there in the mouse, which is why I started working on flies as well, around 2002, because there the technology did exist.

Of course, with the advent of newer techniques like optogenetics in 2005 and more sensitive calcium indicators, that's totally changed. Now the field is as crowded as the stem cell field was when I left it.

Ann Goldstein: And in working with both flies and with mice, you found surprising similarities.

Dr. Anderson: We found similarities at both the circuit level and the molecular level. The fly brain is anatomically completely different from the mouse brain.

Yet—and this is one of the things I find very surprising—in the two systems, the mouse and the fly, using completely independent approaches, we have discovered a small population of neurons that controls both fighting and mating behavior.

In the case of the mouse, it's about 2000 neurons in the hypothalamus. In the case of the flies, it's about three to five neurons in the structure of the brain where the neurons are male-specific. That was a complete surprise. This is still preliminary, but in both cases it seems as if these neurons control state-dependent properties that are associated with these behaviors—levels of arousal, for example—rather than just functioning as switches that determine whether the output is mating or fighting.

That's an example of what seems to be some kind of a circuit motif even though in both systems the structures concerned have no relationship. At the molecular level, I certainly never expected to find genes that were involved in behaviors like mating and fighting and are used in the same way in these two species, but in an unbiased screen recently in flies we turned up a group of cells that release a neuropeptide called tachykinin, which is very important in controlling male-specific aggression in flies.

It turns out that homologs or orthologs of tachykinin in mammals, which are called Substance P or tachykinins, control aggression in rats, in mice, and in cats. There's even a paper from two years ago describing a study of humans by a psychiatrist which showed there was a weak correlation between levels of self-reported aggressiveness and levels of this neuropeptide measured in the cerebral spinal fluid by radioimmunoassay in patients with personality disorders. That's just a correlation, but there are a growing number of examples of neuropeptides that play a conserved role in survival behaviors in vertebrates and invertebrates, like neuropeptide Y, which controls feeding in mammals and in flies, and some evidence that oxytocin, which has been implicated in reproductive behavior in mammals, controls mating in *C. elegans* as Cori Bargmann has shown.

These behaviors, these survival behaviors, which I call the four Fs—feeding, freezing, fighting, and mating—

 doi: 10.1101/sqb.2014.79.01

seem to be controlled to a first approximation by molecules, at least neuropeptides, that have a conserved function. Why that should be, I think, is a deep question and the answer is not clear yet, but it was very surprising to me to find that.

Ann Goldstein: Is that the reason why you decided to start with studying aggression?

Dr. Anderson: I started by working on fear because I was interested in emotional behaviors. And fear—certainly learned fear—from studies of Joe LeDoux and others, was one of the best-characterized emotion circuits, certainly in mice.

The problem was it was difficult to get traction on that in flies. We've continued to work on it, but it's still a matter of debate whether the avoidance responses that flies show to something like a looming shadow involve some sort of persistent internal state, that might be likened to a fear-like state, or whether they're just reflexes, stimulus-response reflexes.

I gravitated towards aggression because at the behavioral level there's just no argument about whether flies are fighting or not. They fight and vertebrates fight, and so when we're starting with obvious conservation at the level of the behavior itself, one could use that as a way of really studying, in parallel, what looks like the same generic behavior in the two systems.

Ann Goldstein: You mentioned that the mouse field now has so many great tools that allow you to ask the kind of questions about aggression in mice. What kind of tools do you want to see more of in the future to help push the field a little further?

Dr. Anderson: Let me start with flies because I think that it's to the point now—and the fly people are going to hate me for saying this—that certain things (experiments at the neural circuit level) are currently easier to do in mouse than in the fly.

It is easier to trace the downstream outputs and the upstream inputs that connect to a given population of genetically identified neurons in mice than in flies, because in mice we have tracer viruses, which cannot be used in flies right now, and which can be genetically targeted to cell populations in mice to map upstream and downstream connections.

We have slice electrophysiological recordings where we can stimulate a particular input from a particular location and record from postsynaptic neurons to confirm synaptic connectivity, and the sign of the connection (whether it's excitatory or inhibitory). And we can record and image neuronal activity in awake, behaving animals, even in freely behaving animals. None of those things are currently possible to do in flies.

Gerry Rubin argues that we won't need viral tracers in flies because in five years we'll have a complete connectome for the fly brain. That may be true, but I do think that having a way of seeing if the neurons you manipulate functionally are actually active during a particular behavior is really important in flies.

I do think having better tracing techniques in flies would be useful. In mice we have such methods, but many of these viral tracing techniques kill the upstream and downstream neurons they infect.

What we want are viral tracers that label those cells but don't kill them so that when we find the next upstream or downstream cell we can see what happens when we stimulate or inhibit them.

Ann Goldstein: And continue down and up the circuit.

Dr. Anderson: Yes, just move down and up the circuit without having to worry about toxicity. I think that's a really important issue. We also need, and this is really a plea, better ways of automatically quantifying behavior in the mouse. Most of the current methods are based on a single mouse moving around in an arena. The kinds of social behaviors we study happen between mice in their home cage and where we're trying to distinguish mating, fighting, sniffing, scratching, and rearing, which is very difficult to do.

For humans to try to score that on a frame-by-frame basis at 30 frames per second, and do it for hundreds of hours of movies, is rate limiting. When we developed those kinds of techniques with Pietro Perona in flies in 2009, automatic behavior recognition was absolutely transformative. We've now done an unbiased screen for aggression-promoting neurons in flies from 3000 different lines together with Gerry Rubin at Janelia Farm. We could never have done that without automated behavior recognition software.

I think that's an area in which the mouse system is lagging behind. Of course the technique everybody wants, which I think is still a way off, is to do the experiment called "playing the piano." Record and play back. Everybody wants to be able to image a group of neurons in the mouse brain under a particular behavioral condition. Watch the spatiotemporal dynamics of neuronal firing and then play back the same spatiotemporal dynamics of firing using multiplexed optogenetic stimulation, with different colors and single cell activation, to see which aspects of those dynamics are really essential for evoking the behavior.

I think that's going to take a little longer. It's coming, and probably in the next three to five years you'll see examples of that. And then we will need the ability to do these kinds of experiments in more than one brain region at a time. (Relative to a human or a monkey) the mouse brain is very tiny and so to try to put even a microendoscopic lens in the hypothalamus and the amygdala, which are half a millimeter apart, is impossible right now. Further miniaturization is critical.

A Conversation with Dora Angelaki

INTERVIEWER: GARY STIX

Senior Editor, Scientific American

Dora Angelaki is the Wilhelmina Robertson Professor and Chair of the Department of Neuroscience, Baylor College of Medicine, with a joint appointment in the Departments of Electrical and Computer Engineering and Psychology, Rice University.

Gary Stix: You gave a talk here yesterday at the conference on how neurons that process sensory information can predict how an animal behaves during a simple perceptual task—I think it was a monkey moved in the world? Can you tell us briefly what the field of multisensory integration is attempting to do? Understanding how one sense works is difficult enough, so I assume that trying to integrate information from multiple senses is even more difficult. Are neuroscientists so attracted by this problem because it's a really tough challenge?

Dr. Angelaki: This is only part of the answer. In the laboratory, we like to simplify the enormous task of understanding how the brain works. Traditionally, neuroscience has studied one sensory system at a time. But in the real world, this is not what happens. As we talk now, I look at you, I see your mouth moving and I hear your voice. This experience is multisensory and our brain is optimized to process all these signals simultaneously.

Gary Stix: You don't stop and say I'm about to look at you.

Dr. Angelaki: No, it all happens together and my perception of you, of the world, of what you say, is based on many sensory signals all acting in unison simultaneously. How that happens in the brain, where it happens in the brain, what computations the brain uses—these are the challenges we seek to identify and solve. There are also different layers of understanding: different people have approached this question from the point of view of psychology and psychophysics, physiology, and/or normative theory.

How can the brain abstractly combine multiple sensory signals into a unified experience? How neurons or a network of neurons actually do this is largely unknown. We work at all levels simultaneously. We take advantage of lessons learned from theory and psychophysics, and try to apply them to neurophysiology.

Gary Stix: In your talk, you described experiments in multisensory integration that looked at the integration of visual and vestibular information as a monkey is moved. Vestibular information is information that we use to sense how our bodies are moving and how we can retain our sense of balance. Can you tell us how you went about doing that experiment?

Dr. Angelaki: When we do simple things like riding a bike or driving a car, we use many sensory signals, including input from our vestibular (also known as the balance system), somatosensory, and visual systems. As we move in the world, our retina sees the velocity of things moving relative to us, our accelerometers in the inner ear are activated, and we receive proprioceptive information from our muscles and joints. For example, when we are in a plane, we feel that it is moving even without looking outside the window.

Gary Stix: Or even just walking around.

Dr. Angelaki: Absolutely. But we don't separate the perceptions—oh, this is my vestibular sense, or this is my proprioceptive sense.

Gary Stix: We wouldn't be able to function in the environment.

Dr. Angelaki: Not without the brain being able to put all of these signals together to try to figure out how we move in the world. This is important, not just to perceive the world but to interact with it. Sometimes when we walk on the sidewalk, we turn to look at a shop window. We do this and still maintain our heading. We can talk to our friend, move our head around, and the brain manages to do all of this because of the interaction of many sensory and motor systems.

In the lab, we have set up virtual reality systems. We study human perception, as well as those of animals, macaques and rodents, to try to understand how neurons in the brain combine multiple sources of complementary, but also often redundant, information. In fact, the kind of equipment we use is very similar to what pilots use during training.

 doi: 10.1101/sqb.2014.79.02

Gary Stix: In a flight simulator?

Dr. Angelaki: It's very much like the flight simulators used for pilot training or those used in amusement parks, in which they accelerate you, as you experience visual stimulation, and you feel like you are moving long distances through space, when you are actually only moving a few centimeters. It is the combination of the stimuli sensed by the ear and body receptors, in combination with the visual stimuli sensed by the retina that gives you the illusion that this is very real. This is what we're doing in the lab. We record brain activity during virtual reality, which allows us to dissociate sensory signals, and thus to understand how we orient and navigate in the world.

Gary Stix: There was one thing that you talked about that I thought was particularly surprising. You looked at the activity of a single neuron and found it could predict what the subject might perceive.

Dr. Angelaki: It is indeed surprising. We knew from other studies that in trained animals, one can find neurons in the brain that are as sensitive as our perception. So if a single cell can do it so well, why do we have so many? In our experiment, rather than recording only from the cortex, we also recorded from the actual receptors from the inner ear, from the vestibular nerve, as well as neurons in the brainstem and cerebellum. We found that even single sensory afferent fibers are as sensitive to motion as our perception. We have 3000 of these neurons carrying information from each motion receptor in the ear to the brain. If one of them is so good, then why do we have so many?

Gary Stix: A single neuron could predict what?

Dr. Angelaki: There are two types of analysis. What I've described so far is how the neuron responses correlate with the stimulus. But, in addition, we record from neurons while the animal performs a task. Then we can also correlate the activity of those neurons with the perceptual choice that the animal makes.

Gary Stix: How the brain decodes the information?

Dr. Angelaki: That's it exactly. The first process is what we call encoding—how the neurons map the sensory world into neural activity. The second process is decoding, which is how the brain takes the activity of all the neurons and generates our perception of the world—critical to orient and be able to initiate an action. We don't know where this happens and any of the details, but that is what we are looking for.

Gary Stix: From that one neuron in the inner ear all the way into the cortex, you could predict what was going to happen?

Dr. Angelaki: Not exactly. It happens only with neurons in the central nervous system. All the neurons are nearly as sensitive as the behavior, but signals about decoding are found only centrally. What is interesting is that neurons with significant correlations with the behavioral choice are not only found in the cortex, but also lower in the brainstem and cerebellum, which are the very early sensory areas that process vestibular signals. You can listen to these neurons and predict what the animal's choice will be. This has many implications about how our brain works and why it doesn't do better. I presented a model, actually several classes of models.

Gary Stix: You're interested in theory and how it might apply to neuroscience and how it might help neuroscience understand the complexities of the brain.

Dr. Angelaki: Absolutely, because I strongly believe that without theory we can never understand how the brain works.

Gary Stix: Some neuroscientists are wedded only to experimentation but you seem to be doing both.

Dr. Angelaki: Yes. For me, it's extremely important to involve theory in all aspects of neuroscience. Not only do I think it's extremely important to use theory to understand how behavior is generated, but I also think that theory is important to connect the different levels of analysis from molecules to the network, to the brain and behavior.

Gary Stix: There is a lot of the emphasis now in neuroscience on gaining this big picture. Do you feel it's necessary to emphasize theory in the projects being set up here and in Europe to try to gain this bigger picture?

Dr. Angelaki: Absolutely. The existing approaches focus primarily on technology development so that we can see as big a piece of the brain at once as possible. That's extremely important and a very valuable goal. Some people make a parallel between neuroscience and astronomy. It was only after the discovery of technically excellent telescopes that we were able to make major advances in our exploration of the cosmos.

This is what is needed in neuroscience. But I don't think astronomy would have been as successful if there hadn't been the appropriate theories able to know what to search for and how to interpret massive amounts of data. Without knowing what to look for and what hypothesis to test, massive data can be overwhelming and not appreciated to their true potential.

Gary Stix: One other thing related to the theory. Turning off a neuron to see whether it causes a deficit in the animal is a standard technique in biology and it's heavily relied upon by neuroscience to prove cause and effect for a particular phenomenon. Yet, you said that, in some cases, that doesn't necessarily tell you that much about the function of a neuron. That seems really surprising.

Dr. Angelaki: That was indeed a provocative statement. I want to emphasize that there are many cases where this approach is extremely helpful, particularly when a causal effect on behavior can be demonstrated. However, for negative results (i.e., those where turning neurons on or off lead to no measurable effects on behavior), we cannot necessarily conclude that these neurons do not participate in the behavior. It's a negative result I emphasized because I presented an example of such a situation in which we inactivated a group of neurons and, using theory, we could predict both neural and behavioral results regardless of whether or not these neurons participated in the particular task we studied.

Gary Stix: I imagine that provoked a lot of discussion after your talk.

Dr. Angelaki: A lot, and many students came to talk to me afterwards and some of them disagreed, which is great. This is what science is about.

A Conversation with Richard Axel

INTERVIEWER: KAREN CARNIOL

Deputy Editor, Cell

Richard Axel is University Professor and Investigator at the Howard Hughes Medical Institute, College of Physicians and Surgeons of Columbia University.

Karen Carniol: You are interested in how odor is represented in the brain. One of the things you've been talking about recently is the difference between innate odors and how they're represented, and learned odors and how they're represented.

Dr. Axel: A small number of odors elicit innate, aversive, appetitive, social, and sexual behaviors, but most odors are not innate. The valence of most odors is learned. I think it's reasonable to believe that innate odors are mediated by neural circuits put in place in response to predictive cues that have existed over long periods of evolutionary time. Learned odors, on the other hand, are elicited, we believe, by ensembles of neurons that are reinforced to acquire meaning over the life of an organism.

What we've been able to demonstrate is that olfactory information, once it leaves the first relay, the olfactory bulb pentafurcates—that is, splits into five. One of the regions it projects to is the cortical amygdala, which we've shown is both necessary and sufficient to elicit innate behaviors. Another of the projections goes to the olfactory cortex, piriform cortex, and we believe that ensemble is not innately determined, and that the ensemble is required for learned olfactory behavior.

Karen Carniol: There's perhaps a clearer degree of functional plasticity in the piriform than would be seen in the amygdala?

Dr. Axel: Yes, that's a very elegant consequence of the way the piriform representation is generated. We believe that this representation is generated by virtue of receiving inputs from random combinations of glomerular low SI in this first relay of the bulb. If the presentation is indeed random, it can have no meaning. Meaning must be imposed by experience. The system is set up to require this plasticity in order to elicit a learned behavior.

Of course, the enormous benefit of this is that a representation is generated and it has no inherent bias with respect to behavioral output. It has no valence. If valence is imposed by learning, this allows the organism not only to respond to the universe of odors, but to be able to respond in a way that allows an appropriate behavioral response to an often very variable and unpredictable world.

Karen Carniol: One of the striking observations on olfactory nerve projections is the topographical organization at the first layer of the olfactory bulb. Was it surprising to you that that's lost when the piriform is reached? Or does it make sense that this organization is preserved globally even though the meaning it might have imposed at the next level is going to be lost?

Dr. Axel: Yes, being an amateur neuroscientist, I was struck by the fact that the piriform discards this order so early in the pathway. I think that inputs from random sources to neurons and sensory cortices will be seen more generally, not just in the piriform cortex. I think we'll see it in higher stations of sensory cortices. In piriform, you see it after two synapses.

But I was certainly surprised by the finding and it forced us to think about what advantages it affords the organism. In fact, if you model a random representation, as we have with Larry Abbott at Columbia, there are many advantages to random inputs that would not be observed with more structured inputs.

Karen Carniol: Have you encountered people who disagree with your interpretation of random inputs and imposition of meaning?

Dr. Axel: I've encountered more people who question our interpretation than agree with it, but gradually that's beginning to shift as more data emerge. Their argument is that in the olfactory bulb, a given odor activates a stereotype spatial pattern of neural activity, and we don't see that in piriform. We see distributive activity. All odors give a dispersed pattern of neural activity which is interdigitating different odors.

There's no discernible order, so critics argue that I do not know the true natural salient odor mixtures that are relevant to the organism. If I could identify those, I would see a maintenance of order in some form. But every effort

 doi: 10.1101/sqb.2014.79.03

we've made to discern odors or mixtures of odors that would restore order have failed. We've tried to address the criticisms.

Karen Carniol: Since the topic of this meeting is cognition, do you think innate cognition exists or is that an oxymoron?

Dr. Axel: I think that the way in which the brain thinks—that is, the material of cognition—is likely to be innate. There's something probably uniquely human about the materials we use. By materials, I mean the neural circuits that we use to think, to carry out what some people call cognition. I'm incapable of it but the rest of the world does it very well.

A Conversation with Cori Bargmann

INTERVIEWER: JAN WITKOWSKI

Executive Director of the Banbury Center
at Cold Spring Harbor Laboratory

Cori Bargmann is Torsten N. Wiesel Professor and Associate Director of the Shelby White and Leon Levy Center for Mind, Brain and Behavior at the Rockefeller University and an Investigator of the Howard Hughes Medical Institute.

Jan Witkowski: Tell us about the system you're interested in.

Dr. Bargmann: My work is with the simplest animal being discussed at this meeting, the nematode worm *Caenorhabditis elegans.* It has only 302 neurons but is nevertheless a real animal. It moves around, it decides what it likes and doesn't like, it learns from past experience and uses that to affect future decisions. And we can understand those kinds of basic processes using this very simple nervous system, and relate what's going on in an individual's brain to its behavior.

Jan Witkowski: So, what aspect of its behavior do you study?

Dr. Bargmann: Most of my lab studies the response to odors, and that's because odors are the things that worms are most interested in. They are so well specialized for detecting odors that 10% of their genome encodes G protein–coupled olfactory receptors. To give you a comparison, humans are thought to be able to smell a trillion different odors, and we do that with about three or four hundred different olfactory receptor proteins. Worms have 2000 olfactory receptor proteins, so we don't even know how many things they can sense. But in any case, this matters to the worm, and because what it smells and what it likes matters to the worm, we study those things to try to gain access to its brain. We have a first-order understanding of the worm's olfactory system. We know what the molecules are that detect odors, we know what neurons those molecules are in, we know which odors the animal finds attractive and why it finds them attractive. So we're now starting to ask the second order of questions, which is why even these simple animals don't act the same every time. When given a particular set of choices, sometimes they'll pick one and sometimes they'll pick the other one. This is true even if you have animals kept under identical circumstances, even if they're getting the identical odor. Even if you have many animals of identical genotypes, you still see this variability. So the question is, where does the brain generate such variability?

Jan Witkowski: So the worms are not robots that always respond in the same way to a given stimulus.

Dr. Bargmann: No, and one of the most striking features of behavior in any animal is that it's variable, that different animals are doing different things. Some of the time they're wandering around, some of the time they seem to be engaged in purposeful action. They're tuning in and out of different behavioral programs.

Jan Witkowski: So what's a worm's response to odor? Presumably, if it's an attractive odor it tends to follow the gradient.

Dr. Bargmann: Yes, that's exactly right. The odors we're studying are attractive odors and one of the responses we look at is whether or not the animal tries to move toward the odor. So, if you take the odor away, for example, it will change its direction. It will look for the odor that is missing, but not always. Sometimes you take the odor away and it doesn't seem to respond at all. It just keeps going on as though it didn't notice. And so, the first question we asked is, well, did it notice? Did its brain even know that the odor was taken away? And since the worm is transparent, we can use fluorescent markers of neural activity in a live animal, without perturbing it at all, to look at the activity of different neurons within the animal's brain. And the answer to the first question is, the worm knows perfectly well that the odor used to be there and now it's not there anymore, because we can see the olfactory neurons generating very strong signals when the odor appears or was removed, and we see the olfactory neurons responding whether or not the animal generates a behavioral response. So, the variability is not because of the ability to detect. The variability is really a choice inside the worm's brain about whether or not to respond to what has just been detected.

 doi: 10.1101/sqb.2014.79.04
Cold Spring Harbor Symposia on Quantitative Biology, Volume LXXIX

Jan Witkowski: With only 302 neurons, what is the decision process that gives rise to this variable response?

Dr. Bargmann: Since you only have 302 neurons, you have to be able to use the property of those neurons to explain the variability of the behavior. And so we've been working inward, on the one hand, from the olfactory neurons, to look at their activity and the activity of their targets. And on the other hand, we've been working back from the motor neurons that generate the changes in direction, the changes in behavior, and asking: Since these two can be far apart, where do we start to see those two signals separating from each other?

What our results suggest is that whether the animal responds or not depends on what is going on in its brain, its own endogenous brain activity when the signal arrives—the internal patterns of brain activity the animal moves through spontaneously, just as part of its locomotor pattern. If signals arrive at certain times, then they're detected. If signals arrive at other times, they're ignored. And we all know that when your mother's shouting at you, sometimes you respond and sometimes you don't. And essentially that is what the nervous system of the worm is doing. At certain times it's available or accessible to external stimuli and at certain other times it's driven by internal stimuli and will, for just a few seconds, not incorporate external stimuli into its decision-making.

Jan Witkowski: Can you explore what those internal stimuli are?

Dr. Bargmann: There are different forms of internal states of the brain that my lab is interested in. One kind is generated by neuromodulators, molecules that represent certain kinds of motivational and emotional states. So, for example, an animal that is hungry responds very differently to a stimulus than an animal that's well fed. Those kinds of differences are to a large degree represented in the neurochemistry of different kinds of neuromodulators operating on certain synapses to make them more or less sensitive. But in the case of this very specific decision, the level at which we're looking is faster than the neuromodulatory systems, and we think it's transient network state. The animal will respond on minute one, respond on minute two, not respond on minute three, respond again on minute four. We think that those variable responses are reflecting shifts in the activity state of the internal circuits. What people have seen for many years in more complex brains, and we are now seeing in worm brains as well, is that different neurons often have collective patterns of activity rather than completely independent activity. One of the terms used to describe this, based on work from John Hopfield and David Tank in the 1980s, is an "attractor state," when groups of neurons tend to become coactive, then maintain each other's activity. So a group of neurons will seize control of the nervous system and maintain that control through their activity. These sorts of activity patterns are what we think allow animals to be responsive or nonresponsive, when temporary patterns of activity representing groups of neurons with collective activities are sensitive or insensitive to other kinds of input. Our work points in that direction. We see that for a smallish number of neurons. Work from other groups, Alipasha Vaziri and Manuel Zimmer's labs, has allowed them to look at the activity of the entire worm brain simultaneously. And they see strong evidence for large groups of neurons that become coactive and then become inactive, perhaps remaining active or in an active state for tens of seconds before flipping down into an inactive state.

Jan Witkowski: What's the functional significance of these states?

Dr. Bargmann: I think that's a question we need to explore. But one influence on us is the idea that at any given time, you want to select one action and suppress other actions. In any given circumstance, the worst thing you can do is essentially seize up and try to do three things at once. So perhaps the nervous system commits to particular kinds of actions, suppresses alternative actions, and then over some period of time will allow new actions to take place. And this concept of action selection can be useful in a variety of different circumstances.

Jan Witkowski: So, changing the subject, you've been involved with the Brain Initiative Project.

Dr. Bargmann: A year ago President Obama announced that there was going to be a grand challenge in neuroscience to understand the brain, analogous to the space program or the war on cancer, or the human genome project. When that was announced, the National Institutes of Health director Francis Collins decided that, rather than simply jump in, he wanted to have a rigorous scientific planning process. So, he asked a group of 15 external scientists in different areas of neuroscience but also engineering, clinical science, chemistry, physics, and genetics to get together and to ask, what are the important problems that we need to study in the brain? What are the best approaches to take to solving them? How would you do that? How much would it cost? I'm co-chair of that planning committee together with Bill Newsome from Stanford University, and we have spent the past year consulting with ourselves and consulting with many, many other neuroscientists and scientists in general to try to answer those questions.

The first year the NIH committed 40 million dollars to starting this new project. This is nothing to sneeze at. The second year the NIH has said that they will spend $100 million on this project, so they like the plan that we came up with in year one enough to double their investment. And we're just about to turn in a final report after a year of planning and we'll see if the NIH continues to show enthusiasm. But I think there's been quite broad support for the idea that we're at a time in neuroscience where we can study brains at a level that we've never been able to study them before. For 50 years we've been able to look at individual neurons and their activity, and for 30 years we've been able to look at whole brains by fMRI, and by imaging. Now it may be possible to look at the networks of neurons and the communicating circuits of neurons that transmit information at high speed, so that we can look at the brain both at high resolution and look at the big picture of many neurons at once. There's a sense that this kind of circuit activity is going to be very important in understanding normal brain function, and also that it may provide a new way to think about what to do for neurological, psychiatric, and brain injury disorders. So, there seems to be broad support for the idea that this is the time to tackle this problem.

Jan Witkowski: And so is it a technical project, in the sense of looking at neurons, looking at their physical connections, at the local level and the brain level, analogous to the way the human genome project was a technical setting of a foundation for future developments in human genetics?

Dr. Bargmann: Yes, I think what inspires the idea of the Brain Initiative now is truly remarkable technical advances that have occurred over the past five to ten years. The first has been the ability to record from many neurons at once, hundreds of neurons at once instead of just one. The second has been the development of optogenetics, methods for perturbing neuron activity and not just watching the activity, but actually linking it to causes. And the third has been remarkable advances in computing technology that let you take very large data sets and make sense of them, and also give you theoretical and modeling capability. But, all of those things are still pretty far from what they would need to be to really address how a circuit works. When you're talking about the circuits that are involved in memory formation, or perception, it's not a hundred neurons you might be thinking about, it might be a million neurons. The first stages of thinking about the Brain Initiative are how to make these methods more powerful, more scalable, less expensive, disseminating them broadly, making them available to a large number of people—developing the technology that would make these large-scale views of brain activity much more powerful. For the first few years, that will be the main investment, and then over time, it will shift to using those tools to asking questions, with some of both at both times.

Jan Witkowski: How long will the project last?

Dr. Bargmann: We were initially told as a working group to think about the first year, then after we handed in the report we were told to keep thinking, and I think that there is a general sense in scientific planning that to really accomplish something in science takes more than 1 or 2 years. Real scientific progress is something that tends to take at least 5 years to get going because you need to set things up, you need to be able to take wrong turns and then correct them. And there's also a sense that if you're thinking out more than about 10 years, you're just going to be hopelessly out of date by the time it's done. And so, I think in real terms people have tended to try to think 10 years ahead, where you think in a lot of detail about what's going to happen in the next couple of years, you think in moderate detail about what might be possible in 5 years, and then you just give a very rough sense of what might be possible at the end of that.

A Conversation with Patricia Churchland

INTERVIEWER: JAN WITKOWSKI

Executive Director of the Banbury Center
at Cold Spring Harbor Laboratory

Patricia Churchland is Professor Emerita of Philosophy at the University of California, San Diego, and an adjunct Professor at the Salk Institute.

Jan Witkowski: You're giving talks here on two rather unusual topics for a science meeting, free will and morality. How did you get involved in these topics in relation to neuroscience?

Dr. Churchland: Many years ago I was a philosophy graduate student, and I was disappointed that much of the philosophy I was learning concerned how we use *words*. But I really wanted to know about the nature of *things*. In particular, I wanted to know about the nature of the mind. By 1969, it was pretty clear that all of our cognitive functions, our feelings and emotions, our perceptions, are really operations of the physical brain. It seemed to me that if I wanted to understand how we think and reason and make decisions, I really needed to know as much as I could about what neuroscience knows about the brain. I was in many ways very fortunate, because I was at the University of Manitoba, where they didn't mind what I did and gave me tremendous freedom, so I was able to teach my classes as well as learn neuroscience at the medical school.

Jan Witkowski: How was this regarded by other philosophers?

Dr. Churchland: It was regarded as sort of crazy, as a failure to understand the nature of philosophy and its view that by analyzing words we would get to deep truths about the nature of the world. That strategy just did not make sense to me. But I wanted to follow the empirical route, and it didn't really matter to me that lots of people thought what I was doing was ultimately futile. Some philosophers still regard paying attention to science irrelevant to the deepest kind of (philosophical) understanding.

Jan Witkowski: The move of psychiatry and psychology into a neuroscience-based discipline has also been very strongly resisted by some psychiatrists.

Dr. Churchland: That's a very interesting parallel. It's absolutely true that the Freudian framework really had a grip on psychiatrists until about the early '70s, when the possibilities for biological psychiatry began to be explored in a systematic way. There was tremendous tension between those who wanted to understand psychiatric conditions as disorders of the physical brain and those who thought it was best to analyze the patient's "deepest" fears and desires in order to cure them.

Jan Witkowski: This is clearly because we attribute special qualities to the brain, compared, say, with our kidneys or our lungs, and free will is an example of such a thing, something thought of as beyond the physical brain.

Dr. Churchland: Certainly Descartes thought that. In the 17th century, Descartes, to the best of my knowledge, was the first person to say that decisions of a genuinely free will had to be uncaused, that such decisions had to be just sort of created by the nonphysical mind. Other people, especially the Scottish philosopher David Hume in the 18th century, said, "Look, that doesn't really make any sense." Hume argued that all of our decisions ultimately must be caused by our motivations, our memories, our drives, our emotions. They must all be caused rather than created without causation. Accordingly Hume thought what we really should be interested in, especially from the point of view of the law, is the difference between those causes that allow for what we call voluntary behavior and those that don't (e.g., the difference between a startle response and turning right at the stop sign).

The way neuroscience has developed exactly follows Hume's point. What we're really interested in are the kinds of habits, the kind of decision-making that operate in a healthy brain, versus those that operate in, for example, an addicted brain or someone who has obsessive compulsive disorder, someone who has lesions to the front of their brain and can't evaluate consequences, and so forth. For those kinds of cases, we realize that the mechanisms for controlled choice and evaluation are not working well.

Jan Witkowski: You would argue then that that demonstrates that there is a biological free will.

Dr. Churchland: Yes, but not the kind that Descartes imagined. It's easiest to consider in the context of self-control. As children are raised, we precisely exploit the fact that the brain is a causal device to enable the child to have self-control. Children can acquire habits that allow them to suppress unwanted behavior, and such acquisition is a causal brain phenomenon. The first thing, of course, they learn is bowel and bladder control. Or more correctly, the very first thing, probably, is that a baby learns not to bite its mother's breast. That's because the mother yelps and says, "Don't you do that again." By interacting with the reward system of the infant's brain, by using approval—which is rewarding—and disapproval—which is not—we encourage good habits, attitudes, norms, and so forth in the developing child. We help them acquire self-discipline and self-control.

Jan Witkowski: To what extent do you have to be conscious of the decision to exercise free will? Haven't there been studies that show that we can make a decision before we're consciously aware of having done so?

Dr. Churchland: I think that consciousness plays some role in voluntary actions, though it's not entirely clear what that role is. One domain where nonconscious habits and nonconscious rationality play a huge role is speech. I can't possibly make a conscious decision in advance of everything I say, because if I do, I'm very quickly tongue-tied. Yet what I am doing—for example, right now in talking to you—is as voluntary as anything I do. Notice that my speech is not impulsive. It is appropriate. It's based on past reflection and knowledge, but I do not make a conscious decision about every word choice and every thought that gets expressed. Of course I have to be awake and paying attention. When do I know exactly what I am going to say? At the very time I am saying it. Remarkably, linguistic expression emerges in this very unconscious, yet very controlled, way. Interestingly, even though the background processing is unconscious, we do hold people responsible for their speech. If in the course of a speech a politician says something that is extremely inappropriate, we hold him responsible. He doesn't get to say, "Oh, I said those terrible things, but . . . "

Jan Witkowski: " . . . I didn't make the decision to do so."

Dr. Churchland: " . . . I didn't make a conscious decision to do that." We don't care if somebody defends his dreadful remarks that way. We say, "You're still responsible." I think that not only speech, but a lot of our behavior involves this interesting mix between intelligent nonconscious processing and taking in consciously the conditions and the stimuli of the present occasion.

Jan Witkowski: Clearly, driving a car is very similar. You're obviously continuously aware of your environment and you're making decisions about what to do without being aware at all of that process.

Dr. Churchland: I think this is true also in very fast sports. Golf, which I play, is very different. We're very deliberate about where we want the ball to go and which club we should use. But if you're playing hockey—and hockey is a spectacularly fast sport—a player may make a decision to shoot rather than pass. Notice that his split-second decision may be unconsciously produced but yet the behavior is no mere reflex. It's based on a veteran player's many hours of skill and knowledge and understanding. It depends on very fast nonconscious processing about where the opposing defenseman is, what the goalie is doing, so forth.

Jan Witkowski: When you talk to people about free will, about this neuroscience biological approach, you must get a lot of kickback from people who just have to believe they have complete free will.

Dr. Churchland: Funnily enough, not much. I often take our dogs to the beach in California and for a while I used to just chat with people and say, "What do *you* mean by free will?" They usually mean something that's very similar to what the courts mean. It means that you weren't forced or coerced to do the action, it means that you weren't involuntarily intoxicated, it means that you weren't insane. But then, as long as you were a normal human being in command of your functions, you have the choice. But they don't mean anything metaphysical and mystical. That's left for philosophers. There are some Cartesian philosophers who still want to think that every time we make a voluntary decision, it's created with no causal antecedents.

Jan Witkowski: That leads on to the next thing I was going to talk about, which is moral values. Don't religious people argue that you have to have free will in order to make moral judgments, that without it we're mere automatons and robots?

Dr. Churchland: Yes, and up to a point I sort of agree, in the sense that you have to have a degree of self-control. If you are so frontally damaged that you act on any impulse that your brain generates, then we don't think of you as having free will. There are some human subjects, perhaps damaged by stroke or by gunshot or sometimes through developmental deficits, who are unable to make appropriate decisions. Whatever impulse arises, they act on it. But normal people with normal brains, people who have been reared to have self-discipline, do have a capacity not to act on every impulse but to evaluate certain actions as unacceptable. They recognize that there can be socially unacceptable impulses or impulses that are personally unacceptable—for example, impulses to try heroin or to start smoking tobacco. You think, "I know that would be a counterproductive thing to do," and you suppress that impulse. Such suppression happens every day to each of us, and I suspect when religious people talk about free will, it's actually that kind of thing that they have in mind. Consequently I don't see a huge gap between my conception of self-control and the idea of free will that some religions espouse. Where there may be a disagreement concerns the idea that the capacity for self-control has a divine source. I think of it as having a brain source.

Jan Witkowski: You used the word *social* a moment ago. One of your arguments is that the biological neuroscience of the way that we are and our sociability is the root of our development of moral values.

Dr. Churchland: Yes, that is essentially right. It's sometimes very tempting, when you think about evolutionary biology and you understand that organisms compete to survive, and you think, "Well, that must mean organisms are one and all essentially selfish." But as Darwin actually recognized, there comes a point where sociality actually helps you compete. The very special case in which it helps you compete is in the case of mammals where there must be caring of parents for dependent offspring, and that's of course certainly true in the case of large-brained mammals, primates, including humans. There seems to be a basic platform of caring for others that's rooted in maternal care, but that can be expanded to care for mates, and probably via genetic changes, to care for kin, to care for friends, and, as long as conditions permit, to care for strangers, other species, and so forth.

It's a basic platform of going beyond caring only for oneself, to deriving pleasure from caring for others and feeling pain when isolated or ostracized. The reason it's there, of course, is that sociality allows us to have communal defense, to have cooperation, to share in foraging strategy, to share in the rearing of children, and so forth. Sociality has been invented many times in evolutionary biology, in insects as well as in mammals and birds.

Jan Witkowski: Does that mean that all societies should have the same moral values or that they have moral values of some sort that specifically help their society?

Dr. Churchland: The way I think of it is that there's a basic motivational platform that manifests itself in our liking to be together, tolerating those we find mildly annoying, and in general enjoying the benefits of cooperation and group defense. Exactly how this social motivation allows for the emergence of different cultural norms and conventions is not well understood. But ecological conditions undoubtedly play some role, as do the specific accidents of history. We find, for example, that the Inuit may value certain things more than, say, certain tribes who live in Brazil, and that may reflect the fact that the tundra is so different from the jungle. But fundamentally, the mammalian capacity for caring for offspring, for mates, for kin, is very deep in our biology. It gives us empathy, generosity, kindness, friendship, and possibly also war. It's part of what mammals are, and it's part of what makes us humans very successful.

Jan Witkowski: This talk seems to be more sociobiology than philosophy. Do you consider yourself a philosopher?

Dr. Churchland: I consider myself, in a way, a very old-fashioned philosopher. What I mean is that throughout history, there have been philosophers such as Aristotle, David Hume, and Adam Smith, who were deeply interested in all of these questions. What is it that distinguishes somebody who makes good choices from somebody who doesn't? In Aristotle's discussion of politics, he emphasizes that we humans are essentially social in our nature. He meant that we typically live in groups—we cooperate, discuss, and thrive in the company of others. He thought much of morality—that is, the social aspect as opposed to the personal aspect of morality—consisted of trying to figure out how best to organize group living, how best to organize ourselves to maximize stability and prosperity. I want to understand where human moral values come from, and how they come to change over time.

I think there are two parts to the story. There's the part of the story that's about the biology of social motivation, and then there's the part of the story that asks how social motivation can lead to our conforming to the norms of our social group. Eventually we want to understand what reasoning really is in neurobiological terms. For example, what are the processes whereby humans can come to realize that a norm, such as trial by ordeal, is causing more trouble than it solves? How can humans figure out that rules of evidence and trial by jury should replace the prevailing system? The kinds of social problems a group faces can change as climate, ecology, and technology change. It may be a question for us, though not for our stone age ancestors, whether we are better off adopting a flat income tax scheme as opposed to a graduated one, or whether organ donation at death should be the default. These are all questions that would have interested Aristotle.

A Conversation with Karl Deisseroth

INTERVIEWER: JAN WITKOWSKI

*Executive Director of the Banbury Center
at Cold Spring Harbor Laboratory*

Karl Deisseroth is the D.H. Chen Professor of Bioengineering and of Psychiatry and Behavioral Sciences at Stanford University.

Jan Witkowski: You are speaking at this meeting about two topics, optogenetics and CLARITY.

Dr. Deisseroth: We've been developing optical technologies for exchanging information with the brain of experimental animals that are carrying out complex behaviors. Optogenetics is one of those. It's a way of using light to control well-defined sets of cells within neurocircuitry during behaviors, to turn them on, turn them off, and assess the causal impact on behavior. Social behavior is in many ways a high-level cognitive task, and it's difficult. It also carries risks. An organism has to be motivated to carry out a social interaction. We have some results on the optogenetic control of circuits in mice that turn up or down the tendency to engage in social interaction.

Jan Witkowski: What sort of social interactions are you looking at?

Dr. Deisseroth: A lot of great work in the past has been done on courtship but we're taking a slightly different tack. Same-sex mice of the same species have a natural willingness or desire to simply be in the same location as each other. If you give animals a choice of being alone or with a novel object of some interest or with another mouse of the same species and gender, they'll prefer to be with the mouse of the same species. There'll be active engagement and approach. It's an interesting built-in process. We're interested in what drives that, what turns that up, what turns that down. We quantify the amount of time the animals choose to spend engaged in this behavior.

Jan Witkowski: What's known about the neural circuits, neurons, or brain areas involved in this behavior?

Dr. Deisseroth: There are different aspects of social behavior. It is a challenge and it also requires motivation. These two aspects map on to certain neural circuits. The prefrontal cortex, which integrates much information and helps generate plans and move towards goals, is certainly involved. But deeper subcortical structures that involve recruitment of reward and motivation pathways are also involved.

Jan Witkowski: Do you think that by using optogenetics and manipulating particular neurons, you'll be able to drive the mice to particular types of behavior?

Dr. Deisseroth: That is a deep question. We're playing with patterns of activity. We're turning up or down certain kinds of behavior, so yes, the short answer is that mice can be modulated in the kinds of behavioral choices they carry out. But we don't want to just play with activity that we think is relevant, we'd also like to see what the animal normally uses. We have a new technique called fiber photometry which lets us collect information on what circuitry the animals normally use in the course of its social decisions and social choices. That's very powerful because then we can see what's normally present and try to mimic it in our experimental interventions.

Fiber photometry is a trick that uses an optical chopper to break the incoming signal at a very specific frequency. We have a detector to pick up signals coming out of the animal that's also tuned to detect signals at just that frequency. The advantage of this is that it lets us detect very small signals that otherwise would be buried in the noise, because only the signal we're interested in modulated at this frequency. So even in a mouse that's freely behaving in a well-lit room, we can pick up activity signals with fluorescent indicators of neural activity that are buried deep within the brain, and some of these neural circuits are involved in social behaviors.

Jan Witkowski: These are free roaming mice with implants to detect these signals?

Dr. Deisseroth: There's a single thin fiber optic that is able to implement either the optogenetic control or the fiber photometry detection of the signal. This is the key advantage: In both cases, it's the same system, same surgery, same location, same site, one thin optic fiber that both delivers the control and can pick up the activity signal. Then in the same animal, we can get both data streams.

Jan Witkowski: Do they use the same fluorescent compounds?

 doi: 10.1101/sqb.2014.79.07

Dr. Deisseroth: For the control, we use proteins encoded by microbial opsin genes that come from algae, from bacteria. They respond to light and open ion channels in the membrane. They themselves don't generate an optical signal but they use light to move ions across the membranes of cells that we're controlling. We pick up the signals and detect the native activity patterns using a different protein. These are genetically encoded reporters of ion flow. Two parallel sets of proteins that help us do the job.

Jan Witkowski: These are transgenic mice. How have you introduced the genes?

Dr. Deisseroth: In both cases, we've injected a virus that carries the gene encoding for these proteins. We take mice that could be transgenic but also can be wild-type mice that have no label.

Jan Witkowski: Eric Nestler has mice that are susceptible to depression. You can see that your system could be used to interrogate what's actually happening in the brains of these mice.

Dr. Deisseroth: Yes, Eric is quite interested in applying the fiber work to his model.

Jan Witkowski: What's the current state of CLARITY at the moment?

Dr. Deisseroth: With CLARITY, we are able to use a chemical engineering trick to make the brain transparent. It looks like it's gone but it's not, just all the light scattering has gone so you see through the brain. We do this by removing the lipids and removing the fats but before we do that, we make the brain robust by building in a hydrogel which keeps all the proteins and cells in place.

We published that about a year ago and the first papers using it are starting to come out. A nice study in spinal cord has just come out from another group. One thing we've been working on is trying to make it faster and easier to use, making the whole process and the data acquisition faster by orders of magnitude, combining CLARITY with light-sheet microscopy. We call this CLARITY-optimized light-sheet microscopy. It lets you collect volumes a 100 or a 1000 times faster than before.

Jan Witkowski: How long does it take to prepare a brain as you were doing 6 months or 9 months ago?

Dr. Deisseroth: For a mouse brain, the actual preparation process takes between a couple of days to a couple of weeks depending on how much you want to stain or label for. The whole process of making the brain transparent and creating a signal is a couple of weeks. What we really sped up is the imaging process. Whole-brain imaging now can be done in tens of minutes or hours, instead of days.

Jan Witkowski: This is imaging with what technique?

Dr. Deisseroth: It's a 100-year old concept called light sheet. With light sheet, you scan a plane of light through your tissue, instead of scanning a point as in confocal and two photon microscopy. These approaches give very nice high resolution but they are quite slow, and also have the disadvantage that you're imaging at a point but illuminating your whole sample which creates bleaching problems and degradation of your signal. With light sheet, you're only illuminating where you're imaging, so it's very efficient and fast. You're not destroying your signal in other parts of the tissue. It required some adaptation to make it well suited for these sorts of intact tissue preparations of the CLARITY-treated brain, or clarified brain as we say.

Jan Witkowski: What's the thickness of these sheets?

Dr. Deisseroth: Traditional resolution in that axis has been considered to be limited to the 10- to 20-micron range. That has to do with the thickness of the light sheet. This is something that can be improved upon with certain kinds of innovations. We can get now, even deep within the tissue, micron-level resolution. We can see fine processes, axons, and dendrites even deep in the middle of the mouse brain, so that's an exciting advance. Not only do we have the speed but we can maintain this high resolution.

Jan Witkowski: What one would like to do is get an insight into dynamics. Do you have a mouse that's been trained to do something or respond in some particular way whose brain you can compare with the brain of a control mouse? What sort of questions can you ask using this technique?

Dr. Deisseroth: What we've been working on is making sure you can do all of this in the same mouse. Not at the same time, but the same mouse at different times. This integration of data streams is a very powerful thing. You can now have a mouse that you've an optogenetic control experiment on. You've turned on or off different kinds of cells. You know that animal's behavior score. In that same animal you might even have done a fiber photometry experiment, or other kinds of imaging to see the native natural patterns of activity in that same mouse, during that same behavior. You can now take that same brain and put it through the CLARITY process. You can register or align the datasets you get at very different times and under different conditions. You can register the datasets and identify regions and cells during behavior. Here is their phenotype, here's their wiring. The goal is to move towards a more integrative understanding of the neural circuitry.

Jan Witkowski: How generally applicable is the clarification technique? Have people tried it with other tissues?

Dr. Deisseroth: There's been an explosion of interest. People have sent us pictures of clarified hearts, pancreata, even whole embryos. There are subtle tweaks to each preparation that are needed. For example, with the mouse brain, to speed up the processes, traditionally we've accelerated the clearance of the tissue by imposing an electric field. You can accelerate the clearance of these lipid-containing micelles out of the tissue. What we're now finding is that for the brain you usually don't have to do this. You can do what we call passive clearing and it's fast enough and good enough. You don't even need the electrophoretic process, which has been a boon for a lot of people who didn't want to get the whole electrophoretic process going. But people are finding for other organs, like the heart, that the electrophoretic method is very important. You can't dispense with that. That has to do with tissue properties and there are different issues for every tissue.

The hard part about working with clarified tissue is representing it in the published page, because there's not really much to see. How do you show a picture of that? We've been trying different strategies, putting the brain over text, over grids that you can read through and so on. In the end people mostly care about the data and that's what we focus on presenting to them.

A Conversation with Robert Desimone

INTERVIEWER: ANN GOLDSTEIN

Scientific Editor, Neuron

Robert Desimone is Director of the McGovern Institute and the Doris and Don Berkey Professor in the Department of Brain and Cognitive Sciences.

Ann Goldstein: As one of the few that study both humans and nonhuman primates, how has your lab's use of both systems helped inform your work in ways that were unexpected or inspiring?

Dr. Desimone: We always start with human behavior. Something in human behavior that we want to try to explain is related to attention these days. Often the human behavioral work has been done—we've done a little bit ourselves, but there's a huge field of human behavioral work and attention. Then we look at how human neuroscience studies have informed us. Sometimes there are open questions in which case we jump in ourselves. In other situations, there's sufficient information that we can go to an animal study because there the goal is to try to link what we see in humans to the underlying neuronal circuitry, which you can't do in human beings. We're getting much better at human techniques like magnetoencephalography (MEG) combined with fMRI. We don't do electrocorticography recordings, but a lot of labs are starting to do that and I think that's super exciting, but you can't get to that level of neural circuits in humans, so we do the monkey studies to provide that link.

Ann Goldstein: How did your combination of fMRI and MEG change the way you approach your work?

Dr. Desimone: When I was at the NIH, over 9 years ago, I pushed for an MEG machine and did some pilot studies. At MIT, we got our machine a little over a year ago, and my lab was one of the first users. One of my postdocs, Daniel Baldauf, who had not worked with MEG before, had to learn not only all that is already known about analyzing MEG data, but had to invent a lot of his own techniques because that's an area that's much less developed than techniques for analyzing fMRI data. He did a huge amount of developmental work and then we started the studies on attention. MEG gives us the exact timing of activity and that's the reason for doing animal work because you need that exact timing.

The fMRI gives you the spatial localization, and the fMRI is usually limited to a second or two. Just think about all the operations your brain can carry out in a second, all the decision-making, reaction, switching, and so on, but MEG is working at the millisecond level, the sort of level at which our brains actually operate.

Ann Goldstein: What do you see as the future for understanding attention and its dysfunction?

Dr. Desimone: One of the things we identified in our latest human study is the important role the prefrontal cortex is playing in the regulation of sensory processing with object feature based attention. The prefrontal cortex always comes at the top of everyone's list for what might play a role in a lot of human psychiatric and neurological disorders. It's the most developed part of the human brain, it's large compared to the prefrontal cortex in the monkey, but compared to the mouse, which many people are using, it's many orders of magnitude larger. The mouse has a tiny prefrontal cortex. What makes us human is really that massive prefrontal cortex. It's clearly very much involved in attention and attention is very much involved in all these psychiatric disorders.

People who are depressed complain about lack of attention, focus, concentration. So do people with lots of learning disorders, most obviously attention deficit disorder. Focusing, keeping on track, blocking out distraction—we all have a problem with these functions, but people with psychiatric disorders often have an even worse problem. It's focused our attention on this part of the brain, and we'd like to be doing studies on that.

 doi: 10.1101/sqb.2014.79.08

Ann Goldstein: Since you have an interest in the effects of overabundant information, how do we deal with this very fast information age we live in now with lots of mobile technology, video games, and so on?

Dr. Desimone: My advice will sound totally trivial, but I recommend trying not to use your brain to block out all that distraction. Instead try to eliminate it from the environment. You will be infinitely better off. Remove the TV or radio in the background and all those other sorts of things, because filtering that out takes a lot of your brain's energy and a lot of your cognitive focus. You'd be better off having that cognitive capacity devoted to whatever you're trying to do, the problem at hand.

A Conversation with James DiCarlo

INTERVIEWER: GARY STIX

Senior Editor, Scientific American

James DiCarlo is Head of the Department of Brain and Cognitive Sciences at the Massachusetts Institute of Technology McGovern Institute for Brain Research.

Gary Stix: Your research is focused on understanding the neuronal representations and computational algorithms that underlie visual object recognition in primates. You've asked some very basic questions recently, such as what is object recognition and why is it challenging?

Dr. DiCarlo: We all have an intuitive feel for what object recognition is. It's the ability to discriminate your face from other faces, this car from other cars, a dog from a camel. But making progress in understanding how our brains are able to accomplish this task is a very challenging problem.

Part of the reason it's challenging is defining what it is; and it's a challenging problem because we're using our brain machinery to solve the problem. It seems effortless to us. When I try to explain, for instance, to my mom, "Oh, I'm working on object recognition," she will reply "Well, I just see and I solve the problem," but of course, she is using the machinery of her brain that we are trying to reverse engineer and to decipher.

What makes the problem challenging is that each object presents an essentially infinite number of images to your retina, so you essentially never see the same object, the exact same image twice. The ability of the brain to deal with all those different images and still know that they're coming from the same object in the environment is a challenge our brain solved, we think, through evolution. But machines are still struggling to understand how to do it.

Gary Stix: This is one of the really big problems both in neuroscience and in machine learning.

Dr. DiCarlo: It's a really great problem because it's at the intersection of neuroscience and machine learning, and of psychology or cognitive science, because objects in the world are what we use to build higher cognition, things like memory or making decisions. Should I reach for this? Should I avoid it? Our brains can't do higher cognition without these foundational elements that we often take for granted unless you're a computer scientist and realize how challenging that problem is.

Gary Stix: Your research is focused on finding some of those foundational elements. There are areas of the brain, such as the inferior temporal cortex, where there might be some clues.

Dr. DiCarlo: It's been known for several decades that there's a portion of the brain, the temporal lobe, that when lost or damaged in humans and nonhuman primates leads to deficits of recognition. So we had clues that that's where the algorithms live. But just saying that that part of your brain solves the problem is not much more specific than saying your brain solves the problem. It's still a very large piece of tissue. But anatomy's told us more about the various visual areas that exist in the brain, a whole network of areas that exist there.

The tools of neurophysiology, and now more advanced tools, allow us to look more closely at neural activity, especially in nonhuman primates. Then we can begin to decipher the actual computations to the level, for example, that an engineer might need to be able to emulate what's going on in our head, or we might need if we wanted to replace those circuits or augment those circuits—to really understand the function at a detailed level, not in just the gross "this part of the brain" sense.

Gary Stix: How are you trying to drill down and find one of these foundational elements?

Dr. DiCarlo: The foundation of any science is the ability to have predictive models of a phenomenon. So for object recognition, as an engineer, if you want to emulate that, you first need to define what are we trying to predict. What would success look like? There are various levels that we might set there, but we've set one, a goal for ourselves, the ability to emulate what we do in core-object recognition.

Put simply, what that means is that I'm going to show you an image for about 200 milliseconds, which is about the time that your eyes dwell as they explore a scene. It's a time that's driven from the biology of how you explore the world with your eyes. We can do a lot with that short time window. We can easily recognize one or more

 doi: 10.1101/sqb.2014.79.09

objects within that one-fifth of a second glimpse. It's not all of vision, but it's a defined space where we can start to get some traction on the problem.

Gary Stix: So you've got a predictive model, and then you want to test that model.

Dr. DiCarlo: In core recognition, images come in and are processed by the eyes, and then through a series of visual areas, processed somehow in ways that are murky, but we can record the neural activity along the processing pathway. Others have done that before us, and now we're doing it at a much larger scale. We can record neural activity, and we're especially interested in a place in the brain called the inferotemporal (IT) cortex, at the end of the temporal lobe, the highest level of this processing chain.

We've found that the population patterns of neural activity there, the firing patterns of all the neurons in that part of the brain, with a very simple model, an algorithm if you will, can predict very accurately the animal's perception and our own perception—our ability to do recognition in that core domain. So we have a model from neural activity in IT cortex to behavioral report.

Gary Stix: So you could predict, say, that a person or a monkey is looking at a tree in the background from examining that neural activity?

Dr. DiCarlo: That's exactly what I mean. The granularity with which we can do that is still being studied but I can tell if you're looking at a tree versus a dog, or I can tell you you're looking at a tree versus a car. I can tell if you're looking at one tree versus another.

We're now trying to see if we can do this on a trial-by-trial basis rather than on an average basis. Already the models are quite good for the tasks that we've tested. Our next step is to look even harder and build better models.

Gary Stix: The challenge with object recognition is that if I'm looking at a tree and then move slightly to the left or the right, the tree changes or I start to see another tree. Will your model still recognize that that's a tree or that it's the same tree?

Dr. DiCarlo: That's the largest thing the model has to deal with. When I say the model deals with that, the neurons up to the IT cortex when we record have dealt with that, so that part of the problem from an engineering perspective is solved.

So once I build a decoder on the IT cortex, a reading of the IT activity, if you will, then a new image of the tree will be properly decoded as a tree. It's a brand-new image, but the model will still make a prediction of what you will say, and the model will be quite accurate.

Gary Stix: What are some of the implications of this for machine learning and perhaps, one day, even for understanding problems that people have with disruptions to neural circuitry?

Dr. DiCarlo: From the machine-learning point of view, this neural activity in IT is something that machine-learning folks would call features. Those are features computed on the image. They're a very powerful set of features for the reason I described. What many people would love to do is to be able to have algorithms that produce those features. Much of machine learning is devoted to finding good features.

The brain, through evolution, has already found some good features, and that's essentially what we're reporting. Here are some nice features. Here's where they are. Here's our evidence that they're nice features. Now the machine-learning community is working, and we're working alongside of them, to help build what are called encoding algorithms that produce those features.

There's a lot of exciting progress in the field, even in the last few years, driven by what are essentially brain-inspired models that are actually now some of the state-of-the-art computer vision algorithms. They were inspired by work of the type done by those who went before us, and I hope from some of our work, to build those kind of models. They turned out to be very powerful in the computer-vision and machine-learning community. Those are the most active and exciting models in computer vision.

Gary Stix: The grand vision of what you're doing is the ability to model this all the way from encoding to neural activation and then to the decoding and perception in the brain?

Dr. DiCarlo: That's exactly the grand vision. If we can do all of that, then we would have a complete end-to-end understanding of this domain of behavior.

Gary Stix: How long until you get there?

Dr. DiCarlo: Within basic-level core recognition, it depends on your level of detail, but in the next 10 years we will have a very good understanding of core basic-level object recognition to the degree that many engineers will be satisfied. We won't know it down to the synapse, but we will know it in the way that the algorithms are very predictive of the neural activity at various levels of the system.

Gary Stix: Do you think that this could provide some insight into what goes wrong when the circuitry in the visual cortex or the temporal cortex goes awry?

Dr. DiCarlo: The most common deficit that affects recognition is major damage to that part of the brain through stroke or lesion. You've taken out those neurons. Maybe our studies could lead to ways that you could sidestep the damages or replace the function.

There are other deficits to the temporal lobe where people have things like deficits in the ability to discriminate among faces or other types of objects. They're not very common, but this kind of work should bear on those deficits, as well. We hope it will also bear on things like how kids learn to read.

At the end of the day, whenever you're doing visual tasks, you're leaning on these kind of representations to do much of your vision. I think this work will help us understand these higher-level issues of, say, social cognition or things like dyslexia. Things that depend on these foundational representations will be better understood because of our understanding of these neural circuits.

Gary Stix: Despite what people see in movies, robots in the real world are still very limited in what they can do. One of their big limitations is the ability to recognize and process information they perceive. Do you think this work could help with that?

Dr. DiCarlo: Certainly. The computer-vision community is already using brain-like algorithms. The next frontier though is expanding the domain of task, not just what you can do in 200 milliseconds, but as you explore a scene with many eye movements or navigate a scene, accumulating information over time. There will be more feedback in the system.

Those are things that we've not yet begun to touch very much, but that's the next frontier as we understand this more core domain, this foundational domain. I won't say that if we do this work we will have robots doing everything you see Data on *Star Trek* doing, but it will be a foundation to enable those next steps.

Gary Stix: Is a larger conceptual framework necessary to create kind of a movie of this whole process, to go beyond 200 milliseconds and create something that can process the full variety of our visual environment?

Dr. DiCarlo: I think what you're asking is what goal we are seeking. Sometimes to know that we've met it, we have to just emulate the system. For instance, if we could build a robot that behaved as well as us and fooled us like Data on *Star Trek*, I think we would all declare that as evidence of success.

This will certainly require, at some point, really embodied systems to test. Vision is not an isolated sense. It has to interact with the motor system, so as these systems get better, you're going to see them on moving things like robots and other autonomous devices. This is already happening. Then the work of my lab and others will have to move beyond vision, and integrate with motor action and other senses.

Gary Stix: The complexity of that seems overwhelming. Don't you think it's going to take many generations to get to these ultimate goals?

Dr. DiCarlo: In that goal, probably. Again, it's hard to make predictions with confidence. In this ventral visual stream that ends in IT cortex is a series of cortical areas where each local bit of cortex looks anatomically very, very similar. One of the exciting ideas that's been in our field for a long time is that there's some interesting local learning algorithm that's of smaller scale in both size and computational power than the global algorithm I've been talking about.

One of the things many of us hope for is that if we could get some insight from this kind of work into that more local algorithm, that could generalize more broadly to things that we're not even thinking about, that might apply to audition or other senses, maybe to some of these higher-level cognitive things, but that's hard to see right now. So there's a hope for a shortcut, if you will, that arises from working on a clearly defined problem.

Gary Stix: You mentioned in a recent paper that there are 40,000 neuron building blocks. Is that what you mean?

Dr. DiCarlo: That's exactly what I'm referring to: the idea that you might have sub-modules of something of that scale, say, a millimeter of cortical tissue. Again, this is an idea that's been around for a while because each piece of cortex looks very similar. What is it doing? What's that fundamental operation?

We are working on it from the problem of vision, but we hope we will get some understanding of that because we set a clear goal of what success looks like. We're building models to emulate what the neurons do at all levels, and so we might start to glean some insight that could generalize outside of vision.

Gary Stix: So these mini modules might be present in other parts of the cortex that are processing incoming sensory information?

Dr. DiCarlo: That's right. They're known to be present. This is a hypothesis that's been in the field for a long time. Like the dream of discovering the DNA code, discovering the cortical processing algorithm, is a long-standing dream.

It will be a big jump and certainly not the whole story. Bits of cortex have to talk over long distances, so everything's not going to be defined as a local small module, but we might get a very big boost from understanding that kind of module.

Gary Stix: When you say an algorithm, is that made up of a kind of code that would be present, say, in the inferior temporal area?

Dr. DiCarlo: People use words like algorithm and models in different ways. If we talk about the algorithm of the ventral stream, that is maybe executed by many subalgorithms, that consist of these building blocks, but when put together, give rise to the global algorithm of the ventral stream. The hope is that the global algorithm of the ventral stream built by subalgorithms that are simpler and more primitive and more understandable and that might generalize well. That subalgorithm in a piece of cortex might take some set of inputs, some learning, local processing, and give some set of outputs. It's essentially a little box that does something clever, and when stacked horizontally across the visual field and then stacked vertically into a deep stack, visual area V1, V2, a series of areas along the temporal lobe, might give rise to the global temporal-lobe or ventral-stream algorithm that I was referring to earlier.

Algorithm is a word computer scientists like, but as neuroscientists, what we're imagining is not just a set of lines or codes that says, "If this, do that," in a way that those of us who write code are used to. It's also going to be a set of neurons that execute that function.

We're building algorithms that are actually already constrained somewhat by the biology. We're not just writing code and hoping that it predicts the responses. We're building systems that are modeled on the biology so that they will translate both to give constraint and to what this neuron does within the algorithm. That's a different level of how deeply you want to look at the system, how you want to describe it.

What's exciting about neuroscience and also challenging is that you have to work at all these levels—what's the goal of the system, how do we describe the algorithm, what are the mechanisms of the neurons and the synapses? We'd like to say we understand all those levels of the system.

Gary Stix: There has been work in the last few years on retinal prosthetics. One approach has been to implant codes into prosthetics so that they can process incoming photons in the same way the retina does. Does the kind of work you're doing connect in some ways with this?

Dr. DiCarlo: This is one of the things we're most excited about right now. There are visual prosthetics for people who, say, have lost a retina. The dominant approach is to try to bypass the retina and reinject a spatial pattern of activity, say, in an early visual area or the subcortical area that comes right behind the retina, called the LGN.

That makes sense from an engineering point of view. The downside is that if you want to get an image in, it's a very high-dimensional space, with many, many pixels that you'd want to play to make it feel like your normal vision would feel like.

We're working at the highest level where your brain has already reduced the dimensionality from millions of pixels to something that's more abstract. If we could have a hundred ways to push the neurons around, we don't know what it would feel like yet, but it might be very functional. It could be a better way to think about brain–machine interfaces, that you only need to have a hundred ways to inject signal, a hundred channels rather than millions, to make a rich perceptual space.

We're now testing what moving neurons around in a monkey does to that monkey's percept right now. This is a long way from a human, but as we start to understand what's happening in an animal, our work might provide a shortcut or a better prosthetic.

Gary Stix: To sum up, what your work is doing is taking something very basic that all of us can relate to and try to find the physical basis in our brains of how we see, how we recognize an object, and come to a fundamental physical and theoretical understanding of that really huge challenge.

Dr. DiCarlo: That's really been the goal of neuroscience since its formation. We believe the brain is a set of mechanisms that give rise to this amazing phenomenology that we feel. We are studying one example of that phenomenology but it's one that many of us can relate to and it would be a foundational success if we can create that end-to-end understanding—a large brick in the foundation of building towards understanding cognition.

A Conversation with Adam Gazzaley

INTERVIEWER: CHARVY NARAIN

Former Senior Editor, Nature Neuroscience

Adam Gazzaley is the founding director of the Neuroscience Imaging Center at the University of California at San Francisco and a Professor in Neurology, Physiology, and Psychiatry.

Charvy Narain: You're studying how to improve the brain.

Dr. Gazzaley: Over the years we've done a lot of basic cognitive neuroscience research trying to understand how our brains function, how they are sensitive to interference, and how that gets worse as we get older. And then around 5 years ago, I started getting frustrated by having only shared the bad news that our brain function decays, without having anything we can do about it. We became driven to see if we could use the tools and the methodology that we have developed to understand the brain, and as a means of improving how the brain processes information.

Charvy Narain: What methods may people already have tried to ameliorate, say, the deficits associated with aging, or even just to improve normal functioning?

Dr. Gazzaley: There's a pretty broad landscape of approaches. The classic is education. That should be discussed because that's the mainstay of boosting how a brain operates. But other approaches have garnered some interest, such as pharmaceuticals. So drugs that are targeted at deficits associated with, let's say, Alzheimer's disease and work at the neurotransmitter system level have at least been considered as approaches to improve cognition. Then there's the general field of cognitive training. What we've been doing at our lab is related to that, perhaps with a little different spin on it.

Charvy Narain: Why should we consider cognitive training rather than popping a pill?

Dr. Gazzaley: In general, the pharmaceutical approach to enhancing how our brain functions has come up pretty short. None of the pharmaceuticals tried for the wide range of cognitive impairments in neurological and psychiatric diseases have worked very well. Some people are unaffected by them. Or they mostly get side effects, or a small effect, and it's generally because the pharmaceutical approach is a pretty blunt one. We can act on neurotransmitter systems, but we can't really act on circuits. So you wind up having to go to very high doses to get effects and then you get lots of side effects. There hasn't been a pill that really corrects these problems. The pharmaceutical approach is a sledgehammer, and the burden of side effects is just so great that the field has been reeling.

Charvy Narain: So what does cognitive training do? How does it work? And how do you actually do it?

Dr. Gazzaley: The drug approach, at least so far, can't activate a circuit. But interacting with a targeted environment—which might be in the form of a training program, in our case a video game—activates a circuit because that's how the brain works. Our idea is to activate that circuit and constantly challenge it by using adaptive algorithms, so that as you get better, the difficulty scales with you. The program keeps pushing on the system until it changes. In physical exercise, repeating a movement means the muscle gets stronger. So that's the general idea in cognitive training. It has selectivity because the interaction activates a circuit and because our brains are plastic, they modify themselves at many different levels in response to a challenge.

Charvy Narain: So what kind of circuits have you been able to get at?

Dr. Gazzaley: We dip back into our more basic science work in cognition and try to understand what systems are vulnerable and how we might design an interaction to really challenge that. In older populations, we know that there are deficits in cognitive control, such as selective attention, interference processing, multitasking, and distraction resistance. So our goals were to challenge prefrontal cortical circuits in older adults who were engaged in a task that had a lot of multitasking and distractions from other tasks.

Charvy Narain: What do these paradigms actually look like?

Dr. Gazzaley: One we've recently had success with and published last year was called NeuroRacer. The player is

 doi: 10.1101/sqb.2014.79.10

travelling through a 3D environment and trying to keep the car they're driving on the road, which is quite difficult, using a joystick so that it turns to the left, right, going up and down the hill. And while that's going on, signs are coming up and they have to respond as rapidly and accurately to target signs. So you have two tasks that are attentionally demanding occurring at the same time, and they both independently adapt. As you get better at them, they get faster and more difficult. Your rewards only occur when they both get better. So the mechanics of the game push you to do what's challenging to do over time, and see if you get better at it.

Charvy Narain: So both speed and accuracy are important in this game and people obviously get better by playing it. But are there generalized benefits and if so, what sort of tasks would people be getting better at?

Dr. Gazzaley: Our hypothesis was that cognitive control abilities have some underlying similarities and commonalities. And that if we improve the ability to deal with multiple tasks, we'd see benefits in other cognitive control abilities that older adults are deficient in. And that's what we're able to show. We see improvement in a very different task, a working memory task, which is essentially holding an image of a face in mind for several seconds even if that face is interrupted by something else. And in a vigilance task—that is, a sustained-attention, boring, repetitive task to identify a rare target—there is also improvement. So it was encouraging to see two tasks that were not explicitly trained in the game improve in the training group, but not in two control groups.

Charvy Narain: And what is the extent of these improvements? If you start with older adults, do they begin to resemble younger people?

Dr. Gazzaley: They do. They resemble the performance of 20 year olds. And neurally they start resembling them. So we use EEG recordings as an outcome measure, preimposed. We measure prefrontal cortical activity, midline frontal theta, that's engaged and bursts right at the most challenging part of the game, a third of a second after a sign comes up when you're also driving. This measure is deficient in 60–70 year olds compared with 20 year olds before the game; then after the game it's not any different. So we can see both neurally and behaviorally a renormalization of a lot of these patterns.

Charvy Narain: How much training does it take, and how long do the benefits last?

Dr. Gazzaley: In this study, it was 12 hours of training, 1 hour a day, 3 days a week, for a month. We tested after 6 months by having subjects play the game itself, and measuring how they did with the two tests at the same time. There was no decline in their abilities 6 months later, which was pretty shocking. Had we known that would happen, we would have done more detailed analysis to see what else remained. Some of the studies we're doing now are to try to understand how skill is retained over such long periods of time.

Charvy Narain: What sort of future work are you planning? Where do you think this technology will lead you?

Dr. Gazzaley: We have a lot of new games that target different cognitive processes. For example, one study is focused on a different type of distraction, internal distraction. It uses the principle of concentrative meditation and then integrates it with the video game dynamics of adaptivity and feedback to create an internal attention video game. We will see if that benefits older adults. We're also moving into different populations. We'd like to see if we can help patients with developmental disorders and ADD, depression, PTSD, traumatic brain injury, or dementia. It's exciting to see if there could be an impact in other populations. We're also working hard to see if we can have the adaptivity of the video game be guided by more than just performance. We think the special ingredient is the closed loop approach, in which you get better as the game gets more challenging. How about if the game doesn't just know how you're performing but knows what's going on in your brain? Maybe even looking at aspects of how your brain is functioning that are not even clearly represented in behaviors. So, the goal is to get real-time EEG recordings and feed those signals into the game mechanics. In essence, the game could diagnose deficits in how your brain processes information in real time and then guide the adaptivity and feedback to target those processes. The video game is not a magic trick, it activates a circuit and we could use other approaches to boost those effects further.

A Conversation with Ann Graybiel

INTERVIEWER: KAREN CARNIOL

Deputy Editor, Cell

Ann Graybiel is an Institute Professor at Massachusetts Institute of Technology.

Karen Carniol: One of your research directions is about value judgments and decision-making in terms of potential trade-offs or rewards or outcomes.

Dr. Graybiel: In everyday life, we all face situations in which our options are mixed—if we do something, the outcome will be partly good and partly bad. This kind of decision-making is called approach–avoidance decision-making, because we have to decide whether to go ahead and do something or to avoid doing it. We face a conflict between approach motivation and avoidance motivation. There are formal ways to test such approach–avoidance conflict, and these tests have been helpful in the diagnosis of problems related to anxiety and depression. Some years ago, Ken-ichi Amemori came to our lab, and he worked out a way that we could use this kind of conflict task in macaque monkeys. This opened up the chance to look at the neural circuits underlying what we think is emotion-related decision-making.

We had been trying to figure out what certain corticostriatal pathways do functionally, and had guessed they might have something to do with weighing different levels of goodness and badness in a given situation. We're interested in compartmental ordering in the striatum. This is a large region deep in the brain that is sort of like a huge ball of cells, but threading through it, like the trains in driftwood, are specialized cellular zones that we thought might be related to something about reinforcement or emotion.

In the search for what those regions might do, Ken-ichi Amemori and I targeted regions of the neocortex that we knew projected to these specialized striatal zones—the so-called "striosomes" or striatal bodies. These cortical regions have been implicated in conditions related to mood disorders such as OCD, addiction, and anxiety disorders. We set out to find out whether we could change the macaques' approach–avoidance decision-making in a task modeled on the tasks used with human subjects by manipulating activity in regions of the anterior cingulate cortex and caudal orbitofrontal cortex, both prefrontal regions that we had found to project to striosomes and that seemed to correspond to cortical regions picked out by a number of human neural imaging studies as abnormal in neuropsychiatric disorders.

We hope this work will have translation value but there's a very deep basic science underpinning it.

The deepest is how does a brain work, and how is it that we have a giant forebrain and a huge neocortex. Second only to the neocortex in the so-called progression index of brain evolution is this giant striatum, the giant ball of cells underneath the neocortex. We know that this deep-brain system communicates with the neocortex in elaborate loop circuits, but in the old days, it was thought to have its primary effects downstream, by way of descending pathways leading to premotor regions that are central pattern generators for movement—regions that control functions like walking and breathing.

Out of this basic anatomy, we developed the idea that maybe the reason the basal ganglia in primates and some other species have such strong output to the neocortex (through a series of steps) is that the basal ganglia loops help us to make new patterns, even cognitive patterns, that we can rely on just as we rely on the brainstem and spinal motor pattern generators for breathing and other elemental activities.

Karen Carniol: Going back to the manipulations, microstimulations in the anterior cingulate cortex that could shift the trade-off values in the macaques, where do you think the projections were that were causing that?

Dr. Graybiel: We found many years ago that those regions, the anterior cingulate cortex and, we think, the caudal orbitofrontal cortex, preferentially project to these distributed compartments in the striatum called striosomes. Dr. Satoko Amemori in our lab did the interesting anatomical tracing of the connections in the very animals in which we did the recording and the microstimulation and found that the region that we manipulated to change the decision-making of the monkeys probably projects striosomes preferentially. This is a recent observation, and we are still doing more experiments to confirm this.

 doi: 10.1101/sqb.2014.79.11

The pathway she works on seems to be related to some balance between pessimism and optimism.

The reason that we are working so hard on this is that maybe this is a clue to how the basal ganglia can affect emotion. Basal ganglia–related pathways definitely are related to whether we can move or not, as witnessed by the problems of patients with Parkinson's disease. But in addition, there's another side of basal ganglia function that somehow relates to a more cognitive or even emotional side of things. Here we are trying to find out whether the striosomes, dispersed though much of the striatum, are functionally connected with regions of the brain that we can relate to mood and emotion. Maybe they help decide what we're motivated to do or not motivated to do. This is such a basic function that it is conserved through evolution, and in fact, the striatum and its striosome-related pathways are too.

Karen Carniol: You talked about shifting the macaques towards more pessimistic decision-making by the microstimulation. How could that interplay with anxiety?

Dr. Graybiel: I don't know whether there really is a "worry circuit," but we do think that anxiety is likely involved in the results. One reason is that when we do this microstimulation, and induce the monkeys to have more pessimistic decision-making, we can completely reverse this effect by giving anxiety-reducing drugs such as diazepam, and the effect is dose-dependent.

Karen Carniol: Another area of your research is how behaviors turn into habits.

Dr. Graybiel: We have been startled to find that that by using optogenetic manipulations of tiny parts of the neocortex of rodents, we can prevent a habit from forming or block a habit that has been formed and then bring it back.

We first found that there is remarkable plasticity in the activity of neurons in the striatum and neocortex as the animals learn habits. We put many electrodes in the sensorimotor part of the striatum in rats and mice, and trained them to run in a T-maze. It turns out that over time, as they learn a habit, the cells become more and more active at the beginning and end of maze runs, but less and less active during the runs. It was as though the brain regions were forming activity markers for the habit as a whole—the entire sequence making up the habit. Then we found that in the nonsensorimotor part of the striatum, it isn't like that at all! Neurons there like to concentrate on the decision-making period of the task. When the animal gets to the T in a T-maze, the activity of those neurons swells up, presumably in relation to the animal's deciding. Then when that behavior becomes a habit, they stop firing as much. This taught us that different cortico-striatal loops can be simultaneously active but can pay attention to quite different aspects of the behaviors being learned.

In psychology there is a longstanding view that decision-making circuits are recruited gradually as the animal learns to make the decisions. Then, as the animal makes the decisions over and over again, the brain can shift a behavior to a habit. This has the great advantage that then, in the habitual mode, we don't have to think out the decision every time. This sequencing of brain activity has been recognized to occur both in the striatum and in the neocortex. And we wondered whether it also had this sort of bracketing of learned, habitual behaviors. Kyle Smith came to our lab and recorded there, and we found that the task bracketing did form, at least in the upper layers of the cortex.

In those upper layers, the cell ensembles at first tended to fire throughout the time as the animals were learning. No great pattern was forming. But when that behavior really got stamped in, after a long training period, all of a sudden, we got the beginning-and-end pattern up in the cortex—very much like the pattern we had seen to develop in the sensorimotor striatum. So maybe we have to have both the striatum and the neocortex developing that pattern until the brain decides "This behavior is a keeper." If it's a keeper, you have a habit.

It is that critical neocortical region that we turn on and off, optogenetically, to turn habits on and off.

A Conversation with Joseph LeDoux

INTERVIEWER: CHARVY NARAIN

Former Senior Editor, Nature Neuroscience

Dr. LeDoux is the Henry and Lucy Moses Professor of Science at NYU in the Center for Neural Science, and he directs the Emotional Brain Institute of NYU and the Nathan Kline Institute. He is also a Professor of Psychiatry and Child and Adolescent Psychiatry at NYU Langone Medical School.

Charvy Narain: You are going to talk about a sort of a new twist in the very old topic on fear and how it works in the amygdala.

Dr. LeDoux: Right. I've been working on fear for roughly 30 years, and after all that time, I decided I hadn't been working on fear at all. It may seem odd, but if you look back over the way I've talked about fear all these years, the idea has been in there the whole time. Basically, I think what we study when we study animals, and also often when we're studying humans, is the processes in the brain that detect and respond to danger. Fear is something additional.

It's the ability of the conscious brain to apprehend that the stuff under the hood has detected and responded to danger and to be aware of all of the consequences that threat detection produces in the brain and in the body, and all of the feedback and all of the other stuff that's going on. It's not just body feedback, but also information from within the brain itself like arousal.

I no longer talk about Pavlovian fear conditioning. I talk about Pavlovian threat conditioning because it's really about how the stimulus has changed in the brain and how that change allows that stimulus now to produce defense responses.

Fear is something we can know and study and talk about in people, but we have no idea what another animal is experiencing, so the language I am proposing is a way to avoid conflation of conscious fear with these threat detection and response mechanisms.

For example, people go to a therapist and seek treatment because they feel bad. They have conscious sensations and feelings that they don't want to have. Those are being influenced but not directly caused by systems underneath. We shouldn't mix those two things up (the nonconscious and conscious), since they might be treatable in different ways.

Charvy Narain: Are you saying that there's threat detection, which is separate, and there's the cognitive sort of stuff which goes with threat detection which deserves the label fear?

Dr. LeDoux: That's pretty close to it, yes. Cognitive behavioral therapy is a common approach, and that approach mixes two things together that need to be separated. Though this therapy is often equated with extinction, it's really much more than stimulus exposure. I think that the cognitive support and stimulus exposure parts need to be separated. I got through this argument in my new book, *Anxious*, but that's not what I'm going to talk about tonight, so maybe I should kind of switch gears.

Charvy Narain: What made you sort of realize that there's this difference? After 30 years of work, what ends up changed?

Dr. LeDoux: It's just been gnawing away at me, kind of like an almost imperceptible itch that I've been wanting to scratch, because I've always said that the conscious feeling of fear is something that involves attention and working memory and all kinds of cognitive processes, but the mistake I made was calling what the amygdala does the fear system. I'm guilty in this mess as well.

Charvy Narain: In that sense almost everybody is.

Dr. LeDoux: I think we're at a point where we've made a lot of progress and a sharper conception of what we're really studying and talking about would help the field advance to the next level.

Charvy Narain: What sort of implications do you think this has for animal models of fear, because you're talking about fear as the sort of conscious perception?

Dr. LeDoux: I would no longer call them animal models of fear. I would call a spade a spade and say that while animal research is very, relevant to human fear and anxiety, when we do so-called fear conditioning we are not

 doi: 10.1101/sqb.2014.79.12

studying those processes that people feel and experience. We do want to understand how those things are happening because they're fundamental to those experiences, but they have to be kept separate.

For example, we can condition someone in the lab to a picture and a shock, and then present that picture without them being aware, right, so they're unconscious of the stimulus. The stimulus produces a response that we would call a fear response, but the person doesn't feel anything. He doesn't even know the stimulus is there. Fear is not what's driving those responses.

Charvy Narain: Do you think these two things are separable in the brain as well?

Dr. LeDoux: They're absolutely separable.

Charvy Narain: Would you talk about some of the work that shows they are separable?

Dr. LeDoux: They're separable because we can separate them in people. And if we don't need conscious fear to explain how threat detection and responding works in a person, we certainly don't need it in a rat. The idea is to understand the stuff that's relevant in rats and people that's the same across species, rather than start by bringing human fear into the rat to explain the human fear. If we want to study human fear, we have to study that.

Charvy Narain: You previously talked about the role of the amygdala in what I guess you'd now say is threat detection. So where is the fear center?

Dr. LeDoux: There's a growing movement in psychology called construction theory of emotion, and I've been a constructionist since the late 1970s when I worked on split brain patients. I've always interpreted fear as the conscious putting together of all of the stuff coming up underneath the hood into a feeling. People like Lisa Feldman Barrett, Jim Russell, and a bunch of others are talking about conscious fear now in this way, as a social or cognitive construction on the basis of unconscious information.

I think fear emerges in cortex, specifically in working memory and attention circuits. If you're not aware of it, you're not feeling it. You have to be aware of something to feel it. You can't have an unconscious feeling, I would say. And to be conscious of something, say a snake, and that you are in danger of being harmed by that snake, the snake and memories about what snakes are and even about what fear is have to be represented in working memory. Otherwise, you can't know that *you* are in danger.

Charvy Narain: So fear is a conscious percept?

Dr. LeDoux: Right. Exactly.

Charvy Narain: There's no such thing as an unconscious feeling of fear. I'm sure you know that there would be many people who would sort of disagree with this idea.

Dr. LeDoux: Most people don't disagree with it in principle, but they're reluctant to change because it's hard to change things.

Charvy Narain: What is the sort of work that would separate out these two things?

Dr. LeDoux: We don't need to do any more work to separate them. It's obvious that they're separable from the research we have now. *Anxious* will reveal all this. I wrote a paper in *PNAS* recently called "Coming to terms with fear," where it's all explained as well.

I'm not doing any research to really test this, because I think it's pretty clear. As I said, most people agree that this is right, but they don't want to change because it's just too much trouble.

Charvy Narain: This idea that what a lot of people call fear is actually not, how does this change things?

Dr. LeDoux: Our research is paid for by public funding. If we tell the public that the amygdala is feeling fear, they're getting the wrong impression.

For example, there was a big splash about a finding that fear could be induced in a woman without an amygdala. But the only reason you'd be surprised that you could have fear in the absence of the amygdala is if you believed when you turned on the amygdala you got the feeling of fear. That leads to interpretations about the role of the amygdala in PTSD or panic disorder. The assumption is feelings of panic and fear and anxiety come from the amygdala. But if that's not true, and I don't think it is, we miss the opportunity to really understand where the awful feelings come from in these people.

The amygdala, when turned on, activates certain behaviors and releases stress hormones. It activates a lot of body systems. It triggers brain arousal. It focuses attention on the threat. It does a lot of stuff that ultimately contributes to the feeling of fear, but is not the feeling of fear.

Fear is not in the amygdala. Fear is a consequence. It's like you've got a lot of ingredients that are nonemotional ingredients that when put together allows the cognitive brain to make the feeling of fear.

Charvy Narain: So fear is a sort of emergent property of the brain?

Dr. LeDoux: You could say that.

Charvy Narain: How would you recontextualize the patients who have got things like PTSD? Or you talked about cognitive behavioral therapy. What sort of implications does this way of thinking have for those things?

Dr. LeDoux: That if you put too much cognition into the process of trying change behavior through extinction, you're putting competing systems at work.

You need to isolate the extinction process to weaken those bonds so that the stimulus is not going to produce the behavioral and physiological responses, and once you've done that, then you can address the cognitive component. I think if you put it all together, you're dealing with systems that should be separated.

Charvy Narain: What are you planning for the future?

Dr. LeDoux: The stuff we've been doing for all these years in rats are defensive reactions. You've got a tone, a shock; the tone elicits the reaction automatically. Now we're looking at how the brain *acts* in the face of danger. We are exploring different ways that the same stimulus, a tone that's been paired with a shock, can have a variety of other consequences that affect instrumental behaviors, the behaviors that are motivated by some goal. But all of this, the reactions and the actions, are implicit, under the hood, defensive activities. Conscious fear is a different matter.

A Conversation with Eric Nestler

INTERVIEWER: JAN WITKOWSKI

Executive Director of the Banbury Center
at Cold Spring Harbor Laboratory

Dr. Nestler is a Nash Family Professor and Chair of the Department of Neuroscience, and Director of the Friedman Brain Institute, at Mount Sinai, and currently a member of the National Advisory Drug Abuse Council.

Jan Witkowski: One of the problems about using mice to study higher level brain functions is finding a model that to some degree reflects what one might expect in humans.

Dr. Nestler: Yes, exactly. Most of the depression field has utilized what are called despair tests, something like the forced swim test where a mouse or rat would be placed in a beaker of water. It would swim for a while, then stop. People say that when it stops it's despairing although we really have no reason to think that. When you give the animal an acute dose of an antidepressant, it swims a little longer. It's hard to imagine a test with so little validity being so widely used in depression research.

Several years ago we set out to try to develop a test that is better validated and has more ethological relevance for a mouse. We figured that social stress was a domain of behavior that really hadn't been studied very much in animal depression models. In this model, which we call social defeat, a normal inbred C57 mouse is placed in the cage of a retired male breeder CD1 mouse. The retired breeder is bigger and meaner and as soon as the test mouse is placed in that cage, the larger mouse attacks it.

We only allow the physical interaction to occur for a couple of minutes before placing a screen between the mice. The test mouse continues to be subjected to all the aggressive cues of the larger mouse but has no more physical contact for the rest of the day. We repeat that process every day, typically for 10 days. Afterwards, we can show that we have induced, in that C57 mouse, a behavioral syndrome that's characterized by anhedonia, which means a loss of interest in pleasurable activities. The mice are less interested in drinking sugar water, eating Reese's Pieces, having sex, or running on a running wheel.

Dr. Nestler: The mice also show anxiety-like behaviors. They develop a metabolic syndrome: they eat too much and get fat and develop a prediabetic state. And they become profoundly socially avoidant. You might think that, after 10 days of being beaten up by different mice, social avoidance is a normal response, but we've shown if these defeated mice are put back in normal caging with the brothers with whom they were raised, for the rest of their lives they remain avoidant and don't play in the cages like their brothers. So we think this behavior is clearly pathologic.

The behaviors are corrected by chronic (not acute) administration of antidepressants that work in human. Interestingly, there's a small subset of C57 mice that don't develop the syndrome and we call them resilient. They've been subjected to the same level of stress yet somehow they manage to avoid most of these deleterious changes.

Jan Witkowski: Only males respond this way?

Dr. Nestler: This is only a male test. We are very interested in adapting social defeat for females but it's something that's still in progress.

Jan Witkowski: Do you know anything about the difference between the resilient and the susceptible mice?

Dr. Nestler: Right now we have no measure that we can apply to a C57 mouse before the stress that predicts accurately whether they're going to be susceptible or resilient, so it's a latent feature in these animals. After the fact, we have described very large differences in brain and peripheral measures that distinguish resilient and susceptible animals and seem to drive those different states but we don't know what initiates them.

Jan Witkowski: How did you get interested in chromatin remodeling in relation to this induced phenotype?

Dr. Nestler: Over the years my lab has been interested in mechanisms by which an environmental stimulus, a drug of abuse, a stress, might produce long-lasting changes. Because we know in humans that, in addiction syndromes or depression syndromes, environmental experiences can change people profoundly for their lifetimes and make them more or less vulnerable to the condition involved. We've thus been struggling to identify the ways in which an environmental stimulus can change the brain permanently. About 10 years ago studies in developmental biology and cancer biology showed that epigenetic

 doi: 10.1101/sqb.2014.79.13
Cold Spring Harbor Symposia on Quantitative Biology, Volume LXXIX

modifications of DNA and histones, and other mechanisms, can cause lifelong changes. We basically applied that insight to the brain, asking whether epigenetic modifications could mediate similarly lifelong changes in brain and behavior.

Jan Witkowski: Did you do a screen for chromatin or epigenetically related molecules?

Dr. Nestler: Yes. Over the last five years we've screened in many different ways. We've screened different histone modifications and different chromatin remodeling proteins. Chromatin remodelers are ATPase-containing complexes that control the movement of nucleosomes along a strand of DNA. We were particularly struck that, when we carried out RNA-Seq to look at global changes in RNA expression in different brain areas that occur as a consequence of susceptibility or resilience, the genes encoding specific types of chromatin remodeling proteins were among the most highly regulated. So in the last couple of years we have focused on this family of proteins.

Jan Witkowski: In particular Acf1?

Dr. Nestler: Yes. Acf1 is an accessory subunit of the ISWI family of chromatin remodeling proteins. There are several families. ISWI includes a critical subunit called Snf2h. As an ATPase, Snf2h can bind to certain subunits and drive mechanisms that activate transcription or bind to other subunits like Acf1 and form complexes that are repressive and decrease gene expression.

Jan Witkowski: They're repressed by altering chromatin structure?

Dr. Nestler: That's right. These complexes repress gene expression by redistributing nucleosomes along a strand of DNA, condensing them and filling in nucleosomes in certain regions. When a gene is active, its promoter region is relatively spared of nucleosomes. Acf1-Snf2h inhibitory complexes would basically fill in nucleosomes in that region and suppress that gene.

Jan Witkowski: The question is, what did you find when you looked in resilient animals? I guess you used the resistant ones as a second type of control?

Dr. Nestler: Yes. We always look at the resilient animals since they provide an interesting counterpoint to how an animal adapts or responds to stress. What we found is that, in a brain area called the nucleus accumbens, which is one of the brain's reward regions, induction of Acf1 and induction of the Acf1-Snf2h complex occurs only in susceptible animals, not in resilient animals, suggesting that the induction may be a factor in causing depression.

Importantly when we look at this nucleus accumbens brain area in human postmortem brain tissue, we see induction of Acf1 in depressed individuals. This is after death but it suggests that the findings in mice are relevant to the human situation.

Jan Witkowski: So now that you've found this difference in the chromatin remodeling between susceptible and resilient mice, what genes is it working on?

Dr. Nestler: The first step is to show directly that the Acf1-Snf2h complex is involved functionally in mediating susceptibility. Using viral vectors, we were able to overexpress Acf1 or knock it down specifically in the nucleus accumbens of an adult mouse. We found that when we overexpressed Acf1 we made the mice more susceptible, and when we knocked down Acf1, preventing the effect seen in susceptibility, we made the mice more resilient.

Dr. Nestler: Then we used chromatin immunoprecipitation sequencing to look genome wide at all the genes that in susceptible mice show increased binding of the Acf1-Snf2h inhibitory complex in nucleus accumbens. We've identified a dozen or so interesting genes. They've never been implicated before in stress responses or depression, so the next step is to determine if these genes act alone or in concert with one another to mediate susceptibility.

Jan Witkowski: Will you do that, again, by overexpression and knockdown?

Dr. Nestler: Yes. We'll use the viral vectors initially and then mutant mice as they are available and explore how manipulating the activity or levels of these genes in the nucleus accumbens controls resilience versus susceptibility.

Jan Witkowski: What are the genetics of acute or severe depression in human beings?

Dr. Nestler: Depression is considered moderately heritable, with about a 40% heritability rate, but it's been very difficult to identify individual genes involved. Accordingly, there is a high discordance among identical twins as well. There is only about a 40% chance that if one identical twin is depressed, his or her identical twin will also be depressed.

Jan Witkowski: So even for identical twins growing up in the same environment, there may be environmental influences, maybe after they left home.

Dr. Nestler: People have talked about difference in position in the uterus and birth factors and things like that.

Jan Witkowski: Right. But your mice are growing up in a controlled environment.

Dr. Nestler: Right. We are using genetically identical animals and we try to raise them in as identical a manner as possible, yet we see this very large divergence between a susceptible response and a resilient response.

Jan Witkowski: The classic definition of epigenetic is a transgenerational change in gene expression that doesn't involve structural changes in DNA. Have you been able to demonstrate such a phenomenon?

Dr. Nestler: In an experiment in my lab, male mice that had been subjected to social defeat stress were bred to normal females. The males were removed as soon as the females were impregnated, and the mothers were allowed to raise their pups. Then we studied the offspring when they were adults. What we found was really quite dramatic. The offspring of defeated fathers demonstrated much greater susceptibility to stress. They had an anxiety-like phenotype at baseline and were much more susceptible to chronic social stress. That was an astounding finding, which gets us into the area of Lamarckian mechanisms that supposedly have been disproved for a century or more.

Jan Witkowski: Have you looked at Acf or the genes affected by Acf1 in the offspring before you stress them?

Dr. Nestler: Not yet. It would be hard to do that before the stress because we have to kill the animal to analyze the state of its genes in brain. However, if we knew that the sons of defeated fathers, for example, showed this increased inherent vulnerability, we could simply analyze their brains and see what changes exist there. It's something we haven't done yet but are keen to do.

A Conversation with Alexandre Pouget

INTERVIEWER: CHARVY NARAIN

Former Senior Editor, Nature Neuroscience

Alexandre Pouget is a Professor in the Department of Basic Neuroscience at the University of Geneva.

Charvy Narain: You'll be talking about how the brain actually encodes information but from an abstract computational perspective. Are you comparing the brain to how a computer works? Is that the sort of approach that you're using?

Dr. Pouget: Yes, there is an interesting kind of analogy that you can draw there. Basically, with any kind of computing device, the first thing you need to figure out is how the device encodes information. If you want to understand how a computer works, you've got to understand first that it encodes information through those strings of bits, binary elements that can be 0s or 1s. Unless you understand that, there is no way you can then understand how it actually computes.

Charvy Narain: How do you approach this problem with respect to the brain?

Dr. Pouget: We have to rely extensively on recordings from our colleagues who are doing experiments in animals and presenting a variety of stimuli. We're asking the animal to move and the same time we record from the neurons. Then we try to establish what the mapping is between the stimulus and the responses.

Charvy Narain: When you establish these sorts of mappings, what do you find?

Dr. Pouget: There are quite a few findings, but the main hypothesis we're pursuing is that the code in the brain must be redundant.

Modern computers can encode billions and billions of numbers with absolute reliability and no errors.

This precision comes with two different costs. One is power. We use about a volt to encode a bit. As a result, you need a lot of power for your computers. A supercomputer may cost $2 million a year just in electricity.

The second issue is reliability. If any piece of hardware breaks in those computers so that the bits can no longer be encoded, you throw away the whole thing or at least, whatever board broke. So a computer is expensive to run and easy to damage, and the contrast with the brain is striking. It's exactly the opposite. In the U.S., you need only a few thousand dollars per year to feed your brain, and it is incredibly good at handling damage. At this point in my life, unfortunately, I lose something in the order of 30,000 neurons per day, and that's only going to get worse. Yet I'm told I continue to function reasonably well. In humans this is called graceful degradations. By the time you're 80, you'll probably have half the brain you had when you were younger, and yet you will continue to function.

Charvy Narain: Even then, the human brain is handling far more data than any supercomputer.

Dr. Pouget: So one of the big questions for us is what's so special about the neural code that allows our brains to have these properties.

We think the key is redundancy. That is, instead of having one bit that's infinitely reliable, you have multiple copies of the same bit that don't have to be perfectly reliable. If you have that, then you can know the value of a bit not just by looking at one of them but by taking the average of the multiple copies that you have. Having multiple copies is what we call the redundant code.

Charvy Narain: Wouldn't you then need a lot of space? The brain is quite compact. If you have multiple bits of information, does that not create space constraints?

Dr. Pouget: It does. You end up having to have more neurons than you would if you were more efficient. But if you want to be robust to damage, you must be able to get the system to function reasonably well even when you've lost half the components in the system. That's the way our brain works. We possibly have way more neurons than we need in principle but this system needs much less energy, which is really important. It takes about a volt to encode a bit in a computer whereas the brain functions on millivolts, a thousand times less.

doi: 10.1101/sqb.2014.79.14

So there are costs to everything. There's no free lunch, but nature seems to have gone for the solution of redundancy, with lots of neurons that are robust to damage.

Charvy Narain: What predictions emerge from this theory?

Dr. Pouget: Whenever you have a redundant code, one of its properties is that the neurons that all encode the same information should share information. We call that being "correlated."

They have to be correlated in a very specific way. It becomes a bit technical, but that's where our work is focusing. We have a very specific prediction about how the neuron should be correlated. We figured out the mathematics behind that about 4 years ago but were badly surprised when the data didn't match the prediction.

Dr. Pouget: People had measured correlation long before we did the mathematical work. When the correlation was first measured, people declared, "Oh, this correlation implies that the brain has a redundant code."

About 10 years ago, another group concluded that from what we can measure in the brain, it doesn't look like a redundant code.

Then we came along and said, "Yes, that's right. For a redundant code, we need another pattern of correlations." The data were already published, and it didn't fit our model, so we had a real puzzle in front of us.

Charvy Narain: How did you square the two?

Dr. Pouget: I'm a theoretician, so I love the data to be wrong, but the data are the data. What we came to realize is that the correlation we see in real neurons is probably a mixture of different factors. Some of them really constrain the code and make it redundant, while others are there because of the way the system is built.

We now have very good reason to believe that the measurements are completely dominated by the correlations that do not constrain the codes. For example, you may have different states of awareness that go up and down.

We know today that this kind of fluctuation in the overall activity of the brain creates a particular pattern of correlations. Those turn out to have almost no impact on the code, but it creates a large correlation that measures between 0 and 1.

Our theoretical prediction is the correlation is about 0.01, 10 times smaller than the correlation that people measure.

Dr. Pouget: Correlations are notoriously difficult to measure. In neuroscience, if we can get about 10 trials, 20 trials, it's a lot and that's hard to get from animals. That's just not enough to find out the value of the correlation coefficient. We think that the correlations that we're looking for are buried under a lot of irrelevant fluctuation in the correlation coefficients.

Charvy Narain: So the correlation you're looking for is much tinier than these and being washed out by irrelevant correlations. Is there any way to get at the correlations you're interested in?

Dr. Pouget: We're not sure yet. It's going to be complicated but there is hope. Ultimately, what we're trying to do is to understand how neural activity controls behavior. That's really our final goal.

The way we do this nowadays is to quantify how much information is captured in your behavior about the stimulus that is making you react, or making you behave. This we can quantify very precisely.

We try to see if, in the brain, the information in different areas involved in behavior has the same order of magnitude as the behavior of your animal or of your subject. We try to match information.

What we want is not the correlations, which are boring, counterintuitive, and hard to measure, we want information. One way to do this is to measure the correlations and compute the information. Another way is to use techniques that bypass the correlation measure altogether and go straight for information.

The problem we've got in doing that is the need to measure all the neurons at the same time while an animal is behaving.

Dr. Pouget: Up until 10 years ago, all we could do was record one neuron in one animal while the animal is performing a task. We know there are millions of neurons involved in any particular task.

What's fantastic in our field is the recent explosion of technology. We are now able to record at least from 100, possibly 1000 neurons at the same time in a behaving animal. In fact, in some systems, we can record all of the neurons at once. We can finally see the entire system at work, live. We now have the techniques needed to try to figure out whether tiny correlations are there or not by measuring information.

Charvy Narain: What sort of techniques would allow you to record from many neurons at once?

Dr. Pouget: In the past, we recorded from neurons using electrical wires that push until they got very close to one neuron. Then we could hear the activity of that neuron and record its electrical activity.

Nowadays, there are stunning new techniques where dyes are put inside the cells. The dye can be expressed in all the cells at the same time. Those dyes are sensitive to calcium. Whenever a cell is active, sending signals, calcium gets inside for a short time, for 10 to 20 seconds.

The dyes expressed in those neurons can sense calcium and light up if they detect it. By looking at how much light comes out, we can figure out how much calcium gets into the cells and therefore how active they are.

Again, one has to keep in mind that we're measuring from cells that are one-tenth of the width of a hair and examining hundreds if not thousands of them at the same time.

Charvy Narain: What sort of work are you planning in the future?

Dr. Pouget: In the next few years, now that we have the technology, we can finally try to establish a precise mapping between the population activity of all the neurons together while the animal is performing a task. We can try to understand what it is in the activity of those neurons that leads to the behavior.

When I was a grad student, 20 or 25 years ago that was the dream. I wasn't sure I would ever see a day where we would be able to do those experiments.

We're going to record all the neurons while the animal is performing the task, and we're going to use all the theoretical tools that we've developed while we were waiting for the technology to establish that mapping. It's going to be the first step toward understanding the mapping between neural activity or thoughts, or behavior, or desire or motivations.

A Conversation with Matthew Rushworth

INTERVIEWER: JAN WITKOWSKI

Executive Director of the Banbury Center
at Cold Spring Harbor Laboratory

Matthew Rushworth is a Research Fellow in the Department of Experimental Psychology and Centre for Functional Magnetic Resonance Imaging of the Brain (FMRIB), University of Oxford.

Jan Witkowski: Tell us about your work in understanding decision-making and the brain mechanisms involved.

Dr. Rushworth: The sort of things I'm interested in are how we make decisions and the brain mechanisms that underlie our ability to make decisions. What we are doing is weighing up one option and considering its value, and comparing that with the value of some alternative course of action, and how we come to a resolution, a decision, about what's the best thing to do.

Jan Witkowski: So what sort of experimental tests do you do?

Dr. Rushworth: We're interested in every type of decision you might make. Perhaps you have a choice of two different wines to drink. And when you're confronted by that choice, what you're going to summon up is a memory of how good, say, that particular white wine and red wine were the last time you tried them. And then you make some comparison between your internal representation of the value of those two different things. And what I'm interested in is how that comparison process works. So when we conduct an experiment we give subjects really simple decisions to make, normally just between two different visual stimuli. And the subject makes the decision in order to win some points and we scan their brain using fMRI while they make that decision, and then we ask them to take a very similar type of decision again a few seconds later, and then again a few seconds later. And by repeatedly asking subjects to make 100 or so simple decisions, we're able to gradually average together the activity that is occurring each time a subject makes a choice.

Jan Witkowski: In this case, you're presenting them with just one alternative but you also add in a third possibility.

Dr. Rushworth: Yes. We think there's a particular part of the brain called the ventromedial prefrontal cortex that carries out a process of value comparison, comparing the value of Choice A with the value of Choice B. And eventually a conclusion is reached about which one would be the better option to take. If we use fMRI to record activity in this brain region, we can see that the activity there is correlated with the difference in value between those two choices. And we think that the difference in value between the two choices is the key decision variable that defines how much better one thing is than the other. So the value difference is what drives our choice and the activity in this brain region covaries with that.

Jan Witkowski: And in the experimental setup, one choice is made more favorable than the other and more points are given for selecting that?

Dr. Rushworth: Sometimes one choice will have more points or a higher probability of points associated with it. And the subjects will have learnt about these contingencies before they come into the experiment. If we add into the mix a third option, we're interested in how this value comparison process gets subtly altered as there are more and more potential options to consider. And it turns out that the value comparison process becomes a bit more inaccurate when there are multiple alternatives.

Jan Witkowski: By "inaccurate" you mean that the subject more frequently chooses a lower value than the higher one?

Dr. Rushworth: Exactly. If, as in our experiments, there is an objectively better option and worse option, on occasions, when we add in a third option, the subjects can become a bit less accurate in choosing between the first two options. So if you were out buying groceries, and you wanted to buy a particular type of bread, and there were just two choices, your preferred option and your less preferred option, you're going to make that choice accurately. But the claim is that when you make that choice when there are multiple different varieties of bread available, somehow your comparison mechanism is confused by the multitude of options. So if you have more opportunities, more potential choices, you might end up being a little less satisfied with the choice that you made.

 doi: 10.1101/sqb.2014.79.15
Cold Spring Harbor Symposia on Quantitative Biology, Volume LXXIX

Jan Witkowski: In the experimental setup, what's the relative value of the third option, the distractor?

Dr. Rushworth: The most distracting type of third option is one that is very low in value. And we think in a situation in which you have two good things to choose between, a very poor third alternative leads to more inaccurate choices.

Jan Witkowski: That seems rather counterintuitive. You'd have thought that if the distractor was higher in, closer in value to the A or B choice, you'd spend more time taking C into account.

Dr. Rushworth: Certainly as that third C option increases in value, you do become more likely to end up choosing it. But there's another type of error that you can look at, and that is inaccuracies in your choice between A and B. So if A's better than B, then B is the worse choice, and the proportion of B choices can suddenly, subtly increase when the C option value decreases. And the reason for this, we think, is because the comparison process is being mediated by an inhibition process. We can imagine two pools of neurons, one representing Choice A and the other Choice B. And they're fighting against one another, so one pool of neurons is inhibiting the other. Ultimately only one of the pools of neurons ends up being active and the other one becomes inactive. It's the inhibition process that mediates this competition process. So A may end up active, or B may end up active, but the other option becomes inactive. The reason why a low-value third option has its effect is because it sends an excitatory influence into the inhibitory pool. If it's a poor-value third option, it doesn't add much to the inhibitory pool, and it's this inhibitory pool that mediates the decision process. So if there isn't much inhibition going on, then the comparison process between those two options doesn't work so effectively. That's the logic behind it.

Jan Witkowski: It sounds a bit like U.K. elections, where the Conservative and Labour parties have now been joined by the Liberal Democrats, who are not a very strong choice, but can spoil the success of one or other party.

Dr. Rushworth: Yes, although other analogies might be less politically contentious. But although there may be something counterintuitive about this process, there are also some situations in which it is a bit more intuitive. Imagine if you're a foraging animal and you don't have long to make a decision because there's a risk of losing the available options. In that situation you don't just want to make the right choice, but to make it quickly as well. So if you have two things together that are both very high in value, and you know there are other things in the environment that are terrible, it's quite a good choice, if not an optimal one, to choose either of those two good things.

Jan Witkowski: So what about the length of time an animal should forage in an increasingly poor environment, rather than risk moving off to what might be a better place?

Dr. Rushworth: That is another line of experimentation, where we're interested in how an animal, or a human in our investigations, has a sense of how well they've been doing recently. We think this is a different type of decision process that depends on different brain areas and different types of mechanisms. And the key idea here is that in some situations, there isn't really a choice between two different things. Instead there's a choice about whether or not to carry on doing something that you're already doing, or to switch to some perhaps more profitable behavior elsewhere. In order to guide this type of decision, we think you need to have an estimate of how well you've been doing recently, and some longer-term, average sense of how well you've been doing over time. And by comparing these two things you can work out whether it's worthwhile carrying on as you were or switching to something else.

Jan Witkowski: You must presume then, that part of this computation must be some estimate of how successful it was in the past to leave this point and risk going some distance away where things may or may not be better.

Dr. Rushworth: You're right that that's an important part of the process, the ability to learn this information, relate it to particular cues in the environment, and recognize those cues when you see them again, telling you that you're again in the same type of situation. But at the moment our investigations are at a slightly earlier stage, where we're just trying to see how it is that our human subjects are trying to estimate how well they're doing now and how well they've been doing over the longer term, and how they compare those two different estimates.

Jan Witkowski: You presumably are not looking at the postchoice period.

Dr. Rushworth: We do sometimes. There are some situations where it seems as if having made a particular choice, we become more likely to commit to that choice again in the future. So it seems as if some of the same brain mechanisms are maybe assessing the success of the choice that we have just made, in order to repeat it, or to switch to some alternative. We know that the same anterior cingulate brain regions that are mediating this second type of decision are evaluating the outcome of the choice, as well as being active before the choice is made.

Jan Witkowski: You're carrying out these experiments in a laboratory setting, or at least in a very carefully controlled setting. Can you devise real-world observational tests, looking at what people do when they're presented with a situation like whether or not to run for a bus or wait for the next.

Dr. Rushworth: Part of the art of designing experiments like that is to drain the decision-making problems of all of their color and interest and make them repetitive, in order to gather consistent data, time after time. And the reason for doing this is that our tools for measuring neural activity are relatively insensitive, so we have to average together activity recorded on multiple different occasions and multiple different simple decisions. But one way we can try to see if these measurements, these rather artificial types of scenario, relate to real-world behaviors is to give our subjects questionnaires about other types of decisions they might make. And then we can try to see if, say, a particular way of performing in our basic, simplistic experimental paradigms might relate to what they habitually do in the real world.

Jan Witkowski: When an individual *C. elegans* is moving along and makes a decision to change direction or go off in a similar direction, is there any relationship between the types of decision and the neural structure system that the *C. elegans* uses to make those sorts of decisions?

Dr. Rushworth: In *C. elegans,* you've got an animal with only 300 or so neurons, nevertheless there are some basic similarities between the models we use and the processes in worms, where neurons or pools of neurons are competing with one another in order to bring about a decision process.

A Conversation with Michael Shadlen

INTERVIEWER: ANN GOLDSTEIN

Scientific Editor, Neuron

Michael Shadlen is a Professor in the Department of Neuroscience, The Kavli Institute for Brain Science at Columbia University and a Howard Hughes Medical Institute Investigator.

Ann Goldstein: What led you to start asking the questions about how decisions are made and decoded in the brain?

Dr. Shadlen: I did my postdoctoral training with Bill Newsome who was the main pioneer in this field, and we had worked out with our colleagues the degree to which you could understand the perceptual limits of monkeys and humans, by relating signal-to-noise relationships in the visual cortex to what you could call noise relationships in behavior—that is, errors and accuracy. We did a lot of computational modeling to reconcile neural recordings and behavior and developed a hypothesis about what downstream structures in the brain were doing to render a choice from the data.

I have a background in neurology. I think the neurobiology of decision-making will ultimately provide insights into neurological disorders affecting higher cognitive function—schizophrenia and autism, for example—because decision-making provides a window on the basic building blocks of cognition. Decisions involve some kind of contingency, using information in a flexible timescale. These are the basic elements of thought in general, and the underlying mechanisms—when they go awry—are probably modes of failure in disease.

The study of decision-making in normal brains is the study of what makes a normal person not confused. That's what led us to study structures in the brain that might be making decisions about something simple like visual stimuli, but we see this in a broader context such as what constitutes a thought or coming to a belief.

Ann Goldstein: How do you study decision-making in a human subject? How do you build a task to test it?

Dr. Shadlen: Every perceptual experiment in psychology has a decision component, because in the end the subject—human, monkey, or rat—is going to look at something, decide what is being seen, and push a button or make an eye movement. I think we must jettison, or at least be skeptical about, the idea that what we know is what we represent in the brain. Instead what we know about the world may come from how the information we receive bears on the hypothesis we have about the world. If I make an answer with a button box, I've got to figure out what it is in my brain that made me move my thumb to push the button. I want to define a decision process backwards, starting with the behavior.

We ask "What's happening in the brain?" and replaced this with a more refined question: "Since I will answer, ultimately, with an eye movement, or a push of a button, what led the brain to form a plan to push this or that button, gaze toward this or that spot, say the word 'right' or 'left'?" So that led us to put our electrodes into a part of the brain that you wouldn't have expected. We worked backwards from, in our case, the ocular motor system, because monkeys reported their answers about decisions by eye movements, and that led us to the deliberation process—all the steps that led to the decision.

Ann Goldstein: There are going to be new studies using microstimulation in humans, following similar investigations in mice. Is this a powerful tool that will help us understand complex cognitive functions such as decision?

Dr. Shadlen: I think it's extremely strong. Most of what we do in systems neuroscience is correlative and I'm a big defender of correlation. Quantitative connections are valuable, but in the end if you're studying neurons, you'd like to say, "Is this circuit involved and can I manipulate it someway?" There are risks to that. There's no guarantee that the neurons you're activating have direct access to behavior. But causal manipulations are really important, and we're seeing new techniques for causal experiments such as optogenetics. What has worked for a long time is electrical microstimulation, especially in the visual cortex.

Ann Goldstein: You are interested in confidence and its neural mechanisms?

Dr. Shadlen: When we make a decision, there are three parts to it: choice, which we can grade as accurate—

 doi: 10.1101/sqb.2014.79.16

correct or incorrect; how long it takes—we can compare with accuracy; and a third more elusive component, which is confidence. You can think of confidence as an internal prediction about whether the decision we've just committed to is correct or not.

So how do you get that? Where does it come from? How do you measure it? A colleague in my lab figured out a way to do this by stealing a trick from the cognitive psychologists, where you get monkeys to bet. You say, "Okay, you've made your decision, but you could opt out of it and get a small reward or stick with it and get a big reward if you're right and nothing if you're wrong." That turns out to be a way to measure confidence.

Confidence in the end is some kind of internal implicit knowledge in the brain of the probability that the decision about to be rendered is correct. The mechanisms that explain how signal-to-noise information in the visual system leads to being correct or incorrect, or faster or slower, also explain how you can predict from your own choice whether you're likely to be correct or not.

In recent experiments, we have stimulated the brain and caused changes in choice, the original experiment that Bill Newsome did 20 years ago. But we could show that there were quantitatively commensurate changes caused in the degree of confidence. That is a bit of a shocker, because if you stimulate the brain artificially through these causal manipulations, we have no idea what it really does to a brain, and we give an animal an opportunity to say, "Wait a second, that was whacky, never experienced anything like that before." But they don't tell us that. They say, "That was like vision."

Ann Goldstein: What are you most excited about in this area? What's the next level for understanding decision and how it relates to other higher cognitive processes?

Dr. Shadlen: What we're doing is exposing what I hope will be principles of cognitive neuroscience. I've just eluded to several of them: integration of evidence, having a reason to stop and therefore trade off speed versus accuracy; doing computations in the language of probability theory; belief that I'm going to be correct or not. I think all of those kinds of elements come into play when we're having a dialogue using language and logic and so forth. What I'd like to think is that as we evolved bigger brains and more of the association cortex that supports these simple kinds of decisions, these operations and more to be discovered, came into play. As I mentioned earlier, I believe that these functions are failure modes in diseases that effect higher brain function such as schizophrenia, autism, and confusion states in general. My dream is that we'll be able to treat diseased brains one day by restoring these kinds of functions—integration of evidence, setting bounds, making decisions, and so on. That's a 20-year horizon, at least! We're nowhere near it yet, but you can almost see the path there.

A Conversation with Karel Svoboda

INTERVIEWER: KAREN CARNIOL

Deputy Editor, Cell

Karel Svoboda is a Group Leader at Howard Hughes Medical Institute's Janelia Farm Research Campus.

Karen Carniol: One of the many things you're interested in is understanding the neural coding that underlies object recognition and decision-making. One of the ways you've chosen to investigate that is through tasks with rodents using their whiskers to perceive. What are you learning?

Dr. Svoboda: Mice and other rodents use their whiskers to explore their environment. It seems like a weird system to study, rather than vision or touch through their paws, but it turns out to have some wonderful advantages and wonderful features for detailed mechanistic analysis of basic cognitive processes, like active sensation, perception, and decision-making. One of the key features is that the system is exquisitely mapped all the way into the cortex. There's a one-to-one correspondence between individual whiskers and small groups of neurons that process information from that whisker. We can zoom in to the relevant neurons. We can record and manipulate the relevant neurons, ask how manipulations have an impact on behavior and be very quantitative about that.

In many other systems, that kind of study is a needle in a haystack problem. Here we know where to look.

The barrel system first had its heyday in studies of experience-dependent plasticity. As the animal develops, information processing is restricted to individual barrels. Beyond that, there are a lot of features in the response of individual neurons to the dynamics of individual whiskers that are learned. Different neurons show different rates of adaptation. In this system we can have a comprehensive understanding of what the neural population is doing that processes information from one whisker.

Karen Carniol: What's the route from understanding the somatosensory perception to a decision-making process in that system?

Dr. Svoboda: We interact with the world by sweeping our eyes over objects or visual scenes to extract the information we want. Through visual search we produce input to the visual system. But the visual system is never studied that way, because it's very difficult to keep track of what exactly a monkey or a mouse looked at. In the whisker system, we can study the decision-making in the barrel cortex in the context of active sensation where the mouse moves its whisker to explore an object, to touch it, to palpate it, to learn about its location or its shape, and so on and so forth.

You can ask questions about what is encoded in the barrel cortex. You can ask about neural coding. Information is carried by action potentials, or spikes, and you can ask questions about how action potentials, trains of action potentials, in different neurons encode information about where the whisker is, what the whisker touches, how the whisker is touched, how it is informed. Then you can go one step further and actually manipulate the activity, to ask not only what is encoded, but what the rest of the brain reads out from that activity, something you can do in this system that is very difficult to do in other systems.

Karen Carniol: What are some of the techniques that you've developed?

Dr. Svoboda: We use electrophysiology extensively to learn what is encoded in the brain. So, what kind of tactile features, what specific types of neurons there are in the barrel cortex. Remember each barrel is still a complex structure with about 10,000 neurons, and at least 20 cell types. We can ask what is encoded in specific neurons using electrophysiology and then with closed loop behavior and photostimulation using optogenetics we can then manipulate these action potential trains in a very precise manner to see if we can push perception one way or another. That is really the arbiter of truth about what actually is extracted about the world by the animal from these action potential trains.

Karen Carniol: The action potentials lead me to my next question, which is another thing you've been looking at: individual synapses, their plasticity, and their stability. Can you talk about that?

Dr. Svoboda: This is an area of study that has been on the wane in the laboratory. Even though we published a paper this year, that experiment was probably done 5 or 6 years ago. The experiments you refer to really go back to my

 doi: 10.1101/sqb.2014.79.17

work at Cold Spring Harbor starting in the late 1900s. We set up some very powerful microscopes that allow us to look at individual synapses in the intact brain. One of the things that had always puzzled me was this literature on enrichment-related plasticity, structural plasticity, of neurons. We know that memories last for a long time, so I thought it was very important to study how stable neurons are, how stable synapses are, what is the lifetime of synapses, or different types of synapses. That's how we got to study structural plasticity in the adult brain and its relationship to experiential manipulations.

What we found was that a substantial fraction of synapses, at least in nonpathological conditions, are remarkably stable. They can last for many months, in some cases for the lifetime of an animal, which is something that I find quite surprising because they are made up of small numbers of fragile, labile proteins. It's still a mystery how that works. But then another fraction of synapses turn over, they appear, and disappear and that is modulated by experience. It's also cell-type-specific. It happens at different rates, and different synapses, thalamocortical synapses, input to the cortex tend to be more stable than intracortical synapses and so on and so forth.

This study of structural plasticity and the cell-type-specific effects really got me very much interested in cortical circuits and how they work, because we could not put this into a functional context without really knowing more about how neurons are hooked up. The logic of information processing is in cortical circuits.

Karen Carniol: Do you see the field of neural circuits and synaptic molecular biology converging at some point in the future?

Dr. Svoboda: I absolutely do and I have a vision of coming back to it. I've dabbled in studying learning at the level of circuits. I've learned that learning is a very, very difficult problem, so we've stepped back and study what I would call function. I want to go back and learn and develop good learning paradigms, for example, good motor learning paradigms and then go in and dissect, look for engrams at the level of defined synaptic pathways. We have to be able to show that those pathways and these changes at synapses are causally related to the memory, which is something that we may be able to get a handle on using new molecular methods over the next decade or so.

A Conversation with Susumu Tonegawa

INTERVIEWER: KAREN CARNIOL

Deputy Editor, Cell

Susumu Tonegawa is the Picower Professor of Biology and Neuroscience at the Massachusetts Institute of Technology (MIT) and the Director of the RIKEN-MIT Center for Neural Circuit Genetics at MIT, as well as the Director of RIKEN Brain Science Institute. He is also an Investigator at the Howard Hughes Medical Institute. Tonegawa won the Nobel Prize in Physiology or Medicine in 1987 for "his discovery of the genetic principle for generation of antibody diversity."

Karen Carniol: Much of your recent work has been focused on memory and in particular on memory engrams.

Dr. Tonegawa: Memory research has a very long history. In the last 100 years, there has been a general idea that when memory information is stored in the brain, the brain cells and the cellular network are altered—biochemically or biophysically—and that those alterations last for a certain period of time corresponding to the length of the memory. However, until recently nobody had pinpointed where exactly a particular memory's information was stored. It's been difficult to study.

There were fragmentary findings in the literature suggesting a synaptic plasticity in which synapse strength was altered or increased, but nobody had shown which cells in the brain had these changes. We wanted to identify the population of cells in which these presumed changes occur, and then to prove that those cells are responsible for the recall of a particular memory. This was very difficult to do until about 2 or 3 years ago, when a technology called optogenetics allowed us to label cells, tag them, and reactivate them to see if their reactivation caused an animal to behave as if it recalled a memory.

With optogenetics, for the first time you can actually see under the microscope the cells that are holding information. In the past, we believed that such cells must exist, in the hippocampus for instance; but the hippocampus has a very heterogeneous population of cells, so we were looking at an averaged picture. Now we can look specifically at the cells which carry memory information. What we see is not a single cell, but a population of cells in the hippocampus which lights up. If we reactivate those cells with an optic fiber implanted deep in the hippocampus, then the animal behaves as if it is recalling a memory. We can even manipulate the cells with optogenetics, and make animals memorize or consolidate memories.

You can also combine such cell manipulation studies with studies of emotion, which has very much to do with memory. A nice experience I had, a terrible experience I had—those kinds of memories are all linked to the brain circuit that controls emotion. The recent work has shown that, in addition to the hippocampus, another area is involved: the amygdala. So we've found that interactions between the hippocampus and amygdala allow us to form memories of different emotional associations.

Though we cannot currently do such optogenetics studies in humans, psychiatrists use a form of psychotherapy that works in a good proportion of depressed patients: They have the patient recall nice, pleasurable experiences they have had in their life. It turns out that in depressed people, memory recall of nice experiences is suppressed, which contributes to the depressed state. Everybody has some nice experiences and some bad experiences in their life, but depressed patients often recall their bad experiences much more easily than their nice experiences.

Empirically, psychiatrists know that helping the patient to recall a nice memory can resolve symptoms in some people, not only for a short time period, but for months, or maybe years. Where in the brain this interaction occurs between the pleasure memory and the fear or anxiety memory is not known. In our most recent work, we are trying to understand the neural circuit for pleasure memory and fear memory, and how one could manipulate these circuits in a mouse in such a way that the positive memory will become dominant.

Karen Carniol: You mentioned that there were two brain areas involved in memory, hippocampus and the amygdala. What's your current interpretation of the functional differences between them in terms of memory coding?

Dr. Tonegawa: The hippocampus has nothing to do with emotion. The primary function of the hippocampus is to store information about the environment and context—what we call episodic memory. The amygdala, on the other hand, can encode emotional association—the plus and the minus of an experience. Since the hippocampus and amygdala are connected, they both contribute to memory recall, and thus both good and bad experiences have associations with environment.

PTSD is an extreme case in which a terrible experience is connected with a specific place or context. A problem for PTSD patients is that they recall the bad memory too easily by encountering an unrelated but similar context, and thus they recall the terrible experience.

We are examining relevant questions in mice by giving them a bad memory and then later giving then a good memory and examining what happened to the bad memory. This we can do by manipulating the memory in both the amygdala and hippocampus. What we find is that there is a circuit which is crucial in switching the dominant emotional association of a memory, either positively or negatively. We've also extended this to animals that are chronically stressed. Often chronic stress leads to depression. So now we are making a mouse model of human depression induced by chronic stress. We've found that manipulation of cells in the hippocampus can alleviate symptoms of depression.

The important point is that we can manipulate the emotional state of a mouse by combining the stressed and depressed state with some very positive experiences in different order, and the mouse shows a behavioral impact of the pleasure memory.

A Conversation with Daniel Wolpert

INTERVIEWER: JAN WITKOWSKI

Executive Director of the Banbury Center at Cold Spring Harbor Laboratory

Daniel Wolpert is a Professor of Engineering at the University of Cambridge and appointed to the Royal Society Noreen Murray Research Professorship in Neurobiology.

Jan Witkowski: What's someone in a department of engineering doing at a biological symposium?

Dr. Wolpert: I would struggle to pass first year engineering! I'm a medical doctor by training, but I got involved in using computational ideas to understand the brain. Over the years I've used control theory, signal processing, and machine learning approaches to understand the brain. When the engineering department wanted someone to do bioengineering they thought they'd hire a biologist with interest in engineering, rather than an engineer with an interest in biology. That's been very productive.

Jan Witkowski: What problems interest you?

Dr. Wolpert: I work on how the brain controls movement. Movements are fundamental to our existence. The only way we can affect the world around us is through the contractions of muscles; there's no other way to improve our survival. It's important to remember that things like sensory, memory, and cognitive processes are important to drive or suppress future movements. The beauty of working on movement is that you have to work on the whole system. It's the final output.

Jan Witkowski: Does that imply then that you feel the primary purpose of the brain and all its systems is controlling movement and that evolutionary pressure on the brain is actually through the control of movements?

Dr. Wolpert: I would say there's no point laying down memories of childhood or perceiving the color of a rose if it doesn't affect how you're going to move later in life, because it has no evolutionary advantage. Although you can study processes like vision in isolation, it's really important to remember that the reason to have these processes is to drive actions.

Another beautiful thing about studying movements is that it's a very hard problem, and we can really appreciate that if we look at how well we do building machines that can do what we can do. We can build machines to play chess and they can beat grand masters, but we can't yet build a robot which can manipulate a chess piece with the dexterity of a 5-year-old child. It's a very challenging mathematical and computational problem, and if we could reverse engineer the brain, not only could we understand it from an intellectual point of view and help people with movement disorders but we could also build better robots.

Jan Witkowski: When I think of robots or robotic arms I think of those big ones welding on a car assembly line. They seem to do okay.

Dr. Wolpert: They do okay. If you want a robot to do a very simple task you can get it to do that one task by hand-tuning a solution. Once you've done that, you can't get it to do another task. What humans are particularity good at is flexibility and manipulation tasks, whereas these big robots are really designed to do one specific thing. What we really hope to do is understand the learning algorithms, which we think are really fundamental to allow machines to learn anything they want to learn.

Jan Witkowski: How do you set about trying to learn those algorithms experimentally?

Dr. Wolpert: Our approach is with humans. We're really interested in the algorithms, and then hope that the neural underpinning of our algorithms will be of interest to other labs who do animal research. We work with undergrads, robotic interfaces, and virtual reality. By controlling what people see and feel, and by giving them new skills they have to learn, we can tease apart the algorithms they use in their brain when they're learning skills.

Jan Witkowski: Can you give an example of what one of those tests looks like?

Dr. Wolpert: We put people into an unusual situation, where they hold a robotic interface which produces a force on the hand as they move, something they've never experienced before. We particularly want to give someone a new skill to learn. I'll mention one of the experiments we've just done. We were fascinated why people follow

 doi: 10.1101/sqb.2014.79.19

through in sport. Why is it that when you throw a ball or hit a ball in tennis or in golf, people tell you the follow-through is important when the action's all over by then? The key to understanding the follow-through is understanding how we activate different motor memories.

When we learn different skills—such as how to ride a bicycle or drive a car—we access different motor memories for these skills. What we've just found out is that the motor memory you activate now depends not only on what you've done in the past, but what you're going to do in the future. If you have a consistent follow-through after an action, you store that in one motor memory. Whereas, if you have different follow-throughs, you store them in different motor memories, and therefore don't consolidate them. What we've shown is that part of the reason you want to follow through is to store everything into one motor memory.

What we've shown is that the half-second before the movement determines which motor memories are accessed. Things happening more than half a second before the movement don't determine which motor memories you can access. Half a second before to half a second after the movement is the critical period. Basically the near future and the near past determine how you activate motor memories, which is an interesting time period. Why that half second is important we don't yet know, and that's an interesting question for us to answer.

Jan Witkowski: How broadly applicable do you think this is across species? Does *C. elegans* make use of this mechanism?

Dr. Wolpert: I don't know if *C. elegans* can learn quite as many things as humans can. I don't know whether they can store many motor skills, but they certainly can switch between skills.

What's really driven me is trying to understand how motor signals work, particularly from a point of view of what we call noise. When I talk about noise I don't mean sound, I mean the variability we have in our senses and in our muscles. One thing that characterizes humans compared with robots is how incredibly variable we are. If I put my hand under a table and try to localize it with my other hand I can be off by centimeters. If I try to play darts I have huge variability, yet we're very skilled. One of the things we've been working on is, how does the brain get around the problems of that variability.

We have so many ways we can do a task. If I asked you to point to your nose, you could do it in many different ways but most people will do such tasks in very similar way. We think the reason people do that is because it's somehow the optimal way to move so as to reduce the bad consequences of this variability. It turns out that this variability makes us move in a particular way. That's rather nice as it means we can generate models to predict how people should move, and that if you move in different ways you're going to be less accurate. I've been driven very much by those noise models, which say what we have to learn about to perceive the world as statistics. As you go around the world you learn about statistics of the outside world, and that's really important to be able to estimate and predict the future.

Jan Witkowski: If you take a thousand people and they're all right handed and they've all got the right number of fingers and you ask them to touch their nose, how variable is the path by which they do it?

Dr. Wolpert: It's not very variable. We all walk roughly the same way, we all reach the same way. You don't have to. Out of the infinite number of ways you could reach to pick something up, there are many ways we don't choose. It's interesting to ask why we choose particular solutions. The way we walk is very stereotypical. People like John Cleese from Monty Python made their early career by violating biological motions, which somehow is funny to us. The fact that we all think so tells us something about the way the brain is organizing these movements.

Jan Witkowski: Presumably a baby learns these things. Is it learned by imitation or is it internally?

Dr. Wolpert: No, I think imitation helps, but I think it's internally learned. I think we start off with a right approximation, but the noise is very big. Variability is much greater in babies. For example, they way they start moving their eyes is very different to the way adults do, and only by the age of 3 or 4 do they start having eye movements similar to the way adults move their eyes. We move our eyes almost all the way to a target but they, because they have so much variability, make small steps, because if they tried to get there in one go they'd have so much variability they'd be hunting around the target all the time. We think these are probably learned throughout experience.

The real question, though, is why I'm not as good as Tiger Woods at golf. It could be that he's found the optimal solutions and I'm just not good and have to keep practicing. What I believe is that we have different levels of variability. I've just got higher level of variability, and therefore whatever I do I can never be as accurate as him. Some people are just born lucky in terms of that variability.

Jan Witkowski: Can you test that?

Dr. Wolpert: It's a little hard to test because it's hard to separate experimentally, the person who is optimal versus those who have noise. We're thinking of ways to do that. It would be wonderful to be able to sort children, to say, "You do piano, while you should be on a big drum." That's a little harder, that's a dream for the future.

A Conversation with Tony Zador

INTERVIEWER: GARY STIX

Editor, Scientific American

Tony Zador is a Professor of Biology and Program Chair in Neuroscience at Cold Spring Harbor Laboratory.

Gary Stix: You are giving a talk at this conference entitled "Sequencing the Connectome." What is a connectome and why is it important?

Dr. Zador: The brain has a lot of neurons. The human brain has 100 billion, the mouse brain maybe 100 million. What we'd really like to understand is how we go from a bunch of neurons to thought, feelings, behavior, all those things. We think that the key to understanding is finding out how the neurons are connected to one another. There have been a lot of traditional techniques for studying connectivity but at a fairly gross level. We can tell that a bunch of neurons here tend to be connected to a bunch of neurons there. There are also techniques for looking at how single neurons are connected but only one at a time. What we would love to be able to do is to tell how every single neuron in the brain is connected to every single other neuron. If you wanted to navigate through the United States, one of the most useful things you could have is a road map. It wouldn't tell you everything about the United States but, without it, it would be very hard to get around.

Gary Stix: You're proposing that sequencing the connectome might be a novel way to understand the connectome.

Dr. Zador: Traditionally, connectivity has been studied as a branch of microscopy. People talk about mapping the brain. Typically, people use one method or another to label a neuron, then observe that neuron at some level of resolution.

The challenge at the core of all microscopy techniques is that neurons can extend long distances—millimeters in a mouse brain, or in a giraffe brain, all the way from the brain to its foot, which must be 10 feet . . . so very long distances.

They're connected with one another at structures called synapses which are below the resolution of light microscopy. So if you really want to understand how one neuron is connected to another, you have to be able to resolve the synapse, which requires electron microscopy. And that means taking incredibly thin sections of brain and imaging them using electron microscopy.

People are doing this impressively well. A big success was in *C. elegans,* an organism with 302 neurons and 7000 synapses. With over 50 person years of work, they were able to reconstruct the entire wiring diagram of this tiny creature. Since then, they've scaled it up. It's working pretty well but it's still extremely challenging.

A few years ago, largely because I'm at Cold Spring Harbor and steeped in sequencing technology, it occurred to me that sequencing has the right scale to handle billions of synapses. If you have a brain that has 100 million neurons and each neuron makes, let's say, 1000 synapses, that's 100 billion synapses.

A sequencing run these days costs about $1000 for one billion reads. The way we propose to do it, one read equals one synapse. That price is coming down. Fifteen years ago, the cost of sequencing the first human genome was approximately $1 billion. Now you can get your genome sequenced for $1000 and within a couple years, it will be well under that.

Sequencing is getting better faster at a rate that's even faster than the rate at which computers get better. My iPhone has more computing power than a computer 20 years ago. That follows something called Moore's Law, and sequencing technology has been improving at a rate substantially faster than Moore's Law, since 2008.

There's every expectation that sequencing will continue to get faster. There's a huge potential benefit to converting the problem of connectivity into a problem of sequencing. Basically, the benefit is cost and speed. Sequencing takes a week to sequence a billion reads.

Gary Stix: A circuit is not a gene. How would you sequence a circuit?

Dr. Zador: The idea is that, first of all, we endow every neuron with a unique, random sequence of DNA. We call it a barcode.

This sounds fanciful but the immune system has solved this problem. B cells and T cells generate novel receptors

 doi: 10.1101/sqb.2014.79.20

through somatic recombination. They scramble pieces of their chromosome to produce, for example, novel antibodies.

We're not using the particular collection of enzymes used in the immune system because they're not really convenient. But we're hijacking similar proteins from other organisms to try to do the same things in neurons.

We're going to put in every neuron a cassette which, when we express the protein, will scramble the pieces and generate a novel sequence in every neuron in the brain. "Every" neuron sounds like a lot, but combinatorics works in our favor.

A sequence of 30 random nucleotides has a potential diversity of 4 to the power of 30 because there are four nucleotides. This is way more than the number of neurons in the brain. If we just put this random cassette in and can cause enough scrambling to occur, then by chance, the probability that two neurons will have the same bar code, is infinitesimally small. If we can do that, the next step is to express little pieces of RNA that encode that little random barcode.

We've already engineered proteins that will drag those RNAs to the synapse. In each synapse, there will be a presynaptic barcode and postsynaptic barcode. With biochemistry, we can join and link them together so you have a single piece of DNA that has the presynaptic barcode and the postsynaptic barcode.

Then all we have to do is read that out and we get, in principle, a huge connectivity matrix that has neurons numbered 1 through 100 million here, 1 through 100 million here, and a dot where neuron 22 is connected to neuron 53. Then another dot where neuron 27 is connected to neuron 68.

Gary Stix: There are lots of different types of neurons and lots of different structures in the brain.

How far do you think you could go with this? Could you look at all of the brain?

Dr. Zador: There's no reason we couldn't do an entire mouse brain. In fact, there's no reason we couldn't do many mouse brains once we get the transgenic mice working. The sequencing costs will be not negligible, but will be well within the range that would make this project worthwhile. We envision getting the connectivity of many individuals, not just a portion of the circuit but the entire circuit.

I've given you the bare-bones version, but there are bells and whistles that we can add that we think will allow us not only to know the connectivity but also the precise location of those cells and their gene expression pattern which, in turn, will tell us about their cell type.

Although it's ambitious, but not beyond what we think is reasonable, we'd like to have the complete connectivity matrix, then associated with each one of those elements, what cell type it is, what genes it expresses. and therefore what cell type it is.

Gary Stix: As with gene sequencing, in principle, this could be done very quickly.

Dr. Zador: Basically, in this form, once we have a transgenic animal, extracting the DNA takes a few days, sequencing takes 2 weeks or less. Per individual, it could take a month. Making the animals we can do that with has taken us a few years and will probably take us a few more years. Once we have them, it won't take much time.

Gary Stix: There's a lot of emphasis in the U.S. and elsewhere on developing new technologies for understanding the brain better. Do you see sequencing as one of them?

Dr. Zador: I've applied for grants. I should be hearing in a few months. What I hope to have soon is a proof of principle that I hope will convince people that this is something worth pursuing. To scale it up to the level I'm talking about is beyond what one lab could do.

After we've shown the proof of principle, I'd love lots of labs to get involved and come up with better ideas for doing a lot of the things we're doing, in the quickest possible way.

I would be thrilled if resources were devoted to this. I think having the complete connectivity of organisms, would be incredibly useful and transform how we do neuroscience.

Gary Stix: Are there competing technologies?

Dr. Zador: There are two other main approaches to figuring out connectivity.

Gary Stix: Those are traditional approaches, aren't they?

Dr. Zador: The people who practice them would say they've extended them to the point that they're not traditional anymore. The main competition, although I see it as complementary, comes from various forms of electron microscopy. That technology has improved so much that for certain questions, it's going to be super useful.

Gary Stix: Could these approaches be combined?

Dr. Zador: It's hard for me to envision how a combination could scale up to the speed and cost required.

Gary Stix: In your recent paper in *PLoS Biology*, you mentioned that this might be a good technique to test out hypotheses such as how the brain circuits go awry in a disorder like autism.

Dr. Zador: Autism is one of my research interests. There's been a lot of progress recently in identifying the genes involved in causing autism. There are dozens and perhaps hundreds of genes which when perturbed, can cause autism. Yet, autism, although it's heterogeneous, is still a meaningful diagnosis. There must be something in common amid the heterogeneity.

There's an appealing hypothesis what is shared is disruption of circuitry. It's possible that the genetic lesions we know cause autism in people could be recapitulated in mice. Then we can ask the question, what goes awry in the circuits of mice which express the same genetic lesions that people have?

The hope is that we could take 20 mouse models of autism, look at their brains, look at their connectivity, and say, "Hey, we noticed that in 17 out of these 20, there's a disruption in this circuit compared with how it is in non-autistic mice." That would lead us to look in the direction of that circuit.

That circuit could be long-range connectivity, let's say from the front of the brain to the back of the brain. Or far more specific things like the inability of one subset of neurons in this part of the brain to connect a particular subset of inhibitory interneurons in some distant region. Or even some local region. There are an enormous number of possible hypotheses. We can test and validate any one of them, but testing them all, at this point, is not possible unless you have a method like the one I describe.

Author Index

Subject Index